Bacterial cellular metabolic systems

Published by Woodhead Publishing Limited, 2013

Woodhead Publishing Series in Biomedicine

1 Practical leadership for biopharmaceutical executives
  J. Y. Chin
2 Outsourcing biopharma R&D to India
  P. R. Chowdhury
3 Matlab® in bioscience and biotechnology
  L. Burstein
4 Allergens and respiratory pollutants
  Edited by M. A. Williams
5 Concepts and techniques in genomics and proteomics
  N. Saraswathy and P. Ramalingam
6 An introduction to pharmaceutical sciences
  J. Roy
7 Patently innovative: How pharmaceutical firms use emerging patent law to extend monopolies on blockbuster drugs
  R. A. Bouchard
8 Therapeutic protein drug products: Practical approaches to formulation in the laboratory, manufacturing and the clinic
  Edited by B. K. Meyer
9 A biotech manager's handbook: A practical guide
  Edited by M. O'Neill and M. H. Hopkins
10 Clinical research in Asia: Opportunities and challenges
  U. Sahoo
11 Therapeutic antibody engineering: Current and future advances driving the strongest growth area in the pharmaceutical industry
  W. R. Strohl and L. M. Strohl
12 Commercialising the stem cell sciences
  O. Harvey
13 Biobanks: Patents or open science?
  A. De Robbio
14 Human papillomavirus infections: From the laboratory to clinical practice
  F. Cobo
15 Annotating new genes: From *in silico* to validations by experiments
  S. Uchida
16 Open-source software in life science research: Practical solutions in the pharmaceutical industry and beyond
  Edited by L. Harland and M. Forster

17 Nanoparticulate drug delivery: A perspective on the transition from laboratory to market
   V. Patravale, P. Dandekar and R. Jain
18 Bacterial cellular metabolic systems: Metabolic regulation of a cell system with ^{13}C-metabolic flux analysis
   K. Shimizu
19 Contract research and manufacturing services (CRAMS) in India
   M. Antani, G. Gokhale and K. Baxi
20 Bioinformatics for biomedical science and clinical applications
   K-H. Liang
21 Deterministic versus stochastic modelling in biochemistry and systems biology
   P. Lecca, I. Laurenzi and F. Jordan
22 Protein folding *in silico*: Protein folding versus protein structure prediction
   I. Roterman-Konieczna
23 Computer-aided vaccine design
   T. J. Chuan and S. Ranganathan
24 An introduction to biotechnology
   W. T. Godbey
25 RNA interference: Therapeutic developments
   T. Novobrantseva, P. Ge and G. Hinkle
26 Patent litigation in the pharmaceutical and biotechnology industries
   G. Morgan
27 Clinical research in paediatric psychopharmacology: A practical guide
   P. Auby
28 The application of SPC in the pharmaceutical and biotechnology industries
   T. Cochrane
29 Ultrafiltration for bioprocessing
   H. Lutz
30 Therapeutic risk management of medicines
   A. K. Banerjee and S. Mayall
31 21st century quality management and good management practices: Value added compliance for the pharmaceutical and biotechnology industry
   S. Williams
32 Sterility, sterilisation and sterility assurance for pharmaceuticals
   T. Sandle
33 CAPA in the pharmaceutical and biotech industries
   J. Rodriguez
34 Process validation for the production of biopharmaceuticals: Principles and best practice.
   A. R. Newcombe and P. Thillaivinayagalingam
35 Clinical trial management: An overview
   U. Sahoo and D. Sawant
36 Impact of regulation on drug development
   H. Guenter Hennings
37 Lean biomanufacturing
   N. J. Smart
38 Marine enzymes for biocatalysis
   Edited by A. Trincone
39 Ocular transporters and receptors in the eye: Their role in drug delivery
   A. K. Mitra
40 Stem cell bioprocessing: For cellular therapy, diagnostics and drug development
   T. G. Fernandes, M. M. Diogo and J. M. S. Cabral

41
42 Fed-batch fermentation: A practical guide to scalable recombinant protein production in Escherichia coli
   G. G. Moulton and T. Vedvick
43 The funding of biopharmaceutical research and development
   D. R. Williams
44 Formulation tools for pharmaceutical development
   Edited by J. E. A. Diaz
45 Drug-biomembrane interaction studies: The application of calorimetric techniques
   R. Pignatello
46 Orphan drugs: Understanding the rare drugs market
   E. Hernberg-Ståhl
47 Nanoparticle-based approaches to targeting drugs for severe diseases
   J. L. A. Mediano
48 Successful biopharmaceutical operations
   C. Driscoll
49 Electroporation-based therapies for cancer
   Edited by R. Sundarajan
50 Transporters in drug discovery and development
   Y. Lai
51 The life-cycle of pharmaceuticals in the environment
   R. Braund and B. Peake
52 Computer-aided applications in pharmaceutical technology
   Edited by J. Petrović
53 From plant genomics to plant biotechnology
   Edited by P. Poltronieri, N. Burbulis and C. Fogher
54 Bioprocess engineering: An introductory engineering and life science approach
   K. G. Clarke
55 Quality assurance problem solving and training strategies for success in the pharmaceutical and life science industries
   G. Welty
56 Nanomedicine: Prognostic and curative approaches to cancer
   K. Scarberry
57 Gene therapy: Potential applications of nanotechnology
   S. Nimesh
58 Controlled drug delivery: The role of self-assembling multi-task excipients
   M. Mateescu
59 In silico protein design
   C. M. Frenz
60 Bioinformatics for computer science: Foundations in modern biology
   K. Revett
61 Gene expression analysis in the RNA world
   J. Q. Clement
62 Computational methods for finding inferential bases in molecular genetics
   Q-N. Tran
63 NMR metabolomics in cancer research
   M. Čuperlović-Culf
64 Virtual worlds for medical education, training and care delivery
   K. Kahol

Woodhead Publishing Series in Biomedicine: Number 18

Bacterial cellular metabolic systems

Metabolic regulation of a cell system with ^{13}C-metabolic flux analysis

KAZUYUKI SHIMIZU

WP

WOODHEAD
PUBLISHING

Oxford   Cambridge   Philadelphia   New Delhi

Published by Woodhead Publishing Limited, 2013

Woodhead Publishing Limited, 80 High Street, Sawston, Cambridge, CB22 3HJ, UK
www.woodheadpublishing.com
www.woodheadpublishingonline.com

Woodhead Publishing, 1518 Walnut Street, Suite 1100, Philadelphia, PA 19102-3406, USA

Woodhead Publishing India Private Limited, G-2, Vardaan House, 7/28 Ansari Road, Daryaganj, New Delhi – 110002, India
www.woodheadpublishingindia.com

First published in 2013 by Woodhead Publishing Limited
ISBN: 978-1-907568-01-5 (print); ISBN 978-1-908818-20-1 (online)
Woodhead Publishing Series in Biomedicine ISSN 2050-0289 (print); ISSN 2050-0297 (online)

© K. Shimizu, 2013

The right of K. Shimizu to be identified as author of this Work has been asserted by him in accordance with sections 77 and 78 of the Copyright, Designs and Patents Act 1988.

British Library Cataloguing-in-Publication Data: A catalogue record for this book is available from the British Library.

Library of Congress Control Number: 2012946405

All rights reserved. No part of this publication may be reproduced, stored in or introduced into a retrieval system, or transmitted, in any form, or by any means (electronic, mechanical, photocopying, recording or otherwise) without the prior written permission of the Publishers. This publication may not be lent, resold, hired out or otherwise disposed of by way of trade in any form of binding or cover other than that in which it is published without the prior consent of the Publishers. Any person who does any unauthorised act in relation to this publication may be liable to criminal prosecution and civil claims for damages.

Permissions may be sought from the Publishers at the above address.

The use in this publication of trade names, trademarks, service marks, and similar terms, even if they are not identified as such, is not to be taken as an expression of opinion as to whether or not they are subject to proprietary rights. The Publishers are not associated with any product or vendor mentioned in this publication.
 The Publishers and author(s) have attempted to trace the copyright holders of all material reproduced in this publication and apologise to any copyright holders if permission to publish in this form has not been obtained. If any copyright material has not been acknowledged, please write and let us know so we may rectify in any future reprint. Any screenshots in this publication are the copyright of the website owner(s), unless indicated otherwise.

Limit of Liability/Disclaimer of Warranty
The Publishers and author(s) make no representations or warranties with respect to the accuracy or completeness of the contents of this publication and specifically disclaim all warranties, including without limitation warranties of fitness of a particular purpose. No warranty may be created or extended by sales of promotional materials. The advice and strategies contained herein may not be suitable for every situation. This publication is sold with the understanding that the Publishers are not rendering legal, accounting or other professional services. If professional assistance is required, the services of a competent professional person should be sought. No responsibility is assumed by the Publishers or author(s) for any loss of profit or any other commercial damages, injury and/or damage to persons or property as a matter of products liability, negligence or otherwise, or from any use or operation of any methods, products, instructions or ideas contained in the material herein. The fact that an organisation or website is referred to in this publication as a citation and/or potential source of further information does not mean that the Publishers nor the author(s) endorse the information the organisation or website may provide or recommendations it may make. Further, readers should be aware that internet websites listed in this work may have changed or disappeared between when this publication was written and when it is read. Because of rapid advances in medical sciences, in particular, independent verification of diagnoses and drug dosages should be made.

Typeset by RefineCatch Limited, Bungay, Suffolk
Printed in the UK and USA

Contents

List of figures xi
List of tables xix
About the author xxv

1 Main metabolism 1

1.1 Introduction 1
1.2 Energy generation in the cell 3
1.3 Carbohydrate metabolism 6
1.4 Respiratory chain pathways 25
1.5 Anaerobic metabolism 30
1.6 Anaerobic respiration 32
1.7 Photosynthesis 33
1.8 Amino acids synthesis 36
1.9 Nucleic acid synthesis and its control 45
1.10 Fatty acid metabolism and degradation 47
1.11 Phosphotransferase system (PTS) 48
1.12 Carbohydrate metabolism other than glucose 48
1.13 ATP generation under aerobic and anaerobic conditions 52
1.14 Metabolic regulation and futile cycle 52
1.15 References 53

2 Brief overview of metabolic regulation of a bacterial cell 55

2.1 Introduction 55
2.2 Metabolic regulation analysis by protein expressions 56
2.3 Metabolic regulation during the time course of batch culture 77
2.4 Reduction of by-product (acetate) formation 84
2.5 References 88

3	**Metabolic regulation by global regulators in response to culture environment**	**95**
	3.1 Introduction	95
	3.2 Carbon catabolite regulation	97
	3.3 Nitrogen regulation	120
	3.4 Phosphate regulation	130
	3.5 Oxygen level regulation	137
	3.6 Acid shock or the effect of pH	154
	3.7 Heat shock response	158
	3.8 Fatty acid metabolic regulation	165
	3.9 Response to nutrient starvation	173
	3.10 References	187
4	**Conventional flux balance analysis and its applications**	**215**
	4.1 Introduction	215
	4.2 Basis for metabolic flux analysis	216
	4.3 Application to photosynthetic bacteria	219
	4.4 Metabolic flux analysis of a single gene knockout *E. coli* under anaerobic conditions	243
	4.5 References	258
5	**^{13}C-metabolic flux analysis and its applications**	**263**
	5.1 Introduction	263
	5.2 Basic principle for flux calculation based on ^{13}C-labeling experiment	265
	5.3 Simple example for flux calculation based on ^{13}C-labeling experiment	270
	5.4 Analytical approach for flux computation	275
	5.5 ^{13}C metabolic flux analysis based on GC-MS	285
	5.6 ^{13}C-MFA using NMR	299
	5.7 ^{13}CMFA for cyanobacteria based on GC-MS and NMR	334
	5.8 Appendix 5.A	351
	5.9 References	354
6	**Effect of a specific-gene knockout on metabolism**	**359**
	6.1 Introduction	359
	6.2 Effect of *ppc* and *pck* gene knockout on metabolism	361

6.3	Effect of *zwf* and *gnd* genes knockout on metabolism	368
6.4	Effect of *pyk* gene knockout on metabolism	377
6.5	Effect of *lpdA* gene knockout on metabolism	382
6.6	Effect of *sucA* and *sucC* gene knockout on metabolism	389
6.7	Effect of *icdA* gene knockout on metabolism	398
6.8	Effect of *pfl* gene knockout on metabolism	412
6.9	Effect of *ldhA* gene knockout on metabolism	426
6.10	References	439

Appendices — 449

- Appendix A: Global regulators and their regulated genes — 449
- Appendix B: Metabolic pathway reactions and the corresponding enzymes and genes — 450
- Appendix C: A table of genes and their associated functions — 452
- Appendix D: Gene names — 454
- Appendix E: Precursor requirements (μmol/g DW) for biomass synthesis of *E. coli* — 460

Index — 463

Figures

1.1	Main metabolic pathways	2
1.2	Metabolic pathways of glycolysis	4
1.3	Structure of ATP	5
1.4	Structure of NADH and NADPH	5
1.5	(a) Pentose phosphate pathway, and (b) the change in carbon numbers	11
1.6	Entner Doudoroff pathway	15
1.7	PDH and TCA cycle	16
1.8	Acetate producing pathways	21
1.9	Anaplerotic pathways	22
1.10	Gluconeogenetic pathways	24
1.11	Modularity of respiratory chains	26
1.12	Proton translocating values per electron in the respiratory chain	26
1.13	Schematic illustration for the electron transfer and proton translocation	28
1.14	Chemical structure of quinones	28
1.15	Pathway of UQ biosynthesis in *E. coli*	30
1.16	Anaerobic pathways	31
1.17	Nitrate respiration	33
1.18	Calvin-Benson cycle	35
1.19	Amino acid synthesis from their precursors	37
1.20	Alanine synthesis from PYR	38
1.21	Valine, leusine, and isoleusine biosynthesis	39
1.22	Glutamate and glutamine synthesis as well as aspartate and asparagine synthesis	39
1.23	Proline biosynthesis from glutamate	40
1.24	Arginine, ornitine, and citorline synthesis pathway	40
1.25	Lysine, threonine, and methionine biosynthesis	41
1.26	Aromatic amino acid synthesis pathways	43
1.27	Several control schemes for aromatic amino acid biosynthesis	44

1.28	Serine, glycine, and cystein synthesis pathways	44
1.29	Histisine synthesis pathways	45
1.30	Nucleic acids synthesis pathways	46
1.31	β oxidation and biosynthesis of fatty acid	47
1.32	Phosphotransferase system (PTS)	49
1.33	Fructose metabolism	50
1.34	Xylose metabolism	51
1.35	ATP balance for aerobic and anaerobic conditions	52
2.1	Batch cultivations of *E. coli* K12 using different DO levels and different carbon sources	58
2.2	2-DE gel maps of the total lysate of *E. coli* cells	63
2.3	The relative expression levels of *E. coli* K-12 proteins of central metabolic pathways under different conditions based on 2DE results	66
2.4	Comparison of the logarithmic protein expression ratios based on 2DE and the corresponding enzyme activity ratios	77
2.5	Growth curves of *E. coli* BW25113, which was grown in minimal media containing 10 g glucose/l as the sole carbon source	78
2.6	Relative gene expression of different global regulators and the metabolic pathway genes known to be regulated by those global regulators during different phases of growth	80
2.7	Concentration of different intracellular metabolites	83
2.8	Total and relative expression of iso-genes of *E. coli* metabolic pathways	84
2.9	Acetate metabolism	86
3.1	Overall metabolic regulation scheme	96
3.2	Outer and inner membrane and periplasm	98
3.3	Inducer exclusion and the activation of adenylate cyclase in the glucose-lactose system	101
3.4	The multiple regulations by Mlc and cAMP-Crp	103
3.5	Batch cultivation of (a) *E. coli* BW25113 and (b) its *cra* mutant	105
3.6	Comparison of enzyme activities of *cra* mutant as compared to the wild type (BW25113)	110
3.7	The effect of dilution rate on the gene transcript levels	113
3.8	Comparison of gene transcript levels of the wild type, *crp* knockout mutant, and *crp*⁺ mutant	116

3.9	Glucose PTS and fructose PTS	118
3.10	Central metabolic pathways and NH_3-assimilation pathways	121
3.11	Ammonia assimilation under diffferent NH_4^+ concentration	122
3.12	Effect of C/N ratio on the fermentation characteristics for the continuous culture at the dilution rate of 0.2 h^{-1}	123
3.13	Schematic illustration of the interaction among several metabolic regulations. Comparison of the transcriptional mRNA levels of the wild type *E.coli* genes cultivated at 100% (C/N = 1.68), 40% (C/N = 4.21), 20% (C/N = 8.42) and 10% (C/N = 1.68) N^- concentration: (a) global regulatory, (b) N^- regulatory, (c) metabolic pathway, (d) respiratory chain	124
3.14	The interaction between nitrogen regulation and catabolite regulation	126
3.15	Overall mechanism of nitrogen assimilation in *E. coli* under C-limited (N-rich) and N-limited conditions	129
3.16	Molecular mechanism of phosphate regulation	131
3.17	Comparison of the transcript levels of the wild type *E.coli* cultivated with different P concentrations of the feed (100%, 55%, 10%)	134
3.18	Schematic illustration of the interaction among several metabolic regulation mechanisms	136
3.19	Comparison of some gene expressions for parent *E. coli* BW25113 and *arcB* mutant at 4 h of batch cultivation along with the gel picture	141
3.20	Comparison of specific enzyme activities of *E. coli* BW25113, its *arcB* and *arcA* mutant at 4 h of batch cultivation	141
3.21	Comparison of the transcript levels between wild type and *fnr* mutant under micro-aerobic continuous culture conditions	144
3.22	Comparison of enzymes activities during micro-aerobic batch culture	145
3.23	Metabolic flux distributions of wild type and *fnr* mutant under micro-aerobic conditions	147
3.24	Comparison of gene expressions	151
3.25	Specific enzyme activities in cell extracts	152
3.26	The role of glutamate decarboxylase for acid resistance	154
3.27	Acid resistance mechanism under acidic conditions	157

3.28	Effect of temperature up-shift on gene expressions in E. coli BW25113 under aerobic continuous culture at the dilution rate of 0.2 h^{-1}	160
3.29	Effect of heat shock on gene and protein expressions and the fermentation characteristics	163
3.30	Metabolic pathways showing levels of enzymes (or proteins) and intracellular metabolite concentrations in the *fadR* mutant E. coli relative to those in the parent at the exponential phase grown in glucose minimal medium under aerobic conditions	171
3.31	Different kinds of sigma factor in E. coli	176
3.32	Schematic diagram on the function of sigma factor as a transcription factor	176
3.33	Various levels of σ^s regulation are differentially affected by various stress conditions	177
3.34	Growth curves of: (a) E. coli BW25113 (parent strain); and (b) E. coli JWK 2711 (*rpoS* mutant)	179
3.35	Intracellular metabolite concentrations of E. coli BW25113 (parent strain) and E. coli JWK 2711 (*rpoS* mutant)	184
4.1	Central reaction network for microalgae metabolism	221
4.2	Cultivation results of C. pyrenoidosa under autotrophic conditions	228
4.3	Cultivation results of C. pyrenoidosa under mixotrophic conditions	229
4.4	Cultivation results of C. pyrenoidosa under cyclic light-autotrophic/dark-heterotrophic conditions	230
4.5	Metabolic flux distribution of *Chlorella* cells	233
4.6	Energy conversion efficiency between the energy supplied to the culture, the energy absorbed by the cells, and the high free energy stored in ATP	238
4.7	Theoretical thermodynamic efficiency of ATP formation from the absorbed energy (Y_{ATP}/AE_{max}) as a linear function of the fraction of the absorbed light energy of the total	239
4.8	Conversion efficiency of: (a) total energy; and (b) light energy during the various growth phases of the mixotrophic cultivation	241
4.9	Energy conversion efficiency between the three energy forms during the first light/dark cycle of the cyclic autotrophic/heterotrophic cultivation	242

4.10	General metabolic pathway of E. coli under oxygen-limited conditions	248
4.11	Factors influencing lactate producing flux	252
4.12	Factors influencing flux through Pta–Ack pathway	255
4.13	The effect of Pyk activity on: (a) Pyk flux; and (b) glycolytic flux	257
5.1	Simple examples of flux calculation	267
5.2	IDV and MAV	268
5.3	Schematic illustration of the data NMR and MS, compared to isotopomer distribution	270
5.4	The fate of labeled carbons for when using: (a) glucose; and (b) acetate or pyruvate, as a carbon source	276
5.5	Transformation of IDVs into MDVs and FMDVs for a C_3 molecule	285
5.6	Schematic illustration of net flux and exchange flux	286
5.7	The relationship between the precursor and amino acid	290
5.8	Derivatization to generate $M\text{-}57^+$ and $M\text{-}159^+$	291
5.9	Net flux distribution in acetate metabolism of E. coli K12 in chemostat cultures at D of 0.11 and 0.22 h^{-1}	295
5.10	Net flux distribution in glucose metabolism of E. coli K12 in chemostat cultures at D of 0.11 and 0.22 h^{-1}	296
5.11	^{13}C isotopomer patterns and NMR multiple spectral patterns	300
5.12	The fate of carbons of precursor metabolites to amino acids	301
5.13	Carbon position and f values	302
5.14	^{13}C–^{13}C scalor coupling	309
5.15	Metabolic flux distribution of the wild type and pckA mutant	311
5.16	Predicted dependencies of the flux through Pck	316
5.17	Bioreaction network of E. coli central carbon metabolism	320
5.18	^{13}C–^{13}C scalar coupling multiplets observed for aspartate from glucose-limited chemostat cultures of E. coli W3110 and the pgi mutant	323
5.19	^{13}C–^{13}C scalar coupling multiplets observed for C-4 of glucose from ammonia-limited chemostat cultures of E. coli W3110 and the pgi mutant	326
5.20	^{13}C–^{13}C scalar coupling multiplets observed for Asp-α from ammonia-limited chemostat cultures of E. coli W3110 and the zwf mutant	330

5.21 Metabolic flux distribution in chemostat cultures of *E. coli* W3110 under glucose-limited conditions — 331

5.22 Specific rates of NADPH production and consumption in glucose (C)- and ammonia (N)-limited chemostat cultures of *E. coli* W3110, the *pgi* mutant, and the *zwf* mutant — 333

5.23 *In vivo* flux distributions in the central metabolism of *Synechocystis* sp. PCC6803 cultivated (a) heterotrophically and (b) mixotrophically — 343

5.24 Synthesis of the precursor for histidine (R5P or X5P) via the reactions of Calvin cycle — 346

6.1 Metabolic flux distribution in the chemostat culture of the wild type *E. coli* (a) and ppc mutant (b) — 365

6.2 Metabolic flux distributions in chemostat culture of glucose-grown *E. coli* parent strain (*upper values*), *gnd* (*middle values*), and *zwf* (*lower values*) mutants at $D = 0.2\ h^{-1}$ — 372

6.3 Metabolic flux distribution in chemostat culture of pyruvate-grown *E. coli* parent strain (*upper values*), *gnd* (*middle values*), and *zwf* (*lower values*) mutants at $D = 0.2\ h^{-1}$ — 373

6.4 Metabolic flux distribution of wild type (upper value) and *pykF*⁻ mutant (lower value) at dilution rate (D) of $0.1\ h^{-1}$ — 380

6.5 Enzyme activities for the *pyk F* mutant at two dilution rates (*D* values) — 381

6.6 Comparison of concentrations of intracellular metabolites in the *pykF* mutant at two *D* values — 381

6.7 Aerobic batch cultivation of (a) *E. coli* BW25113 and (b) *E. coli lpdA* mutant using glucose as carbon source in LB medium — 383

6.8 Enzyme activities of *E. coli* BW25113 and *E. coli lpdA* mutant under aerobic conditions in LB medium — 384

6.9 Comparison of the intracellular metabolite concentrations of *E. coli* BW25113 and *E. coli lpdA* mutant in the batch cultivation — 385

6.10 Metabolic flux distributions of wild type (upper values) at dilution rate of $0.2\ h^{-1}$ and *lpdA* mutant (lower values, underlined) at dilution rate of $0.22\ h^{-1}$ — 387

6.11 Comparison of enzyme activities of *E. coli* BW25113 and
E. coli poxB mutant under aerobic conditions using
synthetic medium   388
6.12 Aerobic batch cultivation of: (a) *E. coli* BW25113;
(b) *E. coli sucA* mutant; and (c) *E. coli sucC* mutant
using glucose as carbon source   391
6.13 Comparison of enzyme activities at a dilution rate of
0.2 h^{-1} in a continuous culture   393
6.14 Intracellular metabolite concentrations of *E. coli*
BW25113, *sucA* and *sucC* gene knockout mutants
at 0.2 h^{-1} specific growth rate in chemostat cultures   394
6.15 Comparison of some gene expression at a dilution rate
of 0.2 h^{-1} in a continuous culture   395
6.16 Metabolic flux distributions of wild-type (upper
values), *sucA* mutant (lower values) at dilution rate
of 0.2 h^{-1} 396
6.17 Batch cultivation results of: (a) parent *E. coli*; and (b) *icd
mutant*, grown on glucose under aerobic conditions   399
6.18 Batch cultivation results of: (a) parent *E. coli*; and
(b) *icd mutant*, grown on glucose under microaerobic
conditions   400
6.19 Batch cultivation results of: (a) parent *E. coli*; and (b) *icd
mutant*, grown on acetate under aerobic conditions   401
6.20 Relative protein and enzyme levels of *icd mutant*
JW1122 from exponential growth phase in comparison
to the parent *E. coli* BW25113 grown on glucose under
aerobic conditions   407
6.21 The cultivation results of: (a) *E. coli* BW25113, and
(b) the *pflB*⁻ mutant, grown on glucose under the aerobic
conditions   413
6.22 The cultivation results of: (a) *E. coli* BW25113; and
(b) the *pflB*⁻ mutant, grown on glucose under the
microaerobic conditions   414
6.23 The specific glucose uptake rates and the specific product
formation rates for: (a) the aerobic conditions; and (b) the
microaerobic conditions   415
6.24 Biomass and metabolite yields on glucose for:
(a) the aerobic conditions; and (b) the microaerobic
conditions   416
6.25 The pyruvate formate lyase activities of *E. coli pfl* mutants
and the parent strain *E. coli* BW25113   417

6.26 Enzyme activities for strains grown on glucose in the microaerobic conditions — 418
6.27 Metabolic regulation mechanisms of lactate production in E. coli pflA and pflB mutants — 423
6.28 Fermentative pathways of E. coli grown on glucose — 427
6.29 Batch cultivation results of: (a) parent E. coli; and (b) ldhA mutant E. coli, grown on glucose under anaerobic conditions — 429
6.30 Intracellular metabolite concentrations in the central metabolic pathways at the exponential growth phase for both ldhA mutant and parent E. coli — 432
6.31 Comparison of gene expressions for parent and ldhA mutant E. coli — 434
6.32 Anaerobic metabolism of glucose in E. coli — 436
6.33 NADH balances in both ldhA mutant and parent E. coli calculated based on the metabolic fluxes — 438

Tables

2.1 Activities of glycolytic and anaplerotic enzymes in response to carbon sources and DO levels — 60

2.2 Activities of PP and E-D pathway enzymes in response to carbon sources and DO levels — 61

2.3 Activities of fermentative enzymes in response to carbon sources and DO levels — 61

2.4 Activities of TCA cycle and glyoxylate shunt enzymes in response to carbon sources and DO levels — 62

2.5 Specific activity of enzymes of *E. coli* metabolic pathways in minimal media under aerobic growth conditions at different phases of growth — 82

3.1 Growth parameters for *E. coli* BW25113 and its *cra* mutant cultivated at the dilution rate of 0.2 h^{-1} where feed glucose concentration was 4 g/l — 106

3.2 Gene expressions of *cra* mutant as compared with the wild type strain — 107

3.3 Effects of dilution rate on fermentation characteristics of wild type *E. coli* — 112

3.4 Effects of the specific gene mutation on the fermentation characteristics at the dilution rate of 0.2 > h^{-1} — 115

3.5 Fermentation characteristics of the wild-type *E. coli* and its *phoB* and *phoR* mutants in the aerobic chemostat culture under different phosphate concentrations at the dilution rate of 0.2 h^{-1} at pH 7.0 — 133

3.6 Growth parameters of *E. coli* BW25113 and *arcB* mutant in aerobic batch cultures — 140

3.7 Specific rate of *soxR* and *soxS* mutants, and parent *E. coli*, grown on glucose under aerobic conditions — 150

3.8 Regulators involved in regulating glutamate-dependent acid resistance — 156

3.9 Fermentation parameters for the aerobic chemostat culture of the wild type *E. coli* BW25113 at the dilution rate of 0.2 h^{-1} — 160

3.10 Batch cultivation characteristics of the parent and the *fadR* mutant *E. coli* in glucose minimal medium under aerobic conditions 165

3.11 Differentially expressed proteins in the *fadR* mutant *E. coli* compared to the parent strain 166

3.12 Specific enzyme activities in cell extracts of the parent and the *fadR* mutant *E. coli* at the exponential phase grown in glucose minimal medium under aerobic conditions 168

3.13 Intracellular metabolite concentrations in the parent and the *fadR* mutant *E. coli* at the exponential phase grown in glucose minimal medium under aerobic conditions 170

3.14 Growth parameters of *E. coli* BW25113 (parent strain) and *E. coli* JWK 2711 (*rpoS* mutant) under aerobic growth conditions in LB media 180

3.15 Growth parameters of *E. coli* BW25113 (parent strain) and *E. coli* JWK 2711 (*rpoS* mutant) under aerobic growth conditions in LB media 181

3.16 Ratio of specific activities of enzymes of *E. coli* BW25113 (parent strain) and *E. coli* JWK 2711 (*rpoS* mutant) during exponential and early stationary phases of growth 183

4.1 Reactions in the networks of three different types of metabolism of *Chlorella* cell 224

4.2 Biochemical reactions for *Chlorella* cell 225

4.3 The consumption of glucose, CO_2 production, and O_2 uptake of *C. pyrenoidosa* under different cultivation conditions 231

4.4 The generation and utilization of ATP in the autotrophic, heterotrophic, and mixotrophic cultures 236

4.5 Theoretical yields of biomass on ATP and ATP maintenance requirements in the autotrophic, heterotrophic, and mixotrophic cultures 236

4.6 Contributions of light energy and glucose to ATP production in the exponential phase of mixotrophic cultures 237

4.7 Biomass yields on the supplied energy ($Y_{X/SE}$) in the autotrophic, mixotrophic, and cyclic autotrophic/heterotrophic culture experiments 240

4.8 Comparison of fermentation results 245

4.9 Comparison of enzyme activities 245

4.10	Comparison of intracellular metabolite concentrations in *E. coli* mutants	247
4.11	Effect of a single-gene knockout on the flux distribution	249
4.12	Effect of a single-gene knockout on flux partitions	250
4.13	Deviation index for LDH flux	253
5.1	Metabolic reactions for acetate metabolism	287
5.2	Metabolic reactions for glucose metabolism	288
5.3	Sensitivity of mass distribution (fragment $[M-159]^+$ of glutamate) upon changes in fluxes of Icl (*aceA*)	292
5.4	Sensitivity of mass distribution (fragment $[M-159]^+$ of glutamate) upon changes in exchange coefficients of Pck (*pckA*)	292
5.5	Experimental determined (exp)* and calculated (cal) fragment mass distribution of TBDMS-derived amino acids from *E. coli* K12 hydrolysates (chemostat culture by using acetate and glucose as the carbon source; D = 0.22 h^{-1})	293
5.6	90% confidence limits for estimated net fluxes and exchange coefficients in acetate metabolism	294
5.7	90% confidence limits for estimated net fluxes and exchange coefficients in glucose metabolism	294
5.8	Growth parameters of *E. coli* K12 at a D of 0.11 and 0.22 h^{-1}, where acetate is used as the sole carbon source	297
5.9	Growth parameters of *E. coli* K12 at a D of 0.11 and 0.22 h^{-1}, where glucose is used as the sole carbon source	297
5.10	Transformation matrix K for calculating f values	304
5.11	Growth parameters of chemostat cultures of *E. coli* wild-type W3110 and *pck* mutant (JWK3366)	307
5.12	Origins of metabolic intermediates in chemostat cultures of *E. coli* W3110 and *pck* mutant JWK3366 determined by flux ratio analysis	308
5.13	Specific enzymatic activities in chemostat cultures of *E. coli* W3110 and *pck* mutant JWK3366	313
5.14	Intracellular metabolite concentrations of *E. coli* W3110 and *pck* mutant JWK3366 in the continuous cultures	314
5.15	Growth parameters of glucose (C)- and ammonia (N)-limited chemostat cultures of *E. coli* wild-type strain W3110, the *pgi* mutant, and the *zwf* mutant	321
5.16	Protein, RNA, and glycogen contents of glucose (C)- and ammonia (N)-limited chemostat cultures of *E. coli* wild-type strain W3110, the *pgi* mutant, and the *zwf* mutant	322

5.17	Origins of metabolic intermediates in glucose (C)- and ammonia (N)-limited chemostat cultures of *E. coli* wild-type strain W3110, the *pgi* mutant, and the *zwf* mutant, as determined by flux ratio analysis	324
5.18	Relative abundances of intact carbon fragments at the carbon positions used for identification of the glyoxylate shunt activity in *E. coli* wild-type strain W3110 and the *pgi* mutant	325
5.19	Relative abundances of intact carbon fragments at the carbon positions used for identification of the ED pathway activity in glucose (C)- and ammonia (N)-limited cultures of the *pgi* mutant	328
5.20	Relative abundances of intact carbon fragments at the carbon positions used for identification of the origin of P5P and E4P pools in glucose (C)- and ammonia (N)-limited cultures of the *zwf* mutant	329
5.21	Stoichiometric reactions for cyanobacteria	335
5.22	Relative intensities of ^{13}C multiplet components of amino acids	338
5.23	Mass isotopomer distribution of ECF-derived amino acids	340
5.24	Independent constraints on the isotopomer distribution of amino acids available from labeling measurements	342
5.25	Growth parameters of exponentially growing *Synechocystis*	344
5.26	Estimated values and 90% confidence regions for estimated free fluxes	345
5.27	Estimated production and consumption of NADPH	349
5.28	Estimated production and consumption of ATP	350
6.1	Cell growth parameters of the wild-type *E. coli* and its *ppc* mutant grown on glucose under aerobic conditions	362
6.2	Specific enzyme activities of the wild-type *E. coli* and its *ppc* mutant grown on glucose under aerobic conditions	363
6.3	Intracellular metabolite concentrations in the wild-type *E. coli* and its *ppc* mutant grown on glucose under aerobic conditions	364
6.4	The NMR spectra of cellular amino acids in the wild-type *E. coli* and its *ppc* mutant	367
6.5	Exponential growth rates of *E. coli* wild-type (WT) and mutant cultures on glucose/pyruvate media	369
6.6	Metabolic parameters of *E. coli* continuous cultures at $D = 0.2\ h^{-1}$	369

6.7	Activities of enzymes located at key branch points and involved in NADPH formation	370
6.8	Absolute metabolic fluxes at several key branch points in the central metabolic pathways, when glucose or pyruvate were used as sole carbon sources	371
6.9	Fragment mass distribution of t-butyldimethylsilyl (TBDMS)-derived amino acids from the *pykF* mutant	378
6.10	Measured and simulated values of the NMR spectra of cellular amino acids	379
6.11	Growth characteristics of E. coli BW25113 at the dilution rate of $0.2\ h^{-1}$ and its *lpdA* mutant at the dilution rate of $0.22\ h^{-1}$ in continuous culture	386
6.12	Growth characteristics of parent strain E. coli BW25113 and its *sucA*, *sucC* mutants in the continuous culture at the dilution rate of $0.2\ h^{-1}$	392
6.13	Specific rate of parent and *icd* mutant grown on different carbon sources under different culture conditions	402
6.14	Summary of MALDI-TOF mass spectrometry data for protein spots showing altered expression levels on 2D gels for parent E. coli (WT) and *icd* mutant	403
6.15	Specific enzyme activities in cell extracts of parent and *icd* mutant grown on glucose under aerobic conditions	405
6.16	Measurement of intracellular metabolites for parent E. coli and *icd* mutant grown on glucose under aerobic conditions	408
6.17	The specific carbon source uptake rates and product formation rates for the E. coli *pfl*A mutant using different carbon sources under microaerobic conditions	416
6.18	The yields (Y) of cell mass (x) and metabolites for different carbon sources for the E. coli *pfl*A mutant grown under the microaerobic and the anaerobic conditions	417
6.19	Enzyme activities for the E. coli *pfl*A mutant grown on different carbon sources under microaerobic conditions	419
6.20	Intracellular metabolite concentrations in cells grown on glucose	421
6.21	Intracellular metabolite concentrations in the E. coli *pfl*A mutant grown on different carbon sources in microaerobic conditions	422
6.22	Specific rates of parent and *ldhA* mutant E. coli grown on glucose under anaerobic conditions	430

6.23 Specific enzyme activities in cell extracts of parent and *ldhA* mutant *E. coli* grown on glucose under anaerobic conditions 431

6.24 Comparison between the ratios of gene expressions, enzyme activities, and metabolic fluxes in *E. coli* grown on glucose under anaerobic conditions 439

About the author

Following a BS and an MS in Chemical Engineering at Nagoya University, Japan, and a PhD in Chemical Engineering at Northwestern University, USA, Kazuyuki Shimizu started his career in 1981 as a research associate at Nagoya University, Japan, and was promoted to associate professor in 1990. He then moved to the Kyushu Institute of Technology, Department of Bioscience and Bioinformatics as a professor in 1991. In 2000, he became an adjunct professor at the Institute of Advanced Biosciences (IAB), Keio University, Japan. Kazuyuki was appointed an associate member of the Science Council of Japan from 2006 to 2011.

He has been working on such topics as.

1. Metabolic flux analysis based on the ^{13}C labeling technique (^{13}C–MFA), where metabolic flux distribution is located on top of omics data, and the rigorous metabolic fluxes obtained by this method are used to analyze metabolic regulation of cells;
2. Integration of the different levels of hierarchical omics information, such as transcriptomics, proteomics, metabolomics, and fluxomics data, where it is critical to understand cell metabolism as a whole cell system; and
3. Systems biology, modeling of the main metabolic pathways of a cell using enzymatic reactions together with metabolic regulation based on global regulator (transcription factor), where metabolic pathway genes relationships may lead to the development of virtual microbes.

He recognizes the importance of uncovering the metabolic regulation mechanism of a cell system, based on both experimental (wet) and computational (dry) approaches. He has also organized a UK–Japan collaboration project on microbial systems biology toward developing virtual microbes.

The author may be contacted at:

Faculty of Computer Science and Systems Engineering
Kyushu Institute of Technology
Iizuka, Fukuoka, 820-8502
Japan
E-mail at shimi@bio.kyutech.ac.jp

or

Institute of Advanced Biosciences
Keio University
Tsuruoka, Yamagata, 997-0017
Japan
E-mail at shimi@ttck.keio.ac.jp

1

Main metabolism

Abstract: The main metabolic pathways, such as the glycolysis (EMP pathway), pentose phosphate pathway, the TCA cycle, the Entner Doudoroff pathway, as well as anaplerotic and gluconeogenetic pathways, are explained. The respiratory chain pathway is also explained in relation to energy generation. Amino acids and fatty acid synthetic pathways and various carbohydrate pathways are also explained.

Key words: Embden-Meyerhoff-Parnas pathway; glycolysis; pentose phosphate pathway; TCA cycle; respiratory chain; catabolism; anabolism; amino acid synthesis; fatty acid synthesis; carbohydrate metabolism.

1.1 Introduction

Metabolism describes the overall chemical or enzymatic reactions that occur in living organisms, which assimilate nutrients with high enthalpy and low entropy and gain free energy during the process of breakdown of nutrients into low enthalpy and high entropy substances, and thus keep the cells alive. All living organisms are in this irreversible state, and if this does not occur, the metabolic processes in the organism become in equilibrium, and the organism can no longer remain alive.

Let us define the **metabolite** as the substrate, intermediate, or product of each metabolic reaction in the cell, and let the **metabolic pathway** be a series of metabolic or enzymatic reactions from the specific substrate to the specific product. Typical central metabolic pathways are shown in Figure 1.1. When studying metabolic regulation or how cell metabolism is regulated, it is important to understand 'catabolism' and 'anabolism'.

Bacterial cellular metabolic systems

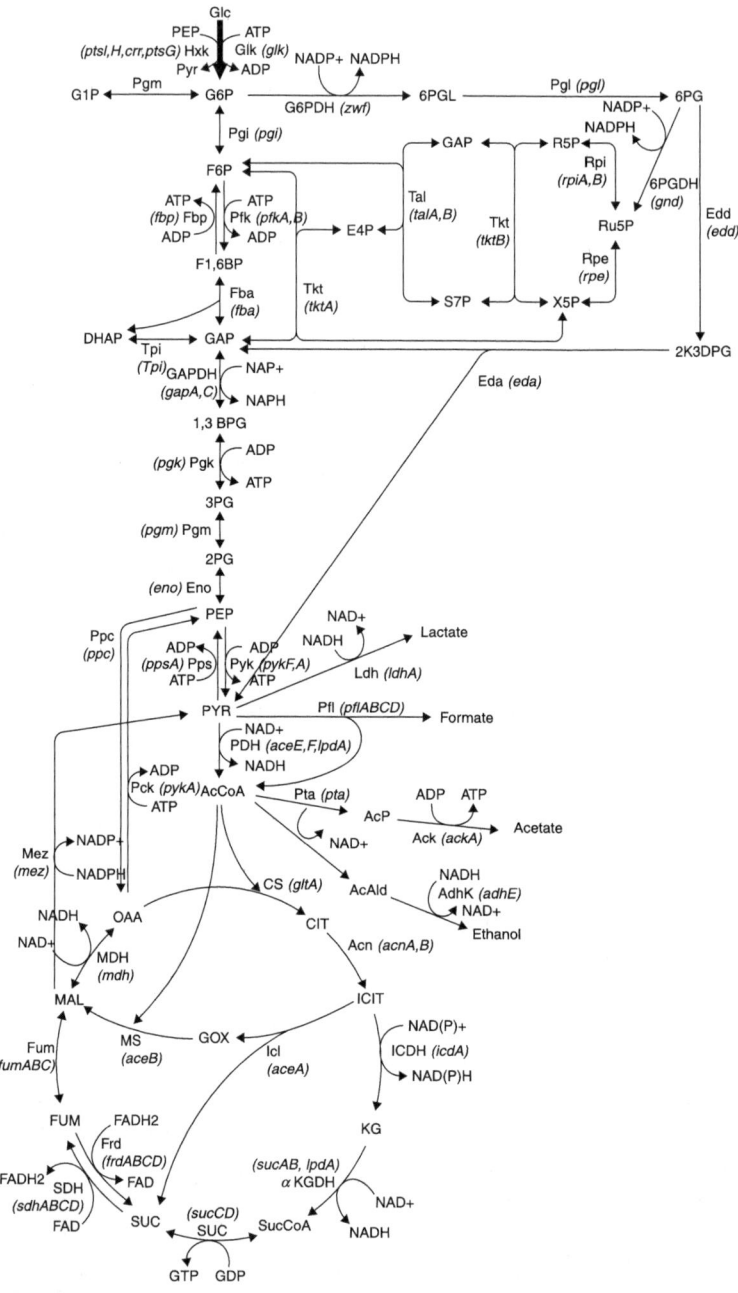

Figure 1.1   Main metabolic pathways

Catabolism is defined as all chemical or enzymatic reactions involved in the breakdown of organic or inorganic materials such as proteins, sugars, fatty acids, etc. in order to obtain energy. **Anabolism** is defined as biosynthetic reactions that lead to the building of cell materials such as proteins, DNA, RNA, lipids, etc. from small molecules, such as pyruvate, produced along the main metabolic pathways using the energy obtained by the process of catabolism.

Briefly, the cell generates energy as ATP (adenosine triphosphate), typically along the glycolysis pathway, from glucose to low molecules such as pyruvate (Figure 1.2). Moreover, the reducing equivalents, such as NADH and $FADH_2$, produced at the glycolysis and TCA (tri carboxylic acid) cycle pathways are oxidized in the respiratory chain, where ATP is produced via the oxidative phosphorylation process. Namely, these energy generating processes are termed **catabolism**. Cell constituents, such as proteins, cell membranes, etc. are formed from their precursor metabolites, such as 3-phosphoglycerate (3PG), phosphoenol-pyruvate (PEP), and pyruvate (PYR) etc., by using the ATP produced by the process of catabolism. This is the **anabolic process**.

1.2 Energy generation in the cell

The source of energy in a cell is ATP, which plays an essential role in a living organism. As shown in Figure 1.3, ATP contains high energy bonds, and if only one phosphate group of ATP is hydrolyzed, the free energy of this hydrolysis is 31 kJ (7.3 kcal), such that:

$$ATP + H_2O \rightarrow ADP + P_i \quad \Delta G = -7.3 \text{ kcal/mol} \quad (1.1)$$

under normal conditions with a pH at 7.0 and the temperature at 25°C. The active ATP often forms complex compounds with Mg^{2+} ions, which can result in a shift of the equilibrium. As stated above, during catabolism, the living organism has to generate ATP from ADP and inorganic phosphate P_i (HPO_3^{2+}) by reversing the above reaction, which is then stored by the cell to keep it alive. Note that ATP analogs, such as guanosine 5′-triphosphate (GTP), uridine 5′-triphosphate (UTP), and cytidine 5′-triphosphate (CTP), are also high energy compounds also comparable to ATP.

An enzyme that uses ATP as a substrate is called **transferase**, and an enzyme that transfers γ-phosphate of ATP to the substrate is called **kinase**:

$$S + ATP \rightarrow S\text{-phosphate} + ADP \quad (1.2)$$

Bacterial cellular metabolic systems

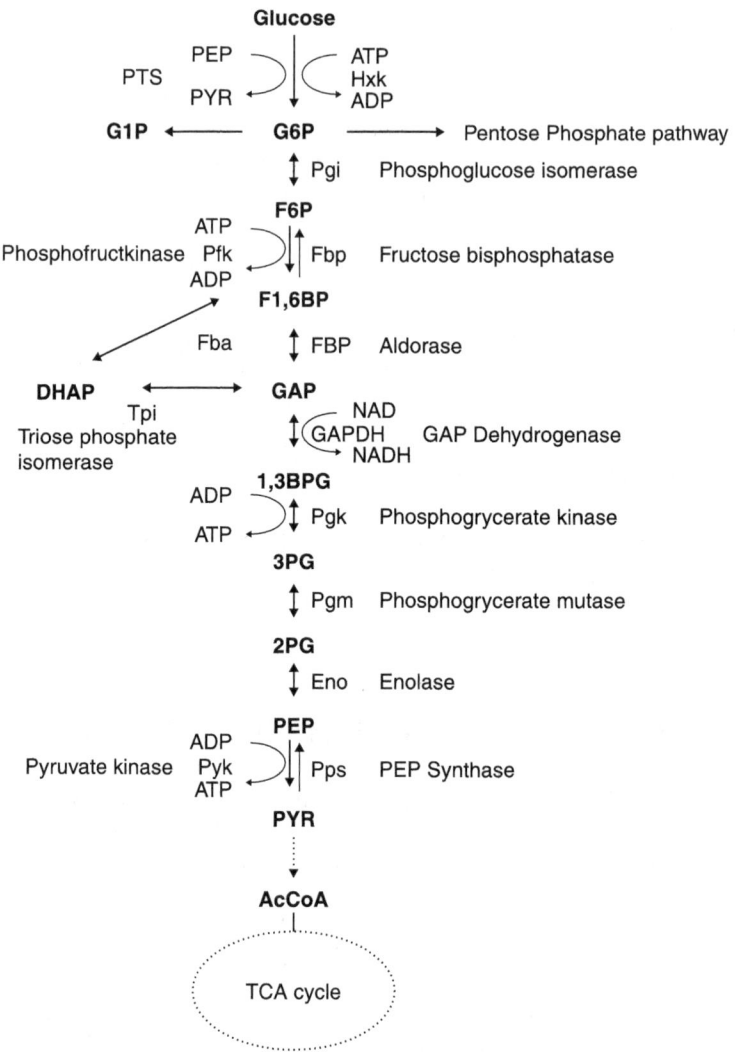

Figure 1.2 Metabolic pathways of glycolysis

where hexokinase (Hxk) and phosphofructokinase (Pfk), etc. belong to this category. Other kinases, such as phosphoglycerate kinase (Pgk), pyruvate kinase (Pyk), and acetate kinase (Ack), are those generating ATP. It may be useful to evaluate the energy charge as (Atkinson, 1968):

$$\text{Energy Charge} = \frac{[ATP] + 0.5[ADP] + 0.0[AMP]}{[ATP] + [ADP] + [AMP]} \quad (1.3)$$

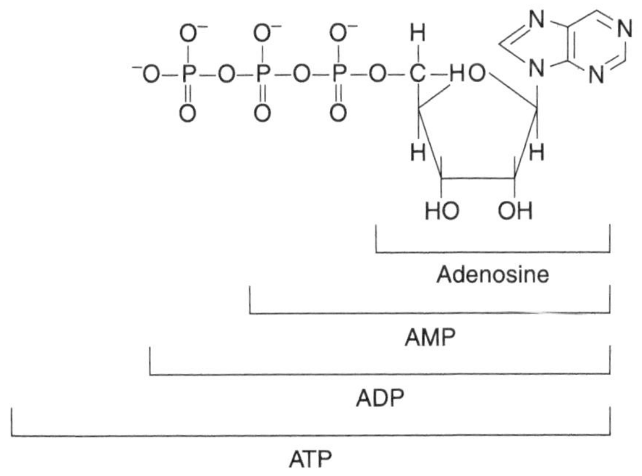

Main metabolism

Figure 1.3 Structure of ATP

where [ ] denotes the concentration.

In relation to ATP, the **reducing equivalents**, such as NADH, FADH$_2$, and NADPH, also play important roles in cell metabolism. Their chemical structures are shown in Figure 1.4, where NADH is utilized for energy generation by oxidative phosphorylation along the respiratory chain, and thus related to catabolism, while NADPH is the reducing power for amino acid synthesis or fatty acid formation, etc., and thus related to anabolism.

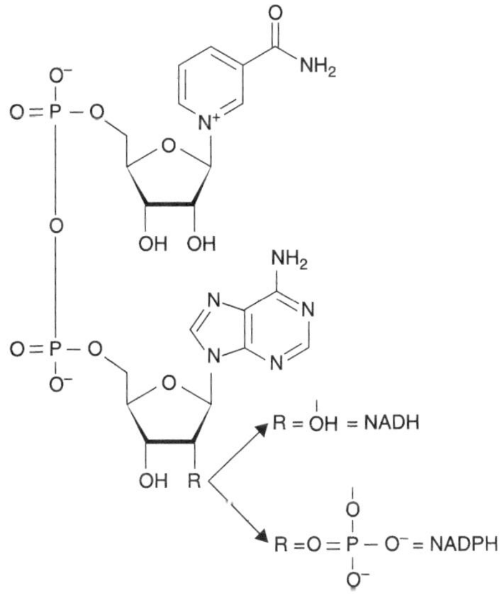

Figure 1.4 Structures of NADH and NADPH

NADH and NADPH can be inter-convertible by transhydrogenase in some bacteria, where Udh converts NADPH to NADH, while Pnt converts NADH to NADPH in *Escherichia coli*, which will be explained in more detail in relation to metabolism in later chapters. Although there are many biochemistry textbooks such as Horton et al. (1996), Lehninger et al. (1993), Stryer (1995), Voet and Voet (1995) etc., let us consider briefly the metabolic pathways one by one from the next section.

1.3 Carbohydrate metabolism

The typical carbohydrate for bacteria is glucose, which is converted to pyruvic acid (PYR) mainly via three different pathways such as the Embden-Meyerhof-Parnas (EMP) pathway, the pentose phosphate (PP) pathway (or hexose monophosphate: HMP) pathway, and the Entner-Doudoroff (ED) pathway. The EMP pathway is often referred to as glycolysis, but the PP and ED pathways can also be called glycolysis in some cases.

1.3.1 EMP pathway

The first step in glucose breakdown is the phosphorylation step that requires 1 mole of ATP catalyzed by hexokinase (Hxk) (ATP: D-hexose phosphotransferase, EC 2.7.1.1) to produce glucose 6-phosphate (G6P):

$$\begin{array}{c}
\text{H}\diagdown\text{C}\diagup\text{O} \\
| \\
\text{H-C-OH} \\
| \\
\text{HO-C-H} \\
| \\
\text{H-C-OH} \\
| \\
\text{H-C-OH} \\
| \\
\text{H-C-OH} \\
| \\
\text{H} \\
\textbf{Glc}
\end{array}
\quad\xrightarrow[\text{ATP}\quad\text{ADP}]{\text{Hxk}}\quad
\begin{array}{c}
\text{H}\diagdown\text{C}\diagup\text{O} \\
| \\
\text{H-C-OH} \\
| \\
\text{HO-C-H} \\
| \\
\text{H-C-OH} \\
| \\
\text{H-C-OH} \\
| \\
\text{H-C-O-P} \\
| \\
\text{H} \\
\textbf{G6P}
\end{array}
\qquad (1.4)$$

Note that enzymatic function relates to the catalytic specificity of an enzyme, which is described by its Enzyme Commission (EC) number

(Webb, 1992). Each enzyme-catalyzed reaction is classified based on a four-digit EC number, where the first number specifies the class of enzymes (1, oxidoreductases; 2, transferases; 3, hydrolases; 4, lyases; 5, isomerases; and 6, ligases). Subsequent digits provide additional function.

The detailed reaction scheme may be expressed as:

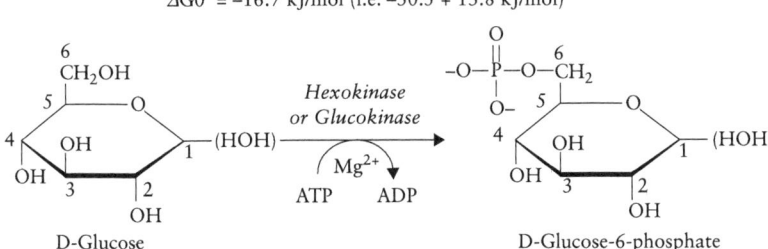

Another phosphorylation system (phosphotransferase system, PTS) using phosphoenol pyruvate (PEP) will be explained later in this and later chapters.

The second step is isomerization of G6P to fructose 6-phosphate (F6P) by the enzyme glucose phosphate isomerase, Pgi (D-glucose-6-phosphate-isomerase, EC 5.3.1.9), where this pathway is highly reversible:

$$
\begin{array}{c}
\text{H}\diagdown\!\!\!\diagup\text{O} \\
\text{C} \\
\text{H}-\text{C}-\text{OH} \\
\text{HO}-\text{C}-\text{H} \\
\text{H}-\text{C}-\text{OH} \\
\text{H}-\text{C}-\text{OH} \\
\text{H}-\text{C}-\text{O}-\text{P} \\
\text{H} \\
\textbf{G6P}
\end{array}
\xrightarrow{\text{Pgi}}
\begin{array}{c}
\text{H} \\
\text{H}-\text{C}-\text{OH} \\
\text{C}=\text{O} \\
\text{HO}-\text{C}-\text{H} \\
\text{H}-\text{C}-\text{OH} \\
\text{H}-\text{C}-\text{OH} \\
\text{H}-\text{C}-\text{O}-\text{P} \\
\text{H} \\
\textbf{F6P}
\end{array}
\tag{1.5}
$$

The third step or the second phosphorylation step requires additional ATP for phosphorylation to produce fructose 1,6-bisphosphate (F16BP or FDP) from F6P. This reaction is catalyzed by the important enzyme, phosphofructokinase (Pfk) (ATP: D-fructose-6-phosphate 1 phosphotransferase, EC 2.7.1.11), which regulates the flow through the EMP pathway, and sometimes becomes a rate-limiting factor of the EMP pathway. It has been shown that Pfk is allosterically inhibited by PEP:

$$\text{F6P} \xrightarrow[\text{ATP} \quad \text{ADP}]{\text{Pfk}} \text{F1,6BP(FDP)} \quad (1.6)$$

The next step is the cleavage of 1,6-diphosphate (F16BP or FDP) to 2 moles of triose phosphate. The enzyme that catalyzes this reaction is fructose diphosphate aldolase (Fba) (fructose-1,6-bisphosphate: D-glyceraldehyde-3-phosphate-lyase, EC 4.1.2.13):

$$\text{F1,6BP} \xrightarrow{\text{Fba}} \text{GAP} + \text{DHAP} \quad (1.7)$$

The cleavage of FDP to 2 moles of triose phosphate, such as glyceraldehyde 3-phosphate (GAP) and dihydroxyacetone phosphate (DHAP), can be interchangeable with the reversible triose phosphate isomerase (Tpi) (D-glyceraldehyde-3-phosphate ketol-isomerase, EC 5.3.1.1). The equilibrium occurs from DHAP toward GAP, assuming the EMP pathway functions properly:

$$\text{DHAP} \xleftrightarrow{\text{Tpi}} \text{GAP} \quad (1.8)$$

Step 6 is a combined oxidation and phosphorylation step, which is catalyzed by glyceraldehyde-phosphate dehydrogenase GAPDH (D-glyceraldehyde-3-phosphate: NAD oxidoreductase, EC 1.2.1.12), and produces NADH and the high-energy component 1,3-bisphospho-D-glycerate (1,3BPG):

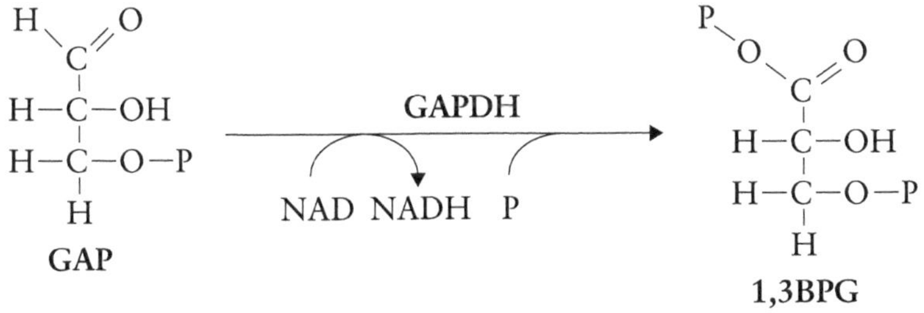

$$\begin{array}{c} H\diagdown\!\!\!\diagup O \\ C \\ | \\ H-C-OH \\ | \\ H-C-O-P \\ | \\ H \\ \textbf{GAP} \end{array} \xrightarrow[\text{NAD NADH P}]{\textbf{GAPDH}} \begin{array}{c} P\diagdown\!\!\!\diagup O\diagdown\!\!\!\diagup O \\ C \\ | \\ H-C-OH \\ | \\ H-C-O-P \\ | \\ H \\ \textbf{1,3BPG} \end{array}$$

(1.9)

In step 7, the high-energy compound 1,3 BPG releases 1 phosphate group as 1 mole of ATP by the reaction catalyzed by phosphoglycerate kinase (Pgk) (ATP: 3-phospho-D-glycerate 1-phospho-transferase, EC 2.7.2.3) to produce 3-phosphoglycerate (3 PG):

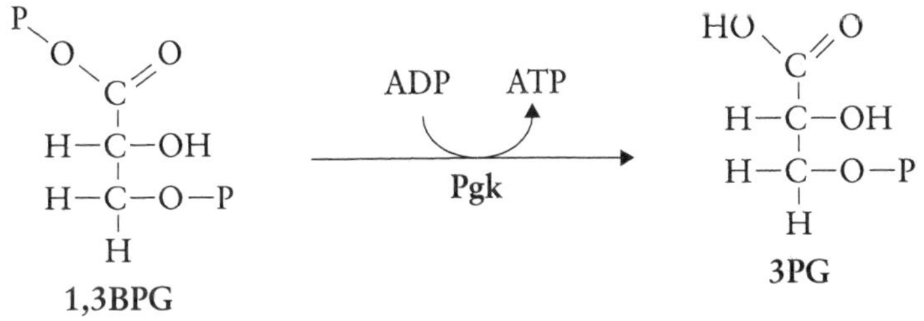

$$\begin{array}{c} P\diagdown\!\!\!\diagup O\diagdown\!\!\!\diagup O \\ C \\ | \\ H-C-OH \\ | \\ H-C-O-P \\ | \\ H \\ \textbf{1,3BPG} \end{array} \xrightarrow[\textbf{Pgk}]{\text{ADP ATP}} \begin{array}{c} HO\diagdown\!\!\!\diagup O \\ C \\ | \\ H-C-OH \\ | \\ H-C-O-P \\ | \\ H \\ \textbf{3PG} \end{array}$$

(1.10)

The conversion of 3 PG to 2 phosphoglycerate (2 PG) is catalyzed by phosphoglycerate mutase (Pgm) (2,3-diphospho-D-glycerate: 2 phospho-D-glycerate phosphotransferase, EC 2.7.5.3). In this reaction, the phosphate group is transferred from the third position to the second position:

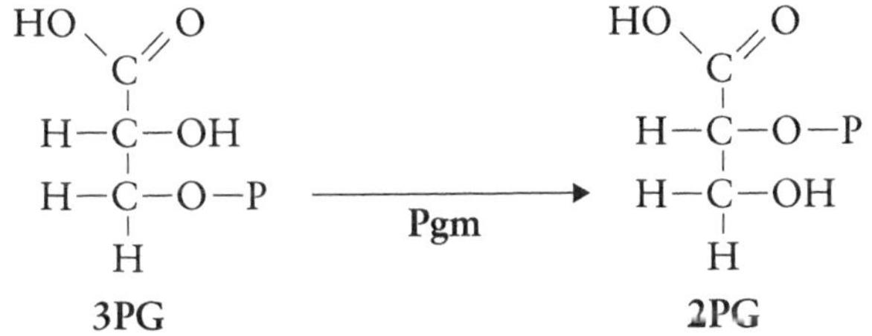

$$\begin{array}{c} HO\diagdown\!\!\!\diagup O \\ C \\ | \\ H-C-OH \\ | \\ H-C-O-P \\ | \\ H \\ \textbf{3PG} \end{array} \xrightarrow{\textbf{Pgm}} \begin{array}{c} HO\diagdown\!\!\!\diagup O \\ C \\ | \\ H-C-O-P \\ | \\ H-C-OH \\ | \\ H \\ \textbf{2PG} \end{array}$$

(1.11)

The next reaction step is catalyzed by enolase (Eno) (2-phospho-D-glycerate hydro-lyase, EC 4.2.1.11) to form phosphoenol pyruvate (PEP) (step 9):

$$\text{2PG} \xrightarrow[\text{H}_2\text{O}]{\text{Eno}} \text{PEP} \quad (1.12)$$

This reaction is connected with an intracellular electron shift, often referred to as the intra-molecular oxidation-reduction reaction. By this reaction, PEP is more energy-rich as compared to 2 PG.

In the final step (step 10) of the EMP pathway, the reaction is catalyzed by pyruvate kinase (Pyk) (ATP: pyruvate phosphotransferase, EC 2.7.1.40) to produce pyruvate (PYR), and the phosphate group of PEP is transferred to ADP to produce ATP. Pyk is the second allosteric enzyme in the EMP pathway:

$$\text{PEP} \xrightarrow[\text{ADP} \quad \text{ATP}]{\text{Pyk}} \text{PYR} \quad (1.13)$$

1.3.2 Pentose phosphate (PP) pathway or hexose monophosphate (HMP) pathway

Figure 1.5 shows the overall pentose phosphate (PP) pathway, which connects to the EMP pathway at G6P, F6P, and GAP. In the PP pathway, G6P is oxidized by the NADP$^+$-linked G6P dehydrogenase (G6PDH) (D-glucose-6-phosphate: NADP oxidoreductase, EC 1.1.1.49) to produce D-glucono-δ-lactone 6-phosphate (PGL). G6PDH is an important regulatory enzyme, where it is inhibited by NADPH:

$$\text{G6P} \xrightarrow[\text{G6PDH}]{\text{NADP}^{2+} \quad \text{NADPH} + \text{H}^+} \text{PGL} \quad (1.14)$$

Main metabolism

The product of the G6PDH reaction is almost immediately hydrolyzed to 6-phosphogluconate (6PG) by gluconolactonase (D-glucono-δ-lactone hydrolase, EC 3.1.1.17):

$$\text{PGL} \xrightarrow[\text{Pgl}]{H_2O} \text{6PG} \quad (1.15)$$

A second NADP$^+$-linked oxidation produces D-ribulose 5-phosphate (Rib5P) from 6 PG this reaction is catalyzed by phosphogluconate dehydrogenase (6PGDH), where the C-1 atom of 6 PG is released as CO_2 to form Rib5P. 6PGDH is also an important regulatory enzyme, where it is inhibited by NADPH:

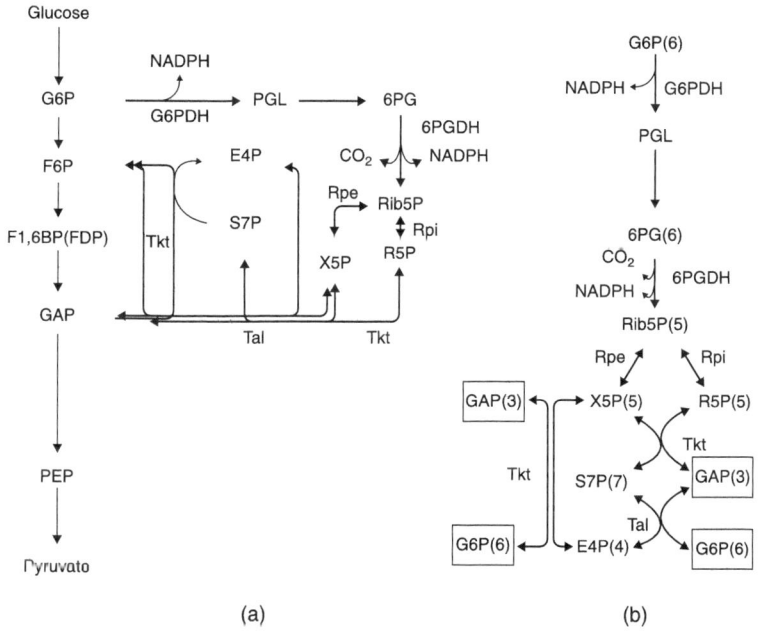

Figure 1.5 (a) Pentose phosphate pathway, and (b) the change in carbon numbers

$$\begin{array}{c} \text{COO}^- \\ | \\ \text{H}-\text{C}-\text{OH} \\ | \\ \text{HO}-\text{C}-\text{H} \\ | \\ \text{H}-\text{C}-\text{OH} \\ | \\ \text{H}-\text{C}-\text{OH} \\ | \\ \text{CH}_2\text{OPO}_3\text{H} \\ \textbf{6PG} \end{array} \quad \xrightarrow[\textbf{6PGDH}]{\text{NADP}^+ \;\; \text{NADPH} + \text{CO}_2} \quad \begin{array}{c} \text{H}_2\text{C}-\text{OH} \\ | \\ \text{C}=\text{O} \\ | \\ \text{HC}-\text{OH} \\ | \\ \text{HC}-\text{OH} \\ | \\ \text{H}_2\text{C}-\text{OPO}_3^{2-} \\ \textbf{Rib5P} \end{array} \quad (1.16)$$

Rib5P is attacked partly by two different enzymes, such as ribose phosphate 3-epimerase (Rpe) (D-ribulose-5-phosphate 3-epimerase, EC 5.1.3.1), which converts Rib5P to xylulose 5-phosphate (X5P), and ribose 5-phosphate isomerase (Rpi) (D-ribose-5-phosphate ketol-isomerase, EC 5.3.1.6), which converts Rib5P to ribose 5-phosphate (R5P):

$$\begin{array}{c} \text{H}_2\text{C}-\text{OH} \\ | \\ \text{C}=\text{O} \\ | \\ \text{HC}-\text{OH} \\ | \\ \text{HC}-\text{OH} \\ | \\ \text{H}_2\text{C}-\text{OPO}_3^{2-} \\ \textbf{Rib5P} \end{array} \quad \begin{array}{c} \xrightarrow{\text{Rpi}} \\ \xrightarrow{\text{Rpe}} \end{array} \quad \begin{array}{c} \begin{array}{c} \text{H}\diagdown\!\!\!\diagup\text{O} \\ \text{C} \\ | \\ \text{HC}-\text{OH} \\ | \\ \text{HC}-\text{OH} \\ | \\ \text{HC}-\text{OH} \\ | \\ \text{H}_2\text{C}-\text{OPO}_3^{2-} \\ \textbf{R5P} \end{array} \\ \\ \begin{array}{c} \text{H}_2\text{C}-\text{OH} \\ | \\ \text{C}=\text{O} \\ | \\ \text{HO}-\text{CH} \\ | \\ \text{HC}-\text{OH} \\ | \\ \text{H}_2\text{C}-\text{OPO}_3^{2-} \\ \textbf{X5P} \end{array} \end{array} \quad (1.17)$$

Both intermediates, such as X5P and R5P, are required for the cleavage reaction catalyzed by transketolase (sedoheptulose-7-phosphate: D-glyceraldehyde-3-phosphate glycolaldehyde transferase, EC 2.2.1.1), which yields GAP and sedoheptulose 7-phosphate (S7P):

$$\begin{array}{c}
\text{R5P} + \text{X5P} \xrightarrow{\text{Tkt}} \text{S7P} + \text{GAP}
\end{array} \tag{1.18}$$

R5P:
H–C(=O), HC–OH, HC–OH, HC–OH, H$_2$C–OPO$_3^{2-}$

X5P:
H$_2$C–OH, C=O, HO–CH, HC–OH, H$_2$C–OPO$_3^{2-}$

S7P:
H$_2$C–OH, C=O, HO–CH, HC–OH, HC–OH, HC–OH, H$_2$C–OPO$_3^{2-}$

GAP:
H–C(=O), HC–OH, H$_2$C–OPO$_3^{2-}$

The second cleavage reaction cleaves both intermediates from the transketolase reaction to produce F6P and erythrose 4-phosphate (E4P), where this reaction is catalyzed by transaldorase (D-glyceraldehyde-3-phosphate dehydroxy acetone transferase, EC 2.2.1.2). Note that E4P and R5P are the important precursors for purine, pyrimidine, and aromatic amino acids:

$$\text{S7P} + \text{GAP} \xrightarrow{\text{Tal}} \text{E4P} + \text{F6P} \tag{1.19}$$

S7P:
H$_2$C–OH, C=O, HO–CH, HC–OH, HC–OH, HC–OH, H$_2$C–OPO$_3^{2-}$

E4P:
H–C(=O), HC–OH, HC–OH, H$_2$C–OPO$_3^{2-}$

GAP:
H–C(=O), HC–OH, H$_2$C–OPO$_3^{2-}$

F6P:
H$_2$C–OH, C=O, HO–CH, HC–OH, HC–OH, H$_2$C–OPO$_3^{2-}$

The third cleavage reaction is carried out by the same transketolase as the first stage reaction, and cleaves E4P and X5P to form GAP and F6P:

$$\begin{array}{c} \text{E4P} + \text{X5P} \xrightarrow{\text{Tkt}} \text{GAP} + \text{F6P} \end{array} \quad (1.20)$$

Note that the pathway reaction from G6P to Rib5P is unidirectional and is known as the **oxidative PP pathway**, while the other reactions in the PP pathway are reversible, so it is called the **non-oxidative PP pathway**.

1.3.3 Entner-Doudoroff pathway

As shown in Figure 1.6, the Entner-Doudoroff (ED) pathway connects to the PP pathway at 6 PG, where 6 PG is converted to 2-keto-3-deoxy-6-phosphogluconate (KDPG) by dehydration reaction catalyzed by phospho-gluconate dehydratase (Edd) (6-phosphogluconate hydro-lyase, EC 4.2.1.12):

$$\text{6PG} \xrightarrow[\text{Edd}]{-H_2O} \text{KDPG} \quad (1.21)$$

Main metabolism

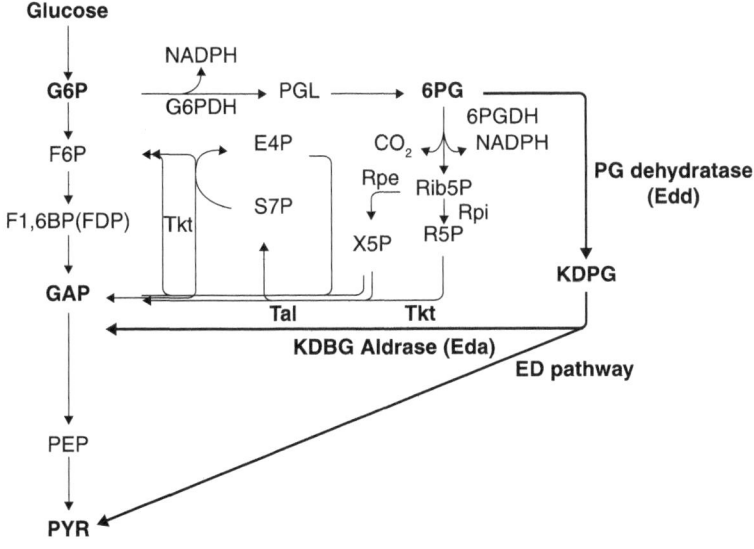

Figure 1.6 Entner Doudoroff pathway

The next step is the cleavage of KDPG to GAP and PYR by phospho-2-keto-3-deoxy-gluconate aldorase (Eda) (6-phospho-2-keto-3-deoxy-D-gluconate D-glyceraldehyde-3-phosphate lyase, EC 4.1.2.14):

$$\text{KDPG} \xrightarrow{\text{Eda}} \text{GAP} + \text{PYR} \qquad (1.22)$$

Bacterial cellular metabolic systems

Although the ED pathway is active in such microorganisms as *Zymomonas mobilis*, this may be induced in other bacteria such as *E. coli*, depending on the genetic and culture conditions, as will be explained later in this book.

1.3.4 PDH and TCA cycle

As shown in Figure 1.7, the terminal product of EMP or ED pathways is PYR, where it is converted to a two carbon acid derivative, such as acetyl CoA (AcCoA), by releasing CO_2 from the first carbon of PYR by the pyruvate dehydrogenase (PDHc) reaction, where this is a multi-enzyme complex consisting of three different enzymes as well as the cofactors, thiamine pyrophosphate (TPP), lipoic acid, and NAD^+.

The three enzymes are pyruvate dehydrogenase (pyruvate: lipoate oxidoreductase, EC 1.2.4.1), lipoate acetyltransferase (acetyl-CoA:

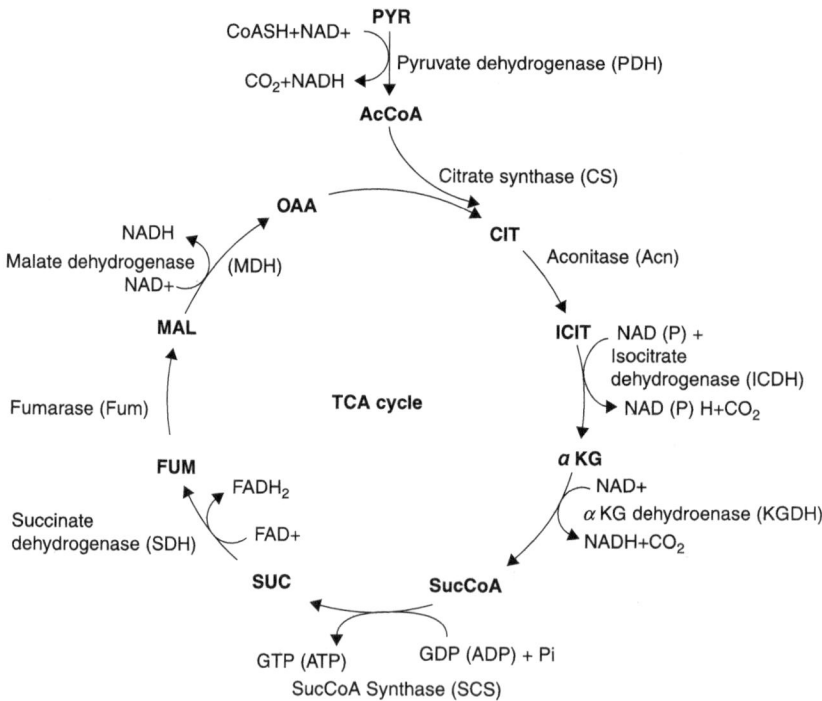

Figure 1.7 PDH and TCA cycle

Main metabolism

dihydrolipoate S-acetyltransferase, EC 2.3.1.12), and lipoamid dehydrogenase (reduced-NAD: lipoamid oxidoreductase, EC 1.6.4.3):

$$\underset{\text{PYR}}{\begin{array}{c} O \\ \| \\ C-O^- \\ | \\ C=O \\ | \\ CH_3 \end{array}} \xrightarrow[\text{PDH}]{\text{CoA-SH + NAD + } \quad \text{NADH + CO}_2} \underset{\text{AcCoA}}{\begin{array}{c} S-CoA \\ | \\ C=O \\ | \\ CH_3 \end{array}} \qquad (1.23)$$

The acetyl CoA thus produced from PYR goes into a series of reactions called either the tricarboxylic acid (TCA) cycle, the Krebs cycle, or the citric acid cycle (Figure 1.7). In the first step of the TCA cycle, AcCoA gives the acetyl group to the four carbon dicarboxylic acids such as oxaloacetate (OAA) to form a six-carbon tricarboxylic acid such as citric acid (CIT). This reaction is catalyzed by citrate synthase (CS) (citrate oxaloacetate-lyase, EC 4.1.3.7), where free CoA is generated, and this can be re-utilized in the formation of AcCoA. Note that the end products of the TCA cycle are NADH and CO_2, where NADH allosterically inhibits the activity of CS:

$$\underset{\text{OAA}}{\begin{array}{c} COO^- \\ | \\ C=O \\ | \\ CH_2 \\ | \\ COO^- \end{array}} + \underset{\text{AcCoA}}{\begin{array}{c} S-CoA \\ | \\ C=O \\ | \\ CH_3 \end{array}} \xrightarrow[\text{CS}]{\text{CoA-SH}} \underset{\text{CIT}}{\begin{array}{c} COO^- \\ | \\ CH_2 \\ | \\ HOC-COO^- \\ | \\ CH_2 \\ | \\ COO^- \end{array}} \qquad (1.24)$$

The next steps are the formation of *cis*-aconitate and then isocitrate by the enzyme aconitate hydratase or aconitase (Acn) (citrate (iso-citrate) hydro-lyase, EC 4.2.1.3):

$$\underset{\text{CIT}}{\begin{array}{c} COO^- \\ | \\ CH_2 \\ | \\ ^-OOC-C-OH \\ | \\ CH_2 \\ | \\ COO^- \end{array}} \xrightarrow{\text{Acn}} \underset{\text{ICIT}}{\begin{array}{c} COO^- \\ | \\ H-C-OH \\ | \\ ^-OOC-C-H \\ | \\ CH_2 \\ | \\ COO^- \end{array}} \qquad (1.25)$$

Isocitric acid (ICIT) is then converted to α-ketoglutaric acid (αKG) or 2-oxoglutarate (2KG) by the reaction catalyzed by isocitrate dehydrogenase (ICDH) (threo-D_s-isocitrate: NADP oxidoreductase, EC 1.1.1.42). NAD(P)H and CO_2 are formed through this reaction, where the microorganisms possess predominantly the $NADP^+$-specific ICDH, whereas fungi and yeasts possess the NAD^+-specific ICDH (Doelle, 1975):

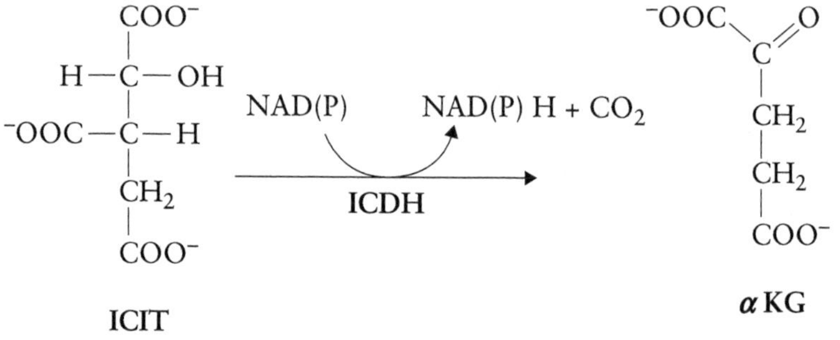

$$\begin{array}{c}
COO^- \\
| \\
H-C-OH \\
| \\
^-OOC-C-H \\
| \\
CH_2 \\
| \\
COO^-
\end{array}
\quad
\begin{array}{c}
NAD(P) \quad NAD(P)H + CO_2 \\
\searrow \quad \nearrow \\
\xrightarrow{\text{ICDH}}
\end{array}
\quad
\begin{array}{c}
^-OOC\diagdown_{C}\diagup^{O} \\
| \\
CH_2 \\
| \\
CH_2 \\
| \\
COO^-
\end{array}$$

ICIT $\qquad\qquad\qquad\qquad\qquad\qquad\qquad$ αKG

(1.26)

The next reaction step is the conversion of αKG to succinyl CoA (SucCoA) by the 2-oxoglutarate dehydrogenase complex (KGDH) (2-oxoglutarate lipoate oxidoreductase, EC 1.2.4.2), a multi-enzyme complex system, similar to PDH. The KGDH complex requires the participation of thiamine pyrophosphate (TPP), α-lipoic acid, CoA, NAD^+, and Mg^{2+}, to produce CO_2 and NADH:

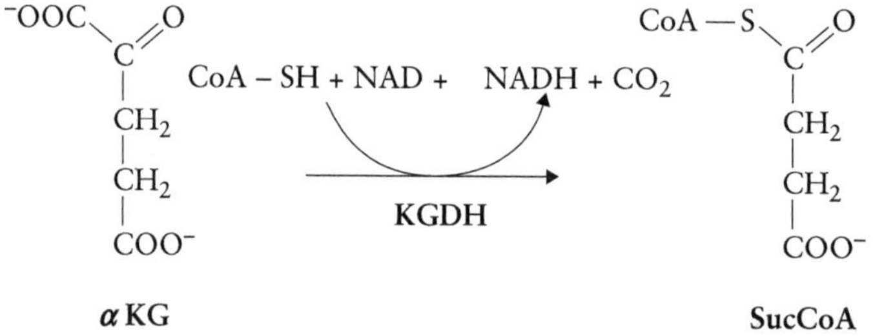

$$\begin{array}{c}
^-OOC\diagdown_{C}\diagup^{O} \\
| \\
CH_2 \\
| \\
CH_2 \\
| \\
COO^-
\end{array}
\quad
\begin{array}{c}
CoA-SH+NAD+ \quad NADH+CO_2 \\
\searrow \quad \nearrow \\
\xrightarrow{\text{KGDH}}
\end{array}
\quad
\begin{array}{c}
CoA-S\diagdown_{C}\diagup^{O} \\
| \\
CH_2 \\
| \\
CH_2 \\
| \\
COO^-
\end{array}$$

αKG $\qquad\qquad\qquad\qquad\qquad\qquad\qquad$ SucCoA

(1.27)

SucCoA is then converted to succinate (SUC) by succinyl-CoA synthetase (SCS) (succinate: CoA ligase, EC 6.2.1.5). Through this reaction step, CoA is released and ATP (GTP) is formed:

$$\underset{\textbf{SucCoA}}{\begin{array}{c} \text{CoA}-\text{S}\diagdown_{\text{C}}\diagup^{\text{O}} \\ | \\ \text{CH}_2 \\ | \\ \text{CH}_2 \\ | \\ \text{COO}- \end{array}} \xrightarrow[\text{SCS}]{\text{ADP (GDP)} + \text{P} \quad \text{ATP (GTP)} + \text{CoA}} \underset{\textbf{SUC}}{\begin{array}{c} \text{COO}- \\ | \\ \text{CH}_2 \\ | \\ \text{CH}_2 \\ | \\ \text{COO}- \end{array}} \quad (1.28)$$

The next step is the dehydrogenation of SUC, where SUC is oxidized to fumarate (FUM) by succinate dehydrogenase (SDH) (succinate: oxidoreductase, EC 1.3.99.1). SDH is closely linked to the electron transport chain, and enters this system at the flavoprotein level, where $FADH_2$ is released by this reaction step:

$$\underset{\textbf{SUC}}{\begin{array}{c} \text{COO}^- \\ | \\ \text{CH}_2 \\ | \\ \text{CH}_2 \\ | \\ \text{COO}^- \end{array}} \xrightarrow[\text{SDH}]{\text{FAD} \quad \text{FADH}_2} \underset{\textbf{FUM}}{\begin{array}{c} \text{COO}^- \\ | \\ \text{CH} \\ || \\ \text{HC} \\ | \\ \text{COO}^- \end{array}} \quad (1.29)$$

Fumarate thus formed is then hydrated at the double bond to form malic acid (MAL) by fumarase or fumarate hydratase (L-malate hydro-lyase, EC 4.2.1.2):

$$\underset{\textbf{FUM}}{\begin{array}{c} \text{COO}^- \\ | \\ \text{CH} \\ || \\ \text{HC} \\ | \\ \text{COO}^- \end{array}} \xrightarrow[\textbf{Fum}]{H_2O} \underset{\textbf{MAL}}{\begin{array}{c} \text{COO}^- \\ | \\ \text{HOCH} \\ | \\ \text{CH}_2 \\ | \\ \text{COO}^- \end{array}} \quad (1.30)$$

The final reaction in the TCA cycle is the dehydrogenation of MAL to oxaloacetate (OAA) catalyzed by malate dehydrogenase (MDH) (L-malate: NAD oxidoreductase, EC 1.1.1.37), where NADH is produced in this reaction step:

$$\underset{\text{MAL}}{\begin{array}{c}\text{COO}^-\\|\\\text{HOCH}\\|\\\text{CH}_2\\|\\\text{COO}^-\end{array}} \xrightarrow[\text{MDH}]{\text{NAD} \quad \text{NADH}} \underset{\text{OAA}}{\begin{array}{c}\text{COO}^-\\|\\\text{C}=\text{O}\\|\\\text{CH}_2\\|\\\text{COO}^-\end{array}}$$

(1.31)

1.3.5 Acetate metabolism

Microorganisms such as *E. coli* produce acetate from AcCoA by the so-called overflow metabolism, while ethanol is formed from PYR in yeast, etc. by a similar mechanism. There are two major acetate producing pathways, the phosphoacetyltransferase/acetate kinase (Pta-Ack) and pyruvate oxidase (Pox) pathways (Figure 1.8). The phosphoacetyltransferase (Pta) [EC 2.3.1.8] converts AcCoA and inorganic phosphate to acetyl phosphate (AceP) and CoA, while acetate kinase (Ack) [EC 2.7.2.1] converts AceP and ADP to acetate (Ace) and ATP in reversible reactions. Note that the cells may require ATP by the Pta-Ack pathway as the second energy source during high growth rates and anaerobic conditions, and that the Pta-Ack pathway may play an important role in the regulation of AceP, where AceP plays a role in metabolic regulation, as will be explained later in this book. The other acetate producing pathway is through pyruvate oxidase (Pox) [EC 1.2.2.2], a peripheral membrane protein that converts PYR, ubiquinone, and H_2O to acetate, ubiquinol, and CO_2, respectively. This pathway is usually induced at the early stationary phase, where acetyl-CoA synthetase (ACS) (acetate: CoA ligase, EC 6.2.1.1) is also activated to convert acetate to AcCoA at the stationary phase.

1.3.6 Anaplerotic pathways

There may exist a close interrelationship between catabolism for energy generation and anabolism for biosynthesis. Some of the intermediates in the main metabolic pathway are necessary as precursors for biosynthesis, as the cell can no longer survive without any of these. Thus the organism must have an ancillary system to take care of replenishment of these intermediates. The routes required for this replenishment are called **anaplerotic** routes (Doelle, 1975). Among the intermediates, OAA is often critical, since its concentration is relatively low due to its utilization by the CS reaction, while it is also a precursor for many amino acids such

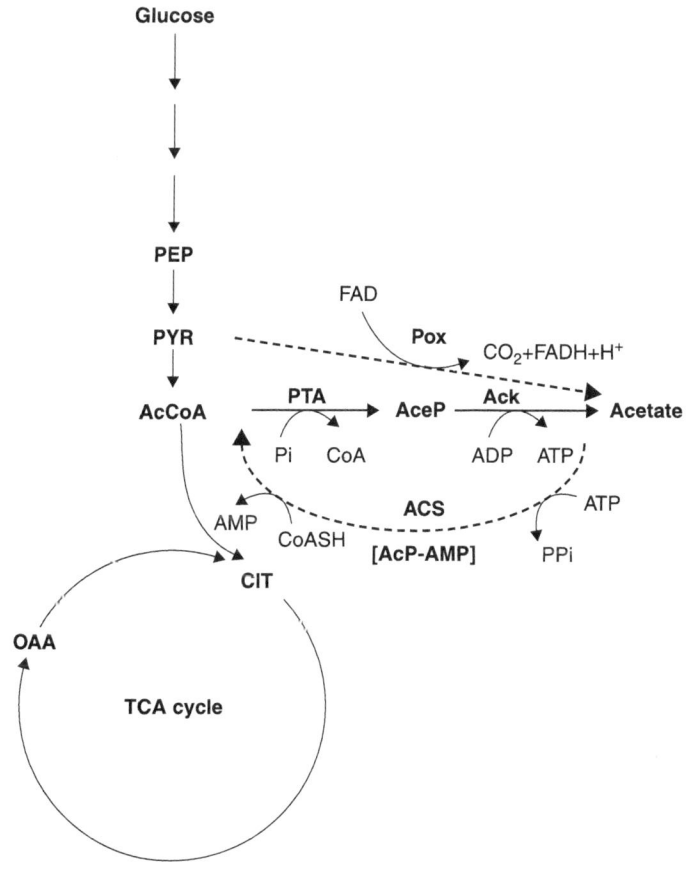

Figure 1.8 Acetate producing pathways

as aspartate and lysine, etc. There are, therefore, several anaplerotic pathways to prevent OAA shortage (Figure 1.9).

The typical anaplerotic pathway is a phosphoenol pyruvate carboxylase (Ppc) (orthophosphate: oxaloacetate carboxy-lyase, EC 4.1.1.31), which catalyzes the reaction of replenishing OAA from PEP as:

$$\text{PEP} + CO_2 \xrightarrow{Ppc} \text{OAA} + H_3PO_4 \qquad (1.32)$$

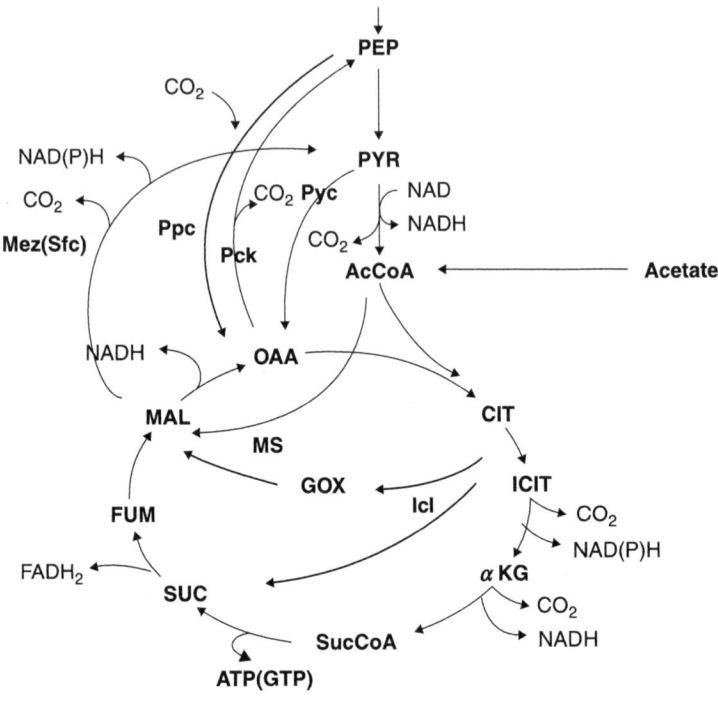

Figure 1.9 Anaplerotic pathways

or the gluconeogenetic pathway, such as PEP carboxykinase (ATP: oxaloacetate carboxy-lyase, EC 4.1.1.49), which converts PEP to OAA and can be also utilized as the anaplerotic pathway by reversing the reaction:

$$\text{OAA} \xrightarrow[\text{Pck}]{\text{ATP} \quad \text{ADP} + CO_2} \text{PEP} \tag{1.33}$$

There is another pathway from PYR to OAA catalyzed by a biotin-dependent pyruvate carboxylase (EC 6.4.1.1); *E. coli* does not have this pathway, while *Corynebacteria*, etc. do have this pathway, i.e.:

Main metabolism

$$\text{PYR} \xrightarrow[\text{Pyc}]{CO_2} \text{OAA} \quad (1.34)$$

Another important anaplerotic pathway is the **glyoxylate pathway**, which consists of isocitrate lyase (Icl) (threo-D_s-isocitrate-glyoxylate-lyase, EC 4.1.3.1) and malate systhase (MS) (L-malate glyoxylate-lyase, EC 4.1.3.2). ICIT undergoes an aldo cleavage to SUC and glyoxylate (GOX) by Icl as:

$$\text{ICIT} \xrightarrow{\text{Icl}} \text{GOX} + \text{SUC} \quad (1.35)$$

The next step is the condensation of AcCoA with GOX to form MAL by MS:

$$\text{GOX} + \text{AcCoA} \xrightarrow[\text{MS}]{\text{CoA-SH}} \text{MAL} \quad (1.36)$$

This glyoxylate pathway has suggested the existence of a cyclic mechanism for replenishing C_4 acids from the TCA cycle for biosynthesis, and this forms the bypass of the TCA cycle and so forms the upper TCA cycle, and plays an important role in the metabolism of short-chain fatty acids (Doelle, 1975) as well as the gluconeogenetic pathway, and is sometimes called the glyoxylate cycle or glyoxylate shunt.

1.3.7 Gluconeogenesis

Acetate or fatty acid can be assimilated and metabolized via the TCA cycle and glyoxylate pathway when glucose is unavailable, and biosynthesis can be made via gluconeogenesis, where acetate is first converted to AcCoA by ACS (Figures 1.8 to 1.10). The MAL thus formed via either

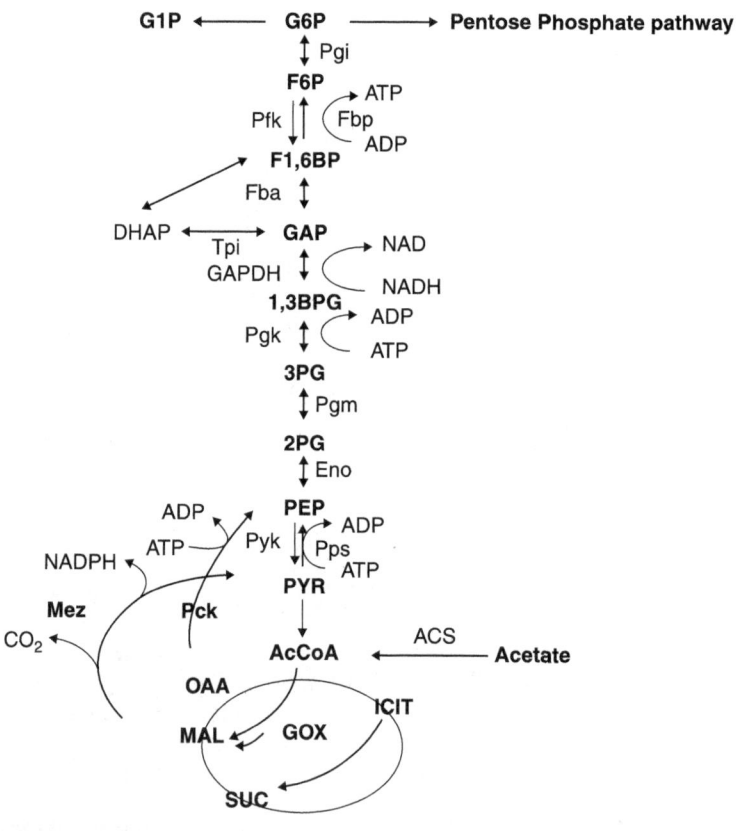

Figure 1.10 Gluconeogenetic pathways

# Main metabolism

the TCA cycle or glyoxylate pathway is converted to PYR by malic enzyme (Mez) (L-malate: NADDP oxidoreductase, EC1.1.1.40), which catalyzes the oxidative carboxylation of OAA forming NADPH + H$^+$, such as:

$$\text{MAL} \xrightarrow[\text{Mez}]{\text{NADP} + \quad \text{NADPH} + CO_2} \text{PYR} \quad (1.37)$$

The PYR thus formed is phosphorylated to form PEP by PEP synthase (Pps) (EC 2.7.9.2), requiring Mg^{2+}, K$^+$, and ATP, such that:

$$\text{PYR} \xrightarrow[\text{Pps}]{\text{ATP} \quad \text{AMP} + Pi} \text{PEP} \quad (1.38)$$

The OAA in the TCA cycle is converted to PEP by Pck, as mentioned above. Most of the EMP pathway reactions are reversible, but Pfk must be replaced by other enzymes, such as fructose bisphosphatase (Fbp) (EC 3.1.3.11), to convert FDP to F6P:

$$\text{FDP} \xrightarrow{\text{Fbp}} \text{F6P} \quad (1.39)$$

1.4 Respiratory chain pathways

Microbial cells can generate energy as ATP under a wide range of redox conditions. The reducing equivalents, such as NADH and FADH$_2$, are re-oxidized in the respiratory chain, where oxygen, nitrate, fumarate, and dimethyl sulfoxide, etc. are the electron acceptors (Figure 1.11).

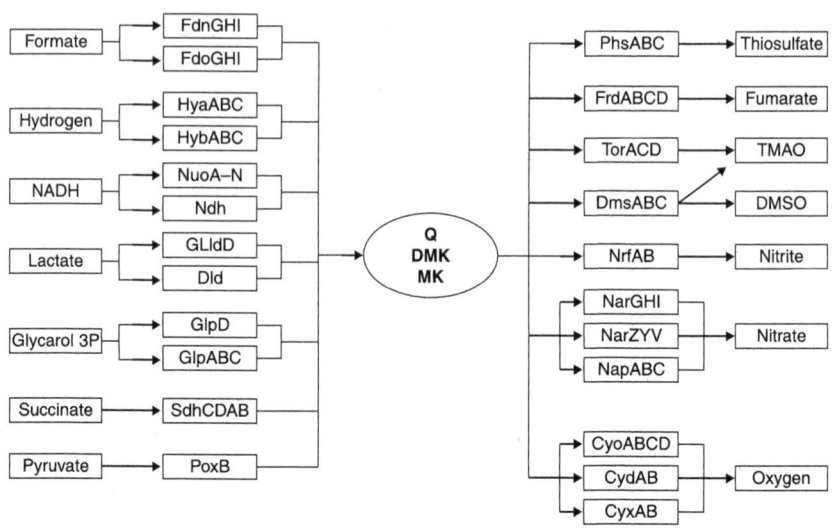

Figure 1.11 Modularity of respiratory chains (Gennis and Stewart, 1996)

This process is coupled to the formation of a **proton motive force** (**PMF**), which is utilized for ATP generation from ADP. *E. coli* can make two different NADH dehydrogenases and two terminal oxidases, as shown in Figure 1.11. Figure 1.12 shows the electron flux by such pathways, where NDH-I is a primary proton pump and results in translocating $2H^+/e^-$, whereas NDH-II is not coupled ($H^+/e^- = 0$). The two cytochrome oxidases are also different in their efficiency of proton translocation, with $H^+/e^- = 2$ for cytochrome bo_3 (Cyo), and $H^+/e^- = 1$ for cytochrome bd (Cyd). This indicates that the number of protons delivered to the periplasm varies from 1 to 4 per electron, depending on which pathways are utilized in relation to oxygen levels. It may be considered that the maximum value

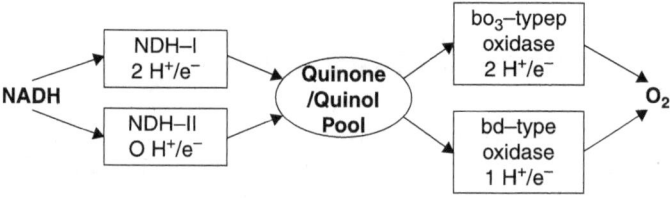

Figure 1.12 Proton translocating values per electron in the respiratory chain

is obtained for the case of NDH-I and Cyo under fully aerobic conditions, and the minimum value is obtained for NDH-II and Cyd under microaerobic conditions, but these regulations are more complex, as will be explained in Chapter 3. Note that the affinity to oxygen is high for Cyd (K_m (O_2) = 0.3 µM) (Mason et al., 2009), while it is low for Cyo (K_m(O_2) = 6.0 µM) (Mason et al., 2009), but the reaction rate such as V_{max} is the reverse (V_m = 218 mol O/mol cytochrome bd/s for Cyd (Bekker et al., 2009), and V_m = 225 mol O/mol cytochrome bd/s Cyo) (Sato-Watnabe et al., 1998). Note also that the important function of the respiratory chain is the maintenance of the redox balance and the regeneration of NAD^+ from NADH, in addition to its bioenergetic efficiency (Gennis and Stewart, 1996).

One of the major functions of the respiratory chain is to generate a proton electrochemical gradient, referred to as the proton motive force (PMF), across the cytoplasmic membrane. The respiratory system is designed to oxidize a wide variety of substrates and to utilize different terminal electron acceptors (Figure 1.11) (Gennis and Stewart, 1996). As shown in Figure 1.12, NDH-II has no transmembrane elements, and all the chemistry occurs on the cytoplasmic side of the membrane, and the bioenergetic contribution of this enzyme to the PMF is negligible. SDH in the TCA cycle contains membrane-spanning subunits, but the protons used in the reduction of ubiquinone come from the cytoplasm and do not generate PMF. The Frd is also unlikely to contribute to the PMF. Figure 1.13 shows the electron transfer and the proton translocation across the membrane (Gennis and Stewart, 1996).

Quinones are widely distributed in nature. These are lipid-soluble components of membrane-bound electron-transport chains, where quinone is present not only in the inner mitochondrial membrane but also in various organs, which implies its role in other than respiratory electron transport (Soballe and Poole, 1999). Quinones may be subdivided into two groups, one of which comprises a benzoquinone called ubiquinone or coenzyme Q, with the latter (CoQ) being mostly used in biochemical and clinical research. Ubiquinone is expressed either as UQ or UQ_n, where n refers to the number of isoprenoid units in the side chain (Figure 1.14). The second group contains the naphthoquinones menaquinone (vitamin K_2, MK, or MK_n) and demethyl/menaquinone (DMK or DMK_n).

Animal cells synthesize UQ only, and MK is obtained from the diet (Soballe and Poole, 1999). In prokaryotes, MK is utilized in pyrimidine biosynthesis under anaerobic conditions (Gibson and Cox, 1993).

Bacterial cellular metabolic systems

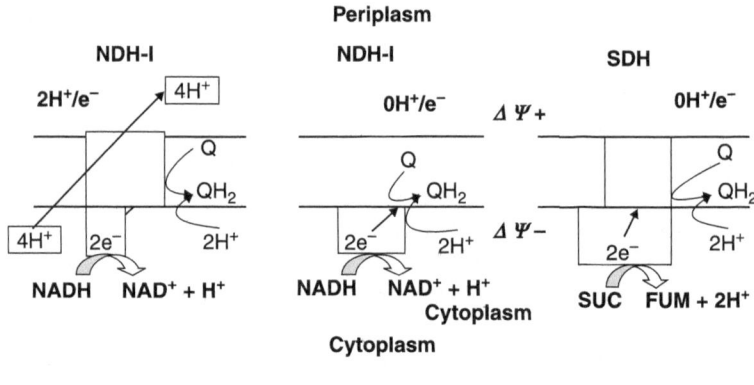
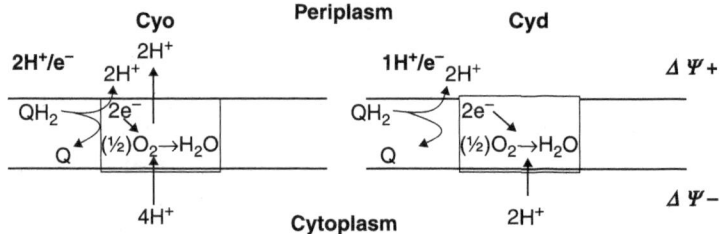

Figure 1.13 Schematic illustration for the electron transfer and proton translocation

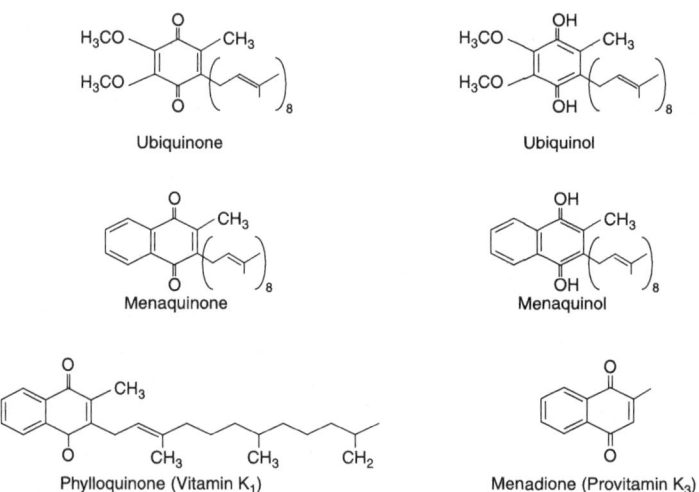

Figure 1.14 Chemical structure of quinones (Mobious et al., 2010)

Humans and some plants, such as tobacco plants, contain CoQ_{10}, while prokaryotes, such as *E. coli* and eukaryotes such as yeast, have UQ_8, MK_8, and DMK_8 as well as other side chains. Most Gram-positive bacteria and anaerobic Gram-negative bacteria contain MK only, whereas most of the aerobic Gram-negative bacteria contain exclusively UQ (Soballe and Poole, 1999). Both types of UQ and MK are found in facultative anaerobic Gram-negative bacteria. Pyloquinone (vitamin K_1), a naphthoquinone with a largely saturated side chain (Figure 1.14), is formed mainly in green plants and plays an important function in blood coagulation (Soballe and Poole, 1999). Plastoquinone (PQ_n) is a benzoquinone that acts as a primary electron carrier in the photosynthetic tissues of higher plant as well as in cyanobacteria.

Plumbagin and juglone are naphthoquinones that are excreted by plants to poison predators. The lack of an isoprenoide side chain increases their quinone solubility and they impose severe oxidative stresses on the cell by intercepting electrons from membrane-bound electron carriers and transferring them to molecular oxygen to reduce superoxide (Soballe and Poole, 1999). Plumbagin and juglone are used to induce oxidative stress or as respiratory inhibitors. Menadione is often used as an electron donor in studies with quinone-dependent oxidoreductase, a redox mediator, or as a superoxide-generating agent (Soballe and Poole, 1999).

UQ was discovered independently, where its role is that of a respiratory hydrogen (or proton plus electron) carrier between NADH dehydrogenase or succinate dehydrogenase and cytochrome systems. Reduction and oxidation of UQ involve 2-electron transfers at the quinone nucleus associated with the addition or release of two single H^+ to form ubiquinol (UQH_8) and UQ, respectively. These reactions are important for both linear electron transfer and transmembrane H^+ translocation (Soballe and Poole, 1999).

Removal or transfer of a single electron and H^+ gives the ubisemiquinone radical (UQ^{*-}). This radical may be stabilized when bound to a protein in association with UQH_2 oxidase, and cytochrome bo' in *E. coli* (Ingledeu et al., 1995; Soballe and Poole, 1999).

An important property of UQ is its hydrophobicity, which allows free movement in the membrane. The biosynthesis of UQ is shown in Figure 1.15, where the nucleus is derived from chorismate, whereas the prenyl side chain is derived from prenyldiphosphate, and the methyl groups are derived from S-adenosylmethionine.

Bacterial cellular metabolic systems

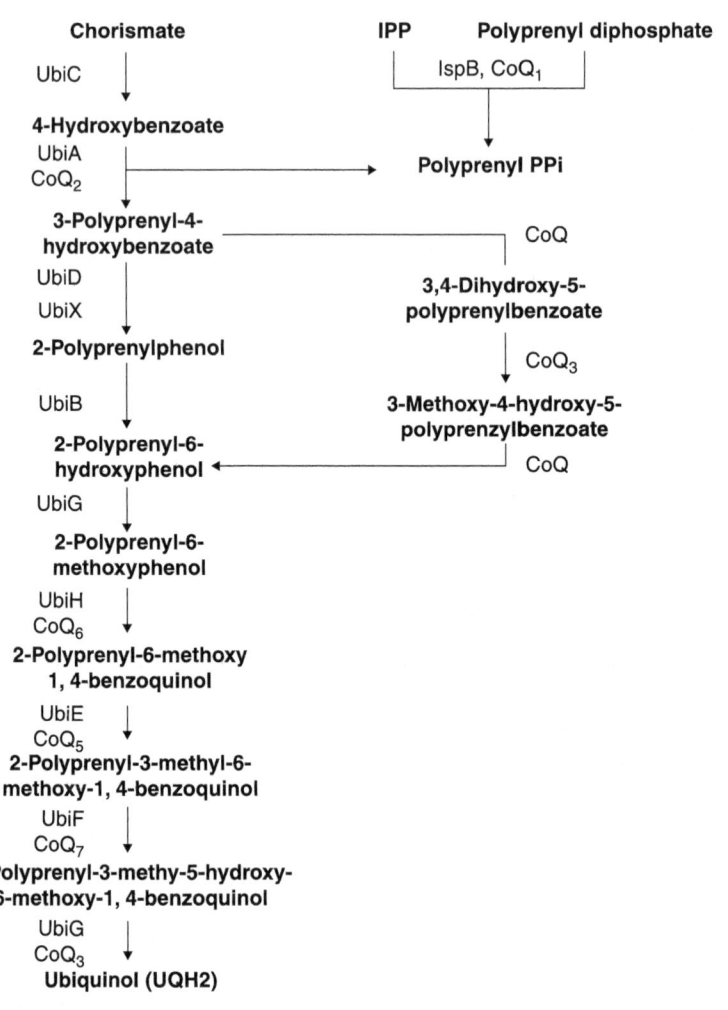

Figure 1.15 Pathway of UQ biosynthesis in *E. coli* (Soballe and Poole, 1999)

1.5 Anaerobic metabolism

In the absence of oxygen or other electron acceptors, the respiratory chain cannot be utilized, and thus ATP is generated via substrate level phosphorylation through the process of degradation of the carbon source in the metabolic pathways. Under such fermentation conditions, cells such as *E. coli* excrete metabolites such as lactate, ethanol, succinate, and formate (also CO_2 and H_2) as well as acetate, where the relative production rates for these metabolites are governed by the demand for redox neutrality (Figure 1.16). The succinate is formed from PEP via Ppc. PYR serves as a common substrate for pyruvate formate-lyase (Pfl) and the pyruvate dehydrogenase complex (PDHc), and this branch point involves the cleavage of PYR. The activity of *pfl*, which encodes Pfl, is under the control of such global regulators as ArcA and Fnr in *E. coli*, and becomes active at lower oxygen concentrations, whereas *aceE,F*, which encode α and β subunits of PDHc, are repressed by ArcA under oxygen limited conditions. At the branch point of AcCoA, the product of both Pfl and

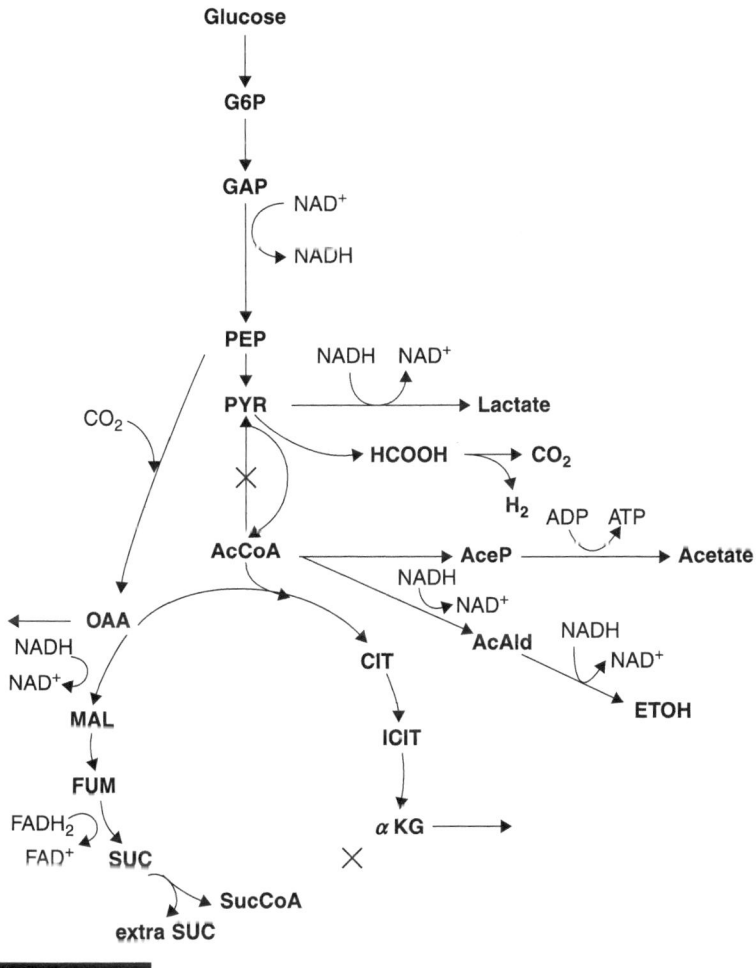

Figure 1.16 Anaerobic pathways

PDHc reactions, AcCoA is converted to either acetate or ethanol, or subsequently undergoes further oxidation in the TCA cycle up to α KG (Figure 1.16).

Anaerobic metabolism is closely related to fermentation, but the meaning of 'fermentation' has been changed during the past hundred years. Pasteur may have used the terms 'cell' and 'ferment' interchangeably when referring to the microbe back in 1857 (Doelle, 1975). The term 'fermentation' originated from wine-making, thus becoming associated with the idea of cells, gas (CO_2) production, and the production of organic by-products (Doelle, 1975). Under anaerobic conditions, energy generation is limited only by substrate level phosphorylation and/or anaerobic respiration, and thus the specific glucose consumption rate is increased to enhance energy production through the EMP pathway (Koebman et al., 2002). The possible fermentation pathways are given in Figure 1.16, where NADH re-oxidation is critical for regeneration of NAD^+ for the metabolism to continue. From research on electron transport systems of microbial metabolism, it may be reasonable to use the term 'fermentation' for the processes that have organic compounds as terminal acceptors (Doelle, 1975). However, with this definition, acetic acid bacteria may not be fermentative but respire aerobically. It may be better to extend the definition of fermentation to also include acetic acid fermentation.

1.6 Anaerobic respiration

The reduction of nitrate to nitrite and then ammonia appears widely in bacteria, archaea, and as plants. *E. coli* and enteric bacteria can combine such reduction processes with electron transport systems (Figure 1.17). Note that nitrate assimilation for anabolism occurs in the cytoplasm, but it can occur in the cytoplasm or periplasm or both, depending on the growth conditions. Moreover, unlike nitrate assimilation, it is strictly an anaerobic process in enteric bacteria, where the expressions of nitrate and nitrite reductase genes are tightly repressed in the presence of oxygen. These are induced under anaerobic conditions and further regulated by the availability of nitrate and nitrite. The regulatory reduction of nitrate to ammonia occurs in many electron-rich environments, such as anoxic marine sediments and sulfide-rich thermal vents, the human gastrointestinal tract, and the bodies of warm-blooded animals.

Main metabolism

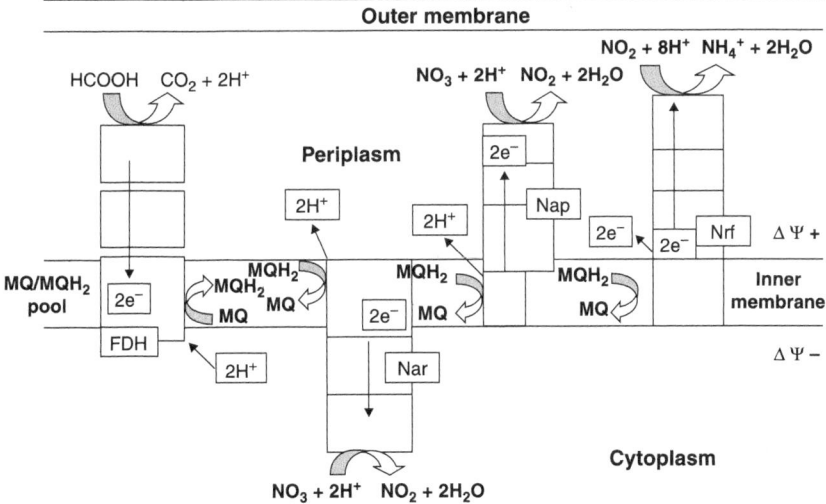

Figure 1.17 Nitrate respiration

1.7 Photosynthesis

In typical plant cells, carbohydrate is formed from CO_2 and water in the atmosphere, with the aid of light energy from the sun. This process is called photosynthesis, where light energy is stored as ATP and NADPH, which are utilized to convert CO_2 to 3-phosphogrycerate (3PG), where 3PG is in turn converted to hexose phosphate. The photosynthetic organisms are not restricted to plant cells but also include photosynthetic bacteria such as cyanobacteria and algae, etc.

Most autotrophic organisms fix CO_2 by the reaction catalyzed by riburose 2-phosphate carboxylase, where CO_2 and H_2O are converted to 3PG. This is an important step in the Calvin (or Calvin-Benson) cycle.

The overall photosynthesis reactions may be expressed as:

$$6CO_2 + 6H_2O \xrightarrow{\text{light}} C_6H_{12}O_6 + 6O_2 \quad (1.40a)$$

or

$$CO_2 + H_2O \xrightarrow{\text{light}} (CH_2O) + O_2 \quad (1.40b)$$

where (CH_2O) denotes carbohydrate.

The first step of photosynthesis is called the light reaction or light phase, where light energy is stored in the form of ATP and NADPH. In this stage, a hydrogen atom is removed from the water molecule and then utilized to reduce NADP$^+$, and the oxygen molecule is left as it is. At the same time, ADP is phosphorylated to form ATP. The reaction during this light phase may be expressed as:

$$H_2O + NADP^+ + P_i + ADP \xrightarrow{light} 0.5 O_2 + NADPH + H^+ + ATP \quad (1.41)$$

In the second stage of photosynthesis, glucose or carbohydrate are formed from CO_2 by utilizing NADPH and ATP produced during the first stage. This stage is known as the dark reaction phase or dark phase, where NADPH is re-oxidized to NADP$^+$, and ATP is hydrolyzed to ADP and P_i as:

$$CO_2 + NADPH + H^+ + ATP \rightarrow (1/6)Glc(CH_2O) + NADP^+ + ADP + P_i \quad (1.42)$$

Photosynthesis occurs in multiple membranes in prokaryotic organisms such as cyanobacteria, while it occurs in organelles as chloroplasts in eukaryotic organisms. The chloroplast forms the complex net structure called the thylakoid membrane, where NADPH and ATP are formed. Variously colored elements are embedded inside the thylakoid membrane to capture photo energy, where green chlorophyll is the most common color element, and light absorption by chlorophyll molecules excites electrons. The excited chlorophyll molecule generates a photon by the fluorescence process, and returns it to its original state. The excited chlorophyll gives the electron a series of enzymes, and ATP is generated through transfer of the electron via this chain of enzymes. This ATP generating process is called photo-phosphorylation. The electron carriers in this process are ferodoxin and some cytochrome. The light phase of photosynthesis consists of two photo-systems. Photo-system I (PS I) is excited by light with a lower wavelength than 700 nm, and produces NADPH. Photo-system II (PS II) requires light with a wavelength lower than 680 nm, and decomposes H_2O into $(1/2)O_2$ and $2H^+$. Here ATP is generated in accordance with the electron flow from PS I to PS II.

In the second stage of photosynthesis, CO_2 is converted to carbohydrate using ATP and NADPH obtained during the light reaction phase. This reaction can be considered as three reactions, namely: i) CO_2 fixation; ii) conversion of fixed CO_2 to carbohydrate; and iii) regeneration of the CO_2 acceptor molecule. In the first process of CO_2 fixation, the

Main metabolism

reduced PP pathway, the C_3 cycle pathway, is so-called because the first intermediate is a three-carbon molecule, typically called the Calvin (or Calvin-Benson) cycle.

In plant cells, CO_2 is fixed by photosynthesis through a structure on the leaf called the stromata. In the first step reaction of the Calvin cycle, 2 moles of 3 PG are formed from 1 mole of CO_2 and 1 mole of riburose 1,5-bisphosphate (Ribu1,5 BP), where this reaction is irreversible and catalyzed by riburose 1,5-bisphosphocarboxylase oxigenase or by the more popular named RubisCo. The Calvin cycle is shown in Figure 1.18, where 3PG is metabolized in the reverse direction to the glycolysis, similar to gluconeogenesis, and 1,3 BPG is formed from 3 PG using ATP by the reaction catalyzed by Pgk, and GAP is formed from 1,3 BPG by the re-oxidation of NADPH (instead of NADH) catalyzed by the isozyme of GAPDH. Then, part of GAP is converted to sucrose or starch, and the rest is used for the regeneration of Rib1,5 BP. In summary, the overall Calvin cycle reaction can be expressed as:

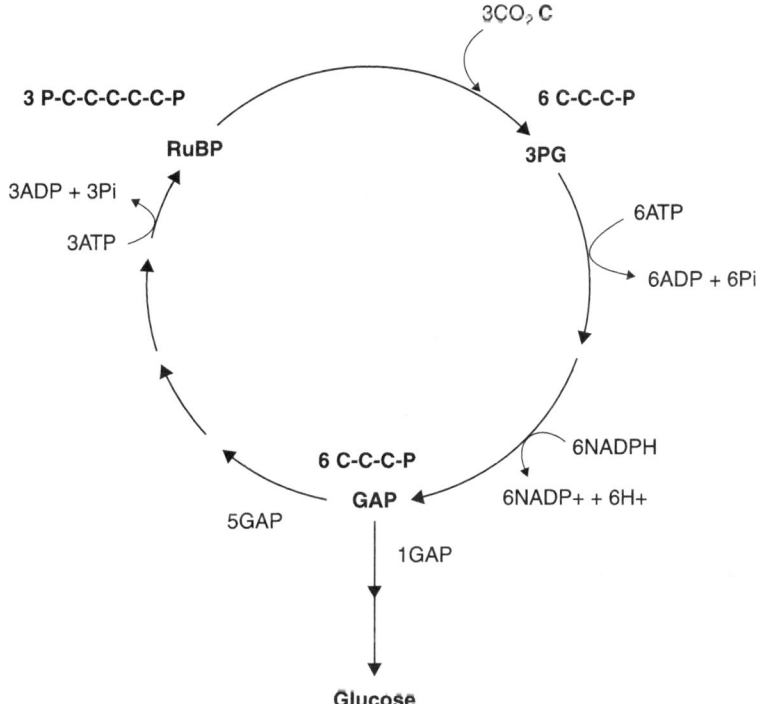

Figure 1.18 Calvin-Benson cycle

$$3CO_2 + 9ATP + 6NADPH + 5H_2O \rightarrow 9ADP + 8P_i \quad (1.43)$$
$$+ 6NADP^+ + GAP \text{ or } DHAP$$

Note that RubisCo catalyzes the carboxylation and oxygen addition, as the alternative name implies. In the reaction of O_2 addition, 1 mole of 3 PG and 1 mole of phosphogrycerate are formed. The 3 PG formed at the oxygen addition reaction of Rib1,5BP enters the Calvin cycle, and 2 moles of 2 PG(C_2) are oxidized to produce 1 mole of CO_2 and 1 mole of 3 PG (C_3), and these again enter the Calvin cycle. During the oxygen addition process, NADH and ATP are used. After light-dependent O_2 uptake is catalyzed by RubisCo, this is released as CO_2 by the phospho-glycerate metabolism. This is called photorespiration.

1.8 Amino acids synthesis

Amino acid synthesis is important for the cell synthesis. Figure 1.19 shows 20 amino acids and their precursors.

Note that the chemical structure of amino acids is:

$$\underset{\alpha\text{-carbon}}{R-\overset{NH_2}{\underset{|}{CH}}-COOH}$$

As can be seen, amino basis ($-NH_2$ or $-NH_3^+$) connects to the α carbon (next to the carboxyl base). An amino acid can produce H^+ from COOH, and thus is an acid. Different amino acids come from different structures of R. Let us now consider how these amino acids are synthesized from their precursors.

1.8.1 Alanine

Pyruvate is the precursor for alanine, where pyruvate is reductively converted to form L-alanine by alanine dehydrogenase, or by the transfer of the amino base to pyruvate by transaminase. L-alanine is converted to D-alanine by a racemase reaction, and is utilized as the constitutent for the cell wall, etc. (Figure 1.20).

Main metabolism

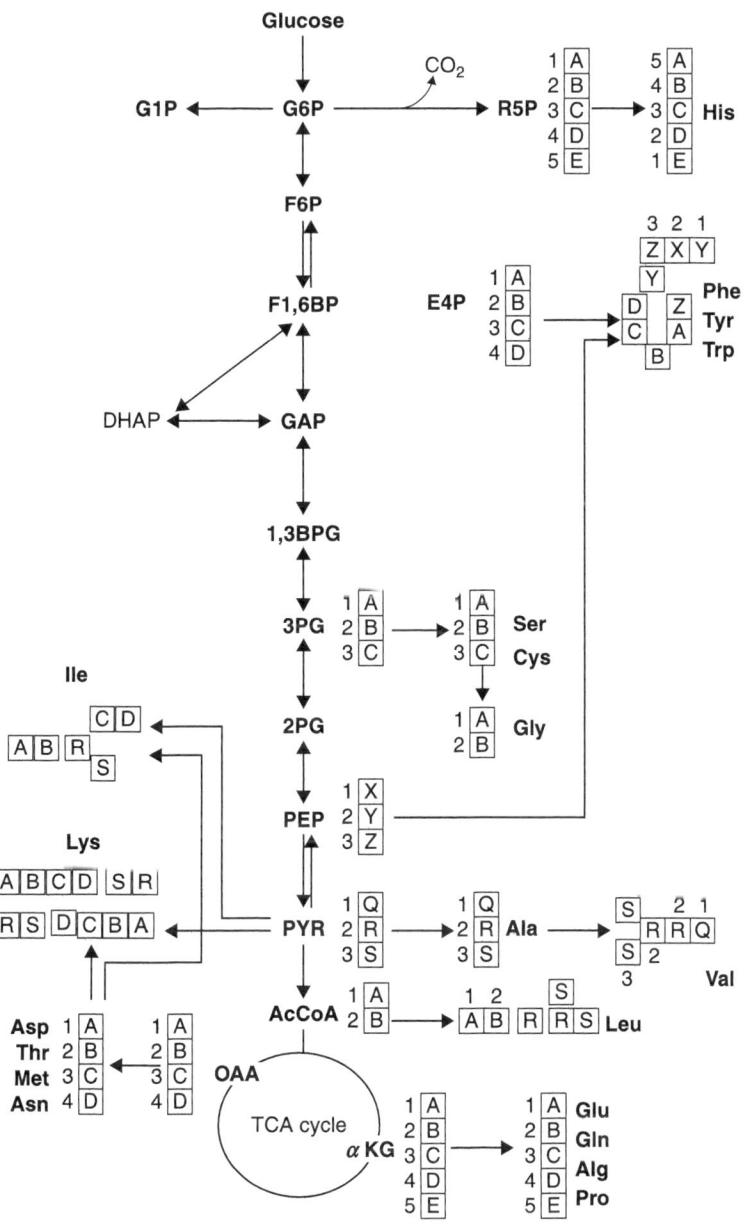

Figure 1.19 Amino acid synthesis from their precursors

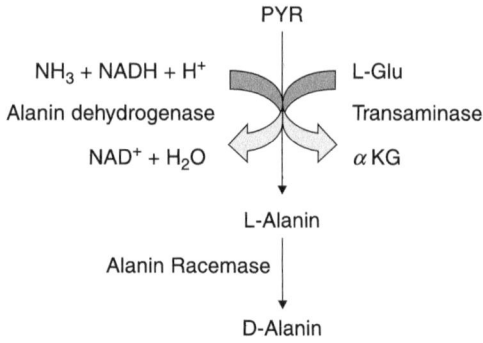

Figure 1.20 Alanine synthesis from PYR

1.8.2 Valine, leusine, isoleusine

Valine, leusine, and isoleusine are the branch chain amino acids, where the isoleusine synthesis pathway from 2-oxobutyrate and the valine synthesis pathway from pyruvate are catalyzed by the common enzyme (Figure 1.21). The final step reactions for valine, leusine, and isoleusine are amine transfer reactions, which are all catalyzed by the branch chain amino-transferase. Moreover, each amino acid synthetic pathway is under feedback control by the final product of amino acids, where isoleusine inhibits threoninehydratase, valine inhibits acetohydroxylic acid synthase, and leusine inhibits isopropyl malic acid synthase. Note that threonine hydratase, acetohydroxylic acid synthase, and branched chain amino acid transaminase are regulated by the multivarent control.

1.8.3 Glutamate, glutamine

Glutamic acid (Glu) is synthesized from α KG in the TCA cycle by glutamate dehydrogenase (GDH) with NH_3 and NADPH (Figure 1.22). The glutamic acid produced inhibits GDH and also controls enzyme synthesis. Note that glutamic acid regulates the synthesis of Ppc and CS. Gutamine is synthesized from glutamate by glutamine synthetase (GS) with NH_3 and ATP (Figure 1.22). Glutamic acid may be also formed from glutamine by glutamate synthase (GOGAT) with NADPH when the NH_3 concentration is low and limiting. The ammonia assimilation or nitrogen regulation pathways are important, which will be explained in more detail in Chapter 3.

Main metabolism

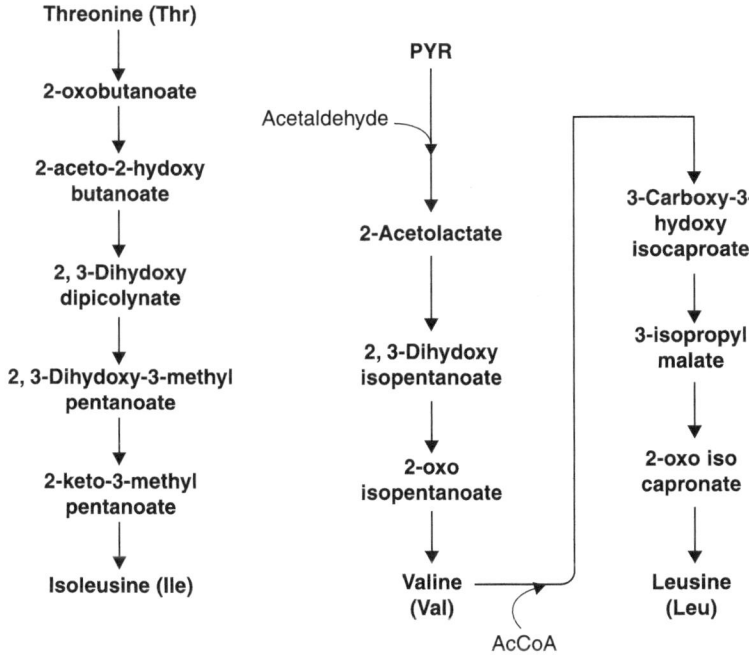

Figure 1.21 Valine, leusine, and isoleusine biosynthesis

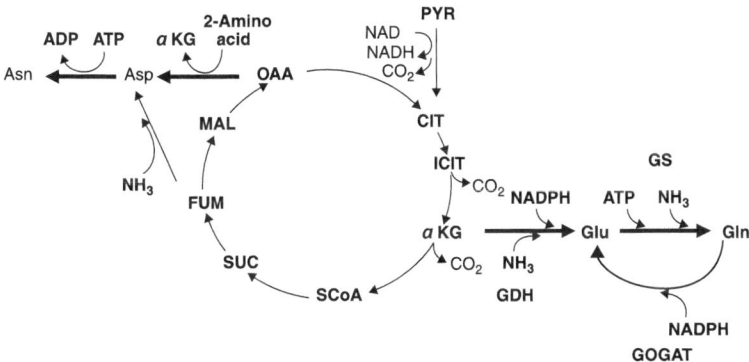

Figure 1.22 Glutamate and glutamine synthesis as well as aspartate and asparagine synthesis

1.8.4 Proline

Proline is synthesized from glutamate, where glutamate is converted to glutamate 5-semialdehyde by glutamate kinase and 5-glutamil phosphate

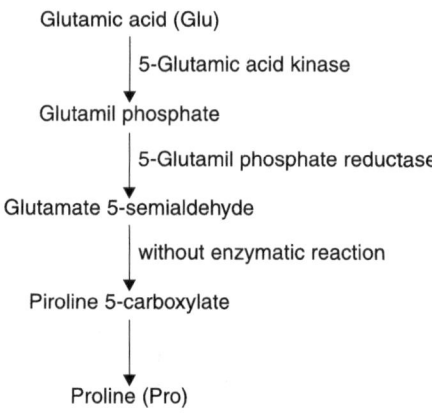

Figure 1.23 Proline biosynthesis from glutamate

reductase, and this becomes proline 5-carboxylic acid and then proline. The first reaction step of these reactions is inhibited by proline (Figure 1.23).

1.8.5 Arginine, ornitine, and citorline

As shown in Figure 1.24, ornitine is formed by five reaction steps, citorline by six steps and arginine by eight steps of reactions from glutamic acid. These sets of reactions are also inhibited by arginine.

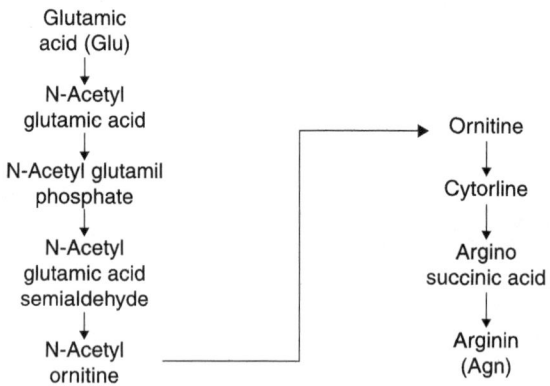

Figure 1.24 Arginine, ornitine, and citorline synthesis pathway

1.8.6 Aspartate and asparagine

Aspartic acid (Asp) is synthesized from OAA in the TCA cycle by aspartate transaminase. As stated before, OAA is synthesized from PEP or PYR by the anaplerotic reaction of Ppc (or Pyc) with fixation of CO_2. In bacteria such as *E. coli*, Ppc is inhibited by Asp. Asparagine (Asn) is synthesized by asparagine synthase from Asp (Figure 1.22).

1.8.7 Lysine, threonine, and methionine

Figure 1.25 shows the lysine (Lys) synthetic pathways, where the bacterial synthesis of lysine takes place via the diaminopimelate (DAP) pathway. L-aspartate (Asp) is formed from OAA by transamination, as stated above. Asp is then activated via phosphorylation by aspartokinase and reduced to form L-aspartate semialdehyde in the first two steps. L-aspartate semialdehyde is at a branch point to enter either threonine, methionine, and isoleusine syntheses or lysine synthesis. Dihydrodipicolinate (DHPS) and dihydrodipicolinate reductase (DHPR) catalyze the third and fourth

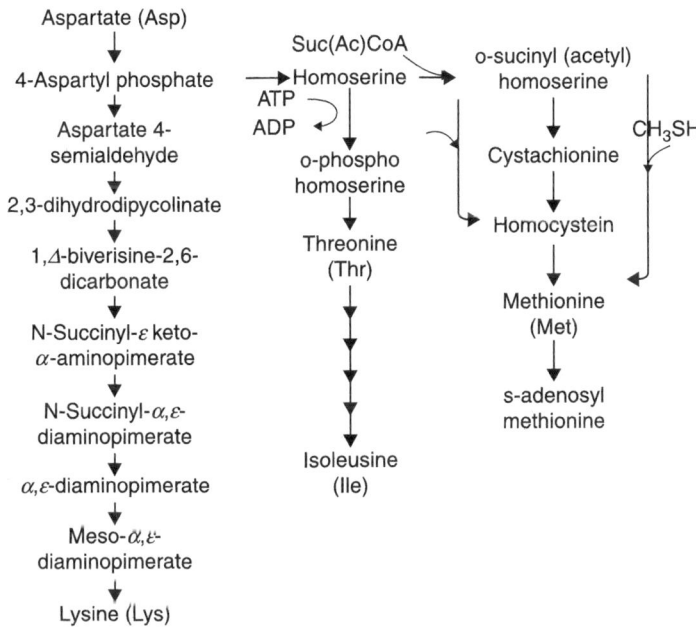

Figure 1.25 Lysine, threonine, and methionine biosynthesis

steps in the lysine synthetic pathway, respectively, and are the enzymes that commit flux to the biosynthesis of meso-diaminopimelate (DAP) and lysine. The synthesis of DAP from L-tetrahydrodipicolinate (THDP) is accomplished by three separate routes: the succinylase and acetylase pathways, in which N-succinylated or N-acetylated intermediates are generated, and the infrequently encountered dehydrogenase pathway. The synthesized meso-DAP can be either used for cell wall synthesis or decarboxylated to L-lysine catalyzed by DAPDC (DAP decarboxylase).

In threonine synthesis, homoserine is first formed from 4-aspartyl phosphate (ASA) by homoserine dehydrogenase. The homoserine is then converted to threonine (Thr) by homoserine kinase and threonine synthase. The limiting pathways for threonine synthesis are the reactions catalyzed by aspart kinase (Ask) and homoserine dehydrogenase, where isozyme I of Ask is threonine sensitive for *E. coli* K12. Moreover, a set of genes form an operon, and a series of enzymes are under multi-varent repression by threonine and isoleusine. In the case of *Corynebacteria*, ASA is under concerted repression when threonine and lysine co-exist, and homoserine dehydrogenase is subject to strong inhibition only by threonine.

Methionine is first formed from homoserine by homoserine-o-succinyl (acetyl) transferase. Then it is converted to cystachionine by cystachionine-γ-synthase by introducing a sulfur molecule. Finally, methionine is synthesized by introducing methyl basis via homocystein. In general, homoserine-o-succinyl (acetyl) transferase is repressed by methionine.

1.8.8 Aromatic amino acids

Aromatic amino acids, such as tryptophane (Trp), phenylalanine (Phe), and tyrosine (Tyr), are formed from E4P in the PP pathway and PEP in the glycolysis. The first reaction of this synthesis is catalyzed by deoxyalabino hepturose phosphate synthase (DAHPS), where this is an important regulatory enzyme. Then shikimic acid (Shik) is formed after four steps from the first reaction, and chorismic acid (CM) is formed after seven steps of reactions from the first reaction (Figure 1.26). From chorismate, antranil acid is formed by antranil acid synthase, and Trp is formed after several reaction steps from this. However, prephenic acid (PA) is formed from CM by corismic acid mutase, and then Trp is formed from PA by prephenic acid dehydrogenase and tyrosine transaminase. Phenylalanine is formed from PA by prephenic acid dehydrogenase and tyrosine transaminase.

It is known that different regulation mechanisms exist, depending on the microorganisms. Figure 1.27 shows three typical regulation patterns

Main metabolism

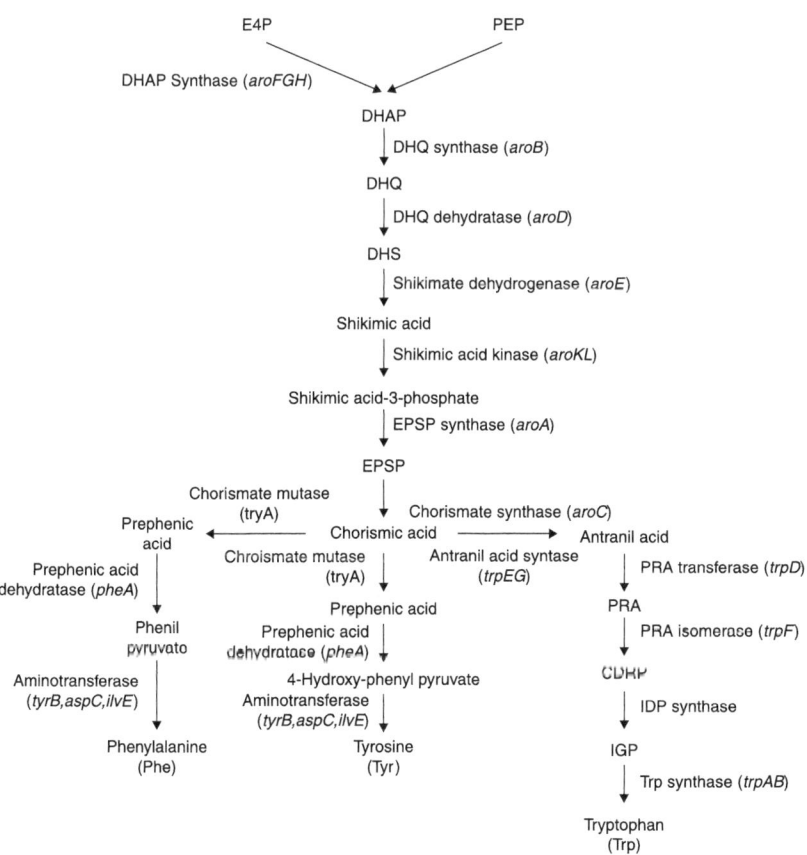

Figure 1.26 Aromatic amino acid synthesis pathways

(Nagai et al., 1996). Figure 1.27a shows the sequential control pattern typically seen in *B. subtilis*, where Phe and Tyr are formed by the branched pathway from PA, and each first enzyme is inhibited by the corresponding products, and the accumulation of CM inhibits DAHPS and shikimate kinase. This control scheme acts sequentially from the end products. Figure 1.27b shows the regulation scheme typically seen in *E. coli.*, etc., where the first enzyme DAHPS from E4P and PEP consist of the three enzymes of I, II, III, each sensitive to Phe, Tyr, and Trp, respectively. Figure 1.27c shows the regulation scheme typically seen in *B. flavum* or *Corynebacteria*, where this scheme is known as the priority synthesis. Namely, antranyl acid synthesis is inhibited by the accumulation of Trp, prephenic acid dehydratase is inhibited by the accumulation of Phe, and DAHPS is inhibited only when both Phe and Tyr have accumulated.

Bacterial cellular metabolic systems

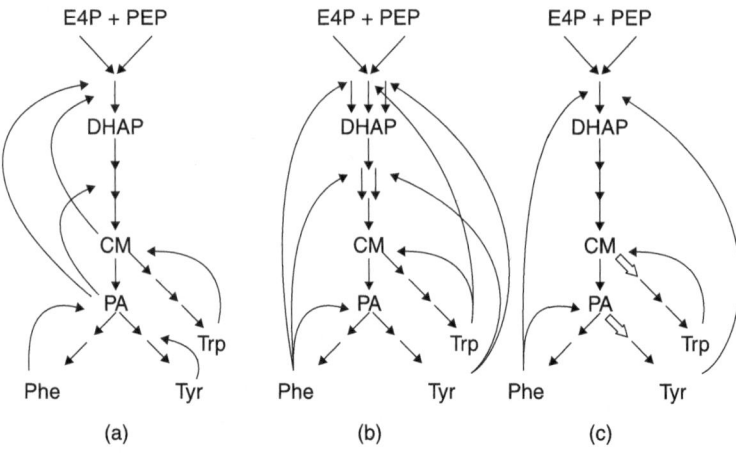

Figure 1.27 Several control schemes for aromatic amino acid biosynthesis

1.8.9 Serine, glycine, and cystein

As shown in Figure 1.28, serine is formed from 3PG in the EMP pathway via three step reactions. The first enzyme phosphoglycerate dehydrogenase

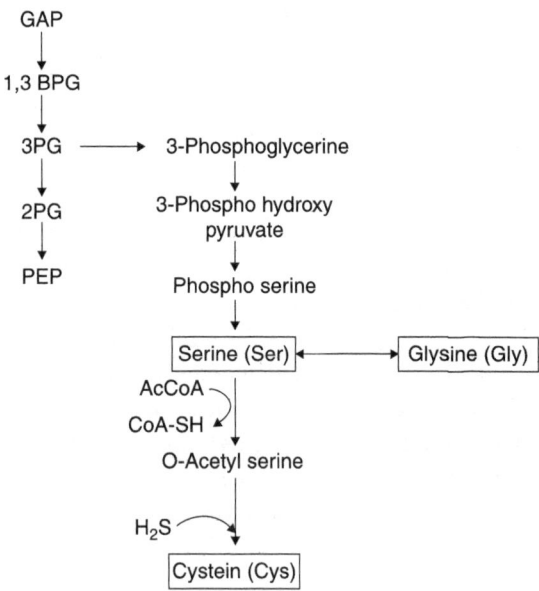

Figure 1.28 Serine, glycine, and cystein synthesis pathways

of these reactions is inhibited by serine. Cystein is formed from serine by the incorporation of H_2S to acetylphosphate formed by acetyltransferase. The mutual exchange takes place between serine and glysine by serine hydroxymethyl transferase. However, glysine can be formed from threonine by transaldorase. Moreover, glycine can be formed from inorganic ammonia salt and 5,10 methylene tetrahydro folate by glycine synthase. Note that cystein contains sulfur.

1.8.10 Histisine

As shown in Figure 1.29, histisine is formed from R5P in the PP pathway, where phospho ribosilpyrophosphate (PRPP) is formed from R5p by the reaction catalyzed by ribose phosphate pyrophosphokinase, and phosphoribosil ATP is formed by ATP-phosphoribosil transferase, where the first step enzyme ATP-phosphoribosil transferase is under feedback inhibition by histisine.

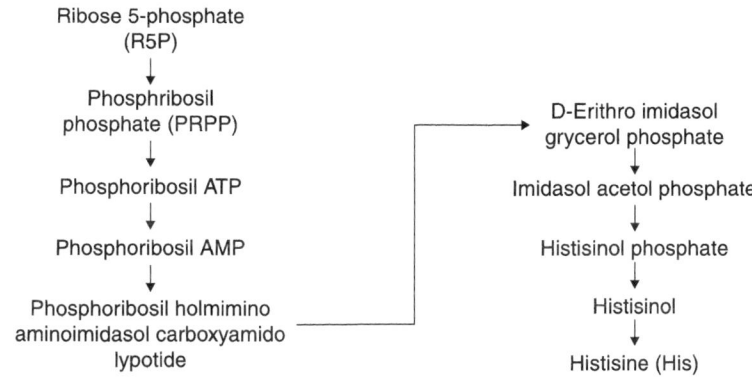

Figure 1.29 Histisine synthesis pathways

1.9 Nucleic acid synthesis and its control

1.9.1 Prine type nucleic acid synthesis

ATP or GTP are synthesized from R5P in the PP pathway, as shown in Figure 1.30, where inosine 5'-monophosphate (IPM) is formed from 5-phosphoribosil pylophosphate (PRPP) by 10 step reactions. It has been

Bacterial cellular metabolic systems

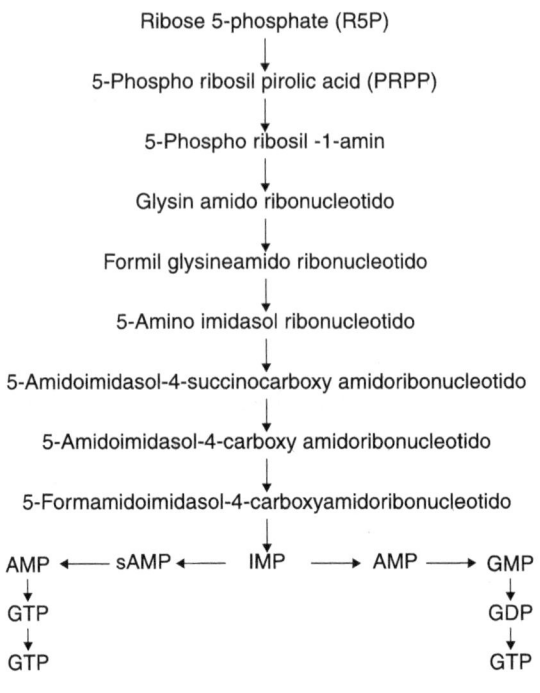

Figure 1.30 Nucleic acids synthesis pathways

known that four bases, such as hypoxantine, adenine, guanine, and xantine and their nucleoside or nucleotide, are interchanged.

1.9.2 Pyrimidine type nucleic acid synthesis

The first enzyme for the synthesis of pyrimidine type of nucleic acid is carbamyl phosphate synthase. The uridine-5′-monophosphate (Urydylate: UMP) is synthesized by six steps. UMP or cytidine acid (cytidine 5′-nucleotidase: CMP) is dephosphorylated by 5′-nucleotidase or phosphatase to become each nucleoside, and decomposed to the corresponding bases by pyrimidine nucleoside phosphorylase. Inversely, bases become nucleosides by the reverse reaction of phospholylase, while nucleotides are synthesized by the salvage synthesis by each kinase. Moreover, cytosine derivatives such as CTP are formed from UTP, where these are converted to urasil derivatives by each corresponding aminase.

1.10 Fatty acid metabolism and degradation

Fatty acid is first activated by ATP and converted to AcylCoA by the action of AcylCoA synthase. Long-chain AcylCoA is decomposed by releasing AcCoA one by one. This degradation pathway is called the β-oxidation pathway, since the β carbon (C-3) of fatty acid is oxidized. Fatty acid is the substance with a highly reductive state. As shown in Figure 1.31, $FADH_2$ or NADH is generated by β-oxidation, and these can be utilized for ATP synthesis by oxidative phosphorylation. Thus, a fatty acid with an even number of carbons is all degraded into AcCoA and goes into the TCA cycle, while that with an odd number of carbons generates propionylCoA, which is carboxylazed and converted to SucCoA via methyl malonylCoA and goes into the TCA cycle.

The synthesis and the degradation by β-oxidation occur along totally different pathways. In mammalian cells, β-oxidation of fatty acid is made

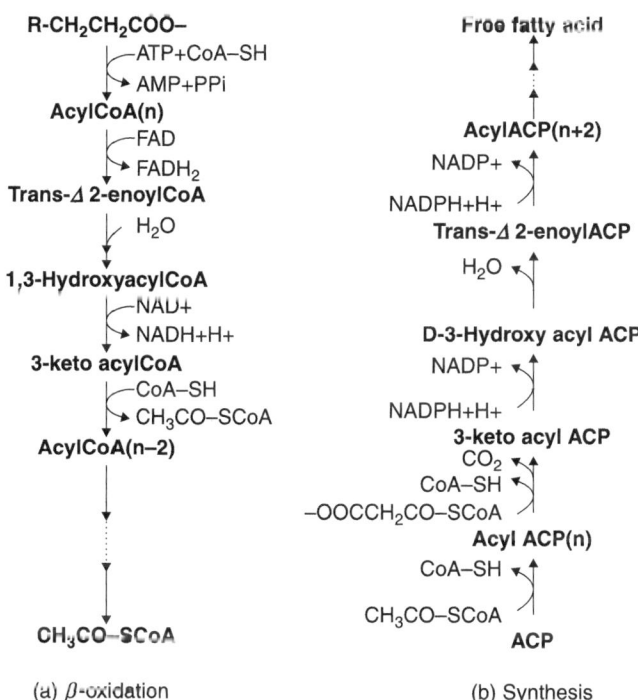

Figure 1.31 β oxidation and biosynthesis of fatty acid

in mytochondria, while the synthesis is made in cytosol. The thioester for β-oxidation of fatty acid is a CoA derivative, while in fatty acid synthesis, the intermediate thioester connects to the acyl carrier protein (ACP). The reaction occurs by two carbons in both synthesis and degradation, and two carbon AcCoA is formed in oxidation, while three carbon malonylCoA is required for the synthetic reaction, and two carbons are added to the chain for extension. In this process, CO_2 is released and the reducing equivalent for the reaction is NADPH.

1.11 Phosphotransferase system (PTS)

As can be seen in Figure 1.32, once inside the periplasm through the outer membrane, various sugars can be internalized into the cytoplasm by the PEP: sugar phosphotransferase system (PTS). This system is widespread in bacteria and absent in Archae and eukaryotic organisms (Postma et al., 1996; Saier, 2000). PTS participates in the transport and phosphorylation of a variety of sugars. The system is composed of the soluble and non-sugar-specific protein Enzyme I (EI), encoded by *ptsI* and the phosphohistidine carrier protein (HPr), encoded by *ptsH*. These proteins relay a phosphoryl group from PEP to the sugar specific EIIA and in turn to EIIBC, where EIIC (in some cases also EIID) is an integral membrane protein permease that recognizes and transports the sugar molecules, which are phosphorylated by EIIB. There are 21 different EII complexes encoded in the *E. coli* chromosome, which are involved in the transport of about 20 different carbohydrates (Deutscher et al., 2006).

1.12 Carbohydrate metabolism other than glucose

Glucose metabolism has been explained in Section 1.3 above. Here, other carbon source metabolisms are briefly explained.

1.12.1 Fructose metabolism

There are three pathways for the utilization of fructose (Figure 1.33) (Kornberg, 2001). In the primary pathway, the fructose (Fru) is

Main metabolism

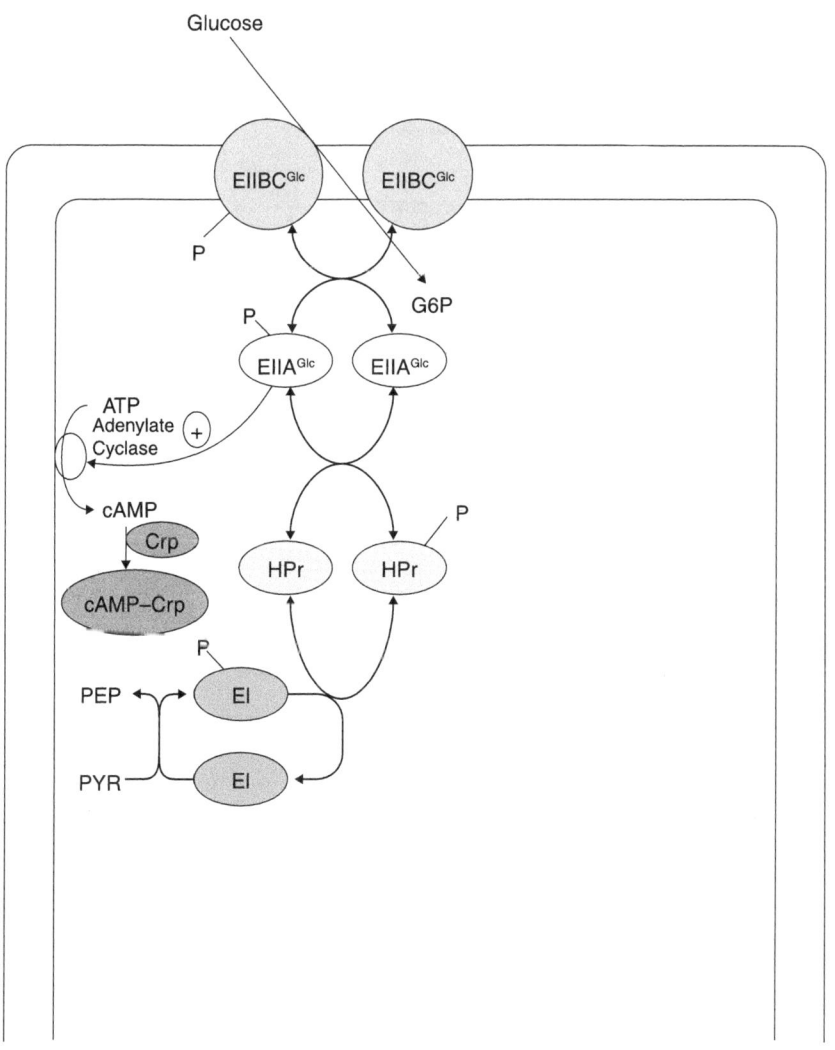

Figure 1.32 Phosphotransferase system (PTS)

transported via the membrane-spanning protein FruA and concomitantly phosphorylated by a PEP: D-fructose 1-phosphotransferase (fructose PTS) system (ATP: D-fructose 1-phosphotransferase, EC 2.7.1.3), which is induced by D-fructose and enters the cell as D-fructose 1 phosphate (F1P), where this process is affected by the transfer of a phosphoryl moiety from PEP to the hexose by the concerted action of two cytoplasmic proteins: EI of PTS and a membrane-associated diphosphoryl protein

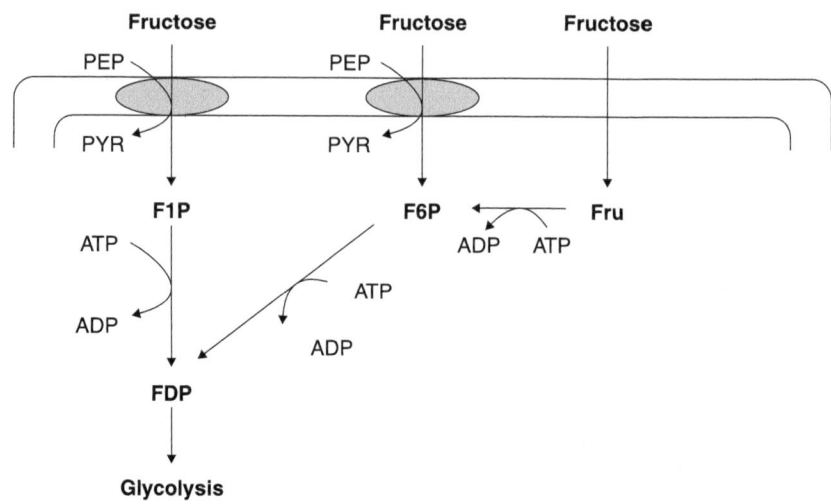

Figure 1.33 Fructose metabolism

(DTP). F1P is then converted to fructose 1,6-diphosphate (FDP) by ATP and by the inducible enzyme D-fructose-1-phosphate kinase (F1PK)(ATP: D-fructose-1-1phosphate 6-phospho-transferase).

In the second pathway, fructose enters the cell via a membrane-spanning protein that has the general ability to recognize sugars possessing the 3,4,5-D-arabino-hexose configuration, which include the permeases for mannose (ManXYZ), glucitol (SrlA), and mannitol (MtlA) (Kornberg, 2001). D-fructose is converted to F6P by a specific sucrose-induced D-fructokinase (ATP: D-fructose 6-phosphotransferase, EC 2.7.1.4), and then converted to FDP by Pfk of the EMP pathway.

In the third pathway, fructose enters the cell by diffusion, using an isoform of the glucose transporter PtsG. Since this mode of entry does not involve the PTS, the free fructose has to be phosphorylated by ATP to become F6P.

1.12.2 Mannose metabolism

The catabolism of mannose can follow two different mechanisms, such as cyclic and non-cyclic mechanisms for the D-isomer, where only the non-cyclic system occurs when L-mannose is the substrate. L-mannose is converted to L-fructose by an isomerase. L-fructose is then phosphorylated at C-1 with ATP by a kinase reaction and L-fructose 1-phosphate is formed, where an aldolase-type cleavage produces DHAP and L-GAP, where this connects to the EMP pathway (Doelle, 1975).

1.12.3 Xylose metabolism

D-xylose is converted to D-xylulose by xylose isomerase (D-xylose keto-isomerase, EC 5.3.1.5) (Figure 1.34). D-xylulose is subsequently phosphorylated by xylulokinase (ATP: D-xylulose 5-phosphotransferase, EC 2.7.1.17) to form D-xylulose 5-phosphate (X5P). Under anaerobic conditions, xylulose reductase (XR) is induced and xylitol and xylitol-5-phosphate are produced, where those may inhibit cell growth (Hausman et al., 1984).

1.12.4 Glycerol metabolism

Glycerol is oxidized to dihydroxyacetone by glycerol dehydrogenase (glycerol: NAD oxidoreductase, EC 1.1.1.6). Dihydroxyacetone is then phosphorylated by a kinase using ATP. Another pathway for glycerol utilization is that glycerol is phosphorylated by glycerol kinase (ATP: glycerol phosphotransferase, EC 2.7.1.30) to form L-glycerol 3-phosphate, which is then converted to GAP.

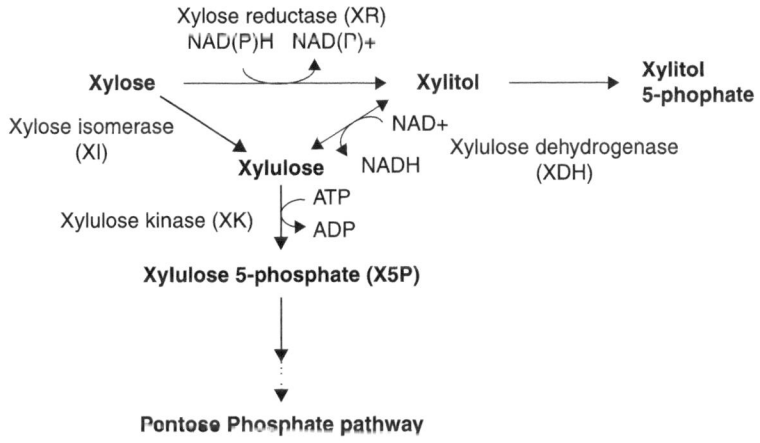

Figure 1.34 Xylose metabolism

1.13 ATP generation under aerobic and anaerobic conditions

Figure 1.35 shows the overall ATP generation under aerobic and anaerobic conditions, which indicates that total ATP generation may be expected to be 38 moles of ATP from 1 mole of glucose by assuming that 1 mole of NADH can be used to generate 3 moles of ATP, while 1 mole of $FADH_2$ can be used to generate 2 moles of ATP generated in the respiratory chain. However, in anaerobic conditions, NADH generated at GAPDH is re-oxidized by producing fermentation metabolites such as lactate, ethanol, and succinate, and only 2 moles of ATP are generated without taking into account ATP generation at Ack for acetate formation (Figure 1.35). Since the specific ATP production rate is correlated with the specific growth rate, the cell growth rate becomes significantly lower under anaerobic conditions than under aerobic conditions. Note that the ATP forming efficiency is a little low in practice, and thus 1.5–2.5 moles of ATP may be formed in practice from 1 mole of NADH. Likewise, the amount of ATP generation for $FADH_2$ is also less than 2 moles of ATP per NADH.

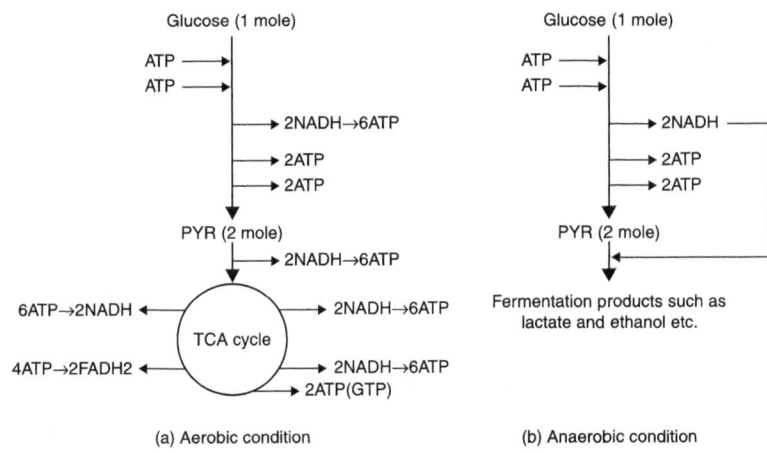

Figure 1.35 ATP balance for aerobic and anaerobic conditions

1.14 Metabolic regulation and futile cycle

Metabolic regulation of a cell achieves cell growth by optimizing ATP generation (catabolism) and cell synthesis (anabolism) during the cell growth phase. Moreover, the cell regulates the metabolism to cope with various kinds of stresses caused by changes in the culture environment, and thus it is not easy to understand the whole metabolic regulation mechanism. Metabolic regulation occurs at both gene and enzyme levels, where enzyme level regulation is typically made by allosteric regulation, which is attained by changing the 3D structure by binding the specific metabolites, etc. For example, G6PDH and 6PGDH are inhibited by NADPH, while Pfk is inhibited by PEP. PDH is also inhibited by NADH, ATP, AcCoA, and so on. Consider the regulation of Pfk and Fbp in the EMP pathway, where Pfk catalyzes the following reaction:

$$F6p + ATP \xrightarrow{Pfk} F1,6BP + ADP \qquad (1.44)$$

while Fbp catalyzes the reverse reaction, such as:

$$F1,6BP \xrightarrow{Fbp} F6P + P_i \qquad (1.45)$$

If these reactions occur at the same time, the energy generated by one pathway is used by the other pathway without efficient use. This phenomenon is called a **futile cycle**, where this occurs due to independent control by each pathway. A similar phenomenon can be seen for Ppc and Pck for anaplerotic and gluconeogenetic pathways, such as:

$$PEP + CO_2 \xrightarrow{Ppc} OAA + ADP \qquad (1.46)$$

and

$$OAA + ADP \xrightarrow{Pck} PEP + ATP \qquad (1.47)$$

Note that these occur depending on the culture conditions.

1.15 References

Atkinson, D.E. (1968) 'The energy charge of the adenylate pool as a regulatory parameter: Interaction with feedback modifiers', *Biochemistry*, 7: 43030–4.

Bekker, M., de Vries, S., Ter Beek, A., Hellingwerf, K.J., and Teixeira de Mattos, M.J. (2009) 'Respiration of *Escherichia coli* can be fully uncoupled via the non-electrogenic terminal cytochrome *bd*- II oxidase', *Journal of Bacteriology*, 191: 5510–17.

Deutscher, J., Franke, C., and Postma, P.W. (2006) 'How phosphotransferase system-related protein phosphorylation regulates carbohydrate metabolism in bacteria', *Microbiology and Molecular Biology Reviews*, 70(4): 939–1031.

Doelle, H.W. (1975) *Bacterial metabolism*, 2nd edition. New York: Academic Press.

Gennis, R.B. and Stewart, V. (1996) 'Respiration', in: Escherichia coli *and* Salmonella: *Cellular and Molecular Biology*, 2nd edition, edited by F.C. Neidhardt et al. Washington, DC: American Society for Microbiology. pp. 217–61.

Gibson, F. and Cox, G.B. (1993) 'The use of mutants of *Escherichia coli* K12 in studying electron transport and oxidative phosphorylation', *Essays in Biochemistry*, 9: 1–29.

Hausman, S.Z., Thompson, J., and London, J. (1984) 'Futile xylitol cycle in *Lactobacillus casei*,' *Journal of Bacteriology*, 160: 211–15.

Horton, H.R., Morgan, L.A., Ochs, R.S., Rawn, J.D., and Serimgeour, K.G. (1996) *Principles of Biochemistry*. Prentice Hall Inc.

Ingledew, W.J., Ohnishi, T., and Salerno, J.C. (1995) 'Studies on a stabilization of ubisemiquinone by *Escherichia coli*. Quinol oxidase, cytochrome bo', *European Journal of Biochemistry*, 227: 903–8.

Koebman, B.J., Westerhoff, H.V., Snoep, J.L., Nilsson, O., and Jensen, P.R. (2002) 'The glycolytic flux in *E. coli* is controlled by the demand for ATP', *Journal of Bacteriology*, 184: 3909–16.

Kornberg, H.L. (2001) 'Routes for fructose utilization by *Escherichia coli*', *Journal of Molecular Microbiology and Biotechnology*, 3: 355–9.

Lehninger, A.L., Nelson, D.L., and Cox, M.C. (1993) *Principles of Biochemistry*. New York: Worth Publishers.

Mason, M.G., Shepherd, M., Nicholls, P., Dobbin, P.S. et al. (2009) 'Cytochrome *bd* confers nitric oxide resistance to *Escherichia coli*', *Nature Chemical Biology*, 5: 94–6.

Mobious, K., Arias-Cartin, R., Breckau, D. et al. (2010) 'Heme biosynthesis is coupled to electron transport chains for energy generation', *Proceedings of the National Academy of Sciences of USA*, 107(23): 10436–41.

Saier, M.H. (2000) 'Vectorial metabolism and evolution of transport systems', *Journal of Bacteriology*, 182: 5029–35.

Sato-Watanabe, M., Mogi, T., Miyoshi, H., and Anraku, Y. (1998) 'Characterization and functional role of the QH site of *bo*-type ubiquinol oxidase from *Escherichia coli*', *Biochemistry*, 37: 5356–61.

Soballe, B. and Poole, R.K. (1999) 'Microbial ubiquinones: Multiple roles in respiration, gene regulation and oxidative stress management', *Microbiology*, 145: 1817–30.

Stryer, L. (1995) *Biochemistry*, 4th edition. New York: W.H. Freeman and Company.

Voet, D. and Voet, J.G. (1995) *Biochemistry*, 2nd edition. New York: John Wiley & Sons.

Webb, E.C. (1992) *Enzyme nomenclature*. Recommendations of the Nomenclature Committee of the International Union of Biochemistry and Molecular Biology. New York: Academic Press.

2

Brief overview of metabolic regulation of a bacterial cell

Abstract: Metabolic regulation of main metabolism is explained based on protein expressions for the batch culture of *Escherichia coli*. The metabolic regulation mechanism is briefly explained for aerobic and anaerobic cultivations of *E. coli* using glucose as a carbon source. Moreover, the metabolic regulations using gluconate, glycerol, or acetate as a carbon source are explained. The metabolic changes during batch cultivation of *E. coli* are also described at the exponential growth phase, early stationary phase, and stationary phase, based on gene expression data.

Key words: metabolic regulation of *E. coli*; proteome; aerobic and anaerobic cultivations; gluconate metabolism; glycerol metabolism; acetate metabolism; stationary phase; RpoS.

2.1 Introduction

All organisms respond to environmental variations, such as nutrient availability, oxygen limitation, etc. Corresponding physiological changes are accompanied by gene and protein expressions, etc. A better understanding of gene and protein expressions for cells under physiological changes is of primary importance. The most important notion is to regard the cell as a whole in its entirety in relation to global metabolic regulation, rather than through individual reactions (Shimizu, 2004; 2009). In this chapter, metabolic regulation of a bacterial cell such as *E. coli*, in response to different carbon sources and lower oxygen concentrations, is explained based on protein expressions, etc. Then we consider how metabolism changes with respect to time during the batch culture, and finally, metabolisms for acetate formation in *E. coli* are described.

2.2 Metabolic regulation analysis by protein expressions

As early as 1975, 2D electrophoresis (2DE) was proposed as the most efficient method of separating complex protein mixtures to analyze global patterns of gene expressions at the protein level (O'Farrell, 1975). Since then, this approach has been widely adopted (Anderson and Anderson, 1998; Gygi et al., 2000a,b; Han and Lee, 2006). One major outcome of these proteome studies is the establishment of 2DE databases for many organisms. These databases provide easy access to the analysis of gene expression in response to genetic or environmental alterations at the protein level. Usually, we choose a condition of interest, and study the global protein responses of the organism to the designated condition, both qualitatively (translational or post-translational modifications) and quantitatively (relative abundance or coordinated expression). In fact, the regulation of metabolic processes ultimately depends upon the control of enzyme activity.

There are three general mechanisms by which the activity of enzymes can be regulated: control by reversible binding of effectors; by covalent modification; and by alteration of enzyme concentration. In the first case, the enzyme is activated or inhibited by binding of a signal molecule, which may or may not be the substrate or product of the enzyme reaction, to the specific regulatory site, producing a conformational change. Substrate effect and allosteric control may be the example. In the case of covalent modification, the structure of the enzyme can be altered by the action of other enzymes. There are a number of such examples, such as the regulation by phosphorylation, i.e. phosphate is incorporated into the enzyme by a protein kinase using ATP, and is removed by a protein phosphatase. The third mechanism that regulates the enzyme activity is the alteration of the concentration of enzyme protein in the cell. The concentration of a protein in the cell is governed by the rate of the synthesis and the rate of degradation. The rate of synthesis of a particular protein may be controlled at several different levels. The rate of transcription of the gene may also be controlled. Other possible sites of control are the processing of the transcript to give mRNA, the transfer of mRNA out of the nucleus, the rate of degradation of mRNA in the cytoplasm, or the rate of translation of mRNA to form the protein on the ribosome. There is strong evidence that the rate of transcription is under rigorous control, and this control is important in determining the enzyme profile of a particular cell type (Martin, 1987). Therefore, to a large extent, enzyme activity reflects the protein expression level.

2.2.1 Fermentation characteristics and protein expressions with enzyme activities

Figure 2.1 shows the time courses of the batch cultivations for different carbon sources and under anaerobic conditions (Peng and Shimizu, 2003). As shown in Figure 2.1, *E. coli* grown in glucose or gluconate medium excreted acetate aerobically. Upon exhaustion of glucose or gluconate, the cells synthesized the enzymes of glyoxylate by-pass, which permits growth on acetate as a sole carbon and energy source. However, glycerol did not promote excretion of acetate aerobically. Under microaerobic conditions, cell growth was restricted concurrently, accompanied by excretion of lactate, acetate, formate, and ethanol (and possibly succinate, though not shown in the figure).

Tables 2.1–2.4 show the enzyme activities involved in the central metabolic pathways of *E. coli*, in which 26 enzymes are listed. Figure 2.2 shows the protein expression maps of 2DE for *E. coli* grown under different conditions. As can be seen from the figure, many proteins were induced or repressed either when the DO level was varied from aerobic to microaerobic conditions in a glucose medium, or the carbon source was changed from glucose to acetate, gluconate, or glycerol under aerobic conditions. In order to see how the protein expression levels changed under different culture conditions, the abundances of each protein under different culture conditions were divided by the corresponding one in the control experiment, such as aerobic conditions using glucose as the sole carbon source. Those ratios are shown in the metabolic pathway maps of Figure 2.3. The thickness of the arrow depends on the magnitude of the change in the protein expression level. Let us analyze the results one by one in the next section.

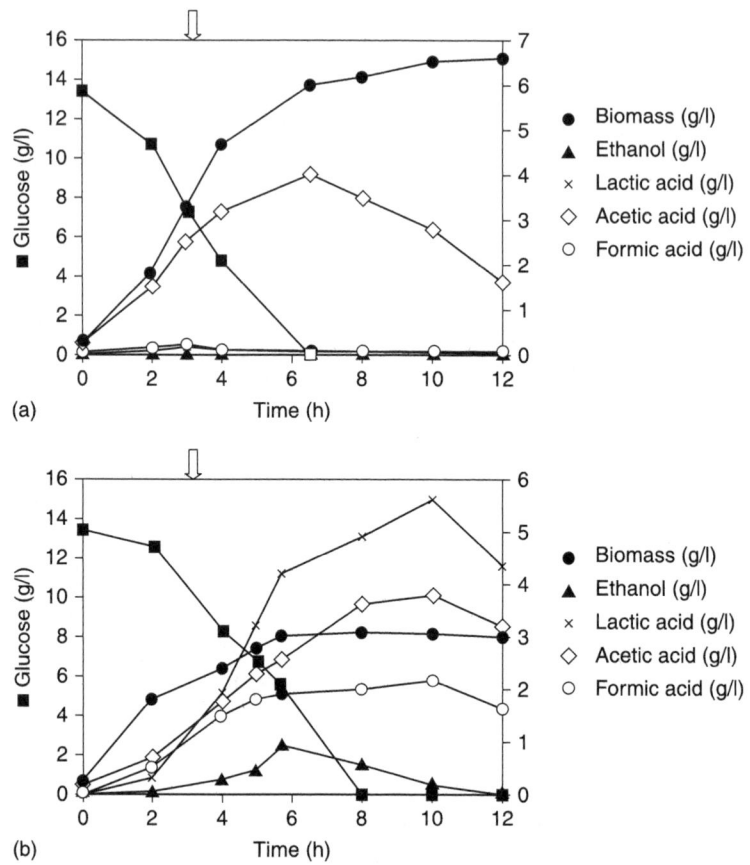

Figure 2.1 Batch cultivations of *E. coli* K12 using different DO levels and different carbon sources: (a) glucose under aerobic conditions; (b) glucose under microaerobic conditions; (c) gluconate under aerobic conditions; and (d) glycerol under aerobic conditions. The arrows indicate that *E. coli* cells were harvested for the proteome analysis and enzyme activity measurement

2.2.2 Metabolic regulation by protein expressions

Glycolysis and anaplerotic pathway

An early investigation showed that glucose transport genes exhibited high basal expression, and *ptsHI* genes were positively stimulated by the

Brief overview of metabolic regulation

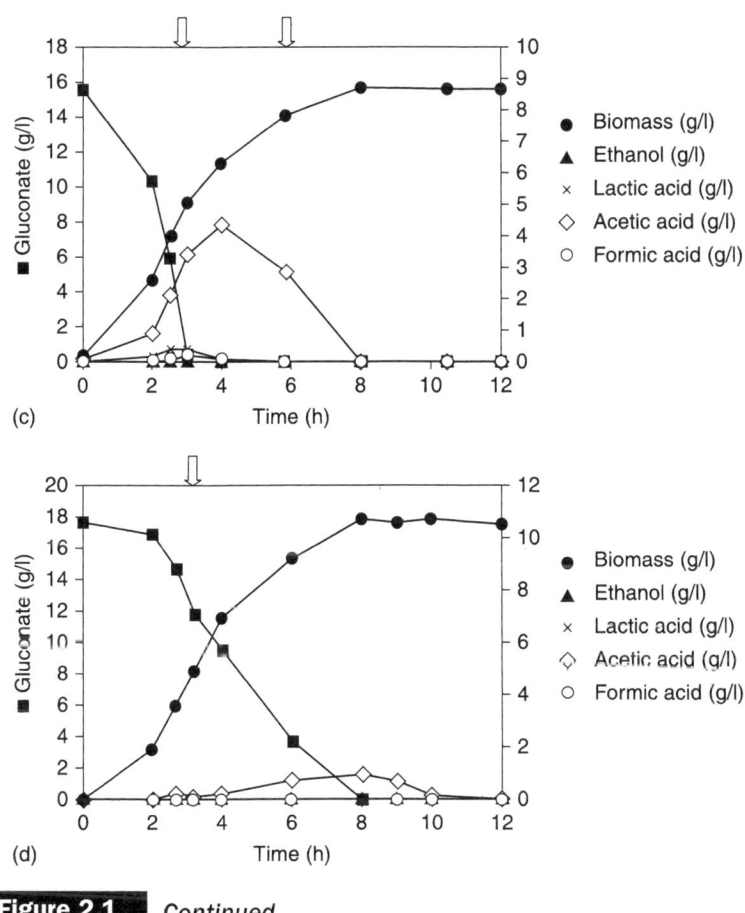

Figure 2.1 Continued

cAMP-CRP receptor protein and also by the growth on glucose, while *crr* promoters within *pts* may be negatively regulated by cAMP-CRP (Danchin, 1988; Fox et al., 1992). Plumbridge (1999) reported that these genes are in the same operon, and the operon is known to be regulated by Mlc. In the absence of glucose, Mlc represses the operons. In the presence of extracellular glucose, the conformation of EIIBC^{Glc} protein changes and bonds strongly with Mlc, which no longer represses the operon. However, an essential feature of the phosphotransferase system (PTS) is that the phosphoryl donor molecule is PEP and not ATP (the regulation of PTS is explained in more detail in Chapter 3). The experimental results show that the glucose transport genes *ptsHI-crr* were slightly up-regulated under microaerobic conditions in comparison to the aerobic growth on glucose, and

Table 2.1 Activities of glycolytic and anaplerotic enzymes in response to carbon sources and DO levels. Ratio calculation was based on the control of aerobic growth on glucose

C source	Glucose					Acetate		Gluconate		Glycerol	
DO level	Aerobic			Microaerobic		Aerobic		Aerobic		Aerobic	
Enzyme	Activity	Control		Activity	Ratio	Activity	Ratio	Activity	Ratio	Activity	Ratio
PTS	0.015±0.001	1		0.017±0.00	1.131	ND	–	ND	–	ND	–
Hxk	0.032±0.00	1		0.039±0.00	1.221	0.022±0.002	0.71	0.027±0.002	0.85	0.014±0.002	0.44
Pgi	3.29±0.05	1		4.66±0.04	1.41	2.88±0.02	0.62	2.86±0.05	0.87	2.18±0.04	0.66
Pfk	0.34±0.01	1		0.54±0.02	1.64	0.19±0.02	0.56	0.17±0.02	0.70	0.22±0.01	0.64
Fbp	0.024±0.00	1		0.033±0.00	1.371	0.052±0.001	2.16	0.12±0.02	5.00	0.077±0.001	3.21
Fba	1.60±0.01	1		1.92±0.01	1.20	0.93±0.02	0.58	1.35±0.04	0.84	1.02±0.02	0.66
Tpi	2.80±0.02	1		3.50±0.02	1.25	2.27±0.02	0.81	3.57±0.02	1.02	4.14±0.02	1.48
GAPDH	0.036±0.001	1		0.051±0.00	1.402	0.018±0.001	0.51	0.039±0.002	1.05	0.045±0.001	1.25
Pgk	0.061±0.01	1		0.080±0.01	1.32	0.28±0.01	0.46	0.068±0.01	1.11	0.076±0.02	1.25
Pyk	0.054±0.00	1		0.084±0.00	1.561	0.016±0.001	0.29	0.058±0.001	1.07	0.036±0.002	0.67
Ppc	0.22±0.02	1		0.13±0.01	0.59	0.050±0.001	0.23	0.26±0.02	1.17	0.40±0.01	1.82
Pck	0.037±0.00	1		0.040±0.00	1.083	0.11±0.03	2.97	0.066±0.003	1.78	0.047±0.004	1.29
Mez	ND	–		ND	–	0.008±0.001	–	ND	–	ND	–

Published by Woodhead Publishing Limited, 2013

Table 2.2 Activities of PP and E-D pathway enzymes in response to carbon sources and DO levels. Ratio calculation was based on the control of aerobic growth on glucose

C source	Glucose						Acetate		Gluconate		Glycerol	
DO level	Aerobic		Microaerobic				Aerobic		Aerobic		Aerobic	
Enzyme	Activity	Control	Activity	Ratio			Activity	Ratio	Activity	Ratio	Activity	Ratio
G6PDH	0.35±0.02	1	0.22±0.01	0.64			0.14±0.01	0.40	0.50±0.01	1.42	0.44±0.01	1.25
6PGDH	0.28±0.02	1	0.21±0.01	0.74			0.19±0.02	0.68	0.45±0.02	1.61	0.33±0.02	1.18
E-D Pathway	0.45±0.02	1	0.15±0.02	0.33			0.26±0.01	0.57	3.01±0.02	6.69	0.38±0.04	0.85

Table 2.3 Activities of fermentative enzymes in response to carbon sources and DO levels. Ratio calculation was based on the control of aerobic growth on glucose

C source	Glucose				Acetate		Gluconate		Glycerol	
DO level	Aerobic		Microaerobic		Aerobic		Aerobic		Aerobic	
Enzyme	Activity	Control	Activity	Ratio	Activity	Ratio	Activity	Ratio	Activity	Ratio
Ack	0.48±0.02	1	0.75±0.01	1.56	0.34±0.02	0.70	0.55±0.04	1.15	0.13±0.02	0.27
LDH	0.66±0.05	1	1.22±0.04	1.85	0.34±0.03	0.51	1.01±0.05	1.53	0.61±0.04	0.92
ADH	0.005±0.002	1	0.062±0.04	12.40	ND	–	0.009±0.002	–	0.005±0.002	–
Pfl	ND	–	0.021±0.02	–	ND	–	ND	–	ND	–

Table 2.4 Activities of TCA cycle and glyoxylate shunt enzymes in response to carbon sources and DO levels. Ratio calculation was based on the control of aerobic growth on glucose

C source	Glucose					Acetate		Gluconate		Glycerol	
DO level	Aerobic			Microaerobic		Aerobic		Aerobic		Aerobic	
Enzyme	Activity	Control		Activity	Ratio	Activity	Ratio	Activity	Ratio	Activity	Ratio
CS	0.051±0.000	1		0.0076±0.0001	0.15	0.25±0.01	4.90	0.32±0.01	6.27	0.23±0.02	4.51
ICDH	1.15±0.02	1		0.14±0.02	0.12	0.26±0.02	0.23	2.20±0.04	1.91	1.88±0.04	1.63
Icl	0.013±0.002	1		0.006±0.002	0.46	0.12±0.02	9.23	0.019±0.001	1.46	0.017±0.003	1.31
KGDH	0.022±0.002	1		ND	–	0.058±0.001	2.91	0.060±0.001	2.73	0.065±0.001	2.95
Fum	0.061±0.004	1		0.017±0.002	0.28	0.20±0.01	3.27	0.11±0.01	1.80	0.10±0.01	1.64
MDH	0.056±0.00	1		0.030±0.001	0.54	0.15±0.01	2.68	1.78±0.01	1.96	0.10±0.01	1.78

Brief overview of metabolic regulation

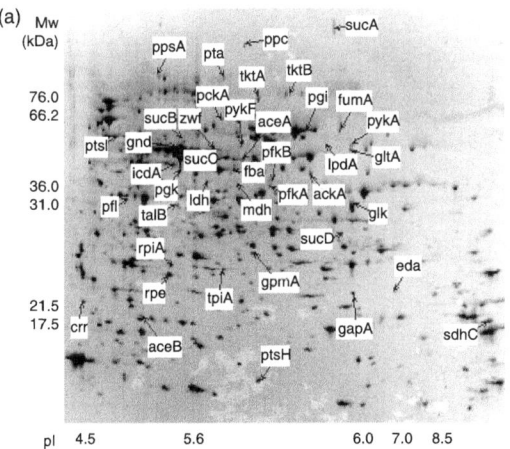

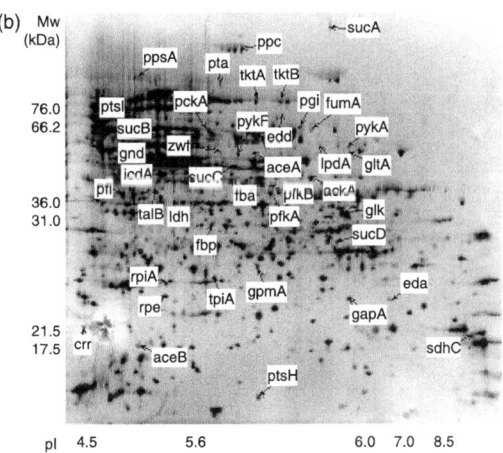

Figure 2.2 2-DE gel maps of the total lysate of *E. coli* cells grown on: (a) glucose under aerobic conditions; (b) glucose under microaerobic conditions; (c) acetate under aerobic conditions; (d) gluconate under aerobic conditions; and (e) glycerol under aerobic conditions.

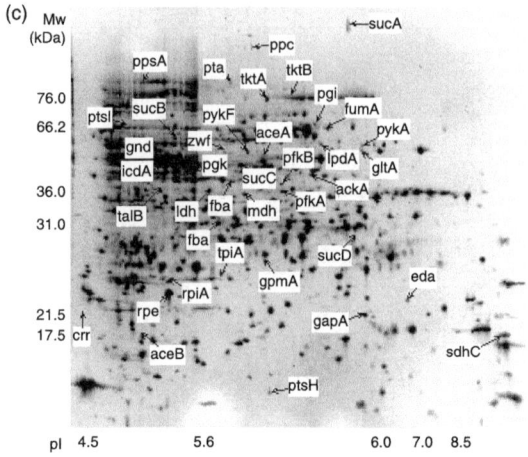
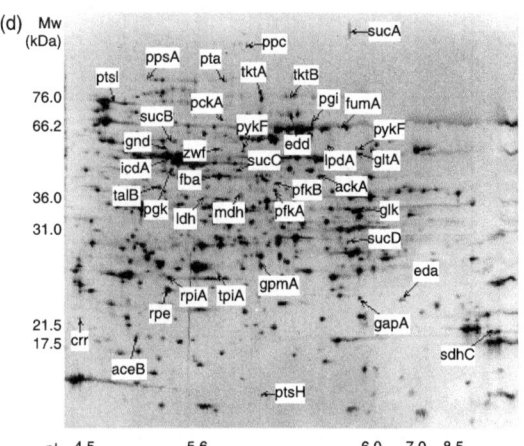
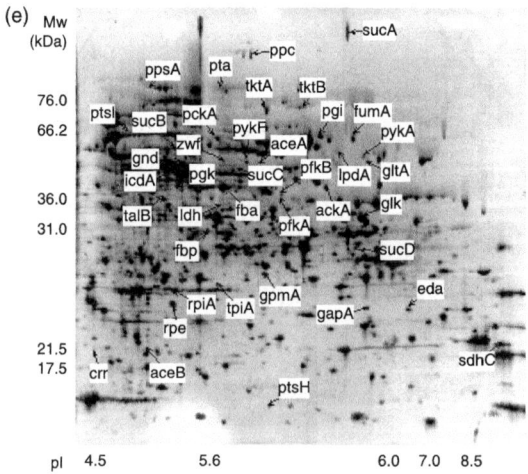

Figure 2.2 *Continued*

significantly down-regulated by 2- to 5-fold when the carbon source was changed to acetate, gluconate, or glycerol. The enzyme activity of PTS was only detected in glucose medium (Table 2.1), confirming the efficient induction by glucose either aerobically or microaerobically. Glk encoding the first enzyme glucokinase, which drives glucose entry into the glycolytic pathway in the cell, was kept relatively unregulated by not more than about 1.2-fold in all cases.

Most of the protein levels for the glycolysis genes, such as *pgi*, *pfkA*, *fba*, *gapA*, *pgk*, *eno*, and *pykF* were also observed to be up-regulated under microaerobic conditions, indicating a 'stepping-up' of anaerobic glucose utilization via glycolysis. The rapid fall of TCA cycle enzyme activities (Table 2.4) and the significant increase in fermentative enzyme activities (Table 2.3) under microaerobic conditions, indicate that glucose utilization was switched toward fermentation under microaerobic conditions.

In contrast, the protein levels for the common glycolysis genes *pgi*, *pfkA*, *fba*, *gapA*, *eno*, *pykF*, and anaplerotic pathway gene *ppc*, were highly repressed by about 2- to 4-fold in acetate medium compared to those in glucose medium. Simultaneously, the protein levels of the gluconeogenic pathway genes *fbp*, *pckA*, and *ppsA* were highly induced by about 2.5-fold, 3.7-fold, and 8.3-fold, respectively, which were thought to be subjected to positive regulation by the catabolite repressor/activator Cra, a global catabolite repression regulator (Saier and Ramseier, 1996), formerly known as the fructose repressor FruR, a global regulator concerned with carbon utilization by transcriptional modulation of the target genes (Ramseier et al., 1995). The enzyme activities given in Table 2.1 agree with the fact that carbon flow will be channeled through the gluconeogenic pathway when acetate is metabolized, and the flux in gluconeogenic direction is much smaller than the glycolytic flux during growth on glucose. Note that in the reaction between F6P and F1, 6BP, PEP and PYR, PEP and OAA, the involved enzymes are tightly regulated. The protein level of *pfkA*, encoding for the main phosphofructokinase, Pfk-1, was repressed by about 2-fold, while *pfkB*, encoding for the minor phosphofructokinase Pfk-2, was kept nearly unchanged. Correspondingly, the activity of Pfk dropped 1.8-fold in acetate medium, while *fbp* gene was up-regulated by 2.5-fold, and the enzyme activity of FBPase varied in a co-ordinate manner, and increased by 2.2-fold (Table 2.1). It has been proved that *pfkA* is subject to the regulation by the catabolite repressor/activator Cra (FruR), while the enzyme Pfk-1 is allosterically activated by ADP and inhibited by PEP and ATP. The inhibitory effect of ATP is opposed

Bacterial cellular metabolic systems

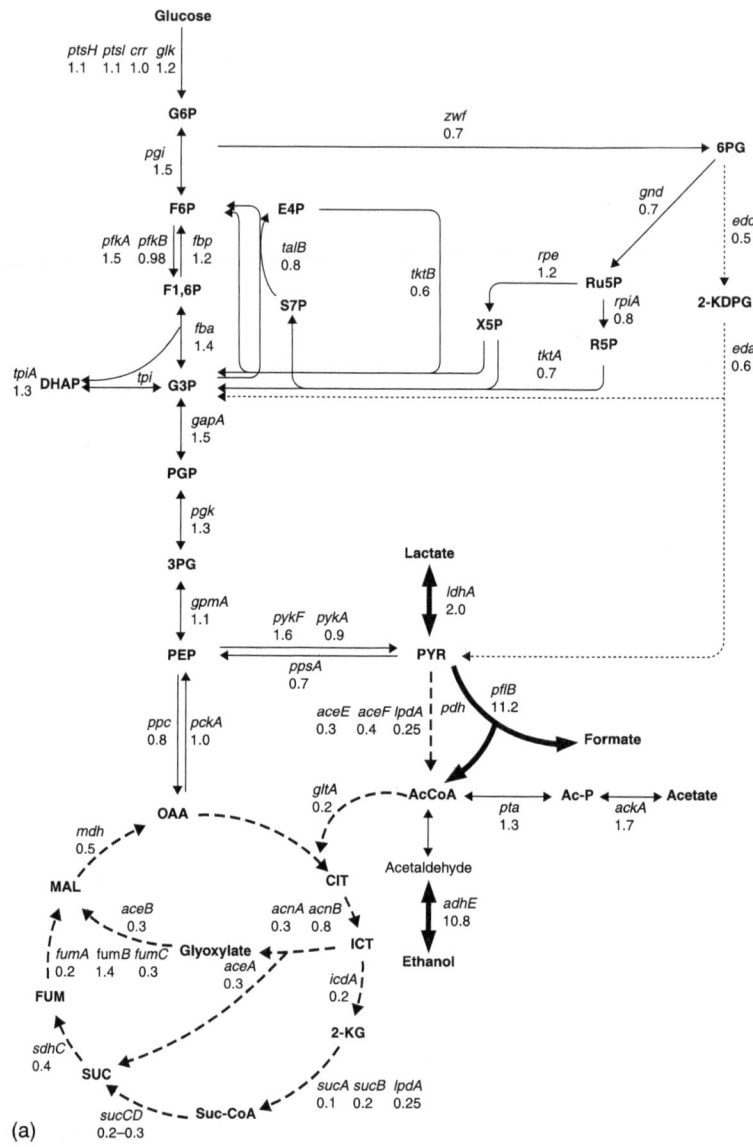

Figure 2.3 The relative expression levels of *E. coli* K-12 proteins of central metabolic pathways under different conditions based on 2DE results: (a) microaerobic growth on glucose; (b) aerobic growth on acetate; (c) aerobic growth on gluconate; and (d) aerobic growth on glycerol. Aerobic growth on glucose was used as the control.

Brief overview of metabolic regulation

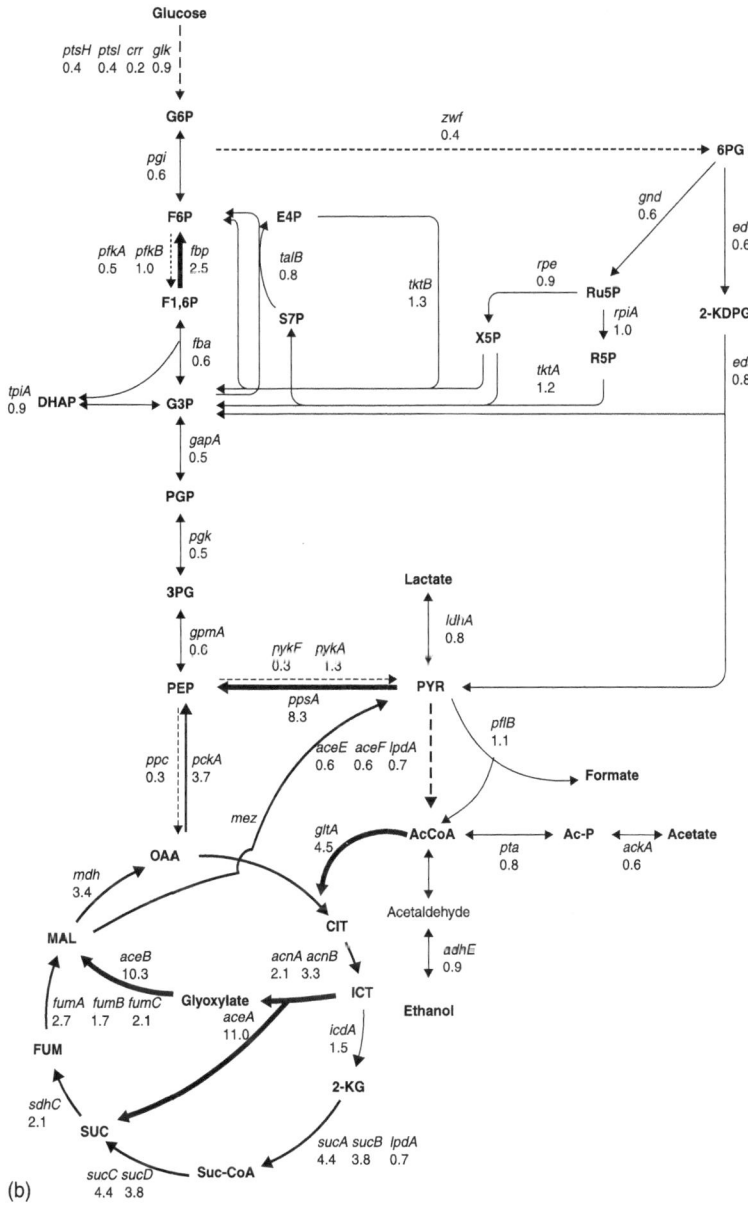

Figure 2.3 (Continued)

Bacterial cellular metabolic systems

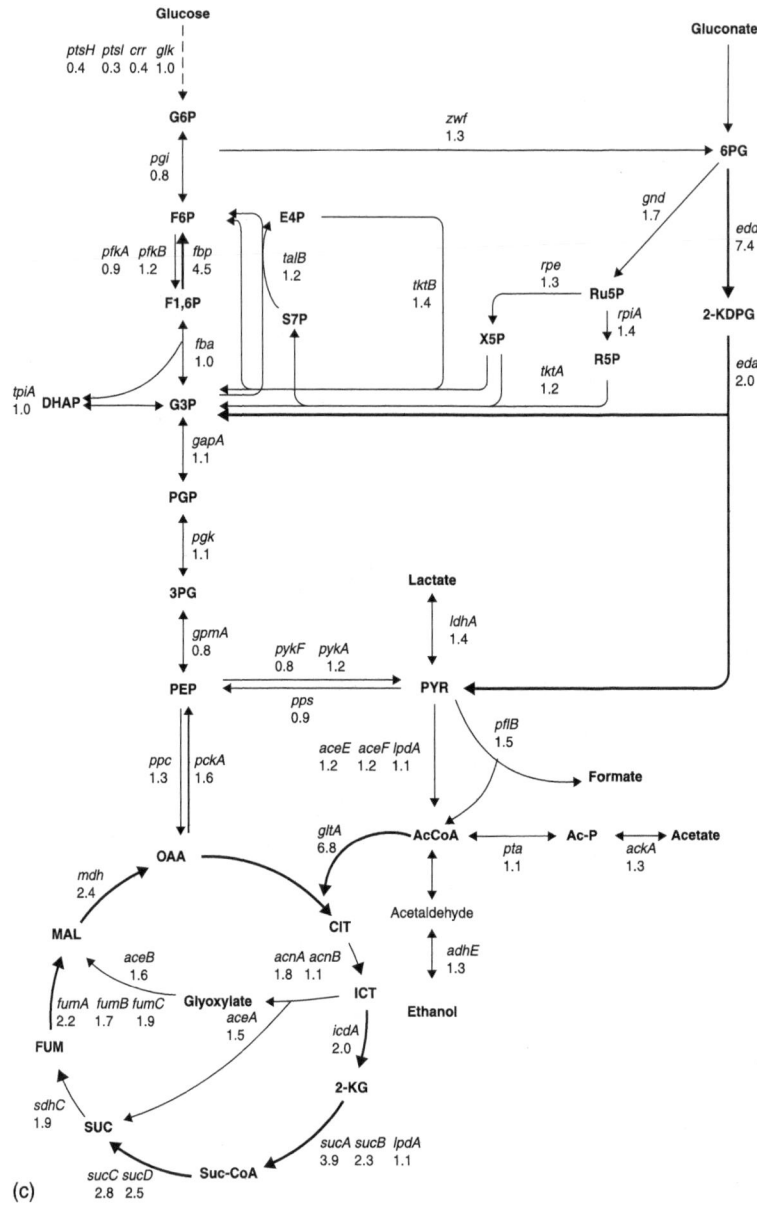

Figure 2.3 Continued

Brief overview of metabolic regulation

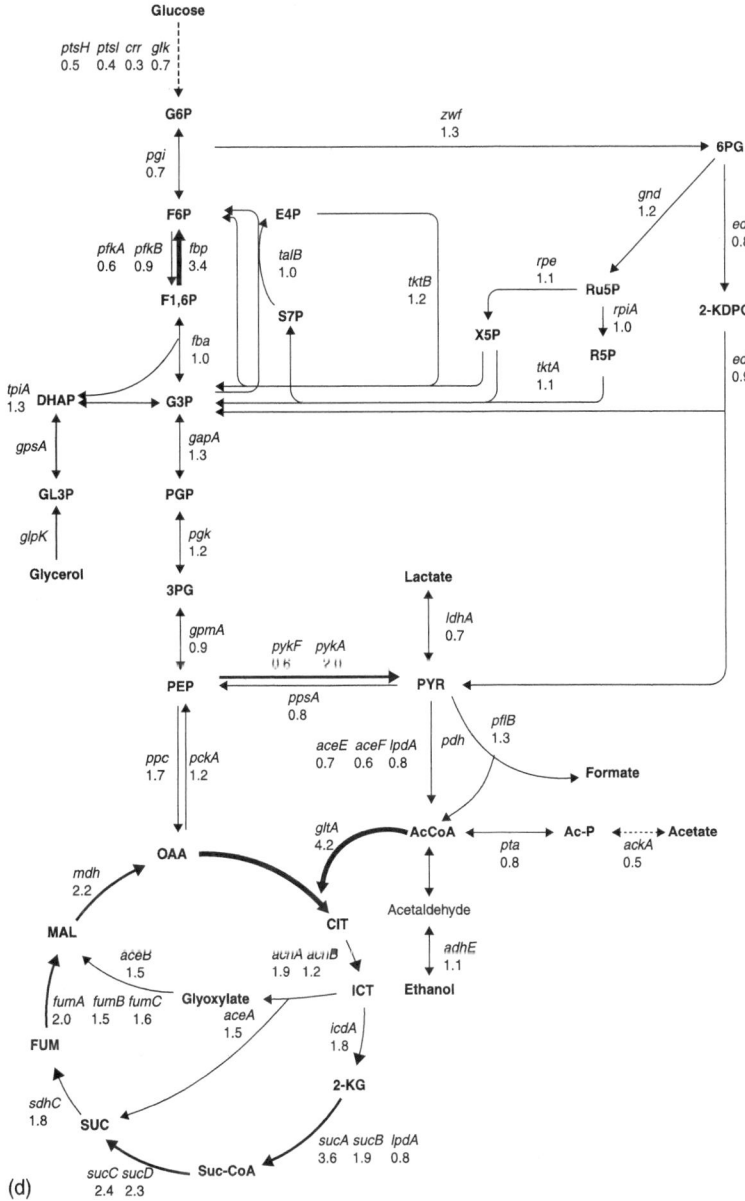

Figure 2.3 Continued

by AMP, and intensified by citrate, which acts as a signal of the availability of alternative sources of ATP (Martin, 1987). This control might link to an increase in the rate of Pfk-1 synthesis under anaerobic conditions. Pfk-2, insensitive to ATP inhibition, may serve as another role (i.e. maintain the futile cycle for the regulatory amplification) (Campos et al., 1984; Guixe and Babul, 1985). The down-regulation of PykF, but not PykA, suggests that these two isoenzymes were differentially regulated. Indeed, *pykF* is negatively regulated by Cra (FruR).

The protein level of *ppc* gene was down-regulated by 3-fold in the acetate medium. This result is consistent with the direction of the metabolic flux. The carbon flux from oxaloacetate (OAA) to phosphenolpyruvate (PEP) will be dominant in such a case, and is catalyzed by Pck. In fact, the protein level of *pckA* gene was up-regulated by 3.7-fold, which is also modulated by Cra at the transcriptional level. The enzyme activities were coordinately regulated, and Pck increased by nearly 3-fold, while Ppc decreased by 4.3-fold (Table 2.1). High protein level expressions of both *ppc* and *pckA* led to futile cycling. Moreover, the protein level of *ppsA* gene was also found to be highly induced by about 8.3-fold in acetate medium. In addition, malic enzyme, Mez, which converts malate to pyruvate, showed some activity only in acetate medium, and was not detectable for other cases. This result suggests that Pps and Mez may play an important role for the gluconeogenic flux during the metabolism of acetate. In fact, the induction of *ppsA* and *maeB* genes of E. coli during growth in acetate medium has been confirmed by DNA microarray (Oh and Liao, 2000; Oh et al., 2002).

In contrast, most of the glycolytic genes were not affected significantly during growth in gluconate medium. An exception is the *fbp* gene, which was induced by 4.5-fold in protein level. The enzyme activity of FBPase increased 5-fold in gluconate medium (Table 2.1). The high induction of the *fbp* gene will drive part of GAP from the ED pathway through gluconeogenesis to supplement G6P. Surprisingly, the protein level of *pckA* gene was found to be induced by 1.6-fold, and the enzyme activity of Pck varied accordingly (Table 2.1). The same phenomenon with *pckA* appeared in cells grown on glycerol. However, this induction is unexpected, since the flux in Pck-mediated reaction is not needed in both cases. The *pckA* may be induced gratuitously by the increased level of cAMP during growth in non-glucose medium. It has been reported that the concentration of cAMP in glycerol-grown cells was much higher than that in glucose-grown cells of E. coli (Epstein et al., 1975; Unden and Duchene, 1987). During growth on glycerol, one significantly affected gene was *pykA*, which was induced by 2-fold at protein level. When *E. coli* was cultivated in glycerol, the flux from PEP to pyruvate is governed by Pyk rather than the PTS. Therefore, it is reasonable to expect that one of the *pyk* genes is up-regulated to satisfy the significantly increased demand for the flux from glycolysis. Apparently, the *pykA* gene rather than *pykF* serves this role. Since Pyk is activated at the protein level by F1, 6BP, whose concentration is relatively low, is not a good indicator of the metabolic state during growth in glycerol (Lowry et al., 1971). Therefore, *pykF* may remain almost inactive due to the low concentration of its allosteric activator. *PykA*, however, is activated by cAMP, which is higher in glycerol medium than in glucose medium (Epstein et al., 1975; Unden and Duchene, 1987), and its activity can still be mediated at the protein level. Therefore, *pykA* is the better choice during growth on glycerol.

Pentose phosphate pathway and Entner-Doudoroff pathway

The two enzymes involved in the oxidative PP pathway, glucose 6-phosphate dehydrogenase (G6PDH) encoded by *zwf* gene and 6-phosphogluconate dehydrogenase (6PGDH) encoded by *gnd* gene, are down-regulated during microaerobic growth in glucose medium and aerobic growth in acetate medium. However, both enzymes are up-regulated when the cells grow in gluconate or glycerol mediums. The enzyme activities varied in a coordinated manner (Table 2.2).

The ED pathway is significantly induced when grown on gluconate, based on the measurements of the overall activity of the ED enzymes and the 2DE result. Table 2.2 shows the activities of the two ED pathway enzymes in *E. coli*. Both enzymes are present at high levels during growth on gluconate, where the protein level of the *edd* gene was highly up-regulated by 7.4-fold, and *eda* was induced by 2-fold, as compared to the case of using glucose as the carbon source, and the overall activity of ED pathway enzymes coordinately increased by about 7-fold (Table 2.2). Note that the overall activity of the ED enzymes, including 6PG dehydratase (edd) and KDPG aldolase (eda), was assayed by measuring 6PG-dependent formation of pyruvate, which is determined colorimetrically as its dinitrophenylhydrazone (Esienberg and Dobrogosz, 1967; Lessie and Whiteley 1969). This assay procedure underestimates the level of KDPG aldolase, which is usually present in excess compared with 6PG dehydratase. It measures the rate-limiting component in the pathway, 6PG dehydatase.

Both enzymes G6PDH and 6PGDH were known to be subject to cellular growth rate regulation, which is proportional to the growth rate influenced by the medium (Wolfe et al., 1979; Pease and Wolfe, 1994). Indeed, the cell growth was slow under microaerobic conditions and in acetate medium, while it was faster during growth on gluconate and glycerol as well as the cell growth on glucose. In addition, an early study reported that 6PGDH was induced by gluconate (Lowry et al., 1971). The other genes involved in non-oxidative metabolism, such as *rpe*, *rpi*, *tal*, and *tkt*, did not differ significantly between growth conditions. It is known that the *edd* gene, containing a regulatory region, is induced by gluconate, and the *eda* gene is probably induced by EDGP, the product of *edd* (Sugimoto and Shiio, 1987; Conway, 1992), and both genes are subjected to negative control by Cra. This result indicates that the ED pathway is predominant in dissimilation of gluconate in *E. coli*, while insignificant in the metabolism of acetate, glucose, or glycerol.

Fermentative pathway

Among the fermentative genes, the protein levels of *pfl* and *adhE* genes were dramatically induced by about 11.2-fold and 10.8-fold, respectively, in response to the shift from aerobic to microaerobic conditions in glucose medium (Figure 2.2). The enzyme activity of Pfl was only detected in microaerobic conditions, and the activity of NADH-dependent ADH also increased by 12.4-fold under microaerobic conditions (Table 2.3).

Regulation of *pfl* synthesis and activity was subjected to control by the Fnr and ArcA/B two-component regulatory systems. Fnr, an oxygen-sensing global regulator, which is an iron sulfur-dependent DNA-binding protein and recognizes a specific sequence motif found in the promoter regions of the genes it regulates, serves as an activator of the transcription of anaerobically regulated genes (Gunsalus and Park, 1996; Kiley and Beinert, 1999). The significant anaerobic induction of the *pfl* operon was thought to account for Fnr and ArcA/B, which mediate the residual transcriptional activation of the operon (Alexeeva et al., 2000; Sawers and Suppmann, 1992). However, the expression of *pfl* was little changed in acetate, gluconate, or glycerol medium, although this fermentative pathway is not used when growing aerobically in acetate medium (Sawers and Suppmann, 1992). Expression of *adhE* is also strongly induced by micro-aerobiosis. However, this induction is independent of the Fnr and ArcA transcription factors. There appears to be a direct correlation between the NADH/NAD⁺ ratio and enzyme synthesis; the higher the ratio, the more ADH is synthesized. Moreover, the ADH protein itself may exert a regulatory function, since the gene expression was enhanced dramatically in the *adhE* mutant (Bock and Sawers, 1999). Therefore, the induction of *adhE* is not surprising, since the previous study showed that the significant differences of the NADH/NAD⁺ ratios between aerobic and anaerobic cultures of 0.02 and 0.75, respectively, have been observed as the DOT (dissolved oxygen tension) of the culture was decreased (Alexeeva et al., 2000). The *ldhA* gene, encoding NAD⁺ linked enzyme LDH, was also highly enhanced by 2-fold during microaerobic fermentation of glucose. This result is consistent with the study by Mat-Jan et al. (1989), who found that the fermentative LDH was cojointly induced by anaerobic conditions and acidic pH (Alexeeva et al., 2000). Since *E. coli* satisfies energy requirements through glycolysis at the accelerated rate under anaerobic conditions, the NADH produced must be re-oxidized to NAD⁺, which is required to maintain glycolysis since it is again the substrate for GAPDH in glycolysis. In the presence of oxygen, the oxidation of NADH occurs through molecular oxygen in the respiratory chain, while in the absence of oxygen it proceeds through reduction of an organic acid, and thereby the conversion of pyruvate to lactate by LDH is preferable. The LDH activity changed in response to carbon sources. Namely, it was repressed in acetate, slightly enhanced in gluconate, and almost unchanged in glycerol.

The protein levels of *pta* and *ackA* genes, involved in the reversible acetyl CoA metabolism in the cell, are up-regulated by 1.3-and 1.7-fold, respectively, in the case of microaerobic conditions (Figure 2.4). The enzyme activity of Ack in microaerobically grown cells was 1.56 times as high as that grown aerobically in glucose medium (Table 2.3). The *pta* and *ackA* genes are constitutively expressed and present in the same operon, but were regulated differentially through different promoters. It has been proposed that the intermediate of this pathway, acetyl phosphate, might be an important effector of gene regulation, while the levels of acetyl phosphate vary dramatically, depending on the carbon source in the growth medium. For example, in the defined medium under limiting phosphate concentrations, very low levels of acetyl phosphate were observed when cells were grown on glycerol (<40 mmol/l), moderate levels on glucose (300 mmol/l), and high levels on pyruvate (Wanner and Wilmers-Riesenberg, 1992; McCleary and Stock, 1994). During growth on glycerol, Ack is down-regulated by almost 3-fold (Table 2.3).

This result is consistent with the observation that acetate production in glycerol medium is much lower than in glucose medium. However, a relatively small change of *pta* was observed. It may perhaps ensure the production of acetyl phosphate for regulatory purposes. During growth on acetate, both *pta* and *ackA* were down-regulated. This result suggests that other genes may be responsible for acetate uptake during growth on acetate. Indeed, it was proposed that the *acs* gene (coding for acetyl-CoA synthetase) is mainly responsible for acetate uptake, whereas the *pta-ackA* pathway is used for acetate excretion during growth on glucose (Oh et al., 2002). In gluconate medium, both genes were observed to be up-regulated. The measured enzyme activities of Ack were roughly correlated with the protein expression levels.

TCA cycle and glyoxylate shunt

It was observed that there was a significant effect of the change in either oxygen level or carbon source on the TCA cycle genes. Under microaerobic conditions, the TCA cycle protein expressions fell by around 2- to 10-fold, as compared to those under aerobic conditions. In contrast, if acetate, gluconate, or glycerol was used as the carbon source instead of glucose, most of the TCA cycle genes were up-regulated by 2- to 6-fold.

Citrate synthase (CS), encoded by the *gltA* gene, the first enzyme to enter the TCA cycle, was suppressed by oxygen limitation and excess glucose, but elevated by oxygen (5-fold) and other oxidized carbon sources (4- to 7-fold), such as acetate, gluconate, or glycerol. The α-ketoglutarate dehydrogenase (KGDH) complex was encoded by *sucA* and *sucB*, and shared the third gene *lpdA* with PDH complex. These genes were strongly suppressed by oxygen limitation and induced by acetate and other oxidized carbon sources. Indeed, the enzyme activity of KGDH was not detectable in glucose medium under microaerobic conditions (Table 2.4). These observations coincide with an early study of Amarasingham et al. (1965). At the important control point of the junction between the TCA cycle and the glyoxylate shunt, *aceA* and *aceB* were cojointly induced by more than 10-fold in acetate medium. The enzyme activity of isocitrate lyase (Icl) increased by more than 9-fold in acetate medium (Table 2.4). The highly regulated glyoxylate shunt was absolutely essential to direct the carbon flow through this bypass to generate a 4-carbon precursor for biosynthesis in acetate metabolism (Chung et al. 1988).

The TCA cycle enzymes (Table 2.4) are shown to be mainly dependent on the gene expression levels. These genes are known to be subjected to cAMP-Crp mediated catabolite repression and ArcA, Fnr, and SoxRS oxidative stress regulatory system mediated aerobic-anaerobic repression (Cronan and Laporte, 1999). Although there was a general reduction in TCA cycle gene expression and enzyme activity related to microaerobic conditions in glucose medium, the variations of individual genes were not coordinated, implying that these genes are subjected to different regulations and different regulatory mechanism. For instance, *fumA*, *fumB*, and *fumC* are regulated differently by oxygen availability, where *fumB* is slightly abundant under microaerobic conditions (1.4-fold) (Figure 2.4), since the expression of the *fumB* gene is controlled by anaerobic transcriptional activator, Fnr, whereas the expressions of *fumA* and *fumC* decreased 3- to 5-fold when the oxygen level was reduced to microaerobic conditions as compared to aerobic conditions (Figure. 2.4). It was proposed that anaerobic expression of *fumA* promoter was repressed by ArcA and Fnr, while expression of *fumC* was repressed by ArcA only, but required the SoxRS regulatory protein, which is a positive regulator for controlling synthesis of proteins in response to oxidative stress (Park and Gunsalus, 1995; Tseng et al., 2001). In addition, the observation of 4-fold higher expression of *fumA* than that of *fumC* suggested that *fumA* is predominant under aerobic conditions. However, *fumC* is relatively less affected by the type of carbon source than *fumA*, indicating that *fumA* is shown to be under catabolite control.

The *acnA* shows 2- to 3-fold lower expressions due to catabolite and anaerobic repression, but *acnB* expression is relatively constant, although 3-fold up-regulated in acetate medium, since *acnB* is used preferably for the glyoxylate shunt, indicating that *acnB* was regulated differently. The *acnA* appeared to be activated by the SoxRS and Fnr (Cronan and Laporte, 1999). It was demonstrated that ArcA functions as a repressor of *gltA* expression during both aerobic and anaerobic growth through a *lacZ* fusion study (Park et al., 1994). The relative pattern of carbon control by *mdh*, *gltA*, and *sdhC* gene expressions are similar. However, the magnitude of oxygen control of *mdh* and *gltA* gene expressions were larger than that of *sdhC*. The difference in magnitude may reflect the function of the *mdh* and *gltA* protein under both aerobic and anaerobic conditions, whereas the *sdhCDAB* proteins are primarily used for aerobic cell growth (Park et al., 1995; Gunsalus and Park, 1996). The protein level of *icdA* gene encoding isocitrate dehydrogenate (ICDH), competing for the common substrate with isocitrate lyase at the node, is up-regulated by 1.5-, 2.0-, and 1.8-fold in acetate, gluconate, and glycerol mediums, respectively. Nevertheless, the activity of ICDH did not agree with this result in the case of acetate, which reduced by nearly 4.3-fold (Table 2.4). This fact may be explained as follows: the *icdA* gene was induced by acetate, and in turn led to an increased amount of the enzyme protein. However, in order to prevent the excess flux of isocitrate through ICDH, the enzyme activity was regulated by reversible inactivation of ICDH, which is catalyzed by bifunctional kinase/phosphatase, encoded by *aceK*, since the phosphorylated form of ICDH has no activity (Holms, 1988). Although regulation of the enzyme ICDH and Icl activities are opposite in acetate medium, the absolute enzyme activity of ICDH, which leads the carbon flow to the TCA cycle, is still dominant over the glyoxylate bypass directing enzyme, Icl (Table 2.4), because of the balance of the rate of supply of NAD(P)H and ATP with the demand for biosynthesis.

Comparison of the protein expression level and enzyme activity

Figure 2.4 shows the correlation of the protein expression levels and the corresponding enzyme activities. The logarithmic ratio of the enzyme activity is plotted versus the logarithmic ratio of the protein expression level. As can be seen in Figure 2.5, most of the spots are close to the 45° line, except ICDH under aerobic acetate conditions, indicating that protein abundances revealed by 2DE are correlated, to a large extent, with enzyme activity. However, it is obvious that there exist deviations from the linear correlation between the logarithmic protein expression ratios and the enzyme activity ratios. The deviations may be caused by the fact that protein products expressed from a single gene can migrate to multiple spots on 2DE gels, and also the proteins from multiple genes can run to the same coordinates on a gel, because of differential protein processing and posttranslational or artifactual modifications (Gygi et al., 2000a,b). Both differential migration and co-migration of proteins complicate comparative and quantitative pattern analysis of 2DE gels.

However, although the enzyme activity is primarily determined by the protein expression level, subtle regulations are exerted by many effectors at the enzyme level to satisfy the need for metabolism. ICDH in acetate medium may be such a case, which was regulated by reversible inactivation catalyzed by bifunctional kinase/phosphatase, as discussed

Brief overview of metabolic regulation

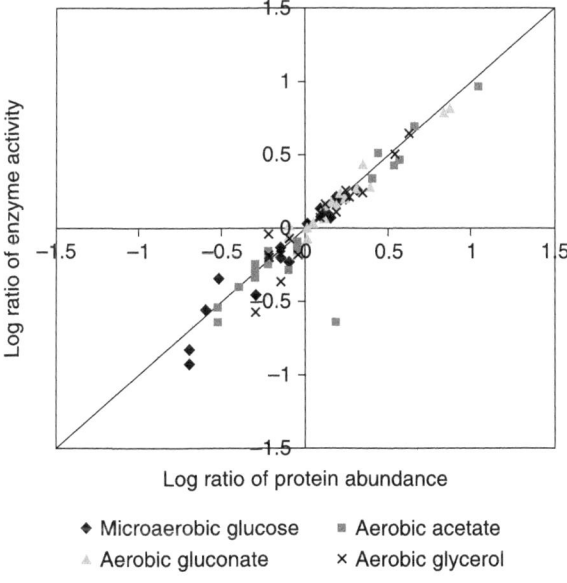

Figure 2.4 Comparison of the logarithmic protein expression ratios based on 2DE and the corresponding enzyme activity ratios. Data were taken from Tables 2.1–2.4 and Figure 2.3. Each set of data was plotted on the figure by the logarithmic ratio value

above. Moreover, the Ppc and Pck enzymes, involved in the anaplerotic and gluconeogenic pathways, are also subject to fine control by some effectors.

2.3 Metabolic regulation during the time course of batch culture

Let us now consider how metabolism changes with respect to time in the batch culture of *E. coli* (Rahman et al., 2007, 2008). The growth curve of *E. coli* BW25113, as shown in Figure 2.5, indicates that the exponential growth of the bacteria continued until about 7.5 h, followed by a short period of growth before the cells entered into the stationary phase. The end of the exponentially growing phase was correlated with almost complete utilization of glucose by the bacteria from the extracellular medium. Therefore, samples collected at 6, 8, and 12 h for gene expression

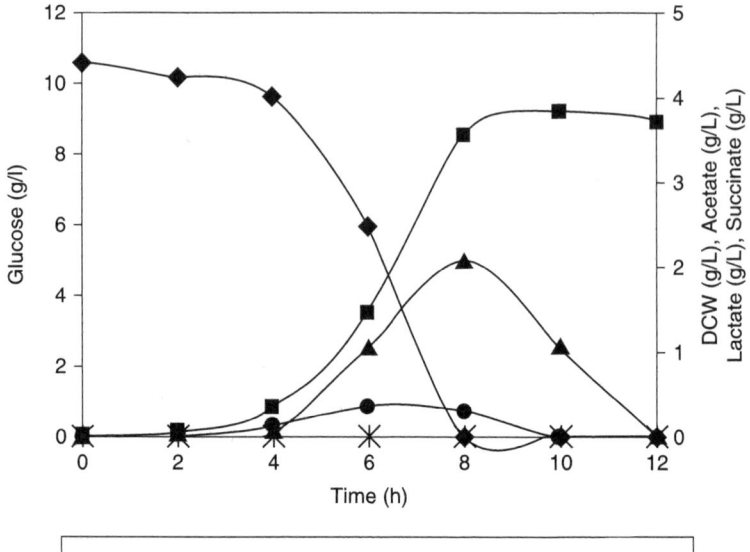

Maximum specific growth rate (m)	0.54 ± 0.03
Specific glucose uptake rate (mM/gDCW/h)	5.52 ± 0.06
Yield of acetate on glucose (g/g)	0.2 ± 0.05
Biomass yield (g DCW/g substrate)	0.36 ± 0.03

Figure 2.5 Growth curves of *E. coli* BW25113, which was grown in minimal media containing 10 g glucose/l as the sole carbon source. Fermentations were carried out at pH 7, at 37°C, with an agitation speed of 350 rpm to ensure 35–40% (v/v) of air saturation to maintain aerobic conditions of growth. Concentrations of ♦ glucose (g/l), ■ DCW (g/l), ▲ acetate (g/l), ✶ succinate (g/l), ● lactate (g/l). Samples for gene expression analyses, enzyme activity measurement, and intracellular metabolite concentration measurements were collected at 6, 8, and 12 h of growth to represent exponential, early stationary, and stationary phases of growth, respectively

and enzyme activity analyses represent the exponential, early stationary, and stationary growth phases, respectively. Gene expression data obtained by RT-PCR were categorized under six global regulatory systems (*rpo*D, *rpo*S, *sox*RS, *fad*R/*icl*R, *cra*, and *arc*A). Figures 2.6a–f shows how the expression of the global regulatory genes changed, and how the metabolic pathway genes in these regulatory networks changed with respect to the growth phases.

The vegetative sigma factor RpoD regulates the expression of a wide range of genes during the exponential phase of growth (Jishage et al., 1996). Gene expression data indicate a slight decrease in the expression of *rpo*D and some of the *rpo*D-dependent genes, such as *pgi* and *ser*A, at the later phases of growth (Figure 2.6a). However, the expression of the general stress regulator *rpo*S increased as the culture proceeded from the exponential growth phase toward the stationary phase, complying with the reports on *rpo*S expression (Aronis, 2002b; Patten et al., 2004). In addition, the RpoS-regulated genes, such as *tkt*B, *tal*A, *acn*A, *suc*A, *fum*C, *acs*, and *sod*C, follow the expression pattern of *rpo*S with approximately 1.5 to 2-fold up-regulation at the later phases of growth, as compared to the expression levels at the exponential phase (Figure 2.6b). Of these genes, *fum*C and *acn*A are also regulated by the oxidative stress regulator, SoxRS. However, the expression of these TCA cycle genes is poorly correlated with the declining expression of *sox*RS from the exponential to the stationary phases of growth (Figure. 2.6c). This indicates that the expressions of *fum*C and *acn*A are predominantly affected by RpoS and not by SoxRS during the later phases of growth.

The induction of the stress regulator, *rpo*S, at the later growth phases is a well-known phenomenon (Aronis 2002; Weber et al., 2005). Expression of *sox*RS is induced by oxidative stress or by the reduction of NADPH/NADP$^+$ ratio (Liochev and Fridovich, 1992; Krapp et al., 2002). At the early stationary phase, although the synthesis of NADPH is expected to be decreased by the lower activities of the pentose phosphate (PP) pathway enzymes, such as G6PDH and 6PGDH, and the TCA cycle enzyme, ICDH, the NADPH level is likely to be maintained by the increased activity of the NADP$^+$-dependent malic enzyme encoded by *mae*B (a gluconeogenic enzyme), resulting in little effect on the NADPH/NADP$^+$ ratio (Table 2.5). Thus, the downward trend of *sox*RS expression at the later phases of growth could be related to the oxidative environment of the cells. The data suggest that SoxRS plays a major role during the exponential phase, when cells experience more oxidative stress resulting from the higher rate of respiration and higher specific growth rate. With the exception of *fum*C and *acn*A, which are related to

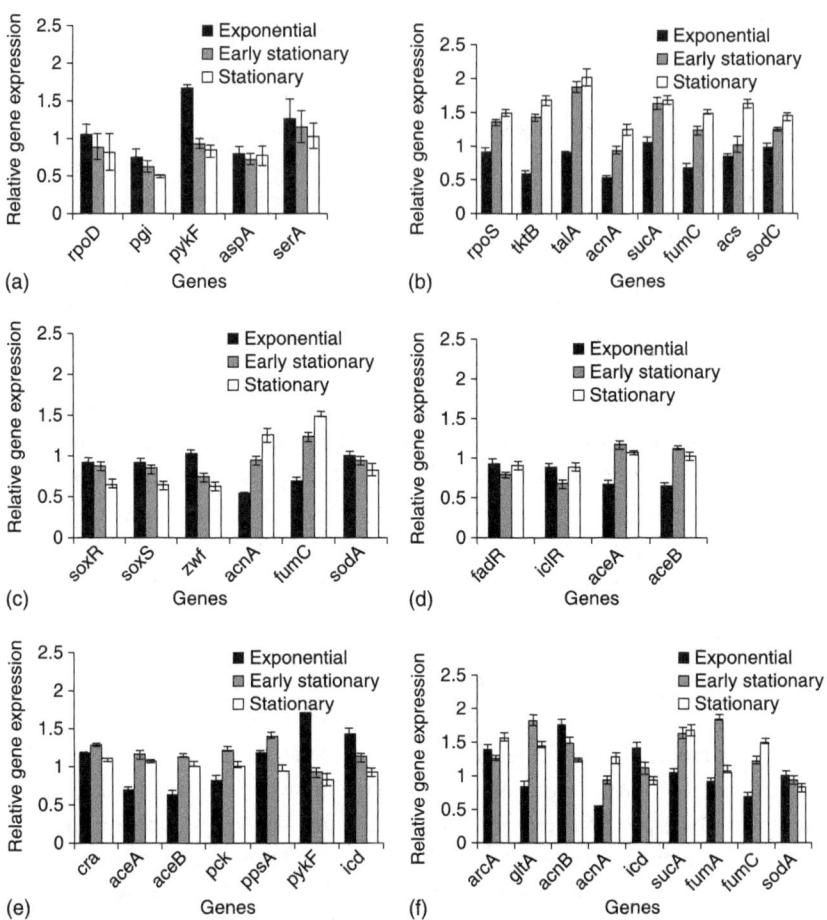

Figure 2.6 Relative gene expression of different global regulators and the metabolic pathway genes known to be regulated by those global regulators during different phases of growth. (a) Relative gene expression of *rpo*D and rpoD-regulated genes; (b) Relative gene expression of *rpo*S and rpoS-regulated genes; (c) Relative gene expression of *sox*RS and soxRS-regulated genes; (d) Relative gene expression of *fad*R/*icl*R and iclR-regulated genes; (e) Relative gene expression of *cra* and cra-regulated genes; and (f) Relative gene expression of *arc*A and arcA-regulated genes. Gene expression data were obtained by RT-PCR of samples collected at different phases of growth

rpoS, the *soxRS*-dependent expression of *zwf* and *sodA* also supports this idea.

Figure 2.6b shows that *acs* is expressed in an *rpoS*-dependent manner during different phases of growth. The higher expression of *acs* accelerates acetate assimilation and increases biomass in the presence of acetate (Kumari et al., 2000; Lin et al., 2006). Again, acetate metabolism leads to the activation of the glyoxylate pathway that requires more AcCoA, as compared to the situation when AcCoA is only used up by the TCA cycle (Stryer, 1988). These facts are consistent with the observations of up-regulations of the glyoxylate pathway genes *aceA* and *aceB* and the low concentration of intracellular Ac-CoA, together with the higher expression of *acs* at the early stationary phase (Figures 2.6b and 2.7). However, *acs* expression shows further increase at the stationary phase, during which the glyoxylate pathway is down-regulated and the extracellular acetate concentration is almost negligible (Figure 2.5). Therefore, the results suggest that *acs* encoded AcCoA synthetase acts as the scavenger of extracellular acetate to maintain the intracellular concentration of AcCoA.

In addition to the RpoS regulated genes, an up-regulation is observed in several Cra regulated genes at the early stationary phase of growth, although the expression of *cra* itself was only slightly changed. A significant reduction of *pykF* expression is observed during the later phases of growth, which can be related to its repression by Cra, or because of its relation with the growth phase-dependent down regulation of *rpoD* (Bledig et al., 1996). The down-regulation of *icd* can be related to acetate metabolism and activation of the glyoxylate pathway (Cronan and LaPorte, 1996). The activities of the glyoxylate pathway enzymes Icl and MS are more than 5-fold higher at the early stationary phase of growth, and the gluconeogenic pathway genes *pck*, *ppsA*, and *fbp* are also significantly increased (Figure 2.6e and Table 2.5). Higher activities of these enzymes led to the reduced level of intracellular AcCoA and PYR, while causing higher levels of OAA and PEP at the early stationary phase as compared to the exponential growth phase (Figure 2.7). The up-regulation of the glyoxylate pathway genes is related not only to extracellular acetate but also to its negative regulator IclR (Gui et al., 1996). The *fadR* and *iclR* changed in a coordinated manner, and their regulated genes, such as *aceA* and *aceB*, changed in the opposite direction, consistent with previous data (Figure. 2.6d) (Gui et al., 1996). The expression patterns of *aceA* and *aceB* are more related to IclR than to Cra (Figures 2.6d, e).

Figure 2.8 clearly shows how the iso-genes *tktA* and *tktB*, *talB* and *talA*, *acnB* and *acnA*, as well as *fumA* and *fumC*, backed up each other

Table 2.5 Specific activity of enzymes of *E. coli* metabolic pathways in minimal media under aerobic growth conditions at different phases of growth

Enzymes	EP (μmol/ min · mg protein)	ES (μmol/ min · mg protein)	STA (μmol/ min · mg protein)
Glycolytic pathway			
Pgi	2.36 ± 0.06	1.59 ± 0.05	0.77 ± 0.05
Pfk	0.93 ± 0.07	0.09 ± 0.04	0.043 ± 0.05
GAPDH	0.11 ± 0.06	0.016 ± 0.05	0.013 ± 0.05
Pyk	0.51 ± 0.04	0.16 ± 0.05	0.09 ± 0.04
Ppc	0.12 ± 0.05	0.054 ± 0.009	0.052 ± 0.06
Pentose phosphate pathway			
G-6PDH	0.33 ± 0.05	0.104 ± 0.04	0.098 ± 0.04
6-PGDH	0.28 ± 0.06	0.12 ± 0.05	0.102 ± 0.06
TCA cycle and acetate metabolism			
CS	0.02 ± 0.01	0.25 ± 0.06	0.13 ± 0.05
Acn	0.16 ± 0.08[a]	0.31 ± 0.04[a]	0.21 ± 0.05[a]
ICDH	0.97 ± 0.05	0.59 ± 0.04	0.36 ± 0.05
SDH	0.09 ± 0.04[b]	0.14 ± 0.05[b]	0.15 ± 0.04[b]
FUM	0.011 ± 0.004[c]	0.021 ± 0.06[c]	0.013 ± 0.03[c]
MDH	0.006 ± 0.008	0.319 ± 0.06	0.19 ± 0.05
Icl	0.05 ± 0.009	0.33 ± 0.06	0.12 ± 0.03
MS	0.043 ± 0.008	0.29 ± 0.05	0.11 ± 0.04
Ack	0.81 ± 0.05	0.57 ± 0.06	0.13 ± 0.04
Acs	0.06 ± 0.005	0.10 ± 0.006	0.36 ± 0.07
Gluconeogenesis and others			
Fbp	0.041 ± 0.06	0.138 ± 0.05	0.088 ± 0.05
Pck	0.015 ± 0.08	0.15 ± 0.007	0.076 ± 0.05
MEZ (NAD^+-dependent)	0.02 ± 0.009	0.026 ± 0.04	0.028 ± 0.04
MEZ ($NADP^+$-dependent)	0.018 ± 0.006	0.075 ± 0.005	0.044 ± 0.008
LDH	0.04 ± 0.006	0.002 ± 0.001	0.002 ± 0.001

[a]Specific activity of total aconitase enzyme
[b]Specific activity of total succinate dehydrogenase enzyme
[c]Specific activity of total fumarase enzyme
EP: exponential phase; ES: early stationary phase; STA: stationary phase

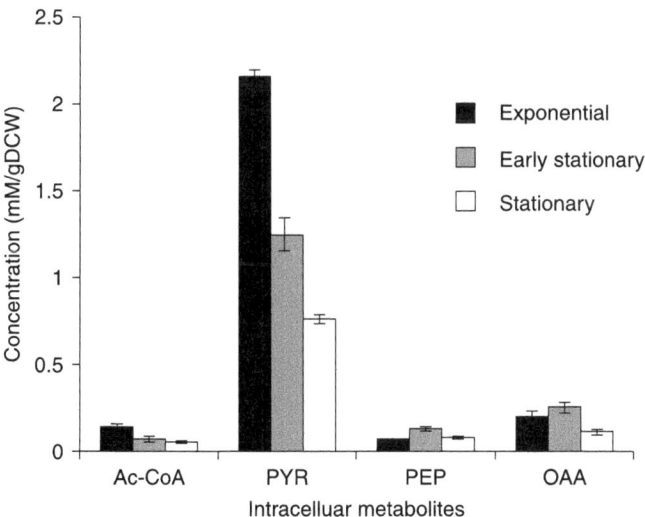

Figure 2.7 Concentration of different intracellular metabolites. Intracellular concentrations of AcCoA, PYR, PEP, and OAA from samples collected at different phases of growth

to maintain the required activities of the corresponding enzymes. With the exception of *fumA* and *fumC*, the results show high correlation between the total gene expression and the corresponding enzyme activities (Table 2.5). Whereas the gene expressions of *acnA*, *tktB*, and *talA* increased and those of *acnB*, *tktA*, and *talB* decreased as the culture proceeded toward stationary phases of growth, the gene expressions of *fumA* and *fumC* changed in a similar pattern (Figure 2.8). The higher expression of both *fumA* and *fumC* at the early stationary phase indicates that despite their similar physiological activities, *fumA* expression is modulated by extracellular acetate, while that of *fumC* is dependent on the growth phase and *rpoS* expression (Figures 2.6b) (Park and Gunsalus, 1995; Chen et al., 2001).

Regarding regulation of the TCA cycle, other global regulator such as ArcA must also be considered. The expression of the *arcA* gene is slightly changed at the early stationary and stationary phases of growth (Figure 2.6f). Although ArcA is known as a negative regulator of the TCA cycle under microaerobic conditions, some of the TCA cycle genes, such as *gltA*, *fumA*, *fumC*, and *acnA*, were up-regulated at the early stationary phase, while other genes, such as *icd*, *sodA*, and *acnB*, showed

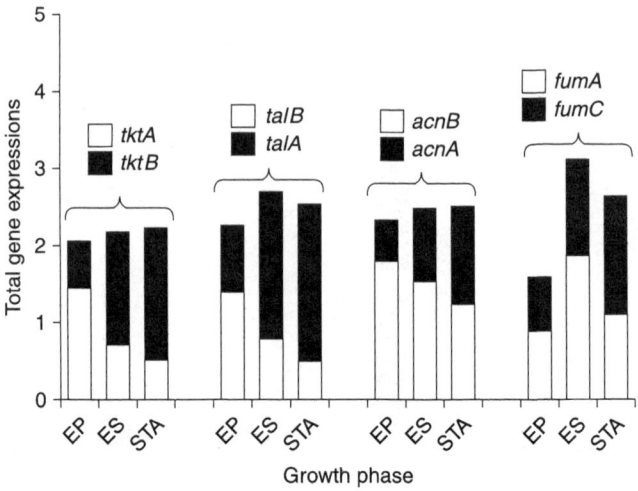

Figure 2.8 Total and relative expression of iso-genes of *E. coli* metabolic pathways. Gene expression data were obtained by RT-PCR of samples collected at different phases of growth. EP: exponential phase; ES: early stationary phase; STA: stationary phase

down-regulation (Figure 2.6f) (Alexeeva et al., 2003; Levanon et al., 2005). Therefore, the effect of ArcA was not dominant under aerobic growth conditions in the present case.

In summary, the above phenomenon demonstrates that the growth phase-dependent changes in *rpoS* expression are the most prominent of global regulators related to *E. coli* metabolism. Increased expression of the *rpoS* gene during the later phases of growth is coordinated with the changes in the expression of other regulators, with the *rpoS* effect at times overriding the effects of other regulators of vital metabolic pathway genes to cope with the changes in extracellular carbon sources and carbon starved conditions.

2.4 Reduction of by-product (acetate) formation

One of the main obstacles for producing the particular metabolite is the formation of by-products such as acetate in *E. coli* or ethanol in yeast.

Rapid aerobic growth of *E. coli* on glucose as well as on gluconate, pyruvate, lactate, glucuronate, and serine but not on glycerol or fructose, and glucose excess conditions are characterized by the formation and excretion of acetate (Veit et al., 2007). This phenomenon is referred to as overflow metabolism or the bacterial Crabtree effect, and its mechanism has been investigated to reduce acetate formation (El-Mansi and Holms, 1989; Rinas et al., 1989; Luli and Strohl, 1990; Holms, 1996; Xu et al, 1999a,b; El-Mansi, 2004; Wolfe, 2005; DeMey et al., 2007; Veit et al., 2007).

It is commonly observed that *E. coli* excretes 10 to 30% of carbon flux from glucose to acetate in glucose-containing media, even when the culture is fully aerated. Acetate excretion is a major limitation in the high cell density culture to obtain high protein concentration (Klenan and Strohl, 1994; van de Walle and Shiloach, 1998). Under aerobic conditions, acetate is generated from pyruvate, either by oxidative decarboxylation by the pyruvate dehydrogenase complex (PDHc), followed by the conversion of acetyl-CoA (AcCoA) to acetate by phosphotransacetylase (Pta) and acetate kinase (Ack) with concomitant formation of ATP at Ack reaction (Hansen and Henning, 1966), or by decarboxylation to acetate directly by pyruvate oxidase (Pox) (Chang et al., 1994) (Figure 2.9). The latter is utilized usually during transition from exponential growth phase to stationary phase under the control of RpoS. After glucose depletion or at low concentration, acetate is assimilated by AcCoA sysnthetase (ACS) with concomitant conversion of ATP to AMP and pyrophosphate (Brown et al., 1977; Kumari et al., 1995), where *acs* is also under control of RpoS. The AcCoA thus formed is metabolized by both the TCA cycle and the glyoxylate pathway for energy generation and cell synthesis. Since acetate is produced mainly through the actions of Pta (encoded by *pta*) and Ack (encoded by *ackA*), genetic techniques have been applied to obtain *pta* mutant, which showed decreased acetate production (Bauer et al., 1990). However, *pta* gene knockout leads to high pyruvate production, which is also undesirable (Diaz-Ricci et al., 1991). It is shown that the Pta-Ack pathway dominates in the exponential growth phase, and the Pox pathway dominates for acetate production in the stationary phase, and that the former pathway is repressed under acidic conditions, whereas the Pox pathway is activated (Dittrich et al., 2005).

E. coli is used for the production of heterologous proteins (Swartz, 2001) since recombinant DNA technology has become available (Cohen et al., 1973). *E. coli* strains were also engineered for the production of chemicals and building blocks (Wendisch et al., 2006). As already

Bacterial cellular metabolic systems

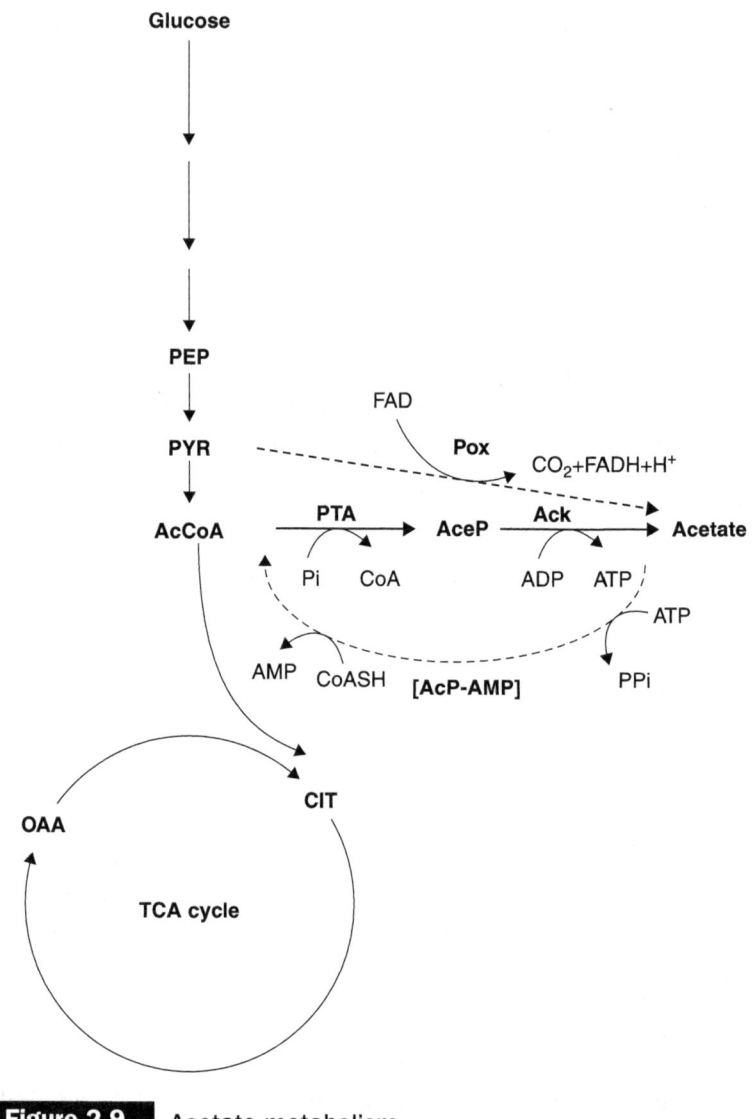

Figure 2.9 Acetate metabolism

mentioned, one of the problems is the acetate formation. Some attempt has, therefore, been made at recombinant protein production (Wong et al., 2008).

Aerobic acetate production is particularly significant at higher growth rates (El-Mansi and Holms, 1989; Majewski and Domach, 1990; Kayser et al., 2005; Wong et al., 2008), and it is a manifestation of the imbalance between glucose uptake and the demands for both biosynthesis and energy production (El-Mansi and Holms, 1989). The most common arguments are that the glucose uptake rate is improperly controlled and that the activity of the TCA cycle is limiting. One way to lower acetate production is to construct mutants with modified glucose uptake rates (Chou et al., 1994). Using this strategy, the TCA cycle can handle all the AcCoA produced by the glycolytic pathway, thus it can eliminate acetate formation. Acetate production may be also reduced by decreasing the glucose uptake rate by decreasing the glucose concentrations in the fermentor in the fed-batch cultivations (Akesson et al., 2001) or by decreasing the expression of *ptsG* that encodes the glucose-specific enzyme II (E II CB^{Glc}) of the phosphotransferase system (PTS) through expression of the *mlc* gene, which encodes Mlc that represses *ptsG* (Hosono et al., 1995; Cho et al., 2005), or by *ptsG* mutation (Flores et al., 2002). Alternatively, enhancement of the TCA cycle and the glyoxylate shunt also apparently reduces acetate production (Farmer and Liao, 1997). It has been reported that the constitutive expression of glyoxylate pathway genes may reduce acetate production by comparing *E. coli B* with *E. coli K* strain (Phue and Shiloach, 2004). The glyoxylate shunt contains two enzymes, isocitrate lyase (Icl) (encoded by *aceA*) and malate synthase (encoded by *aceB*), of which genes are located on the *aceBAK* operon, with *aceK* coding for ICDH kinase/phosphatase. The transcriptional regulation of this operon involves many factors, including IclR, FadR, Cra, ArcAB, and HimAB (Cronan and Laporte, 1999). The expression of *aceBAK* is induced during growth on either acetate or fatty acids, but induction is repressed in the presence of glucose, glycerol, or pyruvate. Mutation in *fadR* results in transcriptionally increased expression of *aceBAK*, even for such carbon sources (Gui et al., 1996) and affects the metabolism (Peng and Shimizu, 2006). Another way of reducing acetate production is to enhance anaplerosis via increasing PEP carboxylase (Ppc) and/or glyoxylate pathway enzyme levels (Farmer and Liao, 1997).

Acetate and other short-chain fatty acids inhibit cell growth and deteriorate metabolite production (Jensen and Carlsen, 1990), as they uncouple the trans-membrane protein potential and thus interfere with efficient energy metabolism. Acetic acid diffuses across the plasma membrane and the cytoplasma protein is released to form the acetate anion, resulting in a decrease of the trans-membrane protein potential (Axe and Bailey, 1995).

Note that carbon flux toward acetate production can be redirected to other less toxic metabolites, such as acetone (Bermejo et al.,

1998), acetoin (Aristidou et al., 1994), lactate (Chang et al., 1999a), polyhydroxybutyrate (PHB) (Chang et al., 1999b), or ethanol (Diaz-Ricci et al., 1991).

2.5 References

Akesson, M., Hagander, P., and Axelsson, J.P. (2001) 'Avoiding acetate accumulation in *Escherichia coli* cultures using feedback control of glucose feeding', *Biotechnology and Bioengineering*, 73: 223–30.

Alexeeva, S., Kort, B., Sawers, G., and Hellingwerf, K.J. (2000) 'Effect of limited aeration and of the *arcAB* system on intermediary pyruvate catabolism in *Escherichia coli*', *Journal of Bacteriology*, 182: 4934–40.

Alexeeva, S., Hellingwerf, K.J., and Joost Teixeira de Mattos, M. (2003) 'Requirement of ArcA for redox regulation in *Escherichia coli* under microaerobic but not anaerobic or aerobic conditions', *Journal of Bacteriology*, 185: 204–9.

Amarasingham, C.R. and Davis, B.D. (1965) 'Regulation of a-ketoglutarate dehydrogenase formation in *Escherichia coli*', *Journal of Biological Chemistry*, 240: 3664–7.

Anderson, N.L. and Anderson, N.G. (1998) 'Proteome and proteomics: new technologies, new concepts, and new words', *Electrophoresis*, 19: 1853–61.

Aristidou, A.A., San, K.Y., and Bennett, G.N. (1994) 'Modification of central pathway in *Escherichia coli* to reduce acetate accumulation by heterologous expression of the *Bacillus subtilis* acetolactate synthase gene', *Biotechnology and Bioengineering*, 44: 944–51.

Aronis, H.R. (2002) 'Stationary phase gene regulation: what makes an *Escherichia coli* promoter sigmaS-selective?' *Current Opinion in Microbiology*, 5: 591–5.

Axe, D.D. and Bailey, J.E. (1995) 'Transport of lactate and acetate through the energized cytoplasmic membrane of *Escherichia coli*', *Biotechnology and Bioengineering*, 47: 8–19.

Bauer, D.A., Ben-Basst, A., Dawson, M., Dela Peunte, U.T., and Neway, J.O. (1990) 'Improved expression of human interleukin-2 in high-cell-density fermentor cultures of *Escherichia coli* K12 by a phosphotrans acetylase mutant', *Applied Environmental and Microbiology*, 56: 1296–302.

Bermejo, L.L., Welker, N.E., and Papoutsakis, E.T. (1998) 'Expression of *Clostridium acetobutylicum* ATCC 824 genes in *Escherichia coli* for acetone production and acetate detoxification', *Applied Environmental Microbiology*, 64: 1079–85.

Bledig, S.A., Ramseier, T.M., and Saier, M.H. Jr (1996) 'FruR mediates catabolite activation of pyruvate kinase (pykF) gene expression in *Escherichia coli*,' *Journal of Bacteriology*, 178: 280–3.

Bock, A. and Sawers, G. (1999) 'Fermentation', in: Escherichia coli *and* Salmonella: *Cellular and Molecular Biology*, 2nd edition, edited by F.C. Neidhardt et al., American Society for Microbiology, Washington DC: Adobe and Mira Digital Publishing. pp. 1–2

Brown, T.D. et al. (1977) 'The enzymic interconversion of acetate and acetyl-coenzyme A in *Escherichia coli*', *Journal of General Microbiology*, 102: 327–36.

Campos, G., Guixe, V., and Babul, J. (1984) 'Kinetic mechanism of phosphjofructokinase-2 from *Escherichia coli*. A mutant enzyme with a different mechanism', *Journal of Biological Chemistry*, 259: 6147–52.

Chang, Y.Y., Wang, A.Y., and Cronan, J.E. Jr. (1994) 'Expression of *Escherichia coli* pyruvate oxidase (PoxB) depends on the sigma factor encoded by the *rpoS* (*katF*) gene', *Molecular Microbiology*, 11: 1019–28.

Chang, D.E., Jung, H.C., Rhee, J.S., and Pan, J.G. (1999a) 'Homofermentative production of D- or L-lactate in metabolically engineered *Escherichia coli* RR1', *Applied Environmental Microbiology*, 65: 1384–9.

Chang, D.E., Shin, S., Rhee, J.S., and Pan, J.G. (1999b) 'Acetate metabolism in a *pta* mutant of *Eshcerichia coli* W3110: importance of maintaining acetyl coenzyme A flux for growth and survival', *Journal of Bacteriology*, 181: 6656–63.

Chen, D.P., Tseng, C.P., Lin, H.T., and Sun, C.F. (2001) 'Oxygen- and growth rate-dependent regulation of *Escherichia coli* fumarase (FumA, FumB, and FumC) activity', *Journal of Bacteriology*, 183: 461–7.

Cho, S., Shin, D., Ji, G.E., Heu, S., and Ryu, S. (2005) 'High-level recombinant protein production by over-expression of Mlc in *Escherichia coli*', *Journal of Biotechnology*, 119: 197–203.

Chou, C.H., Bennett, G.N., and San, K.Y. (1994) 'Effect of modified glucose uptake using genetic engineering techniques on high-level recombinant protein production in *Escherichia coli* dense cultures', *Biotechnology and Bioengineering*, 44: 953–60.

Chung, T., Klumpp, D.J., and LaPorte, D.C. (1988) 'Glyoxylate bypass operon of *Escherichia coli*: Cloning and determination of the functional map', *Journal of Bacteriology*, 170: 386–92.

Cohen, S.N., Chang, A.C., Boyer, H.W., and Helling, R.B. (1973) 'Construction of biologically functional bacterial plasmids *in vitro*', *Proceedings of the National Academy of Sciences of USA*, 70: 3240–4.

Conway, T. (1992) 'The Entner-Doudoroff pathway: history, physiology, and molecular biology', *FEMS Microbiology Reviews*, 103: 1–27.

Cronan, J.E. Jr and Laporte, D. (1999) 'Tricarboxylic acid cycle and glyoxylate bypass', in: Escherichia coli *and* Salmonella: *Cellular and Molecular Biology*, 2nd edition, edited by F.C. Neidhardt et al., American Society for Microbiology, Washington DC: Adobe and Mira Digital Publishing. pp. 1–16.

Cronan, J.E. and LaPorte, D.J. (1996) 'Tricarboxylic acid cycle and glyoxylate bypass', in: Escherichia coli *and* Salmonella: *Cellular and Molecular Biology*, 2nd edition, edited by F.C. Neidhardt et al. American Society for Microbiology, Washington DC: Adobe and Mira Digital Publishing., pp, 208–11.

De Reuse, H. and Danchin, A. (1988) 'The *ptsH*, *ptsI*, and *crr* genes of the *Escherichia coli* phosphoenolpyruvate-dependent phosphotransferase system: A complex operon with several modes of transcription', *Journal of Bacteriology*, 170: 3827–37.

DeMey, M., De Maeseneire, S., Soetaert, W. and Vandamme, E. (2007) 'Minimizing acetate fermentation in *E. coli* fermentations', *Journal of Industrial Microbiology and Biotechnology*, 34: 689–700.

Diaz-Ricci, J.C., Regan, L., and Bailey, J.E. (1991) 'Effect of alteration of the acetic acid sysnthesis pathway on the fermentation pattern of *Escherichia coli*', *Biotechnology and Bioengineering*, 38: 1318–24.

Dittrich, C.R., Bennett, G.N. and San, K.Y. (2005) 'Characterization of the acetate-producing pathways in *Escherichia coli*', *Biotechology Progress*, 21: 1062–7.

El-Mansi, E.M. and Holms, W.H. (1989) 'Control of carbon flux to acetate excretion during growth of *Escherichia coli* in batch and continuous cultures', *Journal of General Microbiology*, 135: 2875–83.

El-Mansi, M. (2004) 'Flux to acetate and lactate excretions in industrial fermentations: physiological and biochemical implications', *Journal of Industrial Microbiology and Biotechnology*, 31: 295–300.

Epstein, W., Rothman-Denes, L., and Hesse, J. (1975) 'Adenosine 3':5'-cyclic monophosphate as mediator of catabolite repression in *Escherichia coli*', *Proceedings of the National Academy of Science USA*, 72: 2300–4.

Esienberg, R.C. and Dobrogosz, W.J. (1967) 'Gluconate metabolism in *Escherichia coli*', *Journal of Bacteriology*, 93: 941–9.

Farmer, W.A. and Liao, J.C. (1997) 'Reduction of aerobic acetate production by *Escherichia coli*', *Applied Environmental Microbiology*, 63: 3205–10.

Flores, S., Gosset, Flores, N., deGraaf, A.A., and Bolivar, F. (2002) 'Analysis of carbon metabolism in *Escherichia coli* strains with an inactive phosphotransferase system by ^{13}C labeling and NMR spectroscopy', *Metabolic Engineering*, 4: 124–37.

Fox, D.K., Presper, K.A., and Adhya, S. (1992) 'Evidence for two promoters upstream of the *pts* operon: regulation by the cAMP receptor protein regulatory complex', *Proceedings of the National Academy of Sciences of USA*, 89: 7056–9.

Gui, L., Sunnarborg, A., and LaPorte, D.C. (1996) 'Regulated expression of a repressor protein: FadR activates *iclR*', *Journal of Bacteriology*, 178: 4704–9.

Guixe, V. and Babul, J. (1985) 'Effect of ATP on fructokinase-2 from *Escherichia coli*. A mutant enzyme altered in the allosteric site for M_qATP^{2-}', *Journal of Biological Chemistry*, 260: 1101–6.

Gunsalus, R.P. and Park, S.J. (1996) 'Aerobic-anaerobic gene regulation in *Escherichia coli*: control by *arcAB* and *fnr* regulons', *12th Forum in Microbiology*, 437–50.

Gygi, S.P., Rist, B., and Aebersold, R. (2000a) 'Measuring gene expression by quantitative proteome analysis', *Current Opinion in Biotechnology*, 11: 396–401.

Gygi, S.P., Corthals, G.L., Zhang, Y., Rochon, Y., and Aebersold, R. (2000b) 'Evalution of two-dimensional gel electropherwsis-based proteome analysis technology', *PNAS*, 97: 9390–5.

Han, M.J. and Lee, S.Y. (2006) 'The Escherichia coli proteome: past, present, and future prospects', *Microbiol. Mol. Biol. Rev*, 70(2): 362–439.

Hansen, H.G. and Henning, U. (1966) 'Regulation of pyruvate dehydrogenase activity in *Escherichia coli*K12', *Biochemistry Biophysics ACTA*, 122: 355–8.

Holms, W.H. (1988) 'Control of flux through the citrate acid cycle and the glyoxylate bypass in *Escherichia coli*', *Biochemical Society Symposium*, 54: 17–31.

Holms, H. (1996) 'Flux analysis and control of the central metabolic pathways in *Escherichia coli*', *FEMS Microbiology Reviews*, 19: 85–116.

Hosono, K., Kakuda, H., and Ichihara, S. (1995) 'Decreasing accumulation of acetate in rich medium by *Escherichia coli* on introduction of genes on a multicopy plasmid', *Bioscience, Biotechnology and Biochemistry*, 59: 256–61.

Jensen, E.B. and Carlsen, S. (1990) 'Production of recombinant human growth hormone in *Escherichia coli*: expression of different precursors and physiological effects of glucose, acetate, and salts', *Biotechnology and Bioengineering*, 36: 1–11.

Jishage, M., Iwata, A., Ueda, S., and Ishihama, A. (1996) 'Regulation of RNA polymerase sigma subunit synthesis in *Escherichia coli*: intracellular levels of four species of sigma subunit under various growth conditions', *Journal of Bacteriology*, 178: 5447–51.

Kayser, A., Weber, J., Hecht, V., and Rinas, U. (2005) 'Metabolic flux analysis of *Escherichia coli* in glucose-limited continuous culture: I. Growth-rate-dependent metabolic efficiency at steady state', *Microbiology*, 151: 693–707.

Kiley, P.J. and Beinert, H. (1999) 'Oxygen sensing by the global regulator, FNR: the role of the iron-sulfur cluster', *FEMS Microbiology Reviews*, 22: 341–52.

Klenan, G.L. and Strohl, W.R. (1994) 'Acetate metabolism by *Escherichia coli* in high cell-density fermentation', *Applied Environmental Microbiology*, 60: 3952–8.

Krapp, A.R., Rodriguez, R.E., Poli, H.O., Paladini, D.H., Palatnik, J.F., and Carrillo, N. (2002) 'The flavoenzyme ferredoxin (flavodoxin)-NADP(H) reductase modulates NADP(H) homeostasis during the soxRS response of *Escherichia coli*', *Journal of Bacteriology*, 184: 1474–80.

Kumari, S., Beatty, C.M., Browning, D.F., Busby, S.J.W., Simel, E.J. et al. (2000) 'Regulation of acetyl coenzyme a synthetase in *Escherichia coli*', *Journal of Bacteriology*, 182: 4173–9.

Kumari, S., Tishel, R., Eisenbach, M., and Wolfe, A.J. (1995) 'Cloning, Characterization, and functional expression of acs, the gene which encodes acetyle coenzyme A synthetase in *Escherichia coli*', *Journal of Bacteriology*, 177: 2878–86.

Lessie, T.G. and Whiteley, H.R. (1969) 'Propertied of threonine deaminase from a bacterium able to utilize threonine as sole source of carbon', *Journal of Bacteriology*, 100: 878–89.

Levanon, S., San, K.Y., and Bennett, G.N. (2005) 'Effect of oxygen on the *Escherichia coli* ArcA and FNR regulation systems and metabolic responses', *Biotechnology and Bioengineering*, 89: 556–64.

Lin, H., Castro, N.M., Bennett, G.N., and San K.Y. (2006) 'Acetyl-CoA synthetase overexpression in *Escherichia coli* demonstrates more efficient acetate assimilation and lower acetate accumulation: A potential tool in metabolic engineering', *Applied Microbiology and Biotechnology*, 71: 870–4.

Liochev, S.I. and Fridovich, I. (1992) 'Fumarase C, the stable fumarase of *Escherichia coli*, is controlled by the *soxRS* regulon', *Proceedings of the National Academy of Sciences of USA*, 89: 5892–6.

Lowry, O.H., Carter, J., Ward, J.B., and Glaser, L. (1971) 'The effect of carbon and nitrogen source on the level of metabolite intermediates in *Escherichia coli*', *Journal of Biological Chemistry*, 246: 6511–21.

Luli, G.W. and Strohl, W.R. (1990) 'Comparison of growth, acetate production, and acetate inhibition of *Escherichia coli* strains in batch and fed-batch fermentations', *Applied Environmental Microbiology*, 56: 1004–11.

Majewski, R.A. and Domach, M.M. (1990) 'Simple constrained-optimization view of acetate overflow in *Escherichia coli*, *Biotechnology and Bioengineering*, 35: 732–8.

Martin, B.R. (1987) 'The regulation of enzyme activity', in: *Metabolic Regulation*, Blackwell Scientific Publications edited by B.R. Martin. Oxford, Boston Palo Alto Melbourne. Pp. 12–27.

Mat-Jan, F., Alam, K.Y., and Clark, D.P. (1989) 'Mutants of *Escherichia coli* deficient in the fermentative lactate dehydrogenase', *Journal of Bacteriology*, 171: 342–8.

McCleary, W.R. and Stock, J.B. (1994) 'Acetyl phosphate and the activation of two-component response regulators', *Journal of Biological Chemistry*, 269: 31576-2.

O'Farrell, P.H. (1975) 'High-resolution two-dimensional electrophoresis of proteins', *Journal of Biological Chemistry*, 250: 4007–21.

Oh, M.K. and Liao, J.C. (2000) 'Gene expression profiling by DNA microarray and metabolic fluxes in *Escherichia coli*', *Biotechnology Progress*, 16: 278–86.

Oh, M.K., Rohlin, L., Kao, K., and Liao, J.C. (2002) 'Global expression profiling of acetate-grown *E. coli*', *Journal of Biological Chemistry*, 277: 13175–83.

Park, S.J. and Gunsalus, R.P. (1995) 'Oxygen, iron, carbon, and supertoxide control of the fumarase *fumA* and *fumC* genes of *Escherichia coli*: role of the *arcA, fnr,* and *soxR* gene products', *Journal of Bacteriology*, 177: 6255–62.

Park, S.J., McCabe, J., Turna, J., and Gunsalus, R.P. (1994) 'Regulation of citrate synthase (*gltA*) gene of *Escherichia coli* in response to anaerobic and carbon supply', *Journal of Bacteriology*, 176: 5086–91.

Park, S.J., Tseng, C.P., and Gunsalus, R.P. (1995) 'Regulation of succinate dehydrogenase (*sdhCDAB*) operon expression in *Escherichia coli* in response to carbon supply and anaerobicsis: role of *ArcA* and *Fnr*', *Molecular Microbiology*, 15: 473–82.

Pease, A.J. and Wolf, R.E. Jr (1994) 'Determination of the growth rate-regulated steps in expression of the *Escherichia coli* K-12 *gnd* gene', *Journal of Bacteriology*, 176: 115–22.

Peng, L. and Shimizu, K. (2003) 'Global metabolic regulation analysis for *E. coli* K12 based on protein expression by 2DE and enzyme activity measurement', *Applied Microbiology and Biotechnology*, 61: 163–78.

Peng, L., and Shimizu, K. (2006) 'Effect of *fadR* gene knockout on the metabolism of *Escherichia coli* based on analyses of protein expressions, enzyme activities and intracellular metabolite concentrations', *Enzyme and Microbial Technology*, 38: 512–20.

Phue, J.N. and Shiloach, J. (2004) 'Transcription levels of key metabolic genes are the cause for different glucose utilization pathways in *E. coli* B (BL21) and *E. coli* K (JM109)', *Journal of Biotechnology*, 109: 21–30.

Plumbridge, J. (1999) 'Expression of the phosphotransferase system both mediates and is mediated by *Mlc* regulation in *Escherichia coli*', *Molecular Microbiology*, 33: 260–73.

Rahman, M. and Shimizu, K. (2007) 'Altered acetate metabolism and biomass production in several *Escherichia coli* mutants lacking *rpoS*-dependent metabolic pathway genes', *Molecular Biosystems*, DOI: 10.1039/b712023k, 1–17.

Rahman, M., Hasa, M.M., and Shimizu, K. (2008) 'Growth phase-dependent changes in the expression of global regulatory genes and associated metabolic pathways in *Esherichia coli*', *Biotechnology Letters*, 30: 853–60.

Ramseier, T.M., Bledig, S., Michotey, V., Feghali, R., and Saier, M.H. Jr (1995) 'The global regulatory protein *fruR* modulating the direction of carbon flow in *Escherichia coli*', *Molecular Microbiology*, 16: 1157–69.

Rinas, U., Kracke-Helm, H.A., and Schugerl, K. (1989) 'Glucose as a substrate in recombinant strain fermentation technology', *Applied Microbiology and Biotechnology*, 31: 163–7.

Saier, M.H. and Ramseier, T.M. (1996) 'The catabolite repressor/activator *(cra)* protein of enteric bacteria', *Journal of Bacteriology*, 178: 3411–17.

Sawers, G. and Suppmann, B. (1992) 'Anaerobic induction of pyruvate formate-lyase gene expression is mediated by the *arcA* and *fnr* protein', *Journal of Bacteriology*, 174: 3474–8.

Shimizu, K. (2004) 'Metabolic flux analysis based on ^{13}C labeling experiments and integration of the information with gene and protein expression patterns', *Advances in Biochemical Engineering/Biotechnology*, 91: 1–49.

Shimizu, K. (2009) 'Toward systematic metabolic engineering based on the analysis of metabolic regulation by the integration of different levels of information', *Biochemical Engineering Journal*, 46: 235–51.

Stryer, L. (1988) *Biochemistry*. New York: W.H. Freeman and Co. -pp. 511–12.

Sugimoto, S. and Shiio, I. (1987) 'Rregulation of 6-phosphogluconate dehydrogenase in *Brevibacterium flavum*', *Agriculture and Biological Chemistry*, 51: 1257–63.

Swartz, J.R. (2001) 'Advances in *Escherichia coli* production of therapeutic protein', *Current Opinion in Biotechnology*, 12: 195–201.

Tseng, C.P., Yum C.C., Linm, H.H., Chang, C.Y., and Kuo, J.T. (2001) 'Oxygen- and growth rate-dependent regulation of *Escherichia coli* fumarase *(fumA, fumB,* and *fumC)* activity', *Journal of Bacteriology*, 183: 461–7.

Unden, G. and Duchene, A. (1987) 'On the role of cyclic AMP and *fnr* protein in *Escherichia coli* growing anaerobically', *Archives of Microbiology*, 147: 195–200.

van de Walle, M. and Shiloach, J. (1998) 'Proposed mechanism acetate accumulation in two recombinant *Escherichia coli* strains during high cell density fermentation', *Biotechnology and Bioengineering*, 57: 71–8.

Veit, A., Polen, T., and Wendisch, V.F. (2007) 'Global gene expression analysis of glucose overflow metabolism in *Escherichia coli* and reduction of aerobic acetate formation', *Applied Microbiology and Biotechnology*, 74: 406–21.

Wanner, B.L. and Wilmes-Riesenberg, M.R. (1992) 'Involvment of phosphotransacetylase, acetate kinase, and acetyl phosphate synthesis in control of the phosphate regulon in *Escherichia coli*', *Journal of Bacteriology*, 174: 2124–30.

Weber, H., Polen, T., Heuveling, J., Wendisch, V.F., and Hengge, R. (2005) 'Genome-wide analysis of the general stress response network in *Escherichia coli*: rS-dependent genes, promoters, and sigma factor selectivity', *Journal of Bacteriology*, 187: 1591–603.

Wendisch, V.F., Bott, M., and Eikmanns, B.J. (2006) 'Metabolic engineering of *Escherichia coli* and Corynebacterium glutamicum for biotechnological production of organic acids and amino acids', *Current Opinion in Microbiology*, 9: 268–74.

Wolf, R.E. Jr, Prather, D.M., and Shea, F.M. (1979) 'Growth-rate-dependent alteration of 6-phosphogluconate dehydrogenase and glucose 6-phosphate dehydrogenase levels in *Escherichia coli* K-12', *Journal of Bacteriology*, 139: 1093–6.

Wolfe, A.J. (2005) 'The acetate switch', *Microbiology and Molecular Biology Reviews*, 69: 12–50.

Wong, M.S., Wu, S., Causey, T.B., Bennet, G.N., and San, K.Y. (2008) 'Reduction of acetate accumulation in *Escherichia coli* cultures for increased recombinant protein production', *Metabolic Engineering*, 10: 97–108.

Xu, B., Jahic, M., Blomsten, G., and Enfors, S.O. (1999a) 'Glucose overflow metabolism and mixed-acid fermentation in aerobic large-scale fed-batch processes', *Applied Microbiology and Biotechnology*, 51: 564–71.

Xu, B., Jahic, M., and Enfors, S.O. (1999b) 'Modeling of overflow metabolism in batch and fed-batch cultures of *Escherichia coli*', *Biotechnology Progress*, 15, 81–90.

Metabolic regulation by global regulators in response to culture environment

Abstract: Basic metabolic regulation mechanisms are explained in terms of catabolite regulation, nitrogen regulation, and phosphate regulation, as well as the effects of acidic pH, heat shock, and nutrient starvation on metabolic regulations. Attention focuses on the effects of global regulators (transcription factors with sigma factors), such as cAMP-Crp, Cra, Mlc, RpoN, ArcA/B, Fnr, SoxR/S, PhoR/B, RpoH, and RpoS on metabolism. The effects of knockout of such genes as cra, crp, mlc, arcA/B, phoR/B, soxR/S, and rpoS, on metabolic regulation are also explained.

Key words: catabolite regulation; nitrogen regulation; phosphate regulation; acidic pH; heat shock; stress oxygen; nutrient starvation.

3.1 Introduction

It is important to clarify the metabolic regulation mechanism and to understand cellular metabolism in response to changes in culture environment. Microorganisms adapt to the changes in the culture, such as carbon sources, nitrogen sources, oxygen availability, by regulating metabolic pathway genes through global regulators and signal transduction. The central metabolic pathways of a cell are controlled by a number of global regulators or transcription factors, depending on culture conditions (Figure 3.1). Biological systems are known to be robust and adaptable to the culture environment. It has become apparent that such robustness is inherent in biochemical and genetic networks. Several

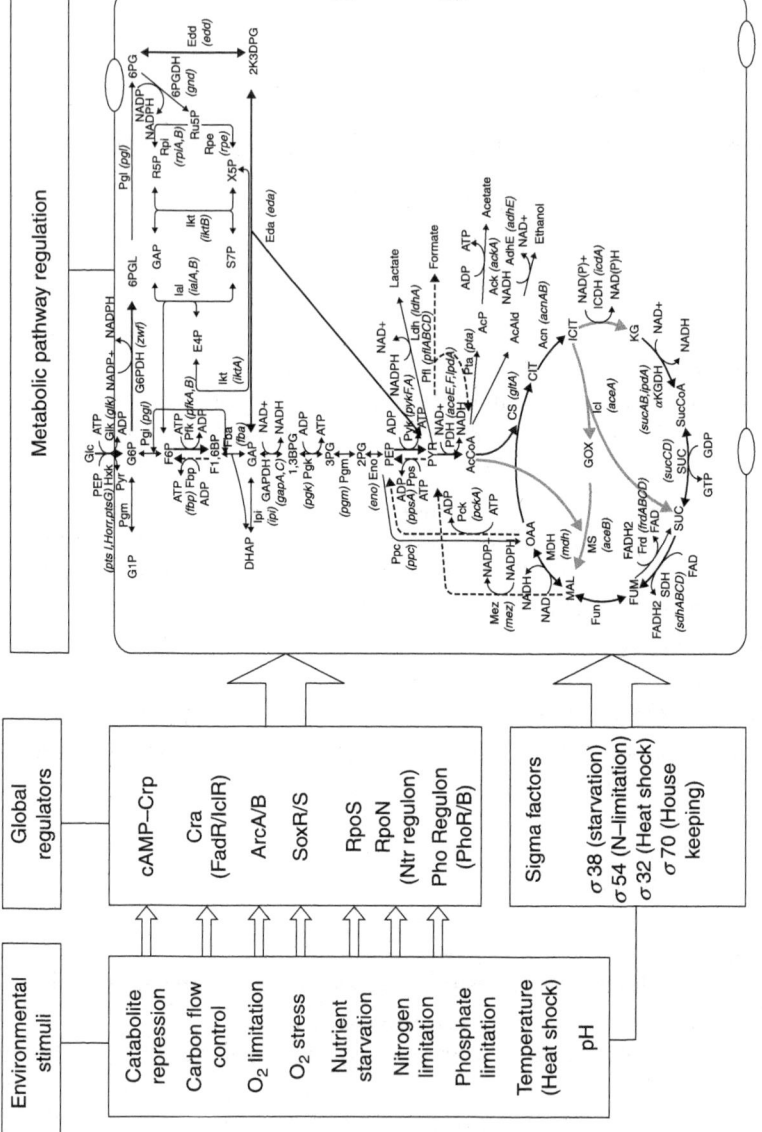

Figure 3.1 Overall metabolic regulation scheme

genes that are necessary to respond to various environmental or nutritional changes require specific recognition by RNA polymerase associated with alternative sigma factors. Here, we consider how the culture environment affects global regulators, and how the metabolic pathway genes are regulated by the corresponding global regulators. The effect of the knockout of the global regulatory gene on metabolism is also considered, so as to understand the roles of global regulators.

3.2 Carbon catabolite regulation

3.2.1 Transport of substrate molecules

The first step in the metabolism of carbohydrates is the transport of these molecules into the cell. In bacteria, various carbohydrates can be taken up by several mechanisms (Gunnewijk et al., 2001). Primary transport of sugars is driven by ATP, while secondary transport is driven by the electrochemical gradients of the translocated molecules across the membrane (Poolman and Konings, 1993), where the secondary transport systems contain the symporters which co-transport two or more molecules, uniporters that transport single molecules, and antiporters that counter-transport two or more molecules. Sugar symporters usually couple the uphill movement of the sugar to the downhill movement of the proton (or sodium ion). Namely, the electrochemical proton (or sodium ion) gradient drives the accumulation of glucose (Gunnewijk et al., 2001). Sugar uptake by group translocation is unique for bacteria and is involved in the phosphotransferase system (PTS), as will be explained in detail in this chapter.

Gram-negative bacteria such as *Escherichia coli* have two concentric membranes surrounding the cytoplasm, and the space between these two membranes is called the periplasm (Figure 3.2). The outer membrane and cytoplasmic membrane constitute a hydrophobic barrier against polar compounds. The outer membrane contains channel proteins, where the specific molecules can only move across these channels. In the outer membrane of *E. coli*, 10^8 channels are formed by the porin proteins (Nikaido and Nake, 1979). The OmpC and OmpF are the most abundant porins present under typical growth conditions, representing up to 2% of the total cellular protein (Nikaido, 1996). Their relative abundance changes, depending on such factors as osmolarity, temperature, and growth phase (Lugtenberg et al., 1976; Hall and Silhavy, 1981; Pratt and Silhavy, 1996). These porins allow glucose to enter the periplasm when

Bacterial cellular metabolic systems

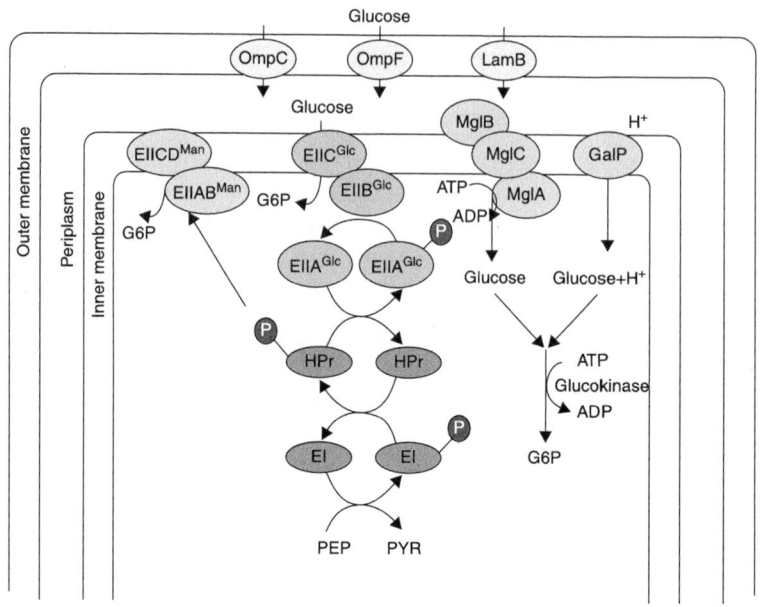

Figure 3.2 Outer and inner membrane and periplasm

glucose is present at higher concentrations than about 0.2 mM (Nikaido and Vaara, 1985; Death et al., 1993). It has been shown that the diffusion rate for glucose is found to be about 2-fold higher through OmpF than through OmpC (Nikaido and Rosenberg, 1983). Under glucose limitation, the outer membrane glycoporin LamB is induced (Death et al., 1993), where this protein permeates several carbohydrates such as maltose, maltodextrins, and glucose (von Meyenburg and Nikaido, 1977). It has also been reported that about 70% of the total glucose import capacity of the cell is contributed by LamB (Deathe et al., 1993). Glucose transport by diffusion through porins of the outer membrane is a passive process (Gosset, 2005).

Once glucose is transported into the periplasm, it can be internalized into the cytoplasm by the PTS. It may be that the glucose concentration in the periplesm is low due to active transport systems in the cytoplasmic membrane (Gosset, 2005). Once inside the periplasm, glucose can be transported into cytosol by PTS, while PTS is widespread in bacteria and absent in Archaea and eukaryotic organisms (Postma et al., 1996; Sair, 2000). PTS is composed of the soluble and non sugar-specific components Enzyme I (EI) encoded by *ptsI* and the phosphohistidine carrier protein (HPr) encoded by *ptsH*, where these transfer phosphoryl group from PEP to the sugar-specific enzyme IIA and IIB. Another component of PTS, enzyme IIC (in some cases also IID) is an integral membrane protein permease that recognizes and transports the sugar molecules, where these are phosphorylated by EIIB. Twenty-one different enzyme II complexes in *E. coli* have been reported, which are involved in the transport of about 20 different carbohydrates (Tchieu et al., 2001). In *E. coli*, EII^{Glc} and EII^{Man} are involved in the transport of glucose. EII^{Glc} is composed of soluble EIIA^{Glc} encoded by *crr*, and the integral membrane permease EIICB^{Glc} is encoded by *ptsG*. The EII^{Man} complex is composed of the EIIAB^{Man} homodimer enzyme and the integral membrane permease EIICD^{Man} (Figure 3.2), where these proteins are encoded in the *manXYZ* operon (Gosset, 2005). In addition to mannose, these proteins can also transport glucose, fructose, N-acetylglucosamines, and glucosamine with similar efficiency (Curtis and Epstein, 1975). In a wild-type strain growing with glucose as a carbon source, *ptsG* is induced, while the *manXYZ* operon is repressed. In the *ptsG* mutant, the glucose can be transported by the EII^{Man} complex, and the cell can grow with less growth rate than the wild type (Chou et al., 1994). When the extracellular glucose concentration is less than 1 µM, this can also be utilized, even at higher glucose concentrations of more than 2 g/L for *pts* mutants (Flores et al., 2005). The induction of these genes is caused by the intracellular galactose that functions as an auto-inducer of the system (Death and Ferenci, 1994). One of the genes induced under glucose limitation is *galP*, which codes for the low affinity galactose: H$^+$ symporter GalP (Figure 3.2).

The genes in the *mglABC* operon encode an ATP-binding protein, a galactose/glucose periplasmic binding protein, and an integral membrane transporter protein, respectively, forming an Mgl system for galactose/glucose (methyl galactoside) import (Gosset, 2005). This high affinity importer belongs to the ATP-binding cassette (ABC) superfamily of the primary active class of transporters (Gosset, 2005). When the extracellular glucose concentration is very low, the Mgl system, together with LamB, attains high-affinity glucose transport (Gosset, 2005). The glucose molecule transported either by GalP or Mgl systems must be phosphorylated by Glk encoded by *glk* from ATP to become G6P (Figure 3.2) (Lunin et al., 2004).

Note that PTS seems to be efficient as it consumes 1 mole of PEP for each internalized and phosphorylated glucose, where 1 mole of PEP is equivalent to 1 mole of ATP, since the conversion of PEP to PYR by Pyk would yield 1 mole of ATP by substrate-level phosphorylation. The high affinity Mgl-glucokinase system is energetically the most expensive, as it consumes 2 moles of ATP per glucose. The GalP-glucokinase system requires 1 mol of H⁺ that is internalized into the cytoplasm and 1 mol of ATP (Figure 3.2).

3.2.2 Carbon catabolite regulation

Among culture environments, carbon sources are by far the most important for the cell, from the point of view of energy generation and biosynthesis. Most living organisms, including bacteria, can use various compounds as carbon sources, where these can be either co-metabolized or selectively used with preference for the specific carbon sources selected from among those available. One typical example of selective carbon-source usage is the diauxie phenomenon observed in *E. coli*, when a mixture of glucose and lactose is used as a carbon source and this phenomenon was first observed by Monod (1942). Subsequent investigations on this phenomenon have revealed that selective-carbon source utilization is common and that glucose is the preferred carbon source in many organisms. Moreover, the presence of glucose often prevents the use of other carbon sources. This preference of glucose over other carbon sources has been named glucose repression, or more generally carbon catabolite repression (CCR) (Magasanik, 1961). CCR is observed in most heterotrophic bacteria, which include facultatively autotrophic bacteria that repress the genes for CO_2 fixation in the presence of an organic carbon source (Bowien and Kusian, 2002). Some pathogenic bacteria, such as *Chlamydia trachomatis* and *Mycoplasma pneumonia*, seem to lack CCR, where these are adapted to nutrient-rich host environments (Nicholson et al., 2004; Halbedel et al., 2007). Another phenomenon can be seen in *Corynebacterium glutamicum*, where co-assimilation of glucose and other carbon sources is made, but it is highly regulated (Wendisch et al., 2000; Frunzke et al., 2008). For some bacteria, such as *Streptococcus thermophilus*, *Bifidobacterium longum*, and *Pseudomonas aeruginosa*, glucose is not a primary carbon source, and the genes for glucose utilization are repressed when preferred carbon sources are available; this phenomenon is known as reverse CCR (Collier et al., 1996; Oarche et al., 2006). CCR is one of the most important regulatory phenomena in many bacteria (Moreno et al., 2001; Blencke et al., 2003; Liu et al., 2005). CCR is important for the cells to compete with other organisms in nature, where it is crucial to select a preferred carbon source in order to improve the growth rate, which then results in higher survival rates than other organisms. Moreover, CCR has a crucial role in the expression of virulence genes, which often enable the organism to access new sources of nutrients. The ability to select the appropriate carbon source that allows fastest growth may be the driving force for the evolution of CCR (Gorke and Stulke, 2008).

3.2.3 Catabolite control of sugar transporters

The *E. coli lac* operon is only expressed if allolactose (a lactose isomer formed by β-galactosidase) binds and inactivates the *lac* repressor. Lactose cannot be transported into the cell in the presence of glucose, because the lactose permease, LacY, is inactive in the presence of glucose (Winkler and Wilson, 1967). As shown in Figure 3.3, phosphorylated E II A^{Glc} is dominant when glucose is absent, and does not interact with LacY, whereas unphosphorylated EIIA^{Glc} can bind and inactivates LacY when glucose is present (Nelson et al., 1983; Hogema et al., 1999). Note that this only occurs if lactose is present (Smirnova et al., 2007). The same mechanism may be seen in the transport of other secondary carbon sources, such as maltose, melibiose, raffinose, and galactose (Misko et al., 1987; Titgemeyer et al., 1994).

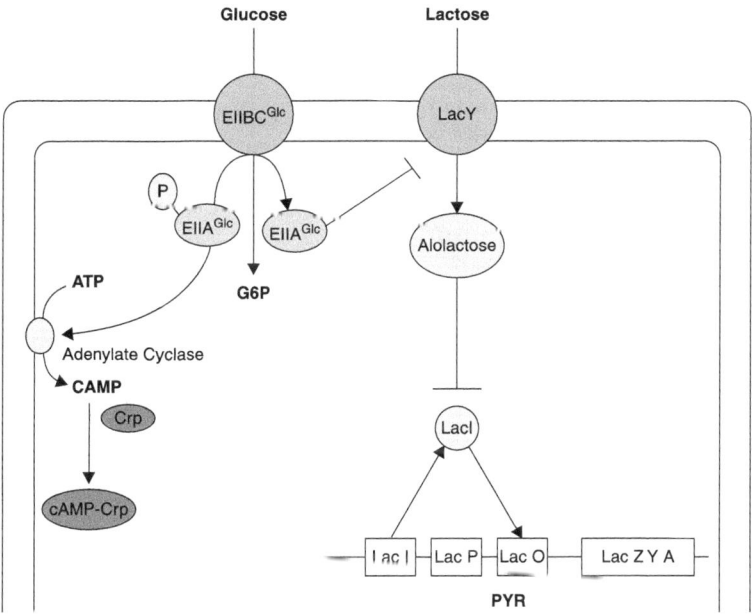

Figure 3.3 Inducer exclusion and the activation of adenylate cyclase in the glucose-lactose system

Inducer exclusion has also been reported for Gram-positive bacteria, and HPr is the major player in these organisms. In *Lactobacillus brevis*, HPr(Ser-P) is formed when glucose is present, and binds and inactivates permease (Djordjevic et al., 2001). By contrast, the lactose permease of *S. thermophilus* is controlled by HPr (His-P)-dependent phosphorylation. In the absence of glucose, HPr (His-P) can phosphorylate a PTS-like domain, thereby activating the permease for lactose transport (Poolman et al., 1995). When glucose is present, HPr becomes phosphorylated on Ser46 and can no longer activate the lactose permease (Gunnewijk and Poolman, 2000).

3.2.4 CCR in E. coli

The central players in carbon catabolite regulation in *E. coli* are the transcriptional activator Crp (cyclic AMP (cAMP) receptor protein, also called catabolite gene-activator protein (CAP)), the signal metabolite cAMP, adenylate cyclase (Cya), the phosphoenol pyruvate (PEP), and carbohydrate phosphotransferase systems (PTSs), where these systems are involved in both transport and phosphorylation of carbohydrates. The PTS in *E. coli* consists of two common cytoplasmic proteins, EI (enzyme I) encoded by *ptsI* and HPr (histidine-phosphorylatable protein) encoded by *ptsH*, as well as carbohydrate-specific EII (enzyme II) complexes (Figure 3.2). The glucose-specific PTS in *E. coli* consists of the cytoplasmic protein EIIA^{Glc} encoded by *crr* and the membrane-bound protein EIICB^{Glc} encoded by *ptsG*, which transport and concomitantly phosphorylate glucose. The phosphoryl groups are transferred from PEP via successive phosphorelay reactions in turn by EI, HPr, EIIA^{Glc}, and EIICB^{Glc} to glucose. The cAMP-Crp complex and the repressor Mlc are involved in the regulation of the *ptsG* gene and the *pts* operon expressions. It has been demonstrated that unphosphorylated EIICB^{Glc} can relieve the expression of *ptsG* gene expression by sequestering Mlc from its binding sites through a direct protein–protein interaction in response to glucose concentration. In contrast to Mlc, where it represses the expressions of *ptsG*, *ptsHI*, and *crr* (Plumbridge, 1998a,b), the cAMP-Crp complex activates *ptsG* gene expression (De Reuse and Danchin, 1988) (Figure 3.4). Since intracellular cAMP levels are low during growth on glucose, these two antagonistic regulatory mechanisms guarantee precise adjustments of *ptsG* expression levels under various conditions (Bettenbrock et al., 2006) (Figure 3.4). It should be noted that unphosphorylated EIIA^{Glc} inhibits the uptake of other non-PTS carbohydrates by the so-called inducer exclusion (de Boris,

Metabolic regulation by global regulators

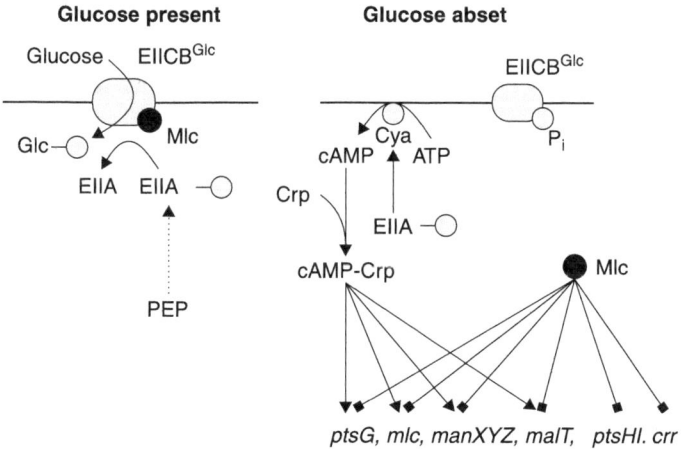

Figure 3.4 The multiple regulations by Mlc and cAMP-Crp

2008), while phosphorylated EIIA^{Glc} (EIIA^{Glc}-P) activates adenylate cyclase (Cya), which generates cAMP from ATP and leads to an increase in the intracellular cAMP level (Park et al., 2006) (Figure 3.4). In the absence of glucose, Mlc binds to the upstream of the *ptsG* gene and prevents its transcription. If glucose is present in the medium, the amount of unphosphorylated EIICB^{Glc} increases due to the phosphate transfer to glucose. In this situation, Mlc binds to EIICB^{Glc}, and thus it does not bind to the operator of *pts* genes (Lee et al., 2000; Tanaka et al., 2000; Bettenbrock et al., 2006). Note that if the concentration ratio between PEP and PYR (PEP/PYR) is high, EIIA^{Glc} is predominantly phosphorylated, whereas if this ratio is low, then EIIA^{Glc} is predominantly dephosphorylated (Hogema et al., 1998; Bettenbrock et al., 2007). EIIA^{Glc} is preferentially dephosphorylated when *E. coli* cells grow rapidly with glucose as a carbon source (Hogema et al., 1998; Bettenbrock et al., 2007). Note also that cAMP levels are low during growth with non-PTS carbohydrates such as lactose, where the PEP/PYR ratio is the key factor that controls phosphorylation of EIIA^{Glc}, which explains dephosphorylation of EIIA^{Glc}, resulting in a low cAMP pool (Hogema et al., 1998; Bettenbrock et al., 2007). As stated above, inducer exclusion is the dominant factor for the glucose-lactose diauxie (Inada et al., 1996a,b; Hogema et al., 1999). The role of cAMP-Crp is then to express the *lac* operon, which is involved in CCR by activating the expression of *ptsG* and EIICB domain of the glucose-specific PTS, and therefore transport of glucose (Kimata et al., 1997).

3.2.5 Carbon flow control in E. coli

In addition to cAMP-Crp, which is dependent on the level of glucose concentration, the catabolite repressor/activator protein (Cra), originally characterized as the fructose repressor (FruR), plays an important role in the control of carbon flow in *E. coli* (Saier et al., 1997; Saier and Ramseier, 1996; Moat et al., 2002). The carbon uptake and glycolysis genes, such as *ptsHI*, *pfkA*, *pykF*, *zwf*, and *edd–eda*, are reported to be repressed, while *ppsA*, *fbp*, *pckA*, *icd*, *aceA*, and *aceB* are activated by the Cra protein (Seier and Ramseir, 1996; Moat et al., 2002) (Appendix A). It has been shown that genes, such as *pfkA*, *pykF*, and *edd–eda*, have Cra binding sites that overlap or follow the RNAP-binding site (Chin et al., 1989; Ramseier et al., 1995; 1996; Lee et al., 2003). It is shown that a mutant defective in the *cra* gene is unable to grow on gluconeogenic substrates such as pyruvate, acetate, and lactate (Saier et al., 1997). This appears to be due to a deficiency in the gluconeogenic enzymes, such as Pps, Pck, some TCA cycle enzymes, the two glyoxylate-shunt enzymes, and certain electron transport carriers (Saier et al., 1997). Molecular level research on *cra* gene expression has been done by several researchers using *lacZ*-transcriptional fusion (Cortay et al., 1994; Ryu et al., 1995; Ramseier et al., 1996; Mikulskis et al., 1997; Prost et al., 1999). The gluconeogenic pathway is deactivated by the knockout of the *cra* gene, and the carbon flow toward catabolism and the glucose consumption rate are expected to increase, since glycolysis pathway genes such as *ptsHI*, *pfkA*, and *pykF* are activated by the *cra* gene knockout. It has been shown that multiple genes knockout, when knocked out together with *cra* gene, and can increase the consumption rate of the substrate and thus improve the rate of metabolite production under certain culture conditions (Sarkar and Shimizu, 2008). However, the regulation mechanism is complex since *icdA*, *aceA*, *B*, and *cydB* genes are repressed, while *zwf* and *edd* gene expressions are activated and thus the ED pathway is activated by the *cra* knockout gene (Sarkar and Shimizu, 2008). The details of such complex regulation networks have not yet been fully investigated for the *cra* mutant (Saier et al., 1997). Phue et al. (2005) also studied the role of the *cra* gene in relation to high density cell cultures of *E. coli* B and *E. coli* K.

3.2.6 Fermentation characteristics and gene expressions of cra gene knockout mutant

Figure 3.5 shows the aerobic batch cultivation result for the wild type and its *cra* gene knockout mutant, where acetate tends to be less consumed

Metabolic regulation by global regulators

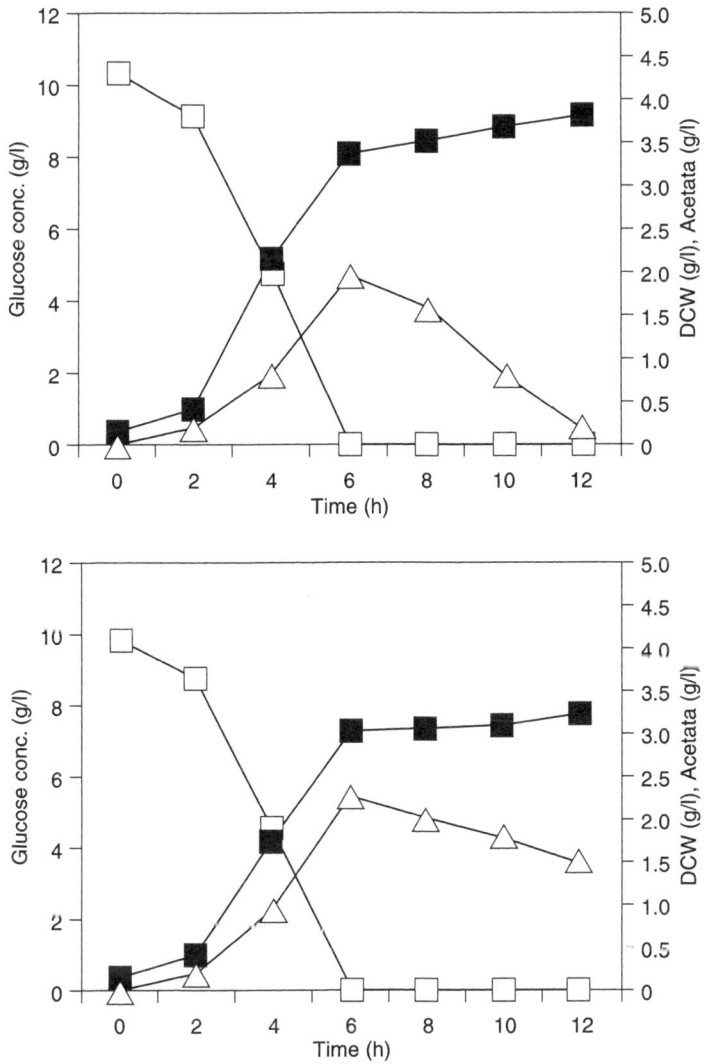

Figure 3.5 Batch cultivation of (a) *E. coli* BW25113 and (b) its *cra* mutant: □, glucose concentration; ■, biomass concentration; △ acetate concentration

at the late growth phase in the case of the *cra* mutant when compared to the parent strain (Sarkar and Shimizu, 2008; Sarkar et al., 2008).

Table 3.1 shows the fermentation characteristics of *E. coli* wild type (BW25113) and its *cra* mutant cultivated in continuous cultures at a dilution rate of 0.2 h^{-1}. The results show the increases in the specific

Table 3.1 Growth parameters for *E. coli* BW25113 and its *cra* mutant cultivated at the dilution rate of 0.2 h^{-1} where feed glucose concentration was 4 g/l

Growth parameters	BW25113	cra mutant
Biomass yield (gg-1)	0.44 ± 0.01	0.30 ± 0.02
Glucose uptake rate (mmole.g^{-1}.h^{-1})	2.54 ± 0.11	3.61 ± 0.03
Acetate production rate (mmole.g^{-1}.h^{-1})	0.02 ± 0.01	0.84 ± 0.02
O_2 uptake rate (mmole.g^{-1}.h^{-1})	10.24 ± 0.52	8.88 ± 0.63
CO_2 evolution rate (mmole.g^{-1}.h^{-1})	8.51 ± 0.35	8.57 ± 0.8

glucose uptake rate and the acetate production rate for the mutant, when compared to the parent strain. Table 3.1 also shows that the cell yield for the mutant is reduced, when compared to the parent strain.

Table 3.2a shows that the expression of the glycolytic pathway gene *pykF* is up-regulated, and those of the gluconeogenic pathway genes, such as *ppsA* and *fbp*, are down-regulated as expected. The gene expression related to the TCA cycle, such as *acnA*, is down-regulated, and the glyoxylate pathway related genes, such as *aceA* and *aceK* were down-regulated in the mutant as expected. The pentose phosphate (PP) pathway related genes such as *zwf* are up-regulated in the mutant. The *adhE* gene is also up-regulated in the mutant, when compared to the parent strain, consistent with the results of Mikulskis et al. (1997). The respiratory pathway related genes, such as *cydA* and *cydB* genes, are down-regulated, which is known to be positively controlled by Cra (Ramseier et al., 1996). The *fru* operon of enteric bacteria is known to be regulated at the transcriptional level by Cra. In the absence of the *cra* gene, the fructose operon genes, such as *fruK* and *fruB*, are up-regulated in the mutant. Other carbohydrate related genes, such as the fucose operon genes, are also up-regulated in the *cra* mutant. Table 3.2b shows that some of the transporter genes, such as *ptsH*, *mglC*, and *xylE*, are up-regulated in the mutant. These are involved in the transport of glucose, galactose, and xylose. Other transporters for the amino acids, such as *gltL* and *proV*, which encode glutamate/aspartate and glycine betaine/l-proline transport proteins, are also up-regulated in the *cra* mutant. Table 3.2c shows that the expression level of the *fabA* gene, which encodes the enzyme that catalyzes the key reaction from unsaturated to saturated fatty acid, is up-regulated in the mutant. The *fabH* (β-ketoacyl-ACP synthase I) involved in the initial step of fatty acid elongation is also up-regulated.

Metabolic regulation by global regulators

Table 3.2 Gene expressions of *cra* mutant as compared with the wild type strain

(a) Carbon and energy related genes

Gene	$\log_2$ (cra/parent)	Description
aceA	−1.43	Isocitrate lyase (EC 4.1.3.1)
aceK	−0.30	Isocitrate dehydrogenase kinase/phosphatase (EC 2.7.1.116) (EC 3.1.3.-).
adhE	1.39	Alcohol_dehydrogenase_(EC__1.1.1.1)
cydB	−1.09	Cytochrome_d_ubiquinol_oxidase_subunit_II_ (EC__1.10.3.-)
cydA	−1.12	Cytochrome_d_ubiquinol_oxidase_subunit_I_ (EC__1.10.3.-)
fbp	−0.81	Fructose-1,6-bisphosphatase (EC 3.1.3.11)
fruB	1.29	PTS_system,_fructose-specific_IIA/FPR_component_(EIIA-Fru)
fruK	1.42	1-phosphofructokinase_(EC__2.7.1.56)_(fructose_1-phosphato_kinase)
fucA	1.14	L-fuculose_phosphate_aldolase_(EC_4.1.2.17)
fucI	1.03	Fucose_Isomerase_FucI_(EC__5.-.-.-)
fucK	1.56	L-fuculokinase (EC 2.7.1.51) (L-fuculose kinase).
fucO	1.48	Lactaldehyde_reductase_(EC_1.1.1.77)
fucP	1.12	Fucose_permease
fucU	0.93	Fucose_operon_FucU_protein
gcl	−1.06	Glyoxylato_carboligase_(EC_4.1.1.47)_(tartronate-semialdehyde_synthase)
gltA	−0.36	Citrate_synthase_(EC__4.1.3.7)
gpsA	−0.58	L-glycerol_3-phosphate_dehydrogenase
mdh	−0.39	Malate_dehydrogenase_(EC__1.1.1.37)
ppsA	−0.84	Phosphoenolpyruvate synthase (EC 2.7.9.2)
pykF	1.43	Pyruvate_kinase_(EC__2.7.1.40)
tktA	1.52	Transketolase_1_(EC__2.2.1.1)_(tk_1)
xylA	1.6	Xylose_isomerase_(EC__5.3.1.5)_(version_1)
zwf	1.76	Glucose-6-phosphate_1-dehydrogenase_(EC_1.1.1.49)

(continued)

Bacterial cellular metabolic systems

Table 3.2 Gene expressions of *cra* mutant as compared with the wild type strain (*continued*)

(b) Metabolic transport related genes

Gene	Log_2 (cra/ parent)	Description
gltL	1.47	Glutamate/aspartate_transport_atp-binding_protein_gltL
malK	1.46	Maltose/maltodextrin_transport_ATP-binding_protein_MalK
manX	1.65	Phosphotransferase_system_enzyme_II_(EC__2.7.1.69),_mannose-specific,_factor_III
mglC	1.34	Galactoside_transport_system_permease_protein
proV	1.41	Glycine_betaine/l-proline_transport_ATP-binding_protein_ProV
ptsH	1.57	PTS_system, Phosphocarrier protein HPr (Histidine-containing protein).
xylE	1.75	D-xylose-proton symporter (D-xylose transporter).

(c) Fatty acids, purine, and pyrimidine metabolism related genes

Gene	Log_2 (cra/ parent)	Description
fabA	2.00	3-hydroxydecanoyl-[acyl-carrier-protein]_dehydratase_(EC__4.2.1.60)
fabB	1.02	3-oxoacyl-[acyl-carrier-protein]_synthase_I_(EC__2.3.1.41)
fabH	1.32	3-oxoacyl-[acyl-carrier-protein]_synthase_(EC__2.3.1.41)_III.
kdsB	−1.00	3-deoxy-manno-octulosonate_cytidylyltransferase_(EC__2.7.7.38)
nth	−2.06	Endonuclease III (EC 4.2.99.18) (DNA-(apurinic or apyrimidinic site) lyase).
purE	0.82	Phosphoribosylaminoimidazole carboxylase catalytic subunit (EC 4.1.1.21) (AIR carboxylase) (AIRC).
purF	0.85	Amidophosphoribosyltransferase_(EC__2.4.2.14)

(d) Amino acid metabolism genes

Gene	Log$_2$ (cra/parent)	Description
aroC	−0.76	Chorismate_synthase_(EC__4.6.1.4).
aroG	−0.78	Phospho-2-dehydro-3-deoxyheptonate aldolase, Phe-sensitive (EC 4.1.2.15) (Phospho-2-keto-3-deoxyheptonate aldolase)
artI	−1.12	Arginine-binding periplasmic protein 1 precursor.
aroL	−0.74	Shikimate_kinase_(EC__2.7.1.71)_II
aroP	−3.06	Aromatic amino acid transport protein aroP (General aromatic amino acid permease).
cysK	−0.78	Cysteine_synthase_A_(EC__4.2.99.8)
hisI	−1.09	Histidine biosynthesis bifunctional protein hisIE [Includes: Phosphoribosyl-AMP cyclohydrolase (EC 3.5.4.19) (PRA-CH); Phosphoribosyl-ATP pyrophosphatase (EC 3.6.1.31) (PRA-PH)].
hisC	−0.92	Histidinol-phosphate_aminotransferase
ilvG	−0.78	Acetolactate_synthase_(EC__4.1.3.18)_II_large_chain
mhpE	−1.06	4-hydroxy-2-oxovalerate_aldolase_(EC_4.1.3.-)
pepD	−2.06	X-his_dipeptidase_(EC__3.4.13.3)_precursor
torA	−1.00	Trimethylamine-n oxide_reductase_precursor_(EC__1.6.6.9)
trpE	1.59	Anthranilate_synthase_component_I_(EC__4.1.3.27)
trpC	0.96	Tryptophan biosynthesis protein trpCF [Includes: Indole-3-glycerol phosphate synthase (EC 4.1.1.48) (IGPS); N-(5′-phospho- ribosyl)anthranilate isomerase (EC 5.3.1.24) (PRAI)].
ybaS	−1.32	Probable glutaminase ybaS (EC 3.5.1.2).

(e) Global and metabolic regulatory genes

Gene	Log$_2$ (cra/parent)	Description
iclR	1.96	Acetate operon repressor.(Repressor_protein_IclR)
fis	1.52	DNA-binding protein fis (Factor-for-inversion stimulation protein) (HIN recombinational enhancer binding protein).
fucR	−0.47	L fucose operon activator.
lysR	1.51	Transcriptional activator protein lysR.
purR	−0.45	Purine nucleotide synthesis repressor.
rpoS	1.62	RNA polymerase sigma-stationary_phase

Table 3.2d shows that most of the genes involved in amino acids metabolism are down-regulated, where most of these genes are related to the non-aromatic amino acids biosynthetic pathway. Although some aromatic amino acids synthetic genes, such as *trpE*, and *trpC*, are up-regulated in the mutant, the *aroL*, *aroG*, and *aroC* genes are down-regulated in the mutant. The histidine biosynthesis related genes, such as *hisI* and *hisC*, are also down-regulated in the mutant. Table 3.2e shows that the *lysR* gene, which encodes the LysR protein that participates in the control of several genes involved in lysine biosynthesis, are up-regulated. The repressor *purR* of the *pur* operon involved in the purine biosynthesis is down-regulated. This is consistent with the up-regulation of purine biosynthesis gene. The *iclR* gene is up-regulated, which is in accordance with the down-regulation of the glyoxylate pathway genes in the mutant. The stress related regulatory genes such as *rpoS* are up-regulated in the mutant. The *fis* gene, which serves as an early signal of nutritional up-shift, is also up-regulated in the mutant as compared to its parent strain. Some of the global regulatory genes, such as *arcA* and *fnr*, do not change in their expression patterns in the mutant when compared to the parent strain.

Some of the enzyme activities are also given in Figure 3.6, where EMP pathway enzymes such as Pfk and Pyk are up-regulated, as well as G6PDH and ED pathway enzyme activities. However, the activity of the gluconeogenic enzymes such as Pck and the activities of ICDH and Icl are down-regulated, and the Ack activity is up-regulated.

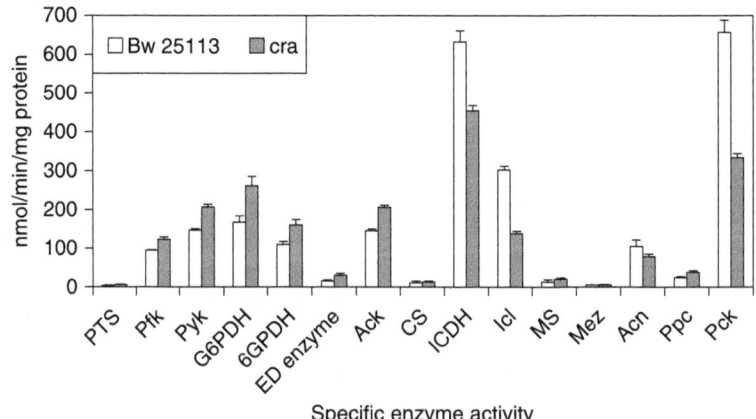

Figure 3.6 Comparison of enzyme activities of the *cra* mutant as compared to the wild type (BW25113)

3.2.7 Effect of glucose concentration on gene expressions in E. coli

Let us consider how culture conditions, such as glucose concentrations, affect global regulators and metabolic pathway genes of wild-type *E. coli* (BW25113) (Yao et al., 2011). Table 3.3 shows the fermentation characteristics of the wild-type *E. coli* for the continuous culture at different dilution rates, where it indicates that the specific glucose uptake rate, acetate production rate, and the specific CO_2 evolution rate (CER) increase as the dilution rate increases. Figure 3.7 shows the effect of the dilution rate (the specific growth rate) on gene transcript levels, where it indicates that in accordance with the increased specific glucose consumption rate, the transcript levels of *ptsG*, *ptsH*, and *pfkA* increase as the dilution rate increases, where the *cra* transcript level decreases and *crp* as well as *mlc* decreases accordingly (Appendix A). The decrease in *crp* is also coincident with a decrease in *cyaA*, which encodes Cya. The transcript levels of *zwf*, *gnd*, *edd*, and *eda* increase as the dilution rate increases in accordance with the decrease in *cra*. The transcript level of *ppc* increases while *pckA* decreases as the dilution rate increases. Moreover, the transcript levels of *fadR* and *iclR* increases, and *aceA* and *aceB* decrease as the dilution rate increases. In accordance with the increase in the specific acetate production rate, the transcript levels of *pta* and *ackA* increase. The TCA cycle genes, such as *gltA*, *acnA*, *fumA*, and *fumC* decrease, while *acnB*, *icdA*, and *lpdA* increase as the dilution rate increases. In accordance with the increase in *soxR/S* transcript levels, *zwf* and *sodA* increase (except the case of dilution rate at 0.7 h^{-1}). In accordance with the decrease in the transcript level of *arcA*, the transcript level of *cyoA* increases, while *cydB* decreases as the dilution rate increases, where the latter phenomenon also coincides with the decrease in *cra* transcript levels (Appendix A). Further observations indicate that in accordance with the decrease in *rpoS* transcript level, *tktB*, *acnA*, and *fumC* decrease as the dilution rate increases.

The effect of the dilution rate (the specific growth rate) on gene transcript levels and fermentation characteristics may be explained as follows: As the dilution rate increases, the glucose concentration increases (though under detectable level in Table 3.3), which causes *cra* transcript levels to be decreased. The *crp* transcript level also decreases as the dilution rate increased, which may be due to a decrease in phosphorylated $EIIA^{Glc}$, causes by higher glucose concentration. This deactivated Cya (as implied by *cyaA* in Figure 3.7b) in turn caused a decrease in cAMP concentrations, and thus cAMP-Crp or *crp* transcript levels decreased (Figure 3.7a). It has

Table 3.3 Effects of dilution rate on fermentation characteristics of wild type E. coli

Dilution rate (h^{-1})	Biomass conc. (g/L)	Glucose conc. (g/L)	Specific glucose uptake rate (mmol/g/h)	Specific acetate formation rate (mmol/g/h)	Biomass yield (g/g)	Specific CER (mmol/g/h)	Carbon recovery
0.2	1.45 ± 0.06	ND*	3.07 ± 0.13	ND*	0.37 ± 0.015	9.15	94%
0.4	1.87 ± 0.09	ND*	4.75 ± 0.23	0.01	0.47 ± 0.023	11.61	98%
0.6	2.0 ± 0.09	ND*	6.67 ± 0.3	0.88 ± 0.04	0.5 ± 0.023	13.17	98%
0.7	1.93 ± 0.08	ND*	8.05 ± 0.34	1.33 ± 0.06	0.48 ± 0.02	15.83	97%

ND*: not detectable, where glucose detectable limit is 0.038 g/l, CER, the CO_2 production rate

Metabolic regulation by global regulators

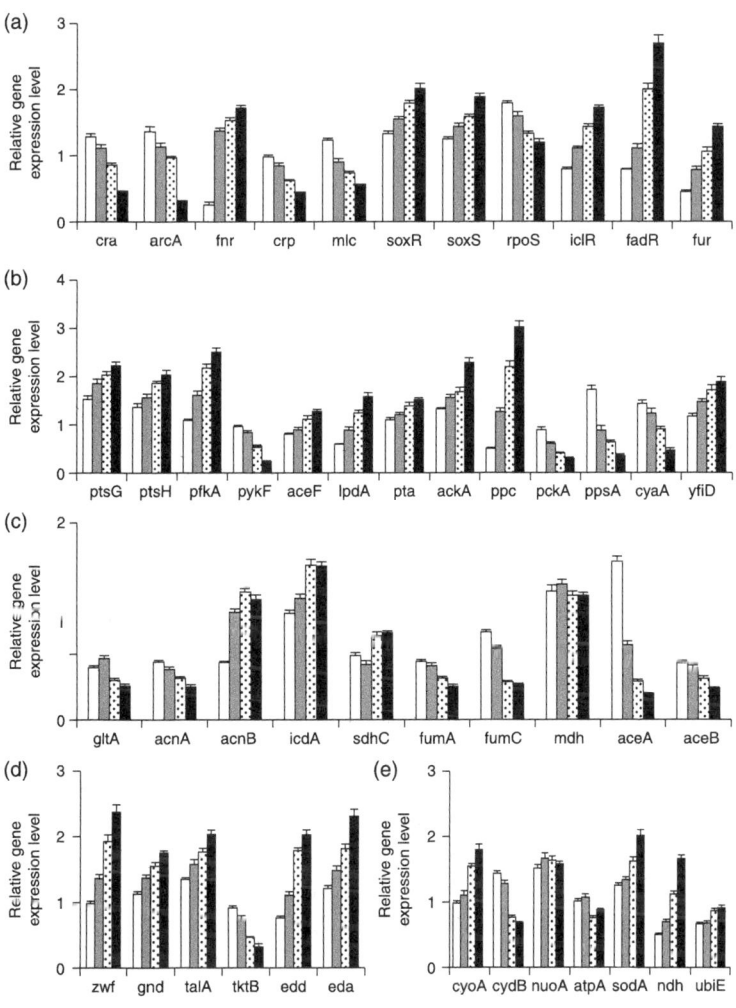

Figure 3.7 The effect of dilution rate on the gene transcript levels. (a) Global regulator genes; (b) PTS, glycolysis, anaplerotic pathway, *cyaA* and *yfiD* genes; (c) TCA and glyoxylate pathway genes; (d) PP pathway genes; (e) Respiratory chain genes. ☐ : 0.2 ; ▨ : 0.4 ; ⊡ : 0.6 ; ■ : 0.7

been reported that cAMP levels start to increase when the glucose concentration becomes less than 0.3 mM. The increases in *fadR* and *iclR* (Figure 3.7a) also coincides with the increase in the glucose concentration as the dilution rate increases. The decrease of *rpoS* (Figure 3.7a) may be also explained by less nutrient stress as the dilution rate increases.

Referring to Figure 3.7 and Appendix A, some of the metabolic pathway transcript levels may be explained with respect to the change in global regulators. The decreases of *aceA* and *aceB* transcript levels coincide with increases of those of *iclR* and *fadR* as the dilution rate increases. The decrease of the transcript levels of *aceA*, *B*, and *pckA*, *ppsA*, and *acnA* (Figure 3.7b,c) coincide with the decrease in the *cra* transcript levels (Figure 3.7a), which also causes an increase in *eda*, *edd*, *pfkA*, and *ptsH* transcript levels. The increase of *cyoA* and decrease of *cydB* coincide with the decrease in *arcA* as the dilution rate increases, where the decrease in *cydB* also coincides with the decrease in *cra*. Note that *ppc* transcript levels increase as the dilution rate increases, which may be due to increased F1,6BP (FDP), where FDP is the effector of Ppc. Although *lpdA* and *aceF* transcript levels increase as the dilution rate increases, TCA cycle gene transcript levels, such as *gltA*, *acnA*, *fumA*, and *fumC* tend to decrease, while *acnB* and *icdA* (*sdhC*) transcript levels tend to increase. The increase in *lpdA* transcript levels may be due to a decrease in *crp*, where cAMP-Crp represses such gene expression. The decrease of *acnA* and *fumC* transcript levels may be caused by the decrease in *rpoS* (Appendix A). The increase in *soxR/S* causes *zwf* and *sodA* to be increased, whereas *acnA* and *fumC* decreases. The latter may be due to a decrease in *rpoS* transcript levels (Appendix A). The increase in *yfiD* may be due to an increase in *fnr* transcript levels, though Fnr may be in an inactive form under aerobic conditions. Upon complex formation of cAMP-Crp, it activates genes encoding the glucose PTS, the TCA cycle, and gluconeogenesis (Appendix A). Although *crp* transcript levels decrease as the dilution rate increases (Figure 3.7a), as stated above, *ptsG* and *ptsH* transcript levels increases (Figure 3.7b). This may be caused by the decrease in *cra*, where *pfkA* gene expression is also increased.

3.2.8 Effect of crp gene mutation on the metabolism

Table 3.4 shows the effect of the *crp* gene knockout and *crp* enhancement (*crp*$^+$) on the fermentation characteristics at the dilution rate of 0.2 h^{-1}, where it indicates that the specific glucose uptake rate is lower and the cell concentration is higher as compared to the wild type, and the specific acetate production rate is higher for the *crp* knockout mutant (Yao et al., 2011). In the case of the *crp*$^+$ mutant, the fermentation characteristics are similar to the wild type by considering the statistical significance.

Table 3.4 Effects of the specific gene mutation on the fermentation characteristics at the dilution rate of 0.2 h^{-1}

Strains	Biomass conc. (g/L)	Glucose conc. (g/L)	Glucose uptake rate (mmol/g/h)	Acetate formation rate (mmol/g/h)	Biomass yield (g/g)
BW25113	1.45 ± 0.06	ND*	3.07 ± 0.13	ND*	0.37 ± 0.015
Δcrp	1.57 ± 0.07	ND*	2.83 ± 0.13	0.35 ± 0.018	0.39 ± 0.018
crp$^+$	1.42 ± 0.07	ND*	3.12 ± 0.14	ND*	0.36 ± 0.018
Δpgi	1.6 ± 0.08	ND*	2.77 ± 0.14	ND*	0.4 ± 0.021
Δmlc	1.65 ± 0.08	ND*	2.69 ± 0.13	ND*	0.41 ± 0.019
ΔmgsA	1.55 ± 0.07	ND*	2.87 ± 0.14	0.1 ± 0.005	0.39 ± 0.018

ND*: not detectable, where glucose detectable limit was 0.038 g/l.

Figure 3.8a shows the transcript levels of the *crp* knockout mutant and the *crp*$^+$ mutant as compared to the wild type, where it indicates that *mlc* is down-regulated in the *crp* knockout mutant, and up-regulated in the *crp*$^+$ mutant. Figure 3.8b indicates that *ptsG* and *ptsH* transcript levels change in a similar fashion to *mlc* and *crp*, which corresponds to the lower glucose uptake rate for the *crp* knockout mutant and similar to the wild type for the *crp*$^+$ mutant (Table 3.4). Figure 3.8 also indicates that *acnA*, *gltA*, *fumA*, *mdh*, *pckA*, and *sdhC* all change in a similar fashion to *crp* (Appendix A). The *cydB* transcript level changes in a similar fashion to *arcA*, while *cyoA* gene changes in a reverse fashion (Appendix A). The *sodA* and *fur* changes in a similar fashion to *soxR/S*. The *tktB* and *fumC* changes in a similar fashion to *rpoS* (Appendix A). The changing patterns of *aceA*, *icdA*, and *pckA* transcript levels are similar to *cra*, while reverse patterns may be seen for *fadR* and *iclR*.

As stated above, the specific glucose consumption rate decreased in the *crp* knockout mutant, as compared to the wild type, is almost the same in the *crp*$^+$ mutant. The *ptsG* gene is activated by Crp, but also repressed by Mlc. Note that the changing pattern of *crp* transcript levels (Figure 3.8a) coincides with that of *mlc* (Figure 3.8a) as well as *cyaA*. It has been reported that there is a Crp binding region in the promoter region of the *mlc* gene.

In the case of the *crp* knockout mutant, the decrease in the glucose consumption rate may be caused by down-regulation of *ptsG* and *ptsH*, whereas *pfkA* and *pykF* are up-regulated. The former may be caused by the *crp* knockout gene, while the latter may be due to down-regulation

Bacterial cellular metabolic systems

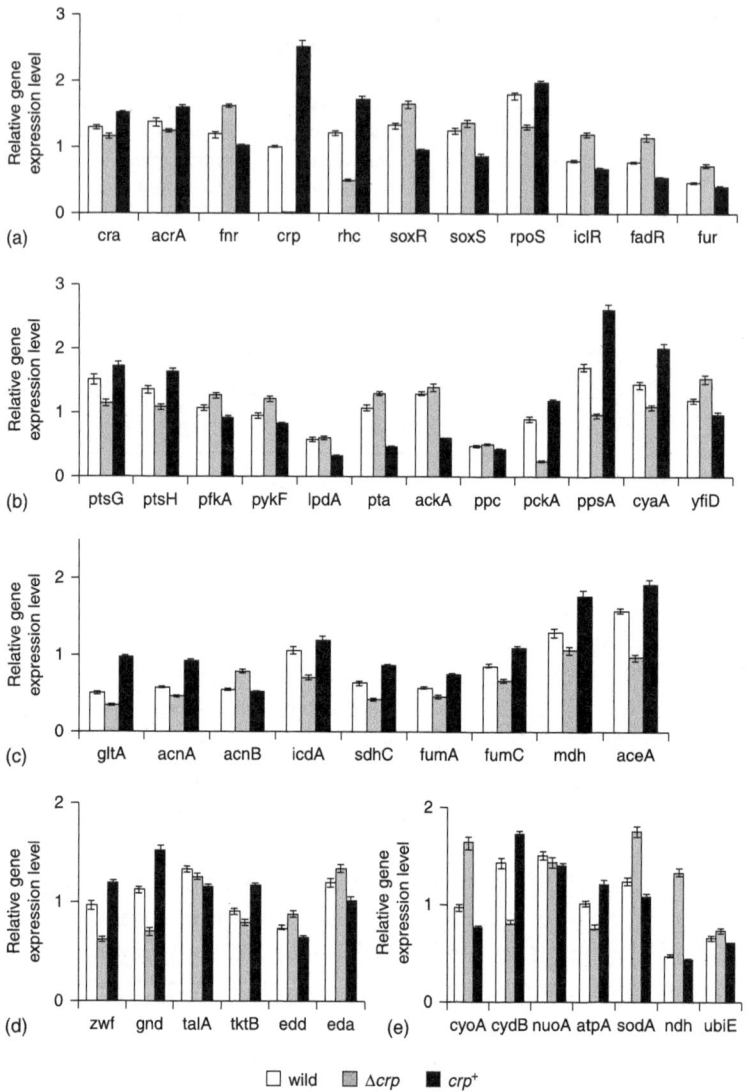

Figure 3.8 Comparison of gene transcript levels of the wild type, *crp* knockout mutant, and *crp*⁺ mutant. (a) Global regulator genes; (b) PTS, glycolysis, anaplerotic pathway, *cyaA* and *yfiD* genes; (c) TCA and glyoxylate pathway genes; (d) PP pathway genes; (e) Respiratory chain genes

of *cra* (Figure 3.8a), where *mlc* may not be dominant. The *crp* gene knockout causes the TCA cycle genes, such as *gltA*, *acnA*, *sdhC*, *fumA*, and *mdh*, to be down-regulated. Figure 3.8a shows *cra* to be down-regulated, which may be caused by a higher glucose concentration, though it is less than a detectable level. The down-regulation of *cra* also causes *icdA* gene as well as gluconegogenic pathway genes *pckA* and *ppsA* to be down-regulated (Figures 3.8a–c). The decrease in the glyoxylate pathway gene *aceA* (Figure 3.8c) may be caused by the increases of *fadR* and *iclR*, which may be due to higher glucose concentrations. The decreased activities of the TCA cycle and glyoxylate pathway may cause acetate to accumulate, where this is also reflected by the up-regulation of *pta* and *ackA* transcript levels (Figure 3.8b). The batch culture of the Δ*crp* mutant indicates that the acetate formed during cell growth phase cannot be consumed during the stationary phase. The decreased transcript levels of *acnA*, *fumC*, and *tktB* may also be due to down-regulation of *rpoS*, which may be due to less nutrient stress caused by an increase in glucose concentration. The *sodA* gene transcript levels increase (Figure 3.8e), consistent with up regulation of *soxR/S* (Figure 3.8a), which might be due to an increase in oxygen concentration. The decrease of *arcA* (Figure 3.8a) is also reflected by this hypothesis, whereas its effect on the TCA cycle genes may be minor and not dominant.

In the case of the *crp*$^+$ mutant, the *ptsG* and *ptsH* transcript levels are increased, whereas *pfkA* and *pykF* are decreased. It should be noted that *crp* gene enhancement causes an increase in the *mlc* gene (Figure 3.7a), and thus Mlc may repress *ptsG* and *ptsH*. This seems to be dependent on the culture conditions. This may happen when glucose concentration is higher, as will be shown for the batch culture. Different from the case of the Δ*crp* mutant, the TCA cycle genes and glyoxyalte pathway genes, as well as gluconeogenic pathway genes, are up-regulated. The activation of the TCA cycle may cause the decrease in cell yield.

3.2.9 Glucose PTS and other PTS such as fructose PTS in E. coli

As stated above, PTS is a transport system that catalyzes the uptake of a variety of carbohydrates and their conversion into their respective phosphoesters during transport (Deutscher et al., 2006). PTS is composed of EI, HPr, and EII, where these accept a phophoryl group from a donor and transfer it to an acceptor, thus cycling between the phosphorylated and unphosphorylated states (Deutscher et al., 2006). EI and HPr are common

Bacterial cellular metabolic systems

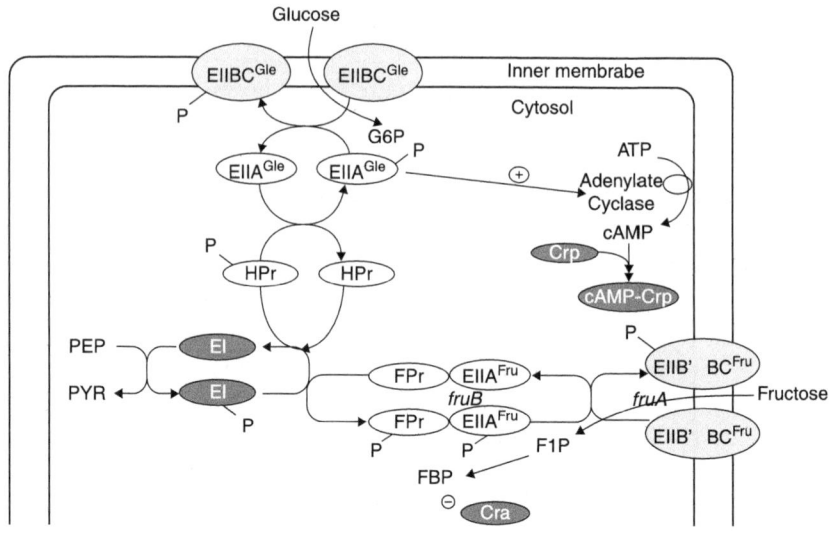

Figure 3.9 Glucose PTS and fructose PTS

to all PTS carbohydrates, while EII is carbohydrate specific. Thus, bacteria usually contain many different E IIs. Each E II complex consists of one or two hydrophobic integral membrane domains (C and D) and two hydrophilic domains (A and B), which together are responsible for the transport of the carbohydrate across the membrane as well as its phosphorylation. *E. coli* contains 15 different E II complexes, where the E II complexes are formed either by distinct proteins that contain EI and/or HPr domains exist (Deutscher et al., 2006). A prominent example for the latter is FPr, which consists of HPr and E IIA domains and mediates the phosphotransfer in the uptake of fructose by *E. coli*. As shown in Figure 3.9, fructose can be transferred and phosphorylated by the fructose PTS (E II BC^{Fru}) or ATP-dependent mannofruct kinase Mak (Aulkemeyer et al., 1991). The E II BC^{Fru} encoded by *fruA* phosphorylate fructose is concomitant with transport to fructose 1-phosphate, which is further converted to FDP by an ATP-dependent Pfk (Kornberg, 2001) (Chapter 1).

3.2.10 CCR in B. subtilis and other Firmicutes

The key players in CCR in *Bacillus subtilis* are the pleiotropic transcription factor CcpA (catabolite control protein A), the Hpr protein of the PTS, the bifunctional Hpr kinase/phosphorylase (HPrK), and the glycolytic intermediates such as FDP and G6P (Henkin et al., 1991; Titgemeyer and

Hillen, 2002; Warner and Lolkema, 2003). Unlike *E. coli*, HPr phosphorylation plays an important role, where phosphorylated HPr serves as the effector for the dimeric CcpA, which controls the expressions of CCR genes (Gorke and Stulke, 2008). The phosphorylation of HPr is catalyzed by HPrK that binds ATP, and its activity is triggered by the availability of FDP as an indicator of high glycolytic activity (Galinier et al., 1998; Reizer et al., 1998; Jault et al., 2000). By contrast, phosphorylase activity prevails under nutrient limitation, and the activation is stimulated by the inorganic phosphate in the cell (Jault et al., 2000; Mijakovic et al., 2002). Under nutrient-rich conditions, HPrK acts as a kinase and phosphorylates HPr, and the cofactor for CcpA is formed. The interaction between CcpA and the phosphorylated HPr is enhanced by FDP and G6P (Seidel et al., 2005; Schumacher et al., 2007).

With the exception of the mycoplasmas, firmicutes also use HPr, HPrK, and CcpA for CCR (Titgemeyer et al., 2002). CcpA in lactic acid bacteria, such as *Lactococcus lactis*, not only represses genes of carbon metabolism but controls metabolic pathway genes such as glycolysis and lactic acid formation pathway genes (Gorke and Stulke, 2008).

3.2.11 CCR in actinobacteria

In *Streptomyces coelicolor* and related species, glucose kinase is the key player of CCR, where it is independent of the PTS (van Wezel et al., 2007). *Corynebacterium glutamicum* is important in the industrial production of amino acids, where it prefers to use multiple carbon sources simultaneously. Diauxic growth is observed when using a mixture of glutamate or ethanol and glucose, where the repressor protein RamB is activated when glucose is present, and binds to the promoter regions of the genes involved in acetate and ethanol catabolism (Gestmeir et al., 2004; Arndt and Eikmanns, 2007). The *ramB* expression is regulated by a feedback of RamB and RamA, where RamA is activated when acetate is present.

3.2.12 CCR in Pseudomonas putida

P. putida can assimilate various aromatic and aliphatic hydrocarbons, where it has been reported that the use of hydrocarbons is repressed by succinate, and this seems to be a general feature of CCR in this organism (Collier et al., 1996; Muller et al., 1996). Under CCR, the translation of operon-specific regulators is inhibited by the binding of an RNA-binding protein Crc to mRNAs of the regulator transcript, and thus CCR seems to be governed by an RNA-binding protein at the level of post-transcriptional control rather than by a DNA-binding transcriptional regulator (Moreno et al., 2007; Moreno and Rojo, 2008).

3.2.13 The impact of CCR on bacterial virulence

CCR is crucial for the expression of virulence genes and for pathogenicity in many pathogenic bacteria. Note that the primary aim of pathogenic bacteria is to gain access to nutrients rather than to cause damage to the host, and that the expression of virulent genes is linked to the nutrient supply of the bacteria (Gorke and Stulke, 2008).

In many firmicules, the mutants devoid of the HPr kinase grow significantly slower than wild-type cells. Therefore, it is suggested that HPr kinase, which generates the cofactor for CcpA, might be a suitable drug target, where the compound that inhibits the kinase activity of HPr has been identified, and this compound inhibits the growth of *B. subtilis* but not of *E. coli*, where *E. coli* does not contain HPr kinase (Gorke and Stulke, 2008).

Crp and cAMP are essential for the expression of virulence genes in enteric bacteria, and therefore, the corresponding *crp* and *cya* mutant strains of *S. enterica* and *Y. enterocolitica* can be used as live vaccines in mice and pigs (Curtiss and Kelly, 1987; Petersen and Young, 2002; Ramströme et al., 2004).

3.3 Nitrogen regulation

Next to carbon (C) source metabolism, nitrogen (N) metabolism is also important in understanding the metabolic regulation. In *E. coli*, assimilation of an N-source, such as ammonia/ammonium (NH_4^+) using α-KG, results in the synthesis of glutamate and glutamine (Figure 3.10). Glutamine synthetase (GS, encoded by *glnA*) catalyzes the only pathway for glutamine biosynthesis. Glutamate can be synthesized by two pathways through the combined actions of GS and glutamate synthase (GOGAT, encoded by *gltBD*) forming GS/GOGAT cycle, or by glutamate dehydrogenase (GDH encoded by *gdhA*) (Yan, 2007). The GS/GOGAT cycle has a high affinity for NH_4^+ (K_m <0.2 mM for GS), and therefore is

Metabolic regulation by global regulators

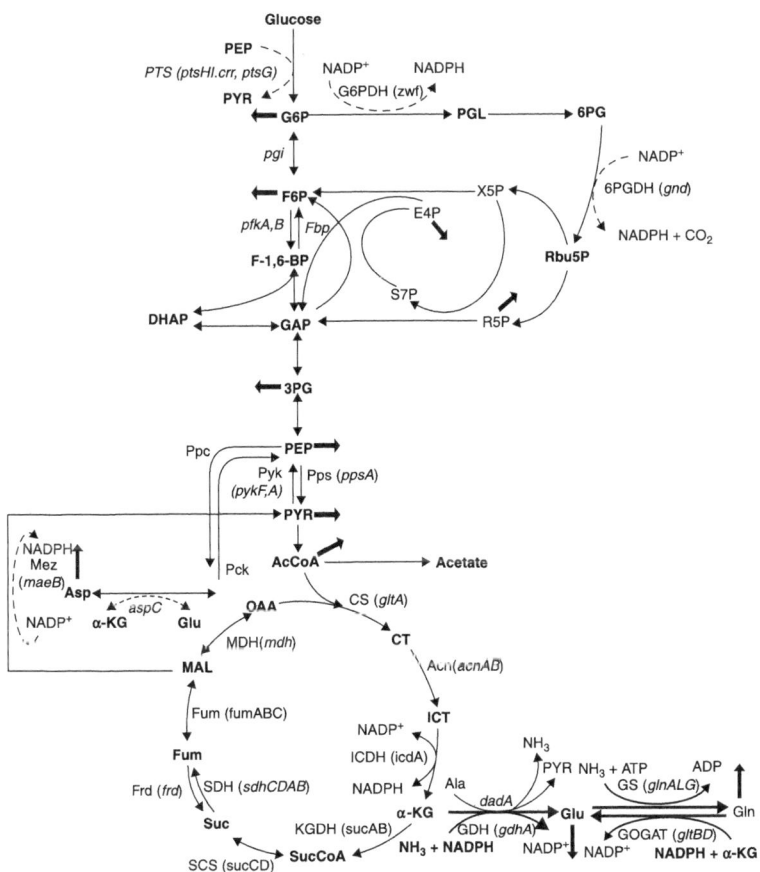

Figure 3.10 Central metabolic pathways and NH_3-assimilation pathways

dominant when nitrogen is scarce in the medium, whereas GDH has a low affinity for NH_4^+ (K_m >1 mM) and is utilized when a sufficient nitrogen source is available in the medium. When extracellular NH_4^+ concentration is low at around 5 µM or less, ammonium enters the cell via AmtB and is converted to glutamine by GS, and UTase uridylylates both GlnK and GlnB (Ninfa et al., 2000) (Figure 3.11). When the extracellular NH_4^+ concentration is more than 50 µM, the metabolic demand for the glutamine pool rises, and UTase deuridylylates GlnK and GlnB. GlnK complexes with AmtB, thereby inhibiting the transporter via AmtB, where GlnB interacts with NtrB and activates its phosphatase activity leading to dephosphorylation of NtrC, and NtrC-dependent gene expression ceases (Ninfa et al., 2000) (Figure 3.11). The nitrogen

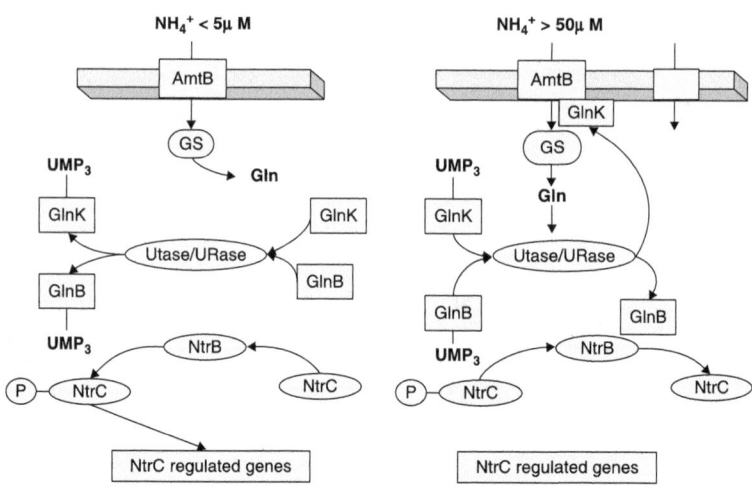

Figure 3.11 Ammonia assimilation under different NH_4^+ concentration

intermediates, such as glutamate and glutamine, provide nitrogen for the synthesis of all the other N-containing components. About 88% of cellular nitrogen comes from glutamate, and the rest from glutamine (Reitzer, 2003). The ATP required for the nitrogen assimilation, using the GS/GOGAT cycle under N-limiting conditions, accounts for 15% of the total requirement in *E. coli*. A significant amount of NADPH is also required for nitrogen assimilation (Reitzer, 2003; Yan, 2007). The other pathways involved in maintaining cellular nitrogen balance under specific conditions include aspartate-oxaloacetate and alanine-pyruvate shunts (Fischer and Sauer, 2003; Zhang et al., 2007).

It should be noted that carbon metabolism is not only controlled by carbon-derived signals, but also by the availability of nitrogen and other nutrients (Commichau et al., 2006). From studies on interdependence of different metabolic routes, two of the major signal transduction systems of nitrogen and carbon metabolism have been identified as P_{II}, a small nitrogen regulatory protein and PTS. Because of the important roles in the regulatory functions, P_{II} and PTS can be regarded as the central processing units of N and C metabolism, respectively. The P_{II} protein senses αKG and ATP, thus linking the state of central carbon and energy metabolism for the control of N assimilation (Commichau et al., 2006). The glucose catabolism is modulated by global regulators, such as Cra, Crp, Cya, Mlc, as stated in the previous sections, while N assimilation is regulated by the P_{II}-Ntr system together with global regulators, such as

Crp, providing a novel regulatory network between C and N assimilation in *E. coli* (Mao et al., 2007). The effects of C and N limitations on *E. coli* metabolism have been investigated for continuous culture (Hua et al., 2003, 2004; Sauer and Eikmanns, 2005; Nanchen et al., 2008; Kumar and Shimizu, 2010).

C and N metabolisms may be linked by energy metabolism, where it has been demonstrated that the glycolytic flux in *E. coli* is controlled by the demand for ATP (Koebman et al., 2002). It has been reported that the P_{II} protein controls N assimilation by acting as a sensor of adenylate energy charge, which is the measure of energy available for metabolism. The signal transduction requires ATP binding to P_{II}, which is synergistic with the binding of αKG. Moreover, αKG serves as a cellular signal of C and N status, and strongly regulates P_{II} functions (Jiang and Ninfa, 2007). The studies on the C and N pathway interdependence have so far focused on the conversion of αKG to glutamate (Ninfa and Jiang, 2005). It is evident that the regulatory mechanism of this conversion is critical for the interdependence of C and N assimilation.

Figure 3.12 shows the effect of the C/N ratio on fermentation characteristics during aerobic continuous culture where the dilution rate was 0.2 h^{-1}, and where the C/N ratio was based on the feed content. Figure 3.12 indicates that the glucose concentration increases, whereas the cell concentration decreases as the C/N ratio increases. Figure 3.12 also shows that the glucose concentration is very low at 100% and 60% of N concentrations (C-limitation), whereas its concentration is high at 20% and 10% of N concentrations (N-limitation). It is also shown that the specific glucose consumption rate, as well as the specific acetate and CO_2 production rates, tends to increase as the C/N ratio increases.

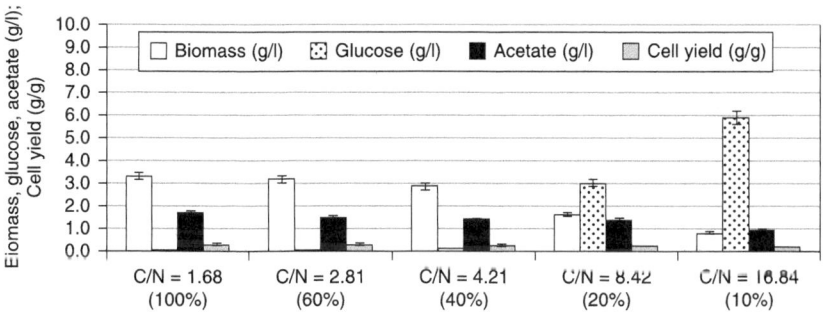

Figure 3.12 Effect of C/N ratio on the fermentation characteristics for the continuous culture at the dilution rate of 0.2 h^{-1}

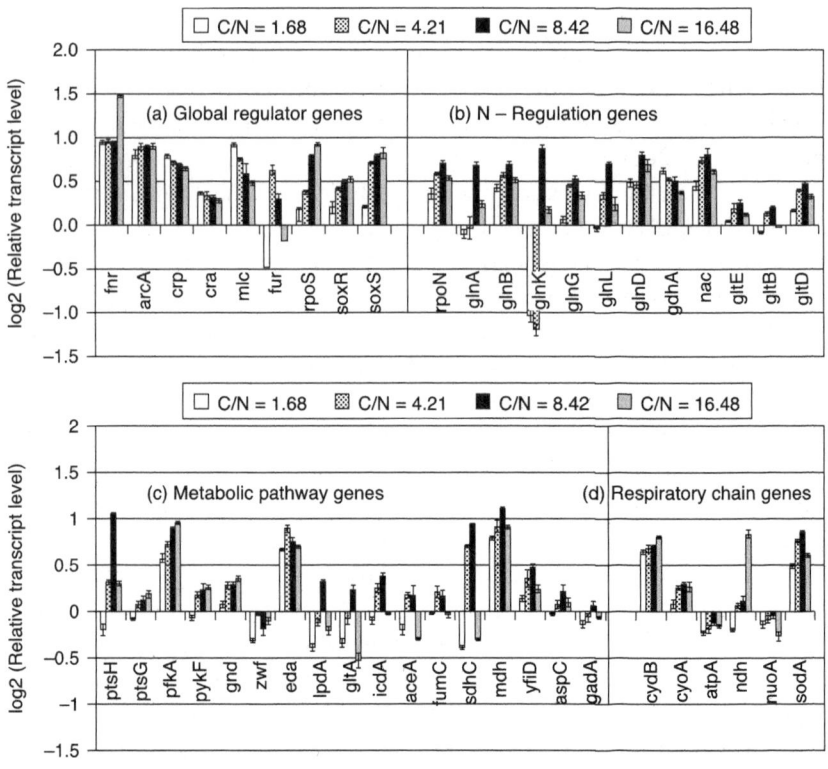

Figure 3.13 Schematic illustration of the interaction among several metabolic regulations. Comparison of the transcriptional mRNA levels of the wild type *E.coli* genes cultivated at 100% (C/N = 1.68), 40% (C/N = 4.21), 20% (C/N = 8.42) and 10% (C/N = 1.68) N⁻ concentration: (a) global regulatory, (b) N⁻ regulatory, (c) metabolic pathway, (d) respiratory chain

In order to interpret the fermentation characteristics, the relative mRNA levels are shown for different C/N ratios in Figure 3.13. Figure 3.13a shows that *crp* transcript levels become lower as C/N ratio increases, which corresponds to the fact that the cAMP-Crp level decreases as glucose concentration increases. In accordance with the change in *crp* transcript levels, the *mlc* level changes in a similar fashion (Gosset et al., 2004). Figure 3.13a also shows that the transcript levels of such genes as *soxR/S* and *rpoS* become higher as the C/N ratio increases, which may be due to oxygen stress caused by higher respiratory activity for the former (Hua et al., 2003), along with nutrient stress for the latter (Rahman et al., 2008). In relation to the up-regulation of *soxR/S*, the *sodA* transcript level increases as the C/N ratio increases, except at the highest C/N ratio (Figure 3.13d). Figure 3.13a shows a high expression of anaerobic regulator *fnr* at the highest C/N ratio, while the transcript level of *arcA* changes little (Kumar and Shimizu, 2010).

The transcript level of *rpoN*, which encodes σ^{54}, increases as the C/N ratio increases (Figure 3.13b). Figure 3.13b also shows that the expressions of *glnA*, *glnL*, *glnG*, and *gltD* genes change in a similar fashion to *rpoN*, indicating the activation of the GS-GOGAT pathway under N-limitation. The *glnB* gene, which codes for P_{II}, also changes in a similar fashion, while *glnD*, which controls uridylylation and deuridylylation, appears somewhat different but the trend seems to be similar (Figure 3.13b). The P_{II} paralogue encoding gene, *glnK*, shows high expressions at 20% and 10% of N-limitation (Figure 3.13b). The expression pattern of *nac* is similar to that of *rpoN*, whereas *gdhA* shows a reverse pattern, implying that *gdhA* is repressed by Nac (Figure 3.13b).

As the C/N ratio increases, the transcript level of the *crp* gene as well as the *mlc* gene decreases, which then causes the transcript level of *ptsG* gene to be increased (Figure 3.13c). In relation to the decrease in the transcript level of *cra*, the transcript levels of such genes as *ptsH*, *pfkA*, and *pykF* increase as the C/N ratio increases (Figure 3.13c). These correspond to the increased specific glucose consumption rate as the C/N ratio increases. Moreover, respiratory chain genes, such as *cyoA*, *cydB*, and *ndh*, together with TCA cycle genes, such as *gltA*, *icdA*, *fumC*, *sdhC*, and *mdh*, show increased expressions as the C/N ratio increases (Figure 3.13d), which corresponds to the increase in the specific CO_2 production rate. Part of the reason for this may be due to the accumulation of α-KG caused by the decreased activity of GDH. Since ferric uptake regulator Fur activates some of the TCA cycle genes, such as *sdh*, *suc*, and *fum* (Zhang et al., 2005), part of the reason may be due to up-regulation of the transcript levels of the *fur* gene (Figure 3.13a).

The glucose concentration in the fermentor increases with the increase in the C/N ratio (Figure 3.12). The glucose uptake is made via PTS in *E. coli*, where phosphate of PEP is transferred by phosphorelation via enzyme I (EI) encoded by *ptsI*, histidine phosphorylatable protein HPr encoded by *ptsH*, glucose specific enzyme II, $EIIA^{Glc}$ encoded by *crr*, and membrane bound $EIICB^{Glc}$ encoded by *ptsG*, as stated in the previous section. When glucose is present in excess, the phosphorylated EIIA^{Glc} transfers phosphate to EIICB^{Glc} for the glucose uptake with phosphorylation, and the unphosphorylated EIIA^{Glc} is dominated in the cytosol (Deutscher et al., 2006). Since unphosphorylated EIIA^{Glc} does not activate Cya, the cAMP level decreases under N-limitation together with the *crp* gene (Figure 3.13a). Since *mlc* is under control of *crp*, the transcript level of *mlc* gene also decreases (Figure 3.13a), which causes up-regulation of transcript levels of *ptsH* and *ptsG* genes (Appendix A). Moreover, an increase in the glucose concentration at a higher C/N ratio may cause down-regulation of *cra*, which causes up-regulation of the glycolysis genes, such as *ptsH*, *ptsG*, *pfkA*, and *pykF*, together with *zwf* (Appendix A).

The GDH pathway is favored when the organism is stressed for energy, because GDH does not use ATP as does the GS pathway (Helling, 1998). Figure 3.13b shows the decreased expression of *gdhA* as the C/N ratio increases. Liang and Houghton (1981) investigated the effect of NH$_4$Cl concentration on GDH and GS activities, and showed the up-regulations of GDH and transhydrogenase activities at lower NH$_4$Cl concentration.

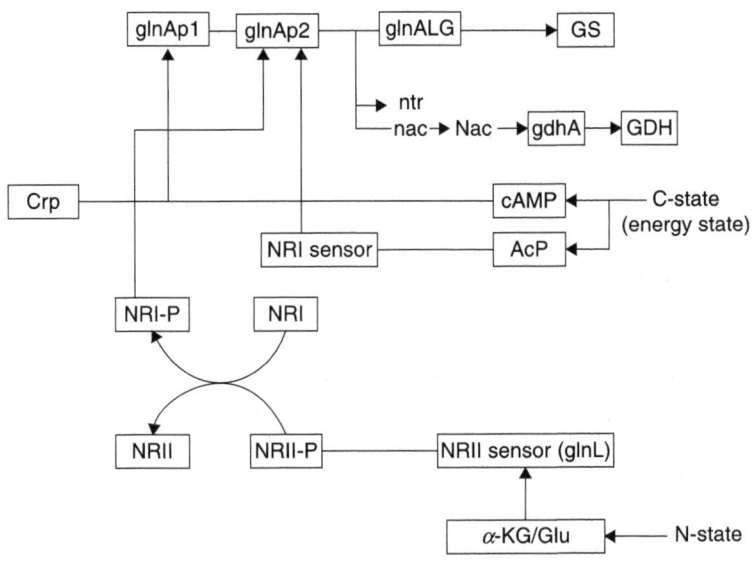

Figure 3.14 The interaction between nitrogen regulation and catabolite regulation

The availability of nitrogen is sensed by the P_{II} protein at the level of intracellular glutamine, where glutamine is synthesized by glutamine synthetase (GS) encoded by *glnA*, and is transported mainly by GlnHPQ. The *glnHPQ* operon is under the control of tandem promoters, such as *glnHp1* and *glnHp2*, where the former is σ^{70}-dependent, and the latter is σ^{54}- and NtrC-P dependent (Willis et al., 1975; Claverie-Martin and Magasanik, 1991). It has been shown that as the major transcriptional effector of the glucose effect, Crp affects nitrogen regulation (Mao et al., 2007). Namely, *glnAp1* is activated by Crp with glutamine as the N-source (Figure 3.14). Through *glnHPQ*-dependent signaling, Crp acts to decrease the amount of the phosphorylated NtrC activator, which in turn causes a decrease in *glnAp2* expression (Mao et al., 2007). However, this regulation is more complex. It has been suggested that σ^{54}-dependent Ntr genes of *E. coli* form a gene cascade in response to N-limitation (Blauwkamp and Ninfa, 2002). The central participants of Ntr response are NR_I or NtrC and NR_{II} or NtrB, and RNA polymerase complexed to σ^{54}. NR_I is the transcriptional activator of σ^{54}-dependent promoters, while NR_{II} is a bifunctional protein that can either transfer phosphate to NR_I or control the dephosphorylation of NR-phosphate. N-limitation results in the phosphorylation of NR_I, which in turn stimulates the expression of *glnALG* operon. The expression of the *glnALG* operon is controlled by tandem promoters, such as *glnAp1* and *glnAp2*, where *glnAp1* is a σ^{70}-dependent weak promoter and its transcription can be activated by Crp and blocked by Ntr-P. However, *glnAp2* is transcribed by RNA polymerase ($E\sigma^{54}$) and is activated by Ntr-P. Therefore, *glnAp2* is responsible for activating *glnA* transcription under N-limitation (Magasanik, 1996). Figure 3.13b shows that the expressions of the *glnA, L,* and *G* genes change in similar fashion as *rpoN* gene expression.

It has been reported that there are no NR_I-P binding sites in the *gdhA* regulatory region (Riba et al., 1988), and it is unlikely for NR_I to directly repress the *gdhA* promoter (Camarena et al., 1998). As it has been shown that Nac is involved in the transcriptional repression of the *gdhA* gene under N-limitation (Camarena et al., 1998), Nac seems to repress *gdhA* gene (Figure 3.13). This figure shows that the transcript level of *gdhA* gene is lower, while *gltB* and *D* genes are higher under N-limitation as compared to C-limitation. NADPH is an important cofactor in GDH and (GS)-GOGAT activities and it has been reported that transhydrogenase plays some role in the regulation of these pathways (Liang and Houghton, 1981). Under N-limitation, the glutamate and glutamine synthetic pathways are expected to be repressed due to shortage of NH_3 for these reactions, and thus NADPH is less utilized, resulting in overproduction of NADPH. Part of this may be converted to NADH by transhydrogenase, and the converted NADH together with other NADH formed may be utilized for ATP production through the respiratory chain. Overproduction of NADPH represses such pathways as G6PDH, 6PGDH, and ICDH in *E. coli*. However, *zwf* is activated in Figure 3.13c, which may be due to SoxR/S encoded by *soxR/S* caused by higher respiratory activity. The ICDH activity is reported to be insensitive to N concentration, where Figure 3.13c also shows little change in *icdA* gene expression.

E. coli possesses two closely related P_{II} paralogues, such as GlnB and GlnK, where GlnB is produced constitutively, and regulates the NtrB (NR_{II})/NtrC (NR_I) two-component system (Ninfa and Atkinson, 2000). It has been shown that the intracellular concentrations of NR_I and NR_{II} increase upon N limitation (Reitzer and Magasanik, 1985; Atkinson and Ninfa, 1993; Atkinson et al., 2002). The phosphorylated NtrC is an activator of various nitrogen-controlled genes, such as *glnA*, which codes for GS (Blauwkamp and Ninfa, 2002) and *glnK*, encoding the second P_{II} paralogues (Atkinson et al., 2002). The increased NR_I, presumably in the phosphorylated form such as NR_I-P, activates the expression of *glnK* and *nac* promoters under N-limitation (van Heeswijk et al., 1996; Pahel et al., 1982). Figure 3.13b shows that the transcript levels of *glnK* and *nac* genes increase as C/N ratio increases, while a slight decrease can be seen at the highest C/N ratio, where it has been reported that *glnK* and *nac* promoters are sharply activated when ammonia is used up (Atkinson et al., 2002).

The *gltBDF* operon has been found to have binding affinity with global regulators such as Fnr and Crp in the promoter region (Paul et al., 2007), while the transcript level of *fnr* gene is higher under N-limitation whereas *crp* gene became lower (Figure 3.13a). The up-regulation of *yfiD* (Fig. 3.13c) may be due to up-regulation of *fnr*.

The Ntr system is composed of four enzymes (Figure 3.15): a uridylytransferase/uridylyl-removing enzyme (UTase/UR) encoded by the *glnD* gene; a small trimeric protein, P_{II} encoded by *glnB*; and the two-component system composed of NtrB and NtrC. GlnD controls the activity of GS by adenylylation/deadenylylation through a bifunctional enzyme adenylyltransferase (ATase), the *glnE* gene product (Shapiro and Stadtman, 1968; Stadtman, 1990; Jaggi et al., 1997). The activity of GlnK becomes high under N-limitation (Figure 3.13b) and contributes to the regulation of NtrC-dependent genes (Maheswaran

Metabolic regulation by global regulators

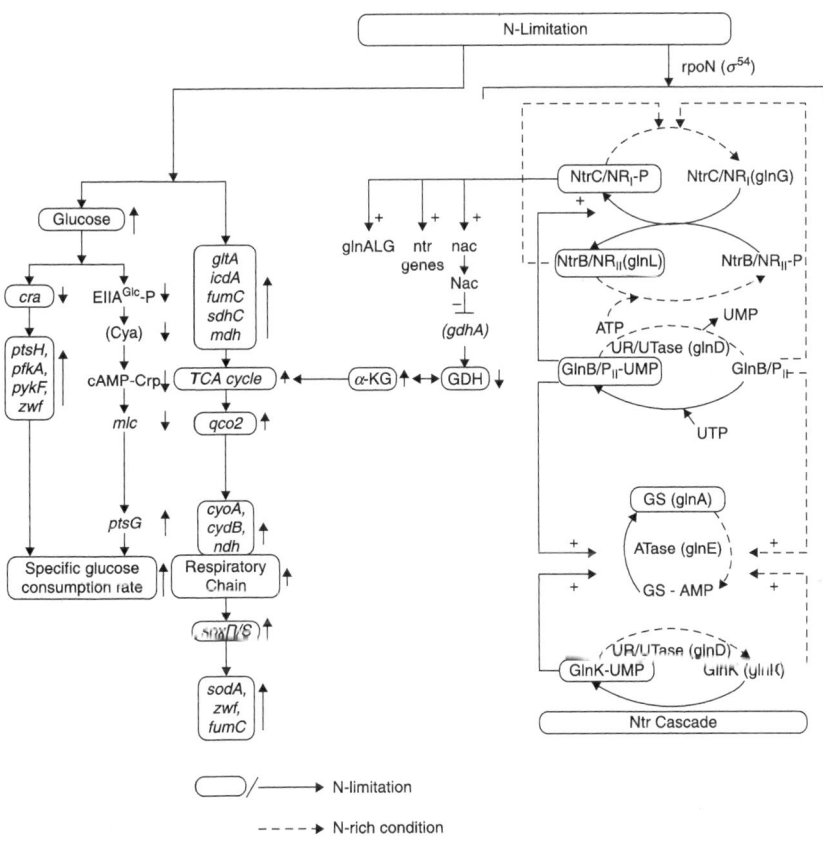

Figure 3.15 Overall mechanism of nitrogen assimilation in *E. coli* under C-limited (N-rich) and N-limited conditions

and Forchhammer, 2003). It has been shown that on GS adenylylation, ATase activity is regulated by UTase/UR and P_{II}, such that upon nitrogen limitation, UTase covalently modifies P_{II} by addition of a UMP group at a specific residue and the resultant uridylylated form of P_{II} promotes deadenylylation of GS by ATase (Figure 3.15). Conversely, under N-rich conditions, the uridylyl-removing activity of GlnD predominates and the deuridylylated P_{II} promotes adenylation of GS by ATase. Adenylylation by ATase is promoted by deuridylylated P_{II}, which is produced by UR action on P_{II} $(UMP)_3$ under higher N-concentration (low C/N ratio) (Figure 3.15). These indicate that UTase/UR and P_{II} acting together sense the intracellular nitrogen status (Merrick and Edwards, 1995). The P_{II} signal transduction proteins, such as GlnB and GlnK, are uridylylated/deuridylylated in response to intracellular glutamine levels, where low intracellular glutamine level, signaling N-limitation, leads to uridylylation of GlnB (Merrick and Edwards, 1995). GlnB is shown to be allosterically regulated by α-KG, and thus GlnB may play a role in integrating signals of C/N status. The NtrB/NtrC two-component system and GlnE, which adenylylates/deadenylylates GS, are the receptors of GlnB signal transduction (Maheswaran and Forchhammer, 2003). It has been suggested that the carbon/cAMP effect is mediated through GlnB uridylylation (Maheswaran and Forchhammer, 2003).

Phosphorylated NR_I/NtrC (NR_I/NtrC-P) activates transcription from N-regulated σ^{54}-dependent promoters by binding to the enhancers (Kustu et al., 1989; Merrick and Edwards, 1995; Jiang et al., 1998; Ninfa et al., 2000). P_{II} and the related GlnK protein control the phosphorylation state of NR_{II}/NtrB by stimulating the phosphatase activity of NR_{II}. The ability of GlnK and P_{II} to regulate the activities of NR_{II} is in turn regulated by the intracellular signals of C and N availability via allosteric control (Ninfa et al., 2000).

3.4 Phosphate regulation

The phosphate (P) metabolism is also important from the energy generation and phosphorelay regulation points of view. The phosphorous compounds serve as major building blocks of many biomolecules, and have important roles in signal transduction (Wanner, 1996). The phosphate is contained in lipids, nucleic acids, proteins, and sugars, and is involved in many biochemical reactions by the transfer of phosphoryl groups (Lamarche et al., 2008). Moreover, phosphate metabolism is closely related to the diverse metabolisms, such as energy and central carbon metabolisms (Ishige et al., 2003). All living cells sophisticatedly regulate the phosphate uptake, and survive even under phosphate-limiting conditions (Baek and Lee, 2006; Wendisch, 2006). *E. coli* contains about 15 mg of phosphate (P) per g (dry cell weight) (Damoglou and Dawes, 1968). Depending on the concentration of environmental phosphate, *E. coli* controls phosphate metabolism through Pho regulon, which forms a global regulatory circuit involved in a bacterial phosphate management (Wanner, 1993, 1996). The PhoR–PhoB two-component system plays an important role in detecting and responding to the changes of the environmental phosphate concentration (Stock et al., 1989; Parkinson, 1993; Baek and and Lee, 2007). It has been shown that PhoR is an inner-membrane histidine kinase sensor protein that appears to respond to variations in periplasmic orthophosphate (P_i) concentration through interaction with a phosphate transport system, and that PhoB is a response regulator that acts as a DNA-binding protein to activate or inhibit specific gene transcription (Smith and Payne, 1992; Wanner, 1996; Harris et al., 2001; Blanco et al., 2002). The activation signal, a phosphate concentration below 4 µM, is transmitted by a phosphorelay from PhoR to PhoB. Phospho-PhoB in turn controls Pho regulon gene expressions. PhoB is phosphorylated by PhoR under phosphate starvation or by PhoM (or CreC) in the absence of functional PhoR (Torriani and Ludke, 1985; Makino et al., 1985, 1988, 1989; Shinagawa et al., 1987; Wanner, 1987; Amemura et al., 1990).

The *E. coli* Pho regulon includes 31 (or more) genes arranged in separate operons, such as *eda*, *phnCDEFGHIJKLMNOP*, *phoA*, *phoBR*, *phoE*, *phoH*, *psiE*, *pstSCAB-phoU*, and *ugpBAECQ* (Hsieh and Wanner, 2010). When P_i is in excess, PhoR, Pst, and PhoU together turn off the Pho regulon by dephosphorylating PhoB. In addition, two P_i-independent controls, which may be a form of cross regulation, turn on the Pho regulon in the absence of PhoR. The sensor CreC, formerly called PhoM, phosphorylates PhoB in response to some (unknown) catabolite, while acetyl phosphate may directly phosphorylate PhoB (Wanner, 1993). When P_i is in excess, P_i is taken up by the low affinity P_i transporter, Pit. Four proteins, such as PstS, PstC, PstA, and PstB, form an ABC transporter

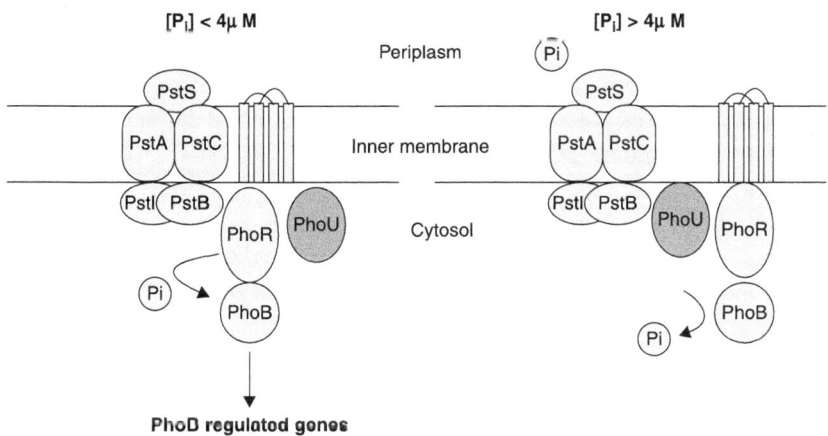

Figure 3.16 Molecular mechanism of phosphate regulation

important for the high-affinity capture of periplasmic inorganic phosphate (P_i) and its low-velocity transport into the cytosol (Van Dien and Keasling, 1998). These proteins are encoded together with PhoU as the *pstSCAB–phoU* operon. PstS is a periplasmic protein that binds P_i with high affinity. PstC and PstA are inner-membrane channel proteins for P_i entry, while PstB is an ATP-dependent permease that provides the energy necessary for P_i transport from periplasm to cytosol (Figure 3.16). When phosphate is in excess, the Pst system forms a repression complex with PhoR, and prevents activation of PhoB. PhoU and PstB are also required for dephosphorylation of phospho-PhoB under P-rich conditions (Wanner, 1997). Indeed, PhoU is essential for the repression of the Pho regulon under high phosphate conditions (Wanner, 1996). It may be considered that PhoU acts by binding to the PhoR, PhoB, or PhoR/PhoB complexes to promote dephosphorylation of phosphorylated PhoB or by inhibiting formation of the PhoR–PhoB complex (Oganesyan et al., 2005).

It has been shown that the *phoB* mutant does not synthesize alkaline phosphatase (*phoA* gene product) (Nesmeianova et al., 1975; Pratt and Torriani, 1977; Zuckier et al., 1980; Guan and Wanner, 1983; Kimura et al., 1989; Yamada et al., 1989) and phosphate binding protein (*pstS* gene product) (Pratt and Torriani, 1977; Kimura et al., 1989; Yamada et al., 1989). It is observed that *phoU* expression changes, depending on phosphate concentration of the *phoB* mutant (Nakata et al., 1984). Since the *phoA* gene mutation leads to the decreased content of membrane proteins or completely lacks them, mutations in the *phoB* gene result in the loss of alkaline phosphate and two membrane proteins (Tsfasman and Nesmeianova, 1981). Nesmeianova et al. (1975) found that *phoB* mutation leads to loss of polyphosphate kinase activity, which catalyzes the synthesis of polyP in *E. coli*. Ault-Riché et al. (1988) also found that the strains with deletion of *phoB* failed to accumulate polyP in response to osmotic stress or nitrogen limitation. Mutations in the *phoB* gene had no effect on *pepN* (Gharbi et al., 1985) and *lky* (*tolB*) expressions (Lazzaroni and Portailer, 1985).

The expressions of the genes under the control of the PhoR–PhoB two-component system are found to be affected by the duration of P-limitation in response to phosphate starvation in *E. coli*. This means that the roles of the PhoR–PhoB two-component regulatory system seem to be more complex (Baek and Lee, 2007). Since phosphate starvation is a relatively inexpensive means of gene induction in practice, the *phoA* promoter has been used for overexpression of heterologous genes (Shin and Seo, 1990).

Table 3.5 Fermentation characteristics of the wild-type *E. coli* and its *phoB* and *phoR* mutants in the aerobic chemostat culture under different phosphate concentrations at the dilution rate of 0.2 h^{-1} at pH 7.0

Fermentation parameters		P-rich (100%) conditions	P-limited (20%) conditions	P-limited (10%) conditions
Biomass concentration (g/l)	Wild	3.86 ± 0.03	3.47 ± 0.05	1.69 ± 0.03
	Δ*phoB*	3.44 ± 0.04	–	3.64 ± 0.01
Glucose concentration (g/l)	Wild	0.660 ± 0.004	0.557 ± 0.001	1.85 ± 0.01
	Δ*phoB*	1.59 ± 0.29	–	1.050 ± 0.001
Acetate concentration (g/l)	Wild	0.046 ± 0.002	0.410 ± 0.001	0.41 ± 0.02
	Δ*phoB*	0.255 ± 0.130	–	0.346 ± 0.010
Specific glucose uptake rate (mmol/gDCW/h)	Wild	2.69 ± 0.05	3.024 ± 0.005	5.36 ± 0.01
	Δ*phoB*	2.72 ± 0.09	–	2.730 ± 0.003
Specific acetate production rate (mmol/gDCW/h)	Wild	0.040 ± 0.002	0.394 ± 0.010	0.81 ± 0.02
	Δ*phoB*	0.247 ± 0.080	–	0.317 ± 0.001

Note: "–" indicates that no data was collected for these conditions. The standard deviation was obtained by triplicate measurements.

A better understanding of the Pho regulon would allow for optimization of such processes (Van Dien and Keasling, 1998).

Table 3.5 shows the effect of P concentration on the fermentation characteristics of the wild-type *E. coli* BW25113, its *phoB* gene knockout mutant (JW0389), and *phoR* mutant (JW0390) in the aerobic continuous culture at a dilution rate of 0.2 h^{-1}, where it indicates that the fermentation characteristics significantly change when feed P concentration become low at around 10% of the M9 medium (Marzan and Shimizu, 2011). In particular, the specific glucose consumption rate and the specific acetate production rate become significantly higher, while cell concentration become significantly lower under such P-limiting conditions.

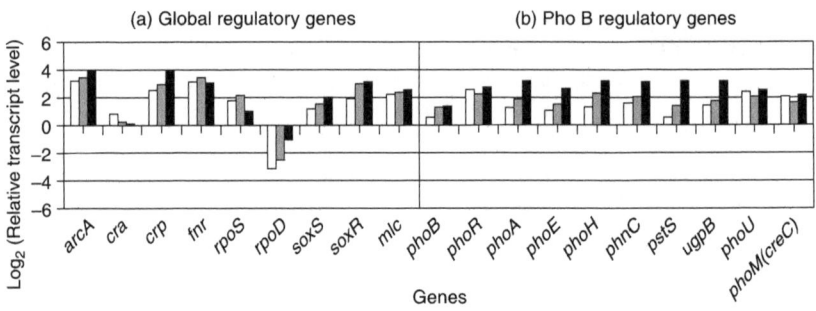

Figure 3.17 Comparison of the transcript levels of the wild type *E. coli* cultivated with different P concentrations of the feed (100%, 55%, 10%): (a) global regulatory genes, (b) PhoB regulatory genes

Figure 3.17 shows the effect of P concentration on the transcript levels, where Figure 3.17b indicates that *phoB* transcript level increases as P concentration decreases, and *phoB* regulated genes, such as *phoA*, *phoE*, *phoH*, *phnC*, *pstS*, and *ugpB*, are all increased in a similar fashion, (Figure 3.17b) and *eda* transcript levels also change in a similar fashion. Note that *phoU* and *phoM* change in a similar fashion to *phoR*, and also that the transcript levels of *rpoD*, which encodes the RNA polymerase holoenzyme containing σ^{70}, increases in a similar fashion to PhoB regulatory genes (Makino et al., 1993). Figure 3.17a also indicates that the transcript levels of *arcA* increase as P concentrations decrease. Figure 3.17a shows that *cra* transcript levels decrease. These are consistent with the increased specific glucose consumption rate (Table 3.4). The decrease in *cra* transcript levels may be due to higher glucose concentration. Moreover, Figure 3.17a also indicates that *soxR/S* transcript levels increase as P concentrations decrease, and accordingly the transcript levels of *rpoD*, (as well as *zwf* and *sodA*) change in a similar fashion. The respiratory chain genes, such as *atpA*, *ndh*, and *nuoA*, also change in a similar fashion, implying that the respiration is activated under P-limitation (Marzan and Shimizu, 2007).

Table 3.5 also shows the effect of the *phoB* gene knockout on the fermentation characteristics under both P-rich and lower P conditions, where it indicates that the glucose concentration increases and cell concentration decreases for the *phoB* mutant as compared to the wild type, and that the specific acetate production rate is higher at P-rich conditions for the *phoB* mutant as compared to the wild type.

Figure 3.17b indicates that *phoB* gene transcript levels increase as P concentration decreases in the wild type, and Figure 3.17a indicates that *rpoD* also increases as P concentration decreases. The *phoA, phoE, phoH, phnC, pstS*, and *ugpB* are all increased in a similar fashion to that of *rpoD*, as mentioned before. Figure 3.17a indicates that the expression pattern of *rpoS* is somewhat different. When cells enter into the P_i-starvation phase in the batch culture, the Pho regulon is activated, and σ^S starts to accumulate in the cytosol (Gentry et al., 1993; Wanner, 1996; Ruiz and Silhavy, 2003). The promoters of the Pho genes are recognized by σ^D-associated RNA polymerase. A mutation in *rpoS* significantly increases the level of AP (Alkaline phosphatase) activity, and the overexpression of σ^S inhibits it (Taschner et al., 2004). It has been reported that in the *rpoS* mutant, the expression of AP is considerably higher than that in the wild-type strain, implying that σ^S is involved in regulation of AP. Other Pho genes, such as *phoE* and *ugpB*, are likewise affected by σ^S. The *rpoS* may inhibit the transcriptions of *phoA, phoB, phoE*, and *ugpB*, but not that of *pstS* (Taschner et al., 2004). Figure 3.17ab indicates that Pho genes are highly expressed as compared to low *rpoS* transcript levels in the case of P-limitation. In contrast, *pst* may be transcribed by both σ^S and σ^D. The Pho regulon is thus evolved to maintain a trade-off between cell nutrition and cell survival during P_i-starvation (Taschner et al., 2004). The previous reports suggest that the Pho regulon and the stress response are interrelated (Spira et al., 1995; Spira and Yagil, 1999; Ruiz and Silhavy, 2003; Taschner et al., 2004; Taschner et al., 2006; Schurdell et al., 2007).

The presence of glucose or mutations in the *cya* or cAMP receptor protein (*crp*) gene leads to induction of the *phoA* gene in the *phoR* mutant. This induction requires the sensor PhoM (CreC) and the regulator PhoB (Wanner et al., 1988). However, PhoM (CreC) may not detect glucose *per se*, where it may detect an intermediate in the central metabolism. Therefore, *cya* or *crp* mutation may indirectly affect PhoM (CreC)-dependent control. In addition to P_i control, two P_i-independent controls may lead to activation of PhoB. These two may be connected to control pathways in carbon and energy metabolism, in which intracellular P_i is incorporated into ATP. One P_i independent control is regulation by synthesis of AcP, where P_i is incorporated into ATP at the Ack (acetate kinase) pathway. AcP may act indirectly on PhoB.

The overall regulation mechanism is illustrated schematically in Figure 3.18 (Marzan et al., 2011).

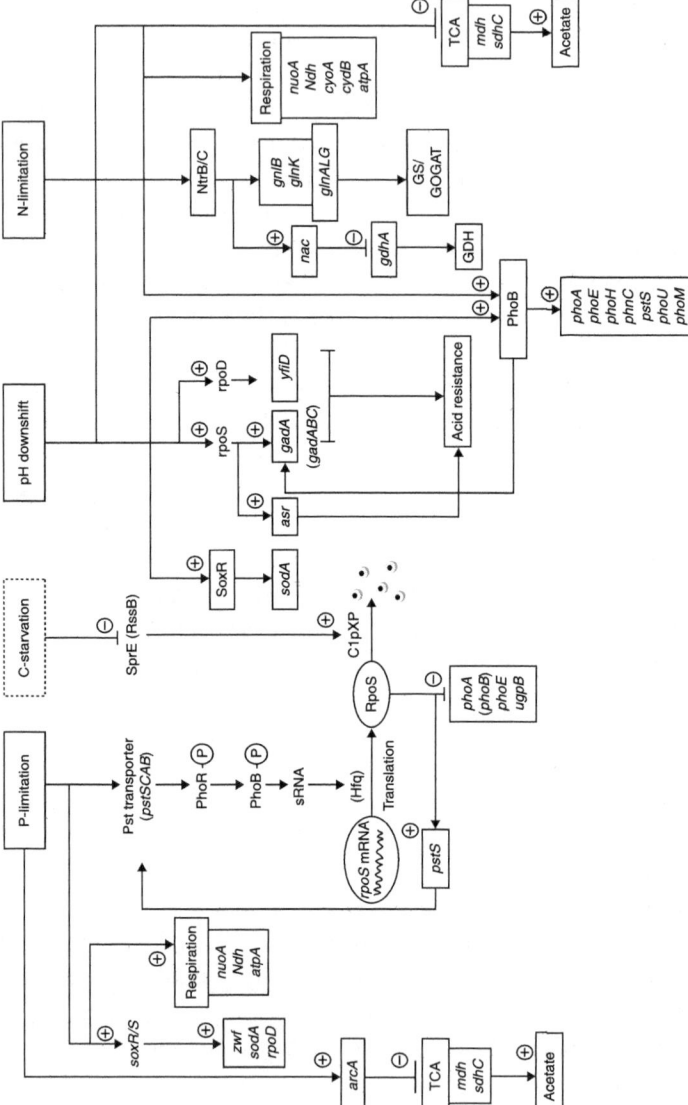

Figure 3.18 Schematic illustration of the interaction among several metabolic regulations

Published by Woodhead Publishing Limited, 2013

3.5 Oxygen level regulation

3.5.1 Effect of oxygen limitation on the metabolism

In addition to nutrient sources, oxygen levels are also important in metabolic regulation. Global regulators, such as Fnr and ArcAB, are mainly responsible for regulation to the availability of oxygen and other electron acceptors in the culture environment, where Fnr regulates the expressions of metabolic pathway genes under anaerobic conditions (Kang et al., 2005), while ArcAB regulates under both anaerobic and microaerobic conditions (Gunsalus, 1992; Alexeeva et al., 2003). Moreover, the genes that encode the primary dehydrogenases, such as *glpD*, *lctPRD*, *aceE,F*, and *lpdA*, are also repressed by ArcA (Iuchi et al., 1990a; Lynch and Lin, 1996b; Cunningham et al., 1998). It has been shown that the ArcA/B system exerts more significant regulation effect on the cell metabolism under microaerobic conditions than under aerobic or anaerobic conditions. The effect of the ArcAB system on the flux distribution at the pyruvate node has been investigated based on the extracellular metabolite concentrations (Shalel-Levanon et al., 2005a,b; Zhu et al., 2006). It is shown that lactate can be overproduced by the *arcA/fnr* double mutant (Zhu et al., 2006) in a similar way to the *pfl* gene knockout (Zhu and Shimizu, 2004, 2005).

Reoxidation of the reducing equivalents, such as NADH generated by the oxidation of the energy source, occurs in the respiratory chain under aerobic or microaerobic conditions. In *E. coli*, NADH is oxidized in the respiratory chain via a coupled NADH dehydrogenage NDH-1 encoded by *nuo* or an uncoupled dehydrogenase NDH-2 encoded by *ndh*, and the electron flows into quinone and the quinol pool. Quinol is then oxidized by either the cytochrome bo or the cytochrome bd terminal oxidase complex, which in turn passes the electrons to the oxygen with concomitant production of water. The two terminal oxidases differ in their affinities for oxygen as well as in their H^+/e^- stoichiometries, where cytochrome bo oxidase has a low affinity for oxygen and translocates two H^+s per e^-, while cytochrome bd has a high affinity to oxygen and translocates one H^+ per e^-. The *cyoABCDE* operon is repressed by both ArcA and Fnr, while *cydAB* operon is activated by ArcA and repressed by Fnr (Alexeeva et al., 2000).

Microbial cells, such as *E. coli*, can generate energy as ATP under a wide range of redox conditions. The reducing equivalents, such as

NADH, are re-oxidized in the respiratory chain, where oxygen, nitrate, fumarate, and dimethyl sulfoxide, etc. are the electron acceptors. This process is coupled to the formation of a proton mortive force (PMF), which is utilized for ATP generation from ADP. In the absence of oxygen, or other electron acceptors, ATP is generated via substrate level phosphorylation through the process of degradation of a carbon source in the metabolic pathways. Under such fermentation conditions, the cell such as *E. coli* excretes such metabolites as lactate, ethanol, succinate, formate (CO_2 and H_2 as well), as well as acetate, where the relative production rates for these metabolites are governed by the demand for redox neutrality. The succinate is formed from PEP via Ppc. Pyruvate serves as a common substrate for pyruvate formate-lyase (Pfl) and the pyruvate dehydrogenase complex (PDHc), and this branch point involves the cleavage of PYR. The expressions of *pfl* genes or the *focApfl* operon, which encode Pfl, are activated by ArcA and Fnr, and become higher at lower oxygen concentrations, whereas *aceE,F*, which encode α and β subunits of PDHc, are repressed by ArcA under oxygen limited conditions. At the branch point of AcCoA, the product of both Pfl and PDHc reactions is converted to either acetate and ethanol, or subsequently undergoes further oxidation in the TCA cycle. The interconversion of Pfl between inactive and active glycyl radical-bearing species occurs at low oxygen concentrations and is controlled by the activities of iron-sulfur protein Pfl activase and the product of the *adhE* gene Pfl deactivase (Kessler and Knappe, 1996; Sawers and Watson, 1998). The active glycyl radical form of Pfl is irreversibly destroyed by molecular oxygen and hence must be either protected from oxygen damage or converted to the inactive, oxygen-stable species during the transition between anaerobiosis and aerobiosis (Wagner et al., 2001; Alexeeva et al., 2000).

3.5.2 Regulation by ArcA/B system

The Arc system, composed of ArcA, the cytosolic response regulator, and ArcB, the membrane bound sensor kinase, regulates the TCA cycle genes, depending on the oxygen level or redox state. The ArcB protein has three cytoplasmic domains: a primary transmitter domain (H1) containing a conserved His292; a receiver domain (D1) containing a conserved Asp576; and a secondary transmitter domain (H2) containing a conserved His717 (Iuchi and Lin, 1992; Ishige et al., 1994; Tsuzuki et al., 1995; Kato et al., 1997). The primary transmitter domain of

ArcB is autophosphorylated at His292 at the expense of ATP (Iuchi, 1993; Georgellis et al., 1997). The phosphoryl group is then sequentially transferred to Asp576 and His717 and from there to Asp54 of ArcA. However, the phosphoryl group on His292 can also be directly transferred *in vitro* to ArcA at a very low rate (Georgellis et al., 1997). The phosphoryl group from His717 can also be transferred to ArcA, but this transfer is regulated by redox conditions (Matsushika and Mizuno, 1998). Namely, upon stimulation by the redox state, ArcB undergoes autophosphorylation, and the phosphoryl group is transferred to ArcA by the His→Asp→His→Asp phosphorelay (Georgellis et al., 1999). Consequently, the phosphorylated ArcA binds to the promoter regions of the TCA cycle and other genes.

It has been reported that ArcA, when phosphorylated, represses the expressions of the genes involved in the TCA cycle and the glyoxylate shunt, such as *glt*A, *acn*AB, *icd*A, *suc*ABCD, *sdh*CDAB, *fum*A, *mdh*, and *aceA,B* (Iuchi and Lin, 1988; Park et al., 1994, 1995, 1997; Lynch and Lin, 1996a). *E. coli* possesses two terminal quinol oxidases in the respiratory chain. The genes *cyo*ABCDE, which encode cytochrome o oxidase that has a low oxygen affinity and mainly functions under aerobic conditions, are repressed by ArcA (Cotter and Gunsalus, 1992). However, the *cyd*AB genes, which encode cytochrome d oxidase that has high oxygen affinity, are activated by ArcA (Iuchi et al., 1990; Drapal and Sawers, 1995; Tseng et al., 1996).

Alexeeva et al. (2003) investigated the effects of different oxygen supply rates on the catabolism in *arc*A mutant. It is shown that the ArcAB system exerts more significant effects on the cell's catabolism under microaerobic conditions than under aerobic or anaerobic conditions. A strong link is demonstrated between redox ratio (NADH/NAD$^+$) and acetate overflow in *E. coli* (Vemuri et al., 2005). It is shown that the commencement of acetate overflow occurs above the critical NADH/NAD$^+$ ratio of 0.06 (Vemuri et al., 2005). Moreover, acetate overflow is delayed by the expression of heterologous NADH oxidase (NOX), an enzyme that serves to reduce the NADH/NAD$^+$ ratio (Vemuri et al., 2005). The redox state has been reported to trigger the Arc regulon (Georgellis et al., 2001; Malpica et al., 2004).

Since phosphorylated ArcA represses TCA cycle genes, the *arcA* gene deletion activates the TCA cycle, resulting in a reduction in the acetate formation (Vemuri et al., 2005). The NADH oxidation by the expression of NOX in the *arc*A knockout gene mutant further reduces the acetate formation, resulting in increased recombinant protein production (Vemuri et al., 2005). Since the TCA cycle is the main source of energy generation and provides important precursors for amino acids, such as glutamate and lysine, it is of practical interest to enhance TCA cycle activity. As stated above, the *arcA/B* genes knockout in *E. coli* transcriptionally activate the TCA cycle and overproduce NADH, which may in turn repress the TCA cycle by allosteric regulation. Moreover, it has been reported that ArcB does not control the TCA cycle under aerobic conditions, due to the fact that oxidized quinone electron carriers inhibit autophosphorylation of ArcB, and it cannot transphosphorylate ArcA (Georgellis et al., 2001).

As expected from regulation, the TCA cycle is activated by the *arcA/B* gene knockout, which then causes a higher NADH/NAD ratio, which in turn represses TCA cycle activity (Nizam and Shimizu, 2009). Vemuri et al. (2006a,b) considered to express the heterologous *nox* gene to oxidize NADH, and in turn activate the TCA cycle, while nicotinic acid and Na nitrate may also activate the TCA cycle (Nizam and Shimizu, 2008).

Although the *arcB* gene knockout may be phenotypically silent under fully aerobic conditions, it may be affected under microaerobic conditions. To investigate the effect of the *arcB* gene knockout on fermentation characteristics, an aerobic batch cultivation was conducted. As shown in Table 3.6, the cell yield of *arcB* mutant is lower as compared to that of the wild type, while the specific glucose consumption rate is higher and the acetate production rate is lower for the former as compared to those of the latter strain. Figure 3.19 shows the gene expressions of the *arcB* gene knockout mutant as compared to its parent strain for the samples taken at 4 h of batch culture, where the TCA cycle genes, such as *gltA*, *icdA*, *fumA*, and *mdh*, are up-regulated compared to the parent strain (Nizam and Shimizu, 2008). The *aceA* gene of the glyoxylate pathway, and the *aceE* and *aceF* genes of PDHc (pyruvate dehydrogenase complex) pathway, are also up-regulated.

Table 3.6 Growth parameters of *E. coli* BW25113 and *arcB* mutant in aerobic batch cultures

	BW 25113	arcB mutant	arcA mutant
Cell yield (g g^{-1})	0.356 ± 0.04	0.344 ± 0.03	0.324 ± 0.04
Specific glucose consumption rate (mmol g^{-1} h^{-1})	1.45 ± 0.02	0.957 ± 0.04	1.38 ± 0.05
Specific acetate production rate (mmol g^{-1} h^{-1})	0.286 ± 0.02	0.14 ± 0.03	0.21 ± 0.03

Metabolic regulation by global regulators

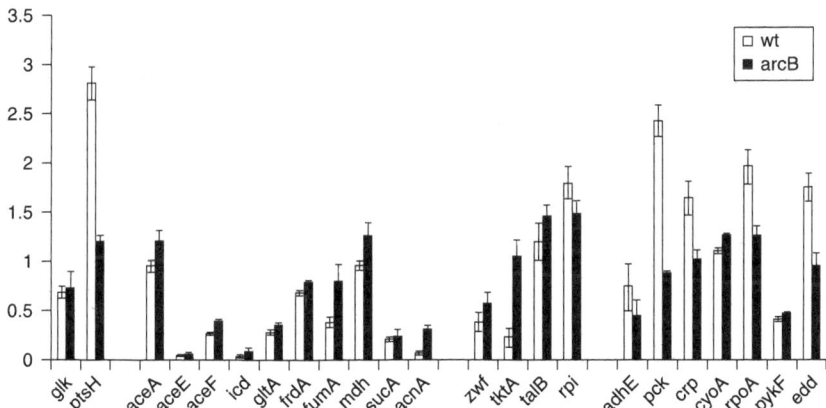

Figure 3.19 Comparison of some gene expressions for parent *E. coli* BW25113 and *arcB* mutant at 4 h of batch cultivation along with the gel picture. Open bars indicate the gene expressions in *E. coli* BW25113 and black bars indicate the gene expressions in *arcB* mutant

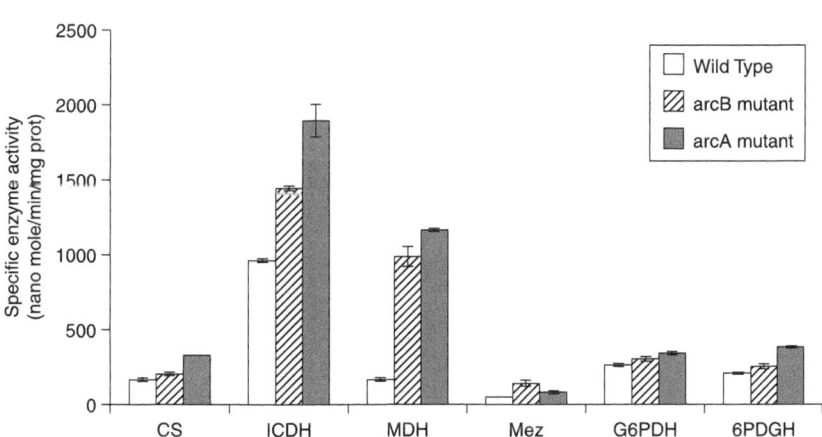

Figure 3.20 Comparison of specific enzyme activities of *E. coli* BW25113, its *arcB* and *arcA* mutant at 4 h of batch cultivation. Open bars indicate the enzyme activities in *E. coli* BW25113, hatched bars indicate *arcB* mutant and black bars indicate *arcA* enzyme activities

Under aerobic or micro-aerobic conditions, TCA cycle related genes, such as *gltA*, *fumA*, *mdh*, and *aceA*, are derepressed, resulting in activation of the TCA cycle enzymes, such as CS, ICDH, and MDH, in the *arcB* gene knockout mutant, as compared with the wild type. The *aceE* and *aceF* genes that code for PDHc, the enzyme that catabolize pyruvate dehydrogenase reactions under aerobic conditions, are up-regulated for the *arcB* mutant (Iuchi et al., 1989; Sawers and Suppmann, 1992; Suppmann and Sawers, 1994; de Graef et al., 1999; Alexeeva et al., 2000), and the increased flux of PDHc for *arcA* mutant. The *zwf* gene is known to be regulated by SoxS and MarA. SoxS can stimulate *zwf* expression for NADPH generation under oxidative stress conditions, and the induction by MarA is related to the superoxide resistance (Li and Demple, 1994; Jair et al., 1995). The reason for the up-regulation of *zwf* gene expression and the up-regulation of G6PDH and 6PGDH in the *arcB* mutant compared to the wild type, may be due to the oxidative stress caused by increased respiration, where SoxRS plays a role under oxidative stress conditions, and regulates such genes as *zwf* (Moat et al., 2002). The increased flux of the PP (Figure 3.23) pathway is consistent with the up-regulation of G6PDH and 6PDGH for the *arcA* mutant (Nizam et al., 2009).

As previously observed, *cyd* gene expression is repressed in *arcA* mutant (Tseng et al., 1996; Shalel-Levanon et al., 2005a,b), which together to TCA cycle activation causes a higher redox state when compared with the wild type. The higher redox state in turn inhibits some of the TCA cycle enzymes, such as ICDH, α-KGDH, and SDH, allosterically.

Since the TCA cycle is the source of energy generation and provides some of the precursors for cell synthesis, the activation of the TCA cycle may lead to improvement of ATP production for the cell growth and/or TCA cycle-related metabolite production in practice. It should be noted that the activation of the TCA cycle reduces the acetate production rate, which is a common obstacle for metabolite production using *E. coli*. However, it should also be noted that the activation of the TCA cycle causes a decrease in cell yield due to higher production rate of CO_2 in the TCA cycle. This may be overcome by activating the glyoxylate pathway by the *fadR* knockout gene (Peng et al., 2006). It is controversial whether cell metabolism is controlled to maximize ATP generation or cell synthesis (Schuetz et al., 2006).

3.5.3 Fnr and Nar systems

Respiration is a fundamental cellular process utilizing different terminal electron acceptors, such as oxygen and nitrate. The ability to sense these electron acceptors is key to the cell's survival. *E. coli* is a metabolically versatile chemoheterotroph grown on a variety of substrates, under various oxygen concentrations with fumarate or nitrate, replacing oxygen as the terminal electron acceptor under anaerobic conditions. Many bacteria utilize oxygen as the terminal electron acceptor, but can switch to other acceptors such as nitrate under oxygen limitation. In *E. coli*, this switch from aerobic to anaerobic respiration is controlled by Fnr (fumarate and nitrate reduction), where it was identified by Sipro et al. (1990). Under oxygen limitation, Fnr binds a $[4Fe-4S]^{2+}$ cluster, and becomes a transcriptionally active dimeric form.

E. coli possesses sensing/regulation systems for a rapid response to the availability of oxygen, redox state, as represented by NADH/NAD⁺ ratio, and the presence of other electron acceptors. These regulation systems channel electrons from donor to terminal acceptors. The pyridine nucleotides, such as NADH and NAD⁺, function as important redox carriers involved in metabolism. These cofactors not only serve as electron acceptors in the breakdown of substrates but also provide the reducing power for the redox reactions in anaerobic and aerobic respirations. A balance for oxidation and reduction of these nucleotides is regulated for catabolism and anabolism, since the turnover of the nucleotides is high compared to their concentrations (de Graef et al., 1999). Under anaerobic conditions, the reoxidation of NADH and the formation of reduced compounds occur, whereas NADH oxidation is coupled to respiration by electron transfer under aerobic or nitrate respiration. In *E. coli*, the genes that code for enzymes specific to respiration and fermentation are mainly under the control of three global regulators, where these exert their effects depending on the redox state of the cell. One of those is Fnr, which is involved in the regulation of gene expressions for fermentation-related enzymes, while the others are the two-component regulatory systems of Nar (nitrate reduction) and Arc (anoxic respiration control).

Metabolic regulation is made by the binding of dimeric Fnr to the promoter regions of the relevant genes with affinities depending on the redox state (Green and Guest, 1993). The ability of Fnr to bind DNA is regulated by a change in equilibrium between monomeric apo Fnr (inactive) and dimeric Fnr (active) *in vivo*. The active form of Fnr binds to DNA to regulate the corresponding genes under anaerobic conditions. Molecular oxygen can oxidize the ion-sulfur cluster of the corresponding region, resulting in monomerization of the protein and subsequent loss of its ability to bind DNA (Kiley and Reznikoff, 1991). Nar plays a role when nitrate is present, and belongs to the two-component redox regulation systems, where it comprises a membrane sensor (NarX), which may act as a kinase causing phosphorylation of the regulator (NarL) under certain conditions. The Nar system activates such genes as nitrate reduction encoding nitrate and nitrite reductases, and represses such genes as fumarate reductase.

As stated above, Fnr and ArcA/B play important roles in regulating the metabolism under anaerobic and microaerobic conditions. However, the detailed regulation mechanism is more complicated, since other global regulators, such as Cra, Crp, RpoS, may play roles.

3.5.4 Effect of fnr knockout gene on the metabolism

In order to clarify metabolic regulation by Fnr, consider the continuous culture at the dilution rate of $0.1\ h^{-1}$ and batch cultures for the wild-type *E. coli* BW25113 and its *fnr* mutant (JW1328) under anaerobic conditions, where a M9 synthetic medium was used (Marzan et al., 2011). The fermentation result indicates that the specific glucose consumption rate is increased, the specific lactate production rate is increased, and the specific production rates of formate, ethanol, and acetate tend to decrease but not as significantly for the mutant as compared to the wild type. Figure 3.21 shows the gene transcript levels for the continuous culture, where it indicates that the *arcA* transcript level decreases, (and the TCA cycle genes, such as *gltA*, *icdA*, *sdhC*, and *mdh*, are increased) for the *fnr* mutant as compared to the wild type. This effect is compounded by the fact that *arcA* transcription

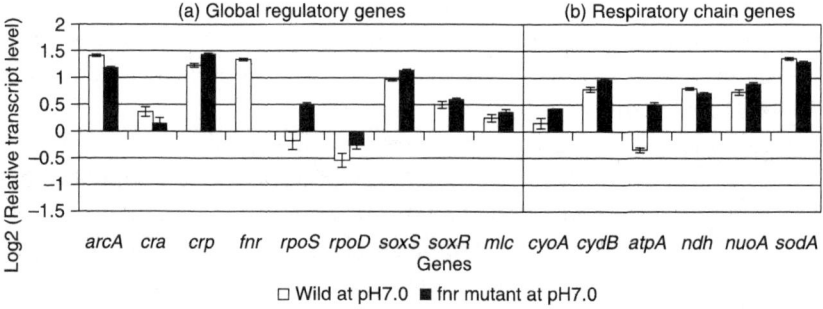

Figure 3.21 Comparison of the transcript levels between wild type and *fnr* mutant under micro-aerobic continuous culture conditions: (a) global regulatory genes, (b) respiratory chain genes

is directly activated by Fnr (Compan and Touati, 1994), which in turn affects the oxidation of quinol (Constantinidou et al., 2006). The increased activities of cytochromes may enhance the oxidation of quinol to quinon, which inhibits the phophorylation of ArcB, and in turn may decrease the phosphorylation of ArcA. The *lpdA* as well as *aceE, F* gene transcript levels significantly increase for the *fnr* mutant as compared to the wild type, which may be due to decreased activity of *arcA*. It was also shown that the *ldhA* gene transcript levels are significantly increased for the *fnr* mutant as compared to the wild type. There might be some relationship between Fnr and Crp. Namely the sequence of the *fnr* gene reveals that it encodes a protein that shows significant homology to CAP/Crp (for catabolic activator protein). However, a number of significant differences between the two proteins have been investigated. Fnr is a monomeric protein, and does not have the conserved group of surface residues that interact with cyclic AMP. It contains an oxygen labile iron-sulfur center as a sensor element for anaerobiosis (Williams et al., 1991; Ziegelhoffer and Kiley, 1995; Lynch and Lin, 1996a,b; Salmon et al., 2003). Several studies have been conducted on the structure and gene sequence for Fnr and Crp proteins. From these studies, it is found that both Fnr and Crp proteins possess almost similar structure and gene sequences. The genes that are controlled by these two global regulators have similar binding sites (Bell et al., 1989; Kiley and Reznikoff, 1991; Williams et al., 1991; Lynch and Lin, 1996a,b; Bo et al., 1998; Unden et al., 2002). Even if some mutation changes the structure of proteins, the mutation in Fnr protein could convert to a Crp protein, and similarly Crp protein could convert to a Fnr protein (Sipro et al., 1990). It may also be considered that both Crp and Fnr proteins can form a heterodimer, which might not allow both of them to function properly

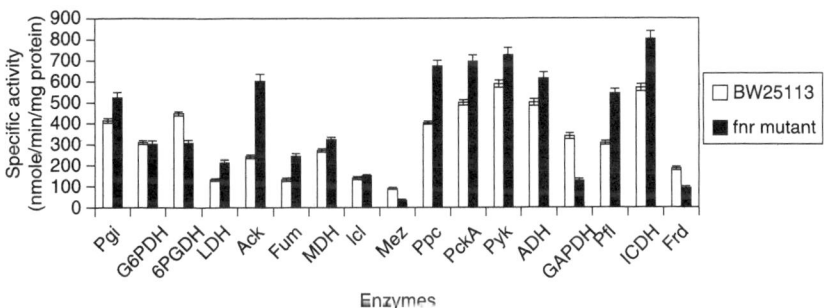

Figure 3.22 Comparison of enzymes activities during micro-aerobic batch culture between (☐) *E. coli* BW25113 and (■) *E. coli fnr* mutant

(Bell et al., 1989; Williams et al., 1991; Ziegelhoffer and Kiley, 1995). Then the absence of an Fnr protein or gene would allow a Crp protein to bind more effectively to the target gene sequence.

Batch fermentation results indicate that the production rate of extra-cellular metabolites, such as acetate, formate, succinate, and ethanol, is reduced, while the production rate of lactate is increased for the *fnr* mutant as compared to the wild type. The enzyme activities are shown in Figure 3.22. Most of the glycolytic enzymes, such as Pgi, GADPH, and Pyk, show higher activities in the *fnr* mutant as compared to its parent strain. The increased activities of GAPDH and ICDH in the mutant are consistent with the results of other researchers (Park et al., 1994, 1995; Chao et al., 1997). Since Fnr is known to act as an activator of the *pfl* gene, significant reduction of Pfl activity is observed for the *fnr* mutant. The reduction of Pfl activity causes increased activity of the other fermentative pathway enzymes, such as LDH in the mutant as compared to the wild type. Other fermentative enzymes, such as Frd which produces succinate from fumarate under anaerobic conditions, are reduced in the mutant, which is consistent with lower succinate production rate in the mutant.

The metabolic flux distributions through the central metabolic pathway of the parent strain and its *fnr* mutant are then estimated based on mass balances. The analysis is performed based on the measurement of the specific rates and the pseudo-steady-state assumption for intracellular metabolites, as explained in Chapter 4. The total number of measured fluxes is 7, and it is higher than the degrees of freedom for the corresponding metabolic network. As a result, the system is an over-determined system. The best estimates for all of the measured and estimated fluxes are then calculated (Tsai et al., 1988), and the comparison of metabolic fluxes between the wild type and the *fnr* mutant is shown in Figure 3.23 (Marzan et al., 2011), where it indicates the lower flux through Pfl, and the increased flux through LDH. The increase in the flux through Ack and decrease in the flux through Frd for the mutant are consistent with enzyme activities. AcCoA has two alternative fates. The energy in the thioester bond can be conserved in the form of ATP by the action of the Pta-Ack pathway, but its formation does not result in the consumption of any reducing equivalents. Alternatively, the energy can be sacrificed by reducing AcCoA to ethanol through two dehydrogenation reactions catalyzed by ADH. The increased activity of LDH in the mutant causes the reduce flux through the ADH pathway.

Since Fnr is known to activate the *frd* and *pfl* genes, the *fnr* mutant produces less succinate and formate. Although *arcA* is known to be activated by Fnr, the regulation mechanism is somewhat complicated. Namely, *cyo* and *cyd* genes are repressed by Fnr, while *cyo* is repressed

Metabolic regulation by global regulators

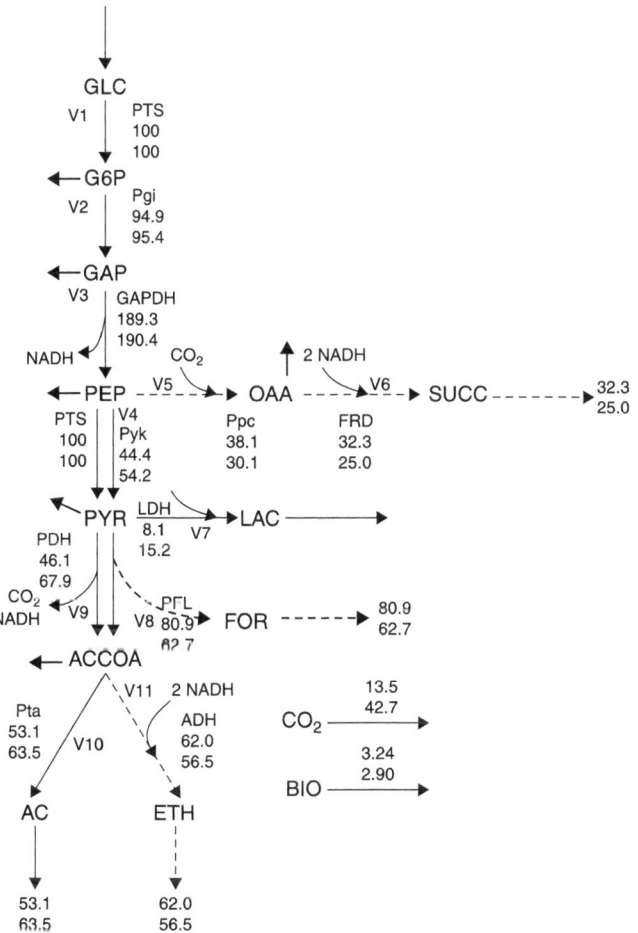

Figure 3.23 Metabolic flux distributions of wild type (upper values) and *fnr* mutant (lower values) under micro-aerobic conditions. The dotted lines indicate the reduced flux. The arrows without destination indicate biomass synthesis

and *cyd* is activated by ArcA. The above result indicates that the *fnr* mutant shows decreased gene expression of *arcA*, and increased gene expressions of both *cyoA* and *cydB*. This implies that the activated cytochrome oxydase increases the quinone pool, which inhibits ArcB phophorylation, and in turn decreases phosphorylation of ArcA, where *arcA* gene expression also decreases due to the *fnr* gene knockout. The down-regulation of *arcA* causes up-regulation of TCA cycle genes as well as *lpdA* and *aceE*, *F*, which code for PDH. Although an indirect effect, the *fnr* mutant causes less growth rate, which causes less biomass concentration, which in turn causes glucose concentration to be increased. This may cause the *cra* gene to be down-regulated and thus activate glycolysis genes, and eventually cause up-regulation of the specific glucose uptake rate. The increased lactate formation may be due to higher NADH/NAD$^+$ ratios caused by the reduced Frd activity, and higher pyruvate concentrations caused by the down-regulation of Pfl and increased flux of Pyk, although PDH activity may be increased to some extent.

3.5.5 Effect of excess oxygen on the metabolism

The microbial cell responds to oxidative stress by inducing antioxidant proteins, such as superoxide dismutase and catarase (Greenberg and Demple, 1989). The well characterized pleiotropic regulators of the antioxidant responses are the OxyR and SoxR proteins (Pomposiello and Demple, 2001). The activation of both proteins results in the transcriptional enhancement of sets of genes whose products relieve the stress by eliminating oxidants and preventing or repairing oxidative damage (Pomposiello and Demple, 2001). SoxR is a member of the MerR family of metal-binding transcription factors, and exists in solution as a homodimer with each subunit containing a [2Fe–2S] cluster. These clusters are in the reduced state in inactivated SoxR, and their oxidation activates SoxR as a powerful transcription factor (Gaudu and Weiss, 1996). The active form of SoxR activates transcription of the *sox* gene. The *sox* gene product, SoxS, belongs to the AraC/XylS family of DNA-binding transcription factors (Amabile-Cuevas and Demple, 1991). SoxS regulates the expressions of more than 17 genes or operons (Liochev and Fridovich, 1992; Gruer and Guest, 1994; Fawcett and Wolf, 1995; Liochev et al., 1999; Gaudu and Weiss, 2000).

Oxygen derivatives, such as superoxide ($O_2^{\cdot -}$), hydrogen peroxide (H_2O_2), and the hydroxyl radical ($\cdot OH$), are usually generated as toxic by-products of aerobic metabolism in a cascade of monovalent reductions from molecular oxygen. Although these are not so reactive *per se*, $O^{\cdot -}$ and H_2O_2 have been reported to cause severe cell damage. H_2O_2, along with Fe^{2+} via the Fenton reaction, produces $\cdot OH$, which can react with any macromolecule such as a protein, membrane constituents, and DNA (Flint et al., 1993; Liochev and Fridovich, 1994). $O_2^{\cdot -}$ exacerbates the Fenton reaction by increasing the intracellular pool of 'free iron', for instance by releasing iron from $O_2^{\cdot -}$-oxidized [4Fe–4S] clusters. $O_2^{\cdot -}$ may also react with nitric oxide (·NO), producing highly reactive peroxynitrite (ONOO⁻), which can generate ·OH. Despite their potential toxicity, reactive oxygen species (ROS) at low concentration have been shown to be actively involved in the cell life and, therefore, should not be entirely eliminated. Potent basic defense systems maintain ROS at harmless levels, but cannot deal with sudden increases in ROS production. This creates an imbalance between production and elimination, referred to as oxidative stress.

Early studies using 2D gel electrophoresis to analyze variations in protein expressions have shown that the synthesis of more than 80 proteins are activated in response to oxidative stress (Greenberg and Demple, 1989). Some of these induced proteins are identified as possessing fundamental antioxidant functions, for example, superoxide dismutase and catalase. The search for mutants with altered antioxidant defenses has led to the isolation and characterization of pleiotropic regulators that operate as redox-regulated genetic switches (Greenberg et al., 1990; Storz et al., 1990; Amabile-Cuevas and Demple, 1991). The best characterized pleiotropic regulators of the antioxidant responses are the OxyR and SoxR proteins (Pomposiello and Demple, 2001). Both proteins have the remarkable ability of directly transducing oxidative signals to genetic regulation. Both proteins are expressed constitutively in an inactive state and are transiently activated in cells under specific types of oxidative stress. Activation of the OxyR and SoxR proteins results in the transcriptional enhancement of sets of genes (regulons), whose products relieve the stress by eliminating oxidants and preventing or repairing oxidative damage (Pomposiello and Demple, 2001).

3.5.6 Effect of soxR/S genes knockout on the metabolism

Here, the effects of *soxR* and *soxS* genes knockout on the central metabolism of *E. coli* are explained, based on fermentation characteristics and gene expressions (Kabir and Shimizu, 2006). Batch aerobic cultivations of *soxR* mutant (JW4024), *soxS* mutant (JW4023), and its parent *E. coli* BW25113 are performed, and the growth parameters are shown in Table 3.7. The specific growth rate is slightly lower for *soxS* mutant (7.9% decrease), but significantly lower for *soxR* mutant (31.6% decrease) compared to the parent strain. The glucose uptake rate is also found to be lower for both *soxR* and *soxS* mutants, as compared to the parent strain. Consequently, biomass yield is lower for mutants, as

Table 3.7 Specific rate of *soxR* and *soxS* mutants, and parent *E. coli* grown on glucose under aerobic conditions. Standard deviations are calculated from four independent measurements

Parameter	BW25113 (Parent)	JW4024 (soxR⁻)	JW4023 (soxS⁻)
μ [h^{-1}]	0.42 ± 0.04	0.27 ± 0.01	0.38 ± 0.03
$Y_{X/S}$ [gDCW.gGLC^{-1}]	0.37 ± 0.01	0.33 ± 0.02	0.34 ± 0.01
q_{GLC} [mM.gDCW^{-1}.h^{-1}]	6.33 ± 0.43	4.50 ± 0.71	6.16 ± 0.55
q_{ACE} [mM.gDCW^{-1}.h^{-1}]	4.16 ± 0.45	4.50 ± 0.44	8.00 ± 0.38
q_{LAC} [mM.gDCW^{-1}.h^{-1}]	0.02 ± 0.01	0.04 ± 0.01	0.09 ± 0.01
q_{CO_2} [mM.gDCW^{-1}.h^{-1}]	0.26 ± 0.03	0.17 ± 0.03	0.22 ± 0.02

μ, specific growth rate; q specific uptake/production rate; $Y_{X/S}$, biomass yield; GLC, glucose; ACE, acetate; LAC, lactate; CO_2, carbon dioxide; DCW, dry cell weight

compared to the wild type. Note that the acetate production rate is significantly increased in *soxS* mutant compared to the parent strain, whereas it is slightly increased in the *soxR* mutant.

The relative expression levels of *pgi*, *crp*, *pfkB*, *fbaA*, *pgk*, *pykA*, *pykF*, *aceE*, and *aceF* involved in the glycolysis are significantly up-regulated in both *soxR* and *soxS* mutants, compared to those in the parent strain (Figure 3.24). Relative expression level of *zwf* is significantly down-regulated in the *soxR* and *soxS* mutants, compared to the parent strain (Figure 3.24). The expression level of the *gnd* gene is significantly down-regulated in the *soxR* mutant, whereas it is slightly regulated in the *soxS* mutant strain. Fermentative genes, such as *pta* and *ackA*, are up-regulated significantly in the *soxS* mutant, whereas the *pta* gene is unchanged and the *ackA* gene is up-regulated in the *soxR* mutant, compared to those in the parent strain.

Among the TCA cycle genes, the relative expression levels of *gltA*, *acnA*, *B*, *icdA*, *sucA*, *sdhC*, and *fumA* are down-regulated, whereas the *mdh* gene is almost unchanged in both *soxR* and *soxS* mutants, as compared to those in the parent strain (Figure 3.29). The down-regulation of these genes is mainly due to the mutation in the *soxRS* system, since the *soxRS* system increases the expression of *fumC* and *acnA* (Hidalgo and Demple, 1996). The expression levels of the genes *secA* and *lepB* are significantly up-regulated, whereas the *dsbA* gene is slightly up-regulated in both the *soxR* and *soxS* mutants (Figure 3.24), indicating that these gene deletions might increase the activity of protein secretion and disulfide bond formation. The transcript levels of *rpoE*, *rpoH*, and *rpoS* (encoding σ^E, σ^{32}, and σ^S, respectively) are up-regulated significantly, whereas *rpoD* (encoding σ^{70}) is almost unregulated in both the *soxR* and *soxS* mutants,

Metabolic regulation by global regulators

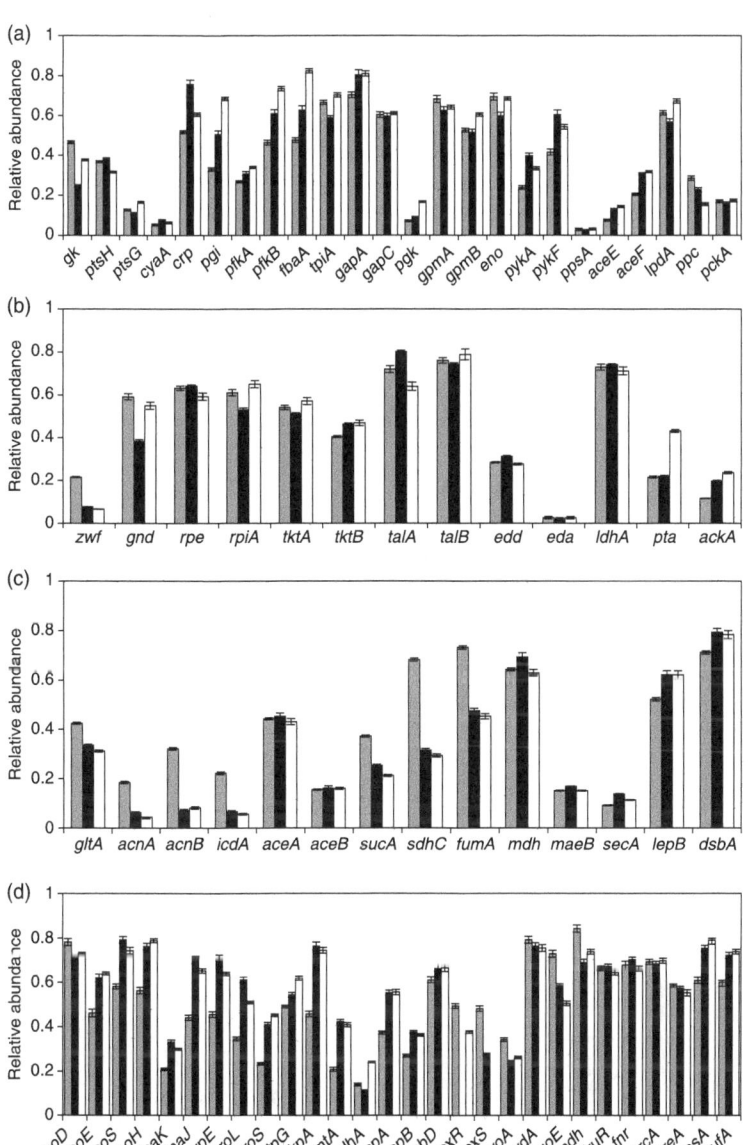

Figure 3.24 Comparison of gene expressions for parent (▨), soxR mutant (■) and soxS (□) mutant E. coli. (a) Transport and glycolytic pathway genes; (b) PP pathway and fermentative; (c) Genes involved in TCA cycle and protein synthesis; and (d) Genes involved in sigma factor, heat shock, NADPH reoxidation, respiration, global, and other regulations

as compared to those in the parent strain (Figure 3.24). The significant up-regulation of heat shock genes, such as *dnaK*, *dnaJ*, *grpE*, *groL*, *groS*, *htpG*, and *fkpA*, in both mutants corresponds to the fact that the heat shock gene expressions are largely proportional to the amount of σ^{32}, though the level of σ^{32} activity is known to be regulated at multiple stages, including translational efficiency and protein stability (Straus et al., 1987). The expression level of the gene *pntA* is also up-regulated significantly in both mutants, and *udhA* is down-regulated in the *soxR* mutant but up-regulated in the *soxS* mutant (Figure 3.2), indicating that deletion of the *soxR* and *soxS* genes affects the transdydrogenase activity. However, expression levels of the genes involved in respiration, such as *cyoA*, *nuoE*, and *ndh*, are down-regulated, except for *cydA* in both the *soxR* and *soxS* mutants.

The specific activities of Pgi and Pfk are up-regulated in both mutants, as compared to the parent strain (Figure 3.25). The enzyme activities of Fba, GAPDH, Pgk, Pyk, and PDH complexes are also up-regulated significantly in the *soxS* mutant, whereas these enzymes are slightly regulated in the *soxR* mutant. The two enzymes involved in the oxidative PP pathway, G6PDH and 6PGDH, which provide NADPH for biosynthesis, are significantly affected in both the *soxR* and *soxS* mutants (Figure 3.25). The activities of G6PDH and 6PGDH decrease in both the *soxR* and *soxS* mutants, compared to the parent stain. The down-regulations of these two enzymes agree with the slower growth rates in both mutants, since these enzymes are known to be under growth rate-dependent regulation (Wolf et al., 1979).

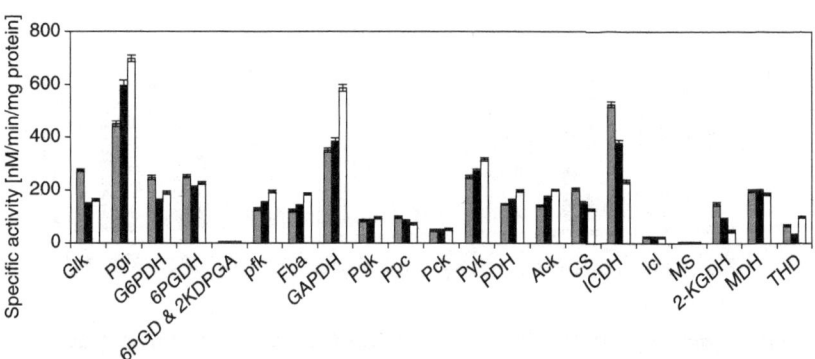

Figure 3.25 Specific enzyme activities in cell extracts of parent (■), *soxR* mutant (■) and *soxS* (□) mutant *E. coli* grown on glucose under aerobic conditions. The mean value from four independent measurements is presented with standard deviation

Metabolic regulation by global regulators

The starting enzyme of the TCA cycle, citrate synthase (CS), which participates in the generation of cellular biosynthetic intermediates and the reduced purine nucleotides used in energy generation via an electron transport-linked phosphorylation reaction, is down-regulated in the *soxR* and *soxS* mutants, as compared to the parent strain (Figure 3.25). This down-regulation causes subsequent decreases in the other TCA cycle enzyme activities, such as ICDH and KGDH, except MDH in both mutants. As a result, a higher specific acetate secretion rate was observed in both mutants, especially in the *soxS* mutant due to the reduced TCA cycle activity. The anaplerotic enzyme Ppc is also found to be down-regulated significantly in both of the mutants compared to the parent strain (Figure 3.25). Ack activity is significantly up-regulated in both mutants compared to the parent strain, consistent with the higher acetate production rate. Moreover, the activity of soluble transhydrogenase (THD) encoded by *udhA* which catalyzes the conversion of NAD^+ and NADPH to NADH and $NADP^+$, is up-regulated in the *soxS* mutant but down-regulated in the *soxR* mutant, compared to those in the parent strain (Figure 3.25), and these results are consistent with the gene expressions (Figure. 3.24).

The activity of Pgi is up-regulated in both the *soxR* and *soxS* mutants, compared to the parent strain, and this result is consistent with the analysis of gene expression levels for *pgi* transcripts. This is mainly due to the down-regulation of the G6PDH enzyme in both of the mutants. The down-regulation of the *zwf* gene in both mutants is also due to the effects of *soxS* and *soxR* genes deletion, since *zwf* is a member of *soxRS* and multiple antibiotic resistance (*mar*) regulons. Thus, unlike *gnd*, *zwf* expression is transcriptionally activated by SoxS during episodes of oxidative stress (Greenberg et al., 1990; Tsaneva and Weiss, 1990).

The *pntA* (membrane bound transhydrogenase) transcripts, which are involved in NADPH generation (Hanson and Rose, 1980), are up-regulated in both the *soxR* and *soxS* mutants. This may be due to down-regulation of NADPH generating enzymes such as G6PDH and 6PGDH in the PP pathway, and ICDH in TCA cycle, since NADPH plays a significant role to reduce oxidative stress (Greenberg et al., 1990). However, *udhA* (soluble transhydrogenase) transcripts, which are involved in the reoxidation of NADPH (Canonaco et al., 2001), are found to be down-regulated in the *soxR* mutant but up-regulated in the *soxS* mutant, compared to that in the parent strain (Figure 3.25), and this result is consistent with the measurement of THD activity (Figure 3.25), indicating that the *soxR* mutant suffers from insufficient re-oxidation of reducing power. This might be one of the reasons why growth of the *soxR* mutant is severely affected.

As mentioned before, acetate production rates are increased significantly in the *soxS* mutant and slightly changed in the *soxR* mutant, as compared to the parent strain. The reason may be due to up-regulation of the glycolytic pathway (down-regulation of PP pathway) and TCA cycle. Moreover, the relative expression levels of *cyoA*, *nuoE*, and *ndh* genes are down-regulated in both mutants, indicating that the *soxR* and *soxS* genes deletion adversely affects the respiratory system and the electron transport chain.

3.6 Acid shock or the effect of pH

The acid barrier of the stomach represents a strong challenge for many pathogenic enterobacteria. Enteric bacteria that cause disease in the human intestine endure transient but extreme acid conditions in the stomach. The normal stomach presents an acid environment at around pH2, with an emptying time of about 2 h (Smith, 2003). Unlike acid sensitive *Vibrio cholerae*, *E. coli*, and *Shigella* have potent acid resistant systems able to withstand a low pH at around 2 for at least 2 h (Lin et al., 1995; Castanié-Cornet et al., 1999). *E. coli* possesses a level of acid resistance rivaling that of the gastric pathogen *Helicobacter Pylori* (Rektorschek et al., 2000). As such, it is important to understand cell metabolism in relation to acidic conditions, from both medical and fermentation points of view. The molecular and physiological response to acid stress has thus been the subject of intense investigation (Foster, 2004; Stincone et al., 2011).

Several acid stress response systems that can protect *E. coli* from acidic conditions have been investigated (Foster, 2004; Richard and Foster,

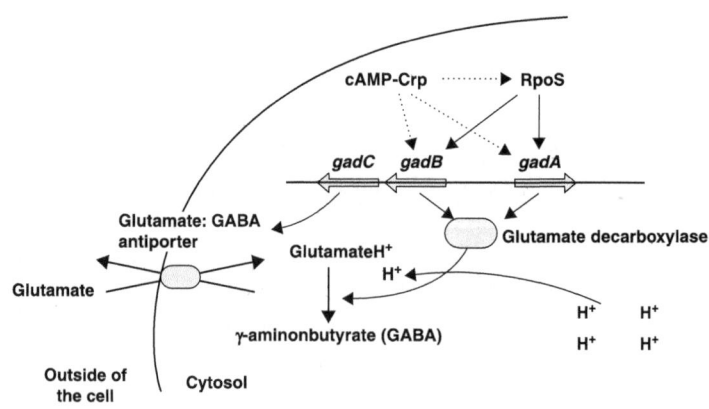

Figure 3.26 The role of glutamate decarboxylase for acid resistance

2003, 2004; Stincone et al., 2011). Some of these depend on the available extracellular amino acids such as glutamate, arginine, and lysine, where the intracellular proton is consumed by the reductive decarboxylation of the amino acid followed by the excretion of the product, such as γ-amino butyric acid (GABA) from cytoplasm to the periplasm by a dedicated antiporter that also imports the original amino acid (Foster, 2004) (Figure 3.26). *E. coli* cells have been demonstrated to exhibit acid resistance by such genes as *gadAB*, which encode glutamate decarboxylase and *gadC*, which encode the Glutamate : GABA antiporter. Glutamate decarboxylase production has been shown to increase in response to acid, osmotic, and stationary phase signals. The *gadA* and *gadB* genes for glutamate decarboxylase isozymes form a glutamate-dependent acid response system, where the process of decarboxylation consumes an intracellular proton and helps maintain pH homeostasis. It has also been shown that there exist similar acid resistant systems using arginine instead of glutamate by arginine decarboxylase, where the antiporter is AdiC in this case (Lin et al., 1995; Castanié-Cornet et al., 1999; Gong et al., 2003; Iyer et al., 2003), and using lysine by lysine decarboxylase (Iyer et al., 2003). Note that cells grown in a media rich in amino acids such as LB are acid resistant (Foster, 2004).

In a typical batch culture, organic acids are most accumulated at the late growth phase or the stationary phase, and it has been known that GadA and GadB proteins increase in response to the stationary phase and low pH (Castanié-Cornet and Foster, 2001). The sigma factor σ^S or RpoS, which increases its amount at the late growth phase and the stationary phase, as well as Crp, are involved in acid resistance (Castanié-Cornet et al., 1999; Foster, 2004). As is implied by the involvement of Crp, the resistance system is repressed when glucose is present. Moreover, it has been shown that FoF_1 proton translocating ATPase is involved in this system (Richard and Foster, 2004). The FoF_1 ATPase is utilized as the protons in the periplasm move into the cytosol across the cell membrane, producing ATP from ADP and P_i by the negative proton motive force (PMF). Since the basic problem of acid stress is the accumulated proton in the cytosol, this proton can be pumped out through FoF_1 ATPase by hydrolyzing ATP and reversed proton movement due to positive PMF at a low pH such as pH2 or 3 (Richard and Foster, 2004). Without amino acids available in the media, this acid response system is activated by utilizing FoF_1 ATPase (Martin-Galiano et al., 2001; Richard and Foster, 2003), where the positive proton motive force (PMF) pumps extrudes protons (H^+) from the cytoplasm with consumption of ATP (Richard and Foster, 2004). Namely PMF is operated in the reverse direction, as compared to producing ATP.

Table 3.8 Regulators involved in regulating glutamate-dependent acid resistance

Protein	Descriptor	Function in acid resistance
RpoD	σ^{70}	Transcription of *gadA/BC*
RpoS	σ^{38}	Transcription of *gadX*
EvgAS	2-component signal transduction	Activates *ydeO* and *gadE* transcription
YdeO	AraC-like regulator	Activates *gadE* transcription
GadE	LuxR-related activator	Required for acid resistance, binds to Gad box, activates transcription of *gadA/BC*, autoactivates *gadE*, represses *ydeO*
GadX	AraC-like regulator	Activator of *gadE*, co-activator of *gadA/BC*, represses *gadW*
GadW	AraC-like regulator	Inhibits RpoS production, activator of *gadE*, can co-activate *gadA/BC* at pH 8
Crp	cAMP receptor protein	Inhibits RpoS production
TrmE	Era-like GTPase	Activates *gadE* mRNA production, stimulates translation of *gadA* and *gadB* mRNA
HNS	Histone-like protein	Negative regulator
TorR	Response regulator of TMAO reductase	Negative regulator of *gadE*

Table 3.8 shows 11 regulators involved in regulating glutamate-dependent acid resistance. In the typical batch culture, the *gadA/BC* loci can be induced during growth in acidic minimal media (pH 5.5) or in the stationary phase regardless of pH (Foster, 2004). However, in complex media such as LB, neither locus is induced until the culture enters into stationary phase.

The expressions of *gadA/BC* genes are under control of GadE and the response regulator RcsB (Castanié-Cornet et al., 2010), where RcsB is part of the RcsCDB phosphorelay, a signal transduction system conserved in members of the *Enterobacteriaceae*. The RcsB can also be activated independently of the phosphorelay, by binding of different co-regulators, such as RcsA (main one), RmpA, TviA, and PhoP. (Castanié-Cornet et al., 2010). As shown in Figure 3.27, EvgS is a sensor kinase and phosphorylates EvgA, where the phosphorylated EvgA activates the *gadE*

Metabolic regulation by global regulators

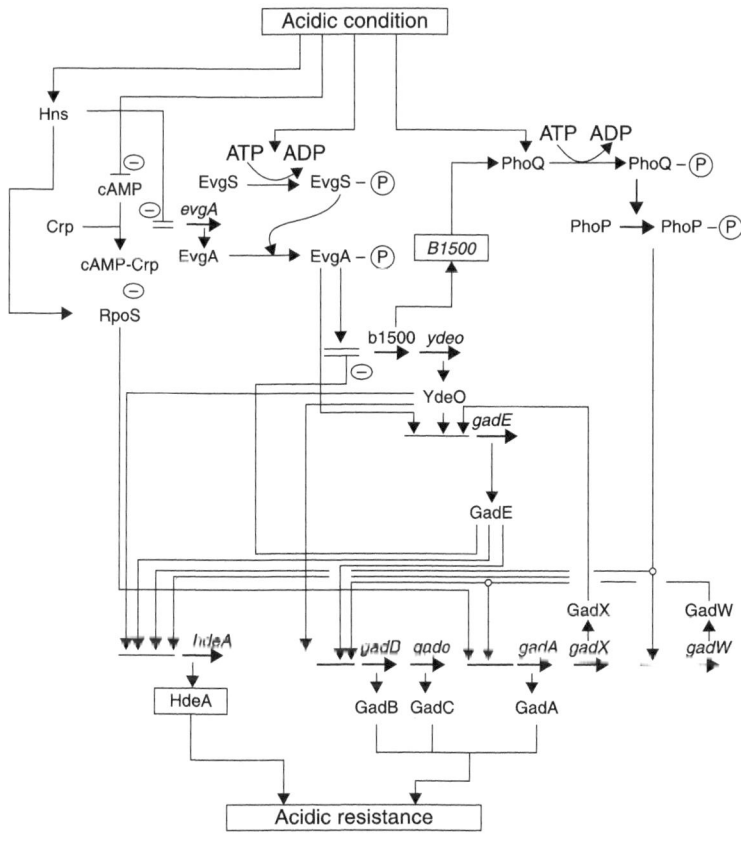

Figure 3.27 Acid resistance mechanism under acidic conditions

gene as well as *ydeO*, where YdeO also regulates the *gadE* gene. It has also been shown that the small membrane protein B1500 connects the signal transduction cascade between EvgS/EvgA and PhoQ/PhoP, where *b1500* is located upstream of *ydeO* and under control of EvgA (Eguchi et al., 2007) and Mg^{2+} turns off the PhoQ/PhoP system (Castelli et al., 2000). Moreover, the phosphorylated PhoP activates *gadW*, and GadW activates *hdeA* and *gadA*, where *hdeA* is under the control of GadW, PhoP-P, and GadE. EvgA regulates at least eight genes related to acid resistance, such as *ydeP*, *b1500*, *ydeO* (Matsuda and Church, 2003).

It has been shown that an acid pH lowers cAMP levels in exponentially growing cells in the minimal glucose medium. This may elevate RpoS that would drive increased expression of *gadX*. However, GadW represses RpoS synthesis under acidic conditions, and in turn GadX synthesis.

GadX, when not repressed by GadW, is acid induced due to changes in cAMP. GadW is also acid induced when it is not repressed by GadX. GadX directly binds to the *gadW* promoter region. GadX and GadW collaborate to repress *gadA* and *gadBC* expressions under alkaline conditions (Ma et al., 2003). The GadX-GadW regulon has also been investigated by DNA microarray (Tucker et al., 2003).

It has also been shown that the two-component system of EnvZ (sensor) and OmpR (regulator) regulates porin expression, where OmpR may be a key regulator for acid adaptation, and thus the *opmR* mutant is sensitive to acid exposure (Stincone et al., 2011). It has been shown that the level of OmpC increases with increased osmolarity when cells are growing in neutral or alkaline media, whereas the level of OmpF decreases at high osmolarity (Sato et al., 2000). It has also been shown that these porin proteins play important roles at acidic conditions.

The acid-inducible *asr* gene is reported to be regulated by the two-component system PhoR/B, which controls the *pho* regulon in response to phosphate starvation, and thus the PhoB-PhoR deletion mutant fails to induce *asr* gene expression (Suziedeliene et al., 1999). It has also been suggested that H$^+$ directly or via its acceptor might activate a sensor protein PhoR in the periplasm (Suziedeliene et al., 1999).

Under anaerobic conditions, additional genes, such as *ackA* and *lpdA*, as well as *hdeA* and *ompT*, are induced. In order to avoid deleterious concentration in the cell caused by the production by the cell at low pH, *ldhA* is induced by acid in order to produce lactate instead of the more harmful acetate plus formate (Bunch et al., 1997).

3.7 Heat shock response

Organisms respond to a sudden temperature up-shift by increasing the synthesis of a set of proteins. This phenomenon is called the heat shock response. The research on the heat shock response of a microorganism contributes to the variety of practical applications, such as temperature-induced heterologous protein production (Hockney, 1994; Hoffmann et al. 2002), and simultaneous saccharification and fermentation (SSF) (Philippidis et al., 2004). Heat shock proteins play a role in the assembly and disassembly of macromolecular complexes such as GroE (Sternberg, 1973), intracellular transport such as Hsp70 (Chirico et al., 1988), transcription such as σ^{70} (Taylor et al., 1984), proteolysis such as Lon (Goff et al., 1984), and translation such as lysil tRNA synthetase (Vanbogelen et al., 1983). The heat shock response in *E. coli* is mediated by $E\sigma^{32}$ (Yura et al., 1993), and it has been shown that at least 26 genes are induced by heat shock (Chuang and Blattner, 1993), where E denotes the RNA polymerase holoenzyme. Among them, *groEL*, *dnaK*, and *htpG* are the genes that code for the major chaperones, such as Hsp 60, Hsp 70, and Hsp 90. ClpP, Lon, and HtrC are involved in proteolysis. DnaK, DnaJ, GrpE, and RpoH are involved in autoregulation of the heat shock response (Tilly et al., 1983; Craig and Gross, 1991; Gamer et al., 1992; Liberek et al., 1992). It has been shown that DnaK prevents the formation of inclusion bodies, by reducing aggregation and promotion of proteolysis of misfolded proteins (Mogk et al., 2003). A bi-chaperone system, involving DnaK and ClpB, mediates the solubilization or disaggregation of proteins (Schlieker et al., 2002). GroEL operates protein transit between soluble and unsoluble protein fractions and participates positively in disaggregation and inclusion body formation. Small heat shock proteins, such as IbpA and IbpB, protect heat-denatured proteins from irreversible aggregation and have been found to be associated with inclusion bodies (Kitagawa et al., 2002; Sorensen and Mortensen, 2005).

Much has been reported on the molecular mechanisms of heat shock proteins (Richmond et al., 1999; Rosen and Ron, 2002). Hoffmann et al. (2002) investigated the metabolic adaptation of *E. coli* during temperature-induced recombinant protein production, and showed that cAMP/Crp-controlled LpdA of the pyruvate dehydrogenase complex (PDHc) and SdhA in the TCA cycle are induced four times, reaching a maximum at 1 h after temperature up-shift. It is also shown that TCA cycle enzymes, such as ICDH and MDH, are initially less produced but regain their respective preshift values about 30 min after the temperature up-shift. Gadgil et al. (2005) investigated the effect of temperature down-shift from 37°C to 33°C and 28°C on gene expressions in *E. coli*. This kind of investigation is useful in analyzing the metabolic changes and investigating the effects of gene modification for strain improvement (Kao, 1999).

Consider how gene expression pattern changes in *E. coli* for the temperature up-shift from 37°C to 42°C in relation to fermentation characteristics (Hassan and Shimizu, 2008). Table 3.9 shows the effect of culture temperature on the fermentation parameters, which indicates that acetate is more accumulated, the cell yield is lower, and the specific glucose consumption rate is lower at 42°C, as compared to the case at 37°C.

Figure 3.28 compares the gene expressions for the two different culture temperatures in continuous culture (Table 3.9), which indicates that the expression of *rpoH* is up-regulated, and the expressions of *dnaK*, *groL*, *groS*, *htpG*, and *ibpB* are up-regulated, which are known to be under the

Table 3.9 Fermentation parameters for the aerobic chemostat culture of the wild type *E. coli* BW25113 at the dilution rate of 0.2 h⁻¹

Fermentation parameters	Culture temperature		% changes
	37°C	42°C	
Specific glucose uptake rate (mmol/gDCW/h)	2.46 ± 0.05	2.28 ± 0.04	−7.31
Specific acetate production rate (mmol/gDCW/h)	0.18 ± 0.02	0.32 ± 0.04	+77.8
Specific CER (mmol/gDCW/h)	5.83 ± 0.05	5.87 ± 0.04	+0.69
Specific OUR (mmol/gDCW/h)	5.74 ± 0.06	5.76 ± .04	+0.35
Biomass yield (gDCW/g substrate)	0.42 ± 0.01	0.34 ± 0.02	−19.0

control of the sigma factor (σ^{32}). Figure 3.28 also shows the up-regulation of *arcA* gene expression, where the *arcA* gene product functions as a repressor of such genes as are involved in the TCA cycle under microaerobic conditions. Figure 3.28 indicates that some of the TCA cycle and glyoxylate pathway genes, such as *icdA*, *sucA*, and *aceA*, are down-regulated. These genes are under the control of ArcA. Figure 3.28 also indicates that the expression of the respiratory chain gene, such as *cydB*, is up-regulated,

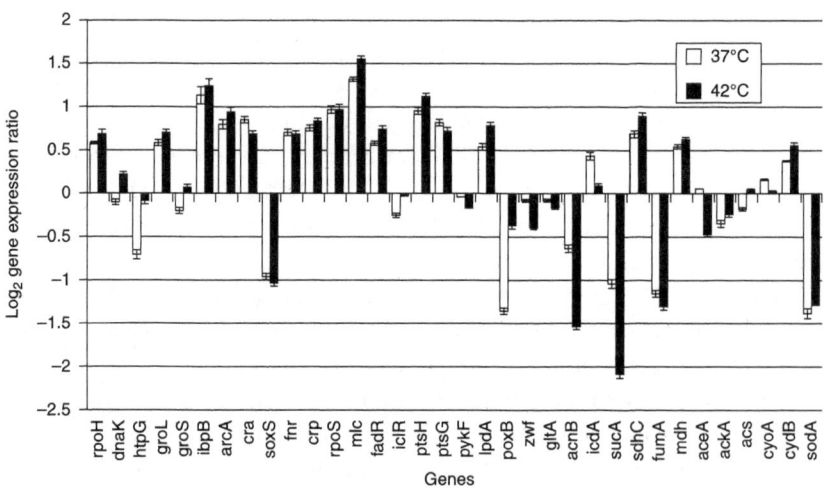

Figure 3.28 Effect of temperature up-shift on gene expressions in *E. coli* BW25113 under aerobic continuous culture at the dilution rate of 0.2 h⁻¹

whereas *cyoA* is down-regulated. This may be due to the up-regulation of *arcA* and its gene product ArcA. Figure 3.28 also indicates that the expression of the *crp* gene, which codes for the cAMP receptor protein Crp, is up-regulated, and the expression of *lpdA*, which is under control of Crp (Appendix A), is also up-regulated (Hoffman et al., 2002). Moreover, *mlc* (making larger colonies) gene expression is higher, and the *ptsG* gene expression is lower. Figure 3.28 also shows that the *cra* gene expression is down-regulated. The gene expressions of *fadR* and *iclR* are also higher, where FadR activates *iclR*, and IclR is known to repress *aceBAK*.

To survive, cells have to control gene expressions precisely in response to changes in their growth environment. Microorganisms, such as *E. coli*, attain this primarily at the transcription level. To control the initiation of the specific transcription, the cell uses diverse mechanisms including various sigma factors. The classical heat shock regulon has been shown to be under the control of the σ^{32} transcription factor, the product of the *rpoH* gene (Gross et al., 1990). The regulation of the sigma factor (σ^{32}) is complex and depends on the feedback control loops involving the *dnaK* chaperone and temperature-induced changes in mRNA secondary structure (Yura and Nakahigashi, 1999). The relative levels of the major heat shock genes, such as *dnaK*, *groS*, *groL*, *ibpB*, *lonA*, and *htpG*, are up-regulated after the temperature up-shift. The expressions of heat shock genes, such as *dnaK*, *groL*, and *ibpB*, increase in the early induction phase (first 10–20 min) and then decline. In *E. coli*, heat shock protein synthesis rates peak at about 5~10 min after the temperature up-shift and then decline to new steady-state levels (Tilly et al., 1986). The heat shock response is made transcriptionally, where it has been known that the RNA polymerase core (E) binds to new initiation subunit σ^{32} (Skelly et al., 1987), and the resulting holoenzyme Eσ^{32} transcribes only heat shock genes (Grossman et al., 1984), which have promoter sequences that differ from those transcribed by E plus σ^{70}, the normal vegetative initiation factor (Cowing et al., 1985). The transcription factor σ^{70} is itself a heat shock protein and the increase in its concentration after heat shock may contribute to its decline in heat shock protein synthesis. Moreover, other heat shock proteins, in particular the *dnaK* gene product, contribute to the shutoff, since the mutations in their genes prolong the high level synthesis of heat shock proteins (Tilly et al., 1983). The heat shock response must be tightly regulated in order to allow rapid changes in heat shock protein synthesis rates. Although the level of mRNA transcribed from the *rpoH* gene increases after heat shock, their increase may be insufficient and too slow to be the sole explanation of the rapid effect of the heat shock. It has been shown that the concentration of active σ^{32} limits the expression of heat shock genes, and that the stability of σ^{32} is modulated (Tilly et al., 1986).

Because of the rapid turn-over (half life <1 min), the cellular concentration of σ^{32} is very low at normal temperature and is limiting for the transcription of the heat shock gene. Upon temperature up-shift, σ^{32} becomes transiently stabilized until the beginning of the shut-off phase of the heat shock response. The heat shock response is induced as a consequence of declining σ^{32} levels and inhibition of σ^{32} activity. Stress-dependent changes in the heat shock gene are mediated by the antagonistic action of σ^{32} and negative modulators, which act upon σ^{32}. These modulators are the DnaK chaperone system, which inactivates σ^{32} by direct association and mediates its degradation by proteases (Arsene et al., 2000). Degradation of σ^{32} is mediated mainly by FtsH and ATP dependent metallo-protease within the inner membrane. The heat shock proteins increased immediately after the temperature up-shift reached a maximum 5–15 min later, and decreased to pre-shift values largely within 1 h, while the maximum induction of many heat-shock proteins, including DnaK and HtpG, was reached at least 30 min later.

The cyclic AMP (cAMP) receptor protein Crp activates transcription for more than 100 promoters. When bound to its allosteric effector cAMP, the Crp homodimer binds to the specific DNA sites near target promoters, enhancing the binding of RNA polymerase holoenzyme (RNAP) and facilitating the initiation of the transcription. The above result indicates that *crp* gene expression increases and *lpdA* gene expression follows a similar pattern after heat shock.

It is shown that *mlc* gene expression follows the same pattern as that of *rpoH* upon heat shock, which confirms that Eσ^{32} is involved in the expression of *mlc* gene. It has been shown that Eσ^{32} plays an important role in balancing the relative concentration of Mlc and EIICB in response to the availability of glucose in order to maintain inducibility of Mlc regulon at higher temperature (Shin et al., 2001). When Mlc is overproduced, it has been known to reduce acetate accumulation (Hosono et al., 1995), and to cause slow growth but give better performance for recombinant protein production (Cho et al., 2005). Mlc is a global regulator of carbohydrate metabolism, and regulates the expression of *pts* operon. It has been shown that Mlc represses *manXYZ* encoding enzyme II of the mannose PTS (Plumbridge, 1998a,b), *malT* encoding the activator of maltose operon, and *mlc* itself negatively (Kim et al., 1999). Moreover, *ptsG* encoding enzyme IICB of the glucose PTS (EIICB^{glc}) and the *pts* operon encoding general PTS proteins, are also known to be repressed by Mlc (Kimata et al., 1998; Tanaka et al., 1999).

The *mlc* promoter is very weak because the nucleotide sequence of the −10 region of the promoter differs from the consensus sequence of the strong promoter of *E. coli*. In addition, Mlc expression is autoregulated by Mlc itself. Therefore, the intracellular concentration of Mlc is limited in *E. coli* (Nam et al., 2001). The *mlc* gene has been known to be transcribed by two promoters, P1 and P2, and have a binding site of its own gene product. It has been shown by *in vitro* transcription assays of *mlc* gene that the P2 promoter could be recognized by RNA polymerase containing the heat shock sigma factor σ^{32} ($E\sigma^{32}$) as well as $E\sigma^{70}$, while the P1 promoter is only recognized by $E\sigma^{70}$. The overall regulation mechanism against heat shock may be expressed as in Figure 3.29.

Let us consider the production mechanism of acetate at higher temperatures. In a typical batch cultivation, cells must switch efficiently from a rapid growth on a favored carbon source such as glucose to a much slower growth on the excreted by-products such as acetate. Acetate excretion occurs through the Pta-Ack pathway, or possibly may be by the Pox pathway. Acetate utilization occurs through AcCoA synthetase (Acs).

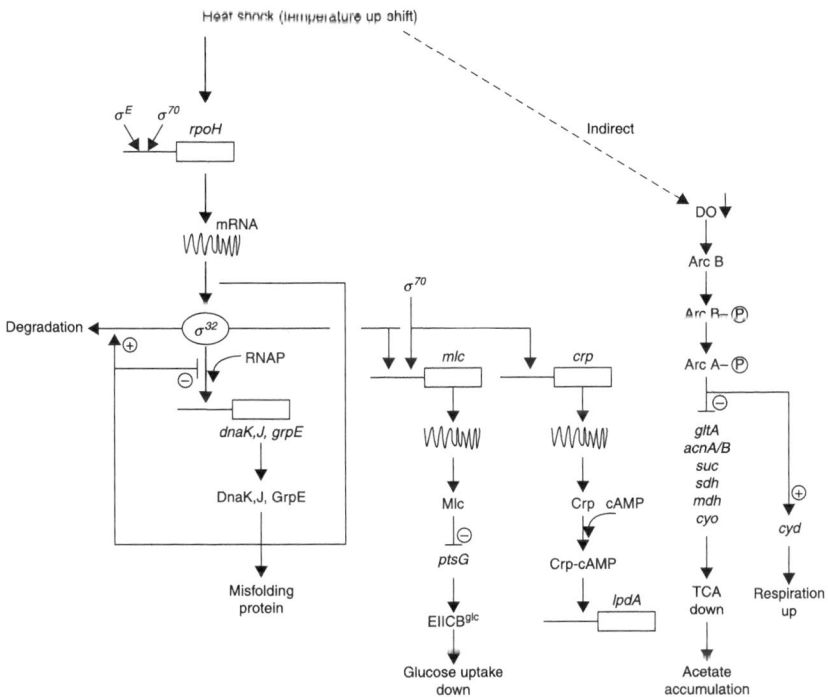

Figure 3.29 Effect of heat shock on gene and protein expressions and the fermentation characteristics

This high-affinity acetate-scavenging enzyme converts acetate to AcCoA, where cells introduce it into the TCA cycle to generate energy and/or the glyoxylate pathway to build cell constituents. The higher expression of *acs* accelerates acetate assimilation in the presence of acetate (Kumari et al., 2000; Lin et al., 2005), which leads to the activation of glyoxylate pathway. Transcription occurs from two σ^{70}-dependent promoters, such as the distal promoter *acs* P1 and proximal promoter *acs* P2 (Kumari et al., 2000; Browning et al., 2002). While multiple factors influence transcription, Crp appears to function directly as the critical transcription factor. Cells control this acetate switch primarily by controlling the initiation of *acs* transcription from the major promoter *acs*P2 (Kumari et al., 2000; Beatty et al., 2003). Activation of *acs* transcription depends on the cAMP-Crp. The cAMP-Crp binds two sites within the *acs* regulatory region. However, it has been shown that Fis and Ihf independently modulate Crp-dependent activation of *acs*P2 transcription (Browning et al., 2004), and the mechanism is not so simple. As such, the activation of *crp* may cause *acs* to be up-regulated. The *acs* gene is also under control of *rpoS*. It has been shown that *acs* is expressed in an *rpoS*-dependent manner during different phases of the growth (Rahman and Shimizu, 2008).

Although cellular ATP may increase for a short period after the temperature up-shift in *E. coli* (Soini et al., 2005), it eventually decreases at higher temperatures (Hoffman et al., 2002; Soini et al., 2005). It has also been reported that the specific CO_2 production rate as well as O_2 consumption rate increases upon temperature up-shift (Hoffmann and Rinas, 2001; Hoffman et al., 2002; Soini et al., 2005). As a result, cell yield decreases and cell maintenance increases (Hoffman et al., 2002; Weber et al., 2002). Although it has been reported that the TCA cycle flux increases upon temperature up-shift at the specific growth rate of 0.08 h^{-1} (Soini et al., 2005), another investigation based on ^{13}C- labeled experiment indicates that the TCA cycle flux becomes low at the dilution rates of 0.45 and 0.32 h^{-1} (Wittmann et al., 2007). It is shown that the repression of TCA cycle genes may be due to up-regulation of the *arcA* gene, while the respiratory activity becomes higher due to up-regulation of the *cyd* gene, which may be caused by up-regulation of the *arcA* gene. The *icdA* and *aceA* genes are known to be repressed by ArcA/B and activated by Cra. The up-regulation of the *arcA* gene expression and down-regulation of the *cra* gene expression both act to repress *icdA* and *aceA* genes, and thus the TCA cycle as well as the glyoxylate pathway is repressed. In accordance with this, the *fadR* and *iclR* expressions are also up-regulated, which are known to repress the *aceBAK* operon. This may cause more acetate accumulation. Moreover, the down-regulation of *cyoA* may limit respiratory activity, while it may be counteracted by the activation of *cydB* gene. It has been reported that respiration is activated during the temperature up-shift (Hoffman et al., 2002). It may be due to activation of *cydB* gene expression, since the K_m value is lower or the affinity to oxygen is higher for Cyd as compared to Cyo. The *arcA* gene shows increased expression after the temperature up-shift (especially first 30 mins) and modulates the expressions of such genes as *cydB*, *cyoA*, and *icdA*. The up-regulation of the *arcA* gene may not be the direct effect of heat shock but indirectly due to lower dissolved oxygen concentration caused by the lower solubility at higher temperature (Soini et al., 2005).

It has been reported that superoxide dismutase gene (*sod*) is induced in response to the oxidative stress imposed by dioxygen or by the redox active compounds such as viologens or quinones caused by the temperature up-shift (Privalle and Fridovich, 1987). It has also been reported that the exposure of a *sodA/B* null mutant *E. coli* to aerobic heat stress caused a profound loss of viability (Benov and Fridovich, 1995; Weber et al., 2002). Moreover, the *sod* gene is under the control of SoxRS, where it becomes significant under dual osmotic and heat stresses (Gunasekera et al., 2008).

3.8 Fatty acid metabolic regulation

The acetate or fatty acid metabolism is also of practical interest. Consider this by looking at the effect of the *fadR* gene knockout on the physiology of *E. coli*, where the parent and its *fadR* mutant are grown on glucose in a minimal medium under aerobic conditions (Peng and Shimizu, 2006). Batch cultivation results are shown in Table 3.10. Compared to the

Table 3.10 Batch cultivation characteristics of the parent and the *fadR* mutant *E. coli* in glucose minimal medium under aerobic conditions

Strain	μ_{max} (h^{-1})	q_{glc} (mmol gDCW^{-1} h^{-1})	$Y_{A/G}$ (g$_{ace}$ gglc^{-1})	$Y_{X/G}$ (g$_{biomass}$ gglc^{-1})
Wild type strain	0.52 ± 0	5.61 ± 0.07	0.33 ± 0.02	0.47 ± 0.01
fadR mutant	0.54 ± 0.01	5.89 ± 0.11	0.24 ± 0.03	0.51 ± 0.01

μ_{max}: the maximum specific growth rate; q_{glc}: the specific glucose consumption rate; $Y_{A/G}$: yield of acetate on glucose; $Y_{X/G}$: yield of biomass on glucose

parent strain, acetate production is reduced in the final concentration, and $Y_{A/G}$ is also reduced, whereas the biomass yield is enhanced in the *fadR* mutant. The specific growth rate and the specific glucose consumption rate of the *fadR* mutant are slightly higher than those of the wild-type strain during exponential growth. This cultivation phenomenon is similar to the results of Farmer and Liao (1997).

Table 3.11 shows the comparison of protein expressions measured by 2DE with MALDI-TOF MS for *fadR* mutant and the wild type, which indicates that the proteins involved in glucose transport and energy metabolism, including PtsH, PtsI, Crr, GapA, Pgk, SucA, FumA, AtpA, AtpC, AtpD, and CyoD, show increased abundance. The levels of amino acids biosynthesis proteins, such as TrpD, AsnB, AroG, and the cell process proteins, such as GreA and RplI, are also higher in the *fadR* mutant. However, the proteins involved in fatty acids biosynthesis pathways, such as AccB and FabD, are down-regulated. Apart from these proteins, the expressions of UspA, the universal stress protein, and DnaK,

Table 3.11 Differentially expressed proteins in the *fadR* mutant *E. coli* compared to the parent strain

Protein name	AC Swiss prot no.	Protein description	Ratio (*fadR*/parent)
Glucose transport and energy metabolism			
PtsI	P08839	Phosphoenolpyruvate-protein phosphotransferase	1.8
PtsH	P07006	Phospho carrier protein HPR	1.5
Crr	P08837	PTS system, glucose-specific IIA component	1.4
GapA	P06977	Glyceraldehyde-3-phosphate dehydrogenase	1.5
Pgk	P11665	Phosphoglycerate kinase	1.3
SucA	P07015	2-oxoglutarate dehydrogenase e1 component	2.0
FumA	P00923	Fumerase hydratase class 1	1.5
AtpA	P00822	ATP synthase alpha chain	1.4
AtpD	P00824	ATP synthase beta chain	1.2
AtpC	P00832	ATP synthase epsilon chain	∞
CyoD	P18403	Cytochrome O ubiquinol oxidase protein cyoD	2.2

Amino acid biosynthesis			
AsnB	P22106	Arsperagine synthetase B	1.4
TrpD	P00904	Anthranilate synthase component II	1.7
AroG	P00886	Phospho-2-dehydro-3-deoxyheptonate aldolase	1.3
Fatty acid biosynthesis			
FabD	P25715	Malonyl CoA-acyl carrier protein transacylase	0.6
AccB	P02905	Biotin carboxyl carrier protein of acetyl coenzyme A carboxylase	0.7
Cell process			
RplI	P02418	50S ribosomal protein L9	1.4
GreA	P21346	Transcription elongation factor greA	1.2
Others			
AceK	P11071	ICDH kinase/phosphotase	1.7
DnaK	P04475	DnaK protein	1.6
UspA	P28242	Universal stress protein	1.4
OppA	P23843	Periplasmic oligopeptide-binding protein	0.6
SgaH	P39304	Probable hexulose-6-phosphate synthase	0.2

the molecular chaperone responsive to heat shock, increases. Both proteins are known to protect the cell from stress conditions. The level of AceK, the bifunctional protein catalyzing phosphorylation/inactivation protein, is higher in the *fadR* mutant. Acetate induced periplasmic transporter OppA shows lower expression levels in the *fadR* mutant as compared to wild type.

Table 3.12 shows the comparison of enzyme activities, where it indicates that Icl and MS, the two enzymes of the glyoxylate shunt, are significantly induced in the *fadR* mutant. CS, Acn, Fum, and MDH were coordinately up-regulated to some extent. However, the activity of ICDH is slightly decreased. Moreover, $NADP^+$-dependent malic enzyme (Mez) is up-regulated, whereas NAD^+-dependent malic enzyme (Sfc) is down regulated in the *fadR* mutant, as compared to the wild type. In addition, Ppc shows lower activity in the *fadR* strain, whereas Pck activity increases.

Table 3.12 Specific enzyme activities in cell extracts of the parent and the *fadR* mutant *E. coli* at the exponential phase grown in glucose minimal medium under aerobic conditions

Enzymes	Activities (nmol mg protein^{-1} min^{-1})		
	Parent	*fadR* mutant	Ratio (*fadR*/parent)
Isocitrate lyase (Icl)	62.3 ± 3.1	231.9 ± 9.5	3.7
Malate synthase (MS)	8.0 ± 0.7	14.9 ± 1.8	1.9
Citrate synthase (CS)	113.0 ± 6.3	144.6 ± 9.1	1.3
Aconitase (Acn)	121.5 ± 7.5	174.7 ± 10.2	1.4
Isocitrate dehydrogenase (ICDH)	1312 ± 59	918.5 ± 50.2	0.7
Fumarase (Fum)	19.7 ± 1.1	64.6 ± 2.8	3.3
Malate dehydrogenase (MDH)	59.3 ± 3.3	75.4 ± 4.3	1.3
NADP+-specific malic enzyme (Mez)	66.3 ± 2.9	112.5 ± 3.6	1.7
NAD+-specific malic enzyme (Sfc)	68.7 ± 1.5	52.8 ± 1.7	0.8
Acetate kinase (Ack)	1075 ± 42.2	767.3 ± 33.5	0.7
Glucokinase (Glk)	123.0 ± 7.1	131.6 ± 7.5	1.1
Glucose phosphate isomerase (Pgi)	2227 ± 56.0	2338 ± 59.2	1.1
Phosphofructose kinase (Pfk)	392.4 ± 13.1	509.7 ± 13.9	1.3
Fructose bisphosphate aldolase (Fba)	857.6 ± 34.3	989.5 ± 47.0	1.2
Triose phosphate isomerase (Tpi)	2474 ± 66.4	2671 ± 59.8	1.1
Glyceraldehyde-3-P dehydrogenase (GAPDH)	136.3 ± 7.4	217.4 ± 8.5	1.6
3-Phosphoglycerate kinase (Pgk)	55.1 ± 4.7	70.6 ± 4.0	1.3
Pyruvate kinase (Pyk)	240.6 ± 13.0	223.8 ± 12.1	0.9
Phosphoenolpyruvate carboxylase (Ppc)	571.5 ± 17.7	473.1 ± 15.2	0.8
Phosphoenolpyruvate carboxykinase (Pck)	78.4 ± 4.2	120.7 ± 9.4	1.5

Glucose-6-P dehydrogenase (G6PDH)	357.6 ± 23.2	439.1 ± 26.1	1.2
6-Phosphogluconate dehydrogenase (6PGDH)	443.2 ± 27.7	465.6 ± 34.0	1.1
Transaldolase (Tal)	218.0 ± 11.5	259.1 ± 15.3	1.2
E-D pathway (Edd and Eda)	6210 ± 110	5490 ± 109	0.9

In the fermentative pathways, Ack activity is reduced in the *fadR* mutant as compared to the parent, which is in agreement with the decrease of acetate production by the *fadR* mutant. These trends of the enzyme activities indicate that the *fadR* mutant utilizes the glyoxylate shunt for the replenishment of OAA for biosynthesis. The activation of the glyoxylate shunt by-passes the TCA cycle and thus prevents the loss of carbons as CO_2 in ICDH and KGDH catalyzed reactions, which leads to the increased utilization of AcCoA.

Table 3.13 shows the comparison of intracellular metabolite concentration for the two strains, where it indicates that PYR and AcCoA concentrations decease, whereas the concentrations of isocitrate (ICIT), αKG, MAL, OAA, and aspartate (ASP) increase in the *fadR* mutant as compared to the parent. These results are in agreement with those reported by van de Walle and Shiloach (1998), who found that the operation of the glyoxylate shunt in *E. coli* BL21 results in accumulation of TCA cycle intermediates and higher biosynthesis fluxes. Similar to PYR, PEP concentration is also reduced in the *fadR* mutant, which may be partially due to the elevated glucose uptake that needs more PEP for PTS. These variations reflect the action of the glyoxylate shunt in the *fadR* mutant. It is also observed that the intracellular concentrations of intermediates in the glycolysis and the PP pathway, such as G6P and 6PG, are reduced in the *fadR* mutant strain, implying the accelerated dissimilation of glucose. Apart from these changes, the ratio of NADPH/$NADP^+$ is lower, while that of NADH/NAD^+ is higher in the *fadR* mutant as compared to wild type.

It has been reported that FadR is a transcriptional regulator with a Helix-turn-Helix motif (van Aalten et al., 2001), regulating metabolic pathways such as the fatty acid biosynthesis and degradation, glyoxylate shunt, and possibly playing a role in the regulation of amino acid biosynthesis directly or indirectly (DiRusso et al., 1992; Gui et al., 1996; Cronan and Subrahmanyam, 1998; DiRusso and Nystrom, 1998). The results show that the *fadR* mutant reduces acetate production and

Bacterial cellular metabolic systems

Table 3.13 Intracellular metabolite concentrations in the parent and the fadR mutant E. coli at the exponential phase grown in glucose minimal medium under aerobic conditions

Metabolites	Intracellular concentrations (mM)		
	Parent	fadR mutant	Ratio (fadR/parent)
AcCoA	0.25 ± 0.026	0.063 ± 0.006	0.3
ICIT	<0.03	0.062 ± 0.007	2.1
α-KG	1.85 ± 0.36	2.22 ± 0.40	1.2
MAL	0.088 ± 0.009	0.38 ± 0.03	4.3
OAA	0.061 ± 0.005	0.17 ± 0.03	2.8
ASP	2.43 ± 0.29	4.47 ± 0.38	1.8
G6P	0.24 ± 0.022	0.16 ± 0.015	0.7
PEP	0.10 ± 0.008	0.08 ± 0.009	0.8
PYR	1.58 ± 0.23	1.027 ± 0.17	0.6
6PG	0.35 ± 0.06	0.33 ± 0.05	0.9
NADH	0.13 ± 0.012	0.12 ± 0.011	0.9
NAD+	2.53 ± 0.188	1.62 ± 0.154	0.6
NADPH	0.032 ± 0.003	0.025 ± 0.003	0.8
NADP+	0.11 ± 0.006	0.10 ± 0.005	0.9

Cell volume: 2.55 µl mg(DCW)$^{-1}$

enhances biomass yield. The activities of Icl and MS, which are involved in the glyoxylate shunt, increase and AceK expression is up-regulated. The overall effect of the *fadR* mutant is illustrated in Figure 3.30 (Peng and Shimizu, 2006).

Induction of the glyoxylate shunt leads to better utilization of AcCoA by increasing the carbon flow through this anaplerotic pathway, which is inferred from the significantly reduced intracellular concentration of AcCoA. The decrease of the intracellular AcCoA pool is therefore suggested to be responsible for the reduced acetate excretion in the *fadR* mutant. In contrast, the higher level of PYR in the parent is directed to acetate due to less TCA cycle enzyme activity. For example, KGDH or accumulation of NADH restricts the carbon flow through the TCA cycle (van de Walle and Shiloach, 1998). Meanwhile, the pools of PEP and PYR are cojointly reduced in the *fadR* mutant due to the draining of carbon into the TCA cycle and the glyoxylate shunt. PEP is known to be a critical

Metabolic regulation by global regulators

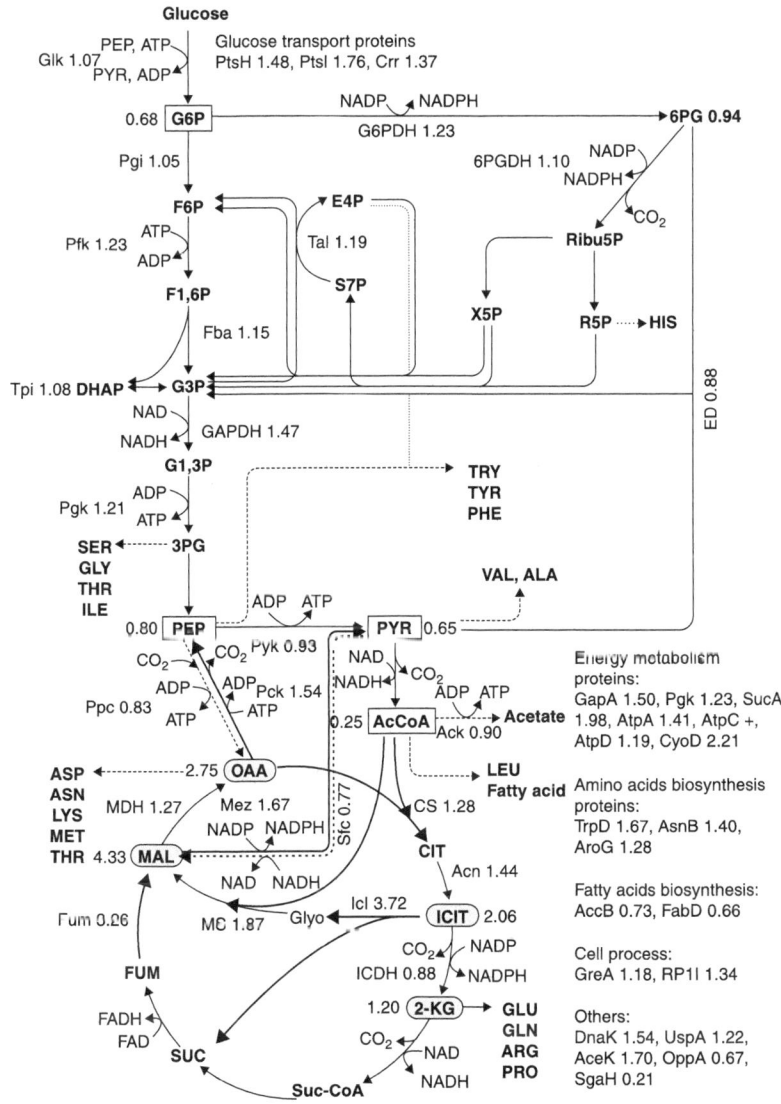

Figure 3.30 Metabolic pathways showing levels of enzymes (or proteins) and intracellular metabolite concentrations in the *fadR* mutant *E. coli* relative to those in the parent at the exponential phase grown in glucose minimal medium under aerobic conditions. The numbers beside the protein names and the metabolites represent the ratios. Oval and boxed metabolites mean increased and decreased intracellular concentrations in *fadR* mutant *E. coli*

metabolite in *E. coli*. It involves not only in the PTS as a phosphoryl donor but also in the regulation of many enzymes as an effector (Fraenkel and Nidhardt, 1999). Therefore it is considered that the up-regulation of the glycolytic enzymes, such as Pgi and Pfk in the *fadR* mutant, are associated with the release from the inhibition due to the lower PEP concentration, since PEP is an inhibitor of both enzymes. Decrease of G6P concentration is responsible for faster glucose uptake in the *fadR* mutant by up-regulation of PTS proteins, since G6P degrades the mRNA of PTS proteins by activating RNaseP enzyme (Morita et al., 2003). Other glycolytic enzymes, such as Fba, Tpi, GAPDH, and Pgk are concurrently up-regulated in the *fadR* mutant to some extent to form more PEP and PYR, which are consistent with D'Alessio and Josse's (1971) results that these enzymes are always regulated proportionally in *E. coli*.

Concomitant with the induction of the glyoxylate shunt, some of the TCA cycle enzymes, such as CS, Acn, Fum, and MDH, are coordinately up-regulated. Besides, SucA, a component of KGDH, and FumA show higher expression levels. These up-regulations are expected to fulfill the role in driving the increased carbon flux due to the action of the glyoxylate shunt. It is reported that CS and Acn, but not ICDH, are regulated in a coordinate mode, which may be due to the fact that citrate is an activator of Acn (Nakano et al., 1998). However, ICDH activity is subject to phosphorylation/inactivation control at the branch of isocitrate to force the carbon flux toward the glyoxylate shunt. The phosphorylation/inactivation of ICDH is exerted by AceK, which is induced in the *fadR* mutant. Decreased carbon flux via ICDH, therefore, restricts the production of NADPH in the TCA cycle, as shown by the NADPH concentration, which is much lower in the mutant than in the parent strain. NADPH is an important cofactor for biosynthesis and mainly formed in the TCA cycle. Up to 60% of the total NADPH is produced in the TCA cycle in the parent strain under aerobic conditions (Hua et al., 2003). To meet the need for biosynthesis, NADPH has to be generated by other NADPH producing pathways. One way is through the NADPH-dependent malic enzyme, Mez, which is up-regulated in the *fadR* mutant. The up-regulation of Mez is probably related to the reduced intracellular AcCoA concentration, as this enzyme is repressed by glucose and AcCoA (Murai et al., 1971). This up-regulation also plays a role in supplying AcCoA from MAL via PYR in the *fadR* mutant. Another route to produce NADPH is via G6PDH and 6PGDH in the PP pathway. Indeed, both enzymes are up-regulated in the *fadR* mutant to some extent. Accelerated cell growth concomitant with the reduction of acetate may cause the shortage of energy in the *fadR* mutant.

Proteome analysis demonstrates that the protein expressions in amino acids biosynthesis, such as AsnB, TrpD and AroG, F_1F_2 proton-translocating ATPase, such as AtpA, AtpC and AtpD, and the ribosomal protein RplI, as well as the transcription elongation factor GreA, are up-regulated in the *fadR* mutant, implying the capacity of biosynthesis of the *fadR* mutant is enhanced. These genes are clustered and show growth rate dependent expression (Yoon et al., 2003). On the contrary, the levels of AccB and FabD involved in fatty acids biosynthetic pathways are negatively affected by the *fadR* mutant, which is consistent with previous studies which indicate that the *E. coli* FadR functions as a repressor of the fatty acid degradative (*fad*) pathways and can also act as an activator of unsaturated fatty acid synthesis (*fab*) (Farewell et al., 1996; Gui et al., 1996; Campbell and Cronan, 2001). In addition, DnaK, the heat shock protein and UspA, the universal stress protein, are induced in the *fadR* mutant. These proteins are known to protect cells from stressful conditions, such as heat shock, starvation, and stress stimulons, thus *fadR* mutation seems to be a stress to the *E. coli* cell. The *uspA* is a member of the *fadR* regulon, and transcription of *uspA* is depressed during exponential growth in *fadR* null mutants (Farewell et al., 1996). Previous studies reveal that the RpoS-regulated genes, periplasmic transporters for amino acids, and peptides and metabolic enzymes, are induced either by acetate or at low pH (Arnold et al., 2001; Kirkpatrick et al., 2001). Of these, it is considered that the down-regulation of OppA is related to the lesser accumulation of acetate in the *fadR* mutant compared to the parent strain.

3.9 Response to nutrient starvation

3.9.1 Metabolic regulation by RpoS

Although many industrial fermentations are conducted in the batch mode, most of the studies have focused on the cell growth phase, and little attention has been paid to the late growth and stationary phases. Since the important metabolites are produced at the early stationary or stationary phases, it is important to clarify the metabolic regulation mechanisms that occur during these phases. During batch fermentation, cultural conditions change from glucose rich to acetate rich conditions, and change further to carbon starvation conditions. The presence of several global regulators, such as RpoD, SoxRS, Cra, FadR, and IclR, have been reported to help *E. coli* to cope with different kinds of metabolic stresses (Cortay et al., 1991; Liochev and Fridovich, 1992; Jordan et al., 1999; Tang et al., 2002; Lu et al., 2003; Varghese et al., 2003; Perrenoud and Sauer, 2005). Apart from these regulators, RpoS, the master regulator of the stationary phase or stress-induced genes in E. coli, regulates such genes as those for the carbohydrate PTS, *crr*, glycolytic pathway genes such as *fbaB* and *pfkB*, the acetate-forming gene *poxB*, the non-oxidative PP pathway genes such as *talA* and *tktB*, and TCAcycle genes such as *acnA* and *fumC*. In addition, some of the genes relating to the amino acid and fatty acid metabolic pathways such as *argH*, *aroM* and *yhgY*, and energy metabolism genes such as *narY*, *appB*, and *ldcC*, have also been identified as being regulated in an *rpoS*-dependent manner (Aronis, 1996; Wei et al., 2000; Aronis, 2002a,b; Lacquor and Landini, 2004; Vijaykumar et al., 2004, Rahman et al., 2006).

E. coli cells are exposed to different stress conditions such as oxidative stress, acid stress, or stresses from particular ion or carbon limitation at different phases of growth. Fortunately, E. coli cells possess several regulatory proteins, which through the regulation of a large number of genes, help the bacteria to cope with a continuously changing environment under different stress conditions, including acid stress, or the other stresses mentioned above (Moat et al., 2002). Of these stress conditions, acid stress, particularly stress from acetate, is one of the major points to be discussed, as this is the major fermentative product of glucose metabolism under aerobic conditions of growth. In addition, acetic acid has been recognized as a problem in recombinant protein production, as it easily passes through the thin lipid layer of the bacterial cell wall and causes damage to protein production (Booth, 1985; Aronis, 1996).

It has been reported that the stress regulatory protein RpoS regulates the expression of approximately 78 genes in E. coli during acid stress (Aronis, 1996; Weber et al., 2005). The glyoxylate pathway is regulated by the global regulators, such as Cra, IclR, and FadR (Gui et al., 1996; Perrenoud and Sauer, 2005). There could be an important relationship among the RpoS and Cra or the other regulators that protect E. coli cells from acid stress, etc.

In general, the bacterial culture medium is considered to be rich in carbohydrate or glucose as the sole carbon source during the exponential growth phase. As the cell such as E. coli utilizes the glucose, acetate is produced as the major fermentative product under aerobic conditions, and the cell exhibits a diauxic shift, which causes termination of the exponential growth phase and stimulation toward the stationary phase. Then E. coli utilizes acetate as a carbon source during the early stationary phase of growth. When acetate is used up, *E. coli* starts to utilize amino acids as carbon and nitrogen sources during the stationary phase. The complex changes occurring among the major metabolic pathway enzymes, their respective genes, and the intermediary metabolites, during a shift from carbon-rich to carbon-limited conditions, have been a major topic of interest in metabolic regulation analyses. RpoS is a well-known global regulator that regulates the expressions of many genes at the onset of the stationary phase or carbon starved conditions and other stress conditions in *E. coli* (Aronis, 2002a; Wei et al., 2000; Vijaykumar et al., 2004).

RpoS is a sigma factor of RNA polymerase. It is known that the core RNA polymerase consists of four subunits, such as two α, one β, and one β'. Part of the RNA polymerase that recognizes the promoter-binding site is generally known as sigma factor (σ). Without this sigma factor, RNA polymerase remains inactive (Maeda et al., 2000). *E. coli* possesses seven different σ factors (Maeda et al., 2000) (Figure 3.31). Depending on the environmental conditions, different sigma units bind with the RNA polymerase so that particular gene expressions are initiated. Of these seven different sigma factors, *rpo*S or σ^{38} has become an important part in bacterial metabolism, as this transcription factor has been shown to be associated with different kinds of stresses in *E. coli* (Maeda et al., 2000; Aronis, 2002).

Unlike other regulators, expressions of which are stimulated by certain effector molecules and these regulators then function by binding to the promoter sites of particular genes, *rpo*S itself is a transcription factor and regulates the expressions of genes at the transcriptional levels. However, once the transcription starts, the sigma factor dissociates from the RNA polymerase (Figure 3.32).

RpoS has been shown to stimulate the expressions of several oxidative stress response genes such as *katE*, *katG*, *sodC*, *dps*, and osmotic stress response genes such as *osmE*, *osmY*. Strains lacking a functional *rpoS* gene also failed to express the genes for acid resistance, such as *gadA* and *gadB*, near-UV resistance gene *nuv*, and acid phosphatase genes *appAR*. (Aronis, 2002). The intracellular level of RpoS itself is regulated by various mechanisms, depending on the stress type and growth conditions. For example, *rpo*S transcription is stimulated by a reduction in growth rate, whereas translation is stimulated by osmotic shock, low temperature, or pH downshifts (Aronis, 2002; Weber et al., 2005) (Figure 3.33). The third mechanism that controls the RpoS level is through proteolysis. While under normal situation, RpoS is rapidly degraded by ClpXP proteases, the proteolytic activity of this enzyme is considerably

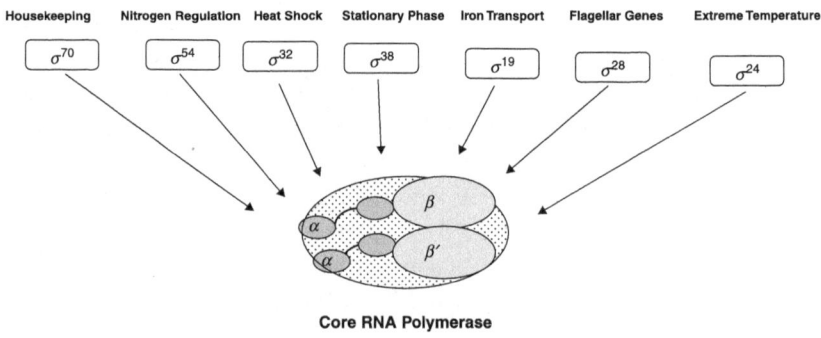

Figure 3.31 Different kinds of sigma factor in *E. coli* (Maeda et al., 2009)

reduced under stress conditions (Lowen et al., 1993; Kusano et al., 1996; Aronis, 2002a).

Although the roles of RpoS are originally described for various types of stress response, it has been demonstrated that the regulatory roles of *rpo*S are not restricted to stress response genes only. In *E. coli*, RpoS-dependent genes are found all over the chromosome, whose function ranges from DNA repair and protein synthesis to the transport, biosynthesis, and metabolism of sugars, amino acids, and fatty acids. Notably, RpoS is found to regulate the expression of DNA repair enzymes, such as the exonuclease encoded by *xth*A and the methyl transferase encoded by *ada*, the gene that determines the cell morphology such as *bol*A, the genes encoding transport and binding proteins, such as *gab*P and *ugp*EC. In addition, a considerably large number of unknown proteins are invariably affected by *rpo*S mutation (Lacour and Landini,

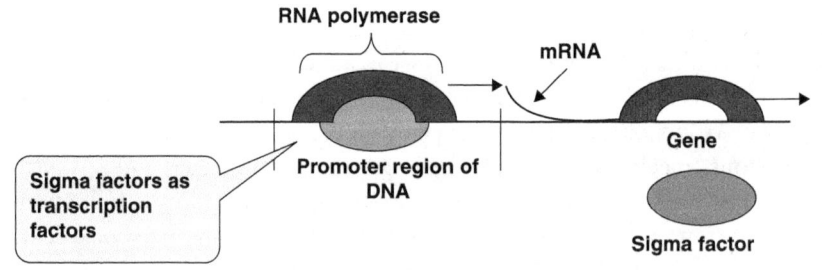

Figure 3.32 Schematic diagram on the function of sigma factor as a transcription factor

Metabolic regulation by global regulators

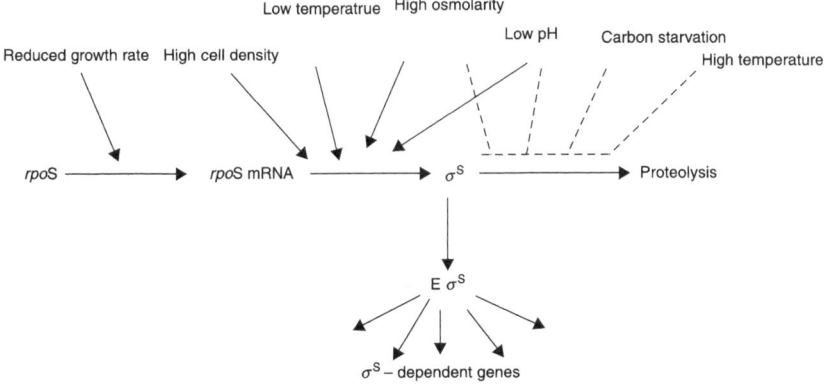

Figure 3.33 Various levels of σ^S regulation are differentially affected by various stress conditions (Aronis, 2002a)

2004; Weber et al., 2005). Considering the wide range of activities of RpoS, it seems obvious that RpoS could make a significant contribution in the maintenance of *E. coli* metabolic pathways at the stationary phase under carbon-starved conditions.

The complexity of the metabolic system of *E. coli* is exemplified by the fact that many metabolic pathway genes are found to be regulated by more than one global regulatory protein. For example, *icd* of the TCA cycle is regulated by RpoD and Cra, *acn*A and *fum*C of the TCA cycle are regulated by SoxRS and RpoS, and *ace*A and *ace*B of the glyoxylate pathways are regulated by Cra and IclR, etc. (Cortay et al., 1991; Liochev and Fridovich, 1992; Jordan et al., 1999; Tang et al., 2002; Lu et al., 2003; Varghese et al., 2003). Moreover, the metabolic pathway of *E. coli* consists of many genes that possess iso-genes. These iso-genes are known to encode back-up enzymes in response to certain environmental stimuli, and the expressions of these enzymes are often regulated by one or more of the global regulators. For example, the *fum*C and *acn*A genes of the TCA cycle are known to encode back-up enzymes at the stationary phase of growth and, as mentioned earlier, they are regulated by both *rpo*S and *sox*RS (Cunningham et al., 1997). It appears that the stress inducible metabolic pathway genes constitute the functional units through which different global regulators co-ordinate metabolic activities in the face of changing culture environment.

In *E. coli*, transketolase is encoded by *tkt*A and *tkt*B genes, and transaldolase is encoded by *tal*A and *tal*B genes, respectively. TktA and TalB are reported to be the major enzymes to catalyze transketolase and transaldolase reactions, and TktB and TalA are the minor enzymes of the non-oxidative PP pathway (Zhao and Winkler, 1994; Sprenger et al., 1995). The non-oxidative PP pathway is important as E4P, the important precursor of the aromatic amino acids, and S7P, which is the cell wall components of *E. coli*, are synthesized only through the non-oxidative branch. E4P and S7P are produced through the consecutive reactions catalyzed by Tkt and Tal (Datta and Racker, 1961; Sprenger et al., 1995). While the physiological roles of the major and minor enzymes have been elucidated, the reports on the positive regulation of the minor transketolase (encoded by *tkt*B) and transaldolase (encoded by *tal*A) of the non-oxidative PP pathway by the stress regulator *rpoS*, indicate that these genes might play a significant role under carbon starved conditions (Lacour and Landini, 2004; Jung et al., 2005; Weber et al., 2005).

However, the TCA cycle genes, such as *fum*C and *acn*A, encode FumC and AcnA enzymes of the TCA cycle, respectively. While fumarase catalyzes the conversion of FUM to MAL by the hydration reaction, Acn catalyzes the conversion of CIT to cis-aconitate to ICIT through dehydration and hydration reactions (Stryer, 1988). However, both FumC and AcnA have iso-enzymes such as FumA and AcnB, which are encoded by the genes *fum*A and *acn*B, respectively. It has been reported that FumA and AcnB play major roles of fumarase and aconitase when the cell grows under optimum growth conditions. Both FumA and AcnB possess Fe-S clusters that make these enzymes vulnerable to oxidative stress, and under such conditions, FumC and AcnA act as back-up enzymes (Jordan et al., 1999; Chen et al., 2001; Tang et al., 2002; Varghese et al., 2003). It has been reported that RpoS regulates the expressions of *fum*C and *acn*A at the stationary phase of growth. It is also reported that the expressions of both *fum*C and *acn*A genes are also regulated by the oxidative stress regulators SoxRS (Park and Gunsalus, 1995a; Krapp et al., 2002).

In summary, RpoS plays an important role at the late growth phase and the stationary phase, as explained in Chapter 2 (see Section 2.3) (Rahman and Shimizu, 2008; Rahman et al., 2008).

3.9.2 Effect of rpoS gene knockout on the metabolism

Figure 3.34 shows the comparison of the batch cultivation characteristics for the wild-type (*E. coli* BW25113) and the *rpoS* mutant (JW2711) strains, where it indicates that the mutant cells enters the stationary phase

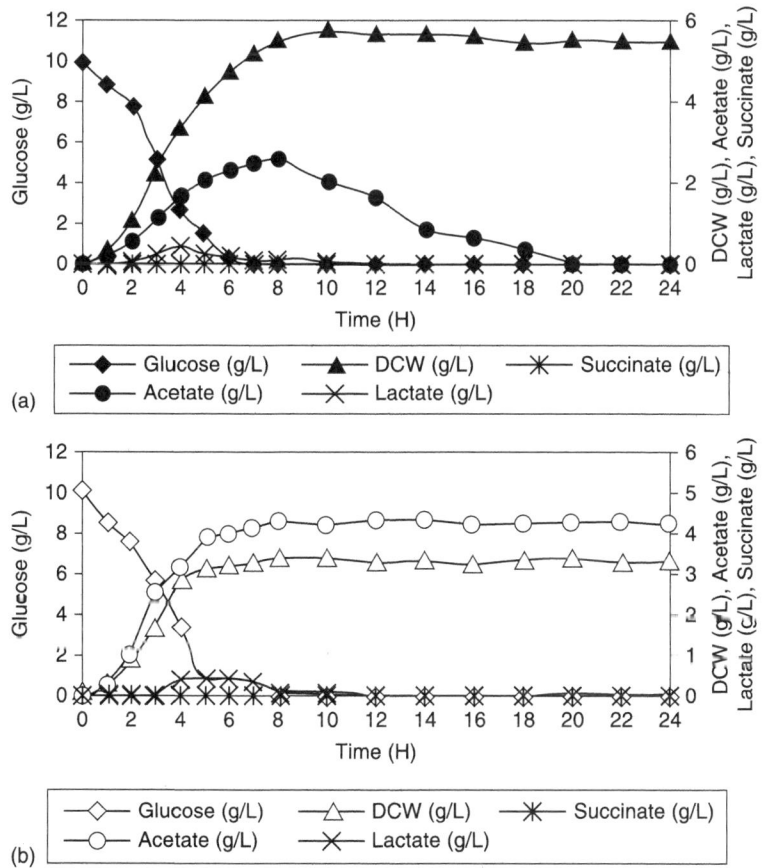

Figure 3.34 Growth curves of: (a) *E. coli* BW25113 (parent strain); and (b) *E. coli* JWK 2711 (*rpoS* mutant). Strains were grown for 24 h in LB media containing 10g/L of glucose at 37°C, at pH 7.0 under aerobic conditions

much earlier than the wild-type cells. Although the glucose consumption by both strains is similar, acetate accumulation/consumption patterns are significantly different. The extracellular acetate concentration at the mid-exponential phase is more than 2-fold higher in the case of the mutant compared to that of the wild type. Moreover, the acetate consumption rate is significantly lower for the mutant throughout the cultivation period. In contrast, lactic acid is found to be utilized by the mutant during the stationary phase. The growth parameters of the wild type and *rpoS* knockout mutant are given in Table 3.14. As shown in the table, biomass yield is significantly lower in the mutant, and the specific O_2 uptake rate

(OUR) and the specific CO_2 evolution rate (CER) are considerably higher for the mutant compared to the wild-type strain.

In order to determine how the mutant copes with the absence of a vital gene such as *rpoS*, it may be identified that such genes are up-regulated at the early stationary phase from the microarray data (Rahman et al., 2006). A total of 208 genes are up-regulated in the mutant, excluding the genes for hypothetical proteins. Among these genes, 25 genes (~12%) are up-regulated in both phases of growth. Microarray data reveals that, of the central metabolic pathways, significant reduction of the expression is observed for several genes during the early stationary phase, with the exception of *fum*C, which is up-regulated. Interestingly, at the exponential growth phase, a substantial up-regulation of the TCA cycle genes is observed. An exceptionally higher level of expression of certain acetate producing genes such as *cys*DEK and down-regulation of acetate utilization pathway genes such as *acs*, *aceBAK*, coincide with the accumulation of acetate throughout the fermentation period. However, the lactate permease gene is up-regulated, thereby making lactate available for use during the stationary phase. Among the genes for amino acid metabolism, the *rpoS* mutation results in significant down-regulation of certain genes involved in the metabolic pathways of valine, isoleucine, serine, asparagine, methionine, and arginine, whereas some of the genes for alanine, leucine, threonine, cysteine, lysine, histidine, and glutamate metabolism are up-regulated during early stationary phase.

A significant down-regulation is also observed for molybdenum uptake transport protein, vitamin B_{12}, and ferric ion transporter proteins, and

Table 3.14 Growth parameters of *E. coli* BW25113 (parent strain) and *E. coli* JW 2711 (*rpoS* mutant) under aerobic growth conditions in LB media

Growth parameters	BW25113	JW2711
Maximum specific growth rate (μ)	0.57 ± 0.05	0.57 ± 0.04
Specific glucose uptake rate (mM/g DCW/h)	5.8 ± 0.03	6.09 ± 0.05
Specific acetate Production rate (mM/g DCW/h)	20 ± 0.05	54.51 ± 0.03
Biomass yield (g DCW/g Substrate)	0.55 ± 0.05	0.34 ± 0.04
Specific CER (mM/g DCW/h)	17.56 ± 0.05	22.34 ± 0.06
Specific OUR (mM/g DCW/h)	18.11 ± 0.06	21.55 ± 0.05

most of the glucose or carbohydrate transport proteins of the phosphotransferase (PTS) system. Expressions of several important enzymes of fatty acid biosynthesis and phospholipid biosynthesis are suppressed in the mutant, whereas some of the genes involved in fatty acid degradation are up-regulated. In contrast to this situation, the key fatty acid degradation regulator *fadR* is also up-regulated during the stationary phase. Table 3.15 shows the expression levels obtained by RT-PCR, which are consistent with the corresponding data in the microarray analysis, as shown in the parentheses.

In order to further investigate the metabolism of the *rpoS* mutant at the protein level, the specific activities of several enzymes of the central metabolic pathways are shown in Table 3.16. The enzymes catalyze three important reactions in the glycolytic pathway, such that Pgi, Pfk, and Pyk are down-regulated in the early stationary phase, whereas these activities are higher in the exponential growth phase compared to the parent strain. However, the relative activity of Ppc is higher, even during the stationary phase. The enzyme activities of two of the TCA cycle enzymes, such as ICDII and MDH of the mutant, are not significantly different from the parent strain at the early stationary phase, but these are

Table 3.15 Growth parameters of *E. coli* BW25113 (parent strain) and *E. coli* JW2711 (*rpoS* mutant) under aerobic growth conditions in LB media

Genes	Ratio (JW2711/BW25113)	
	Exponential phase	Early stationary phase
Glycolytic pathway		
ptsH	1.5 ± 0.036 (1.789)	0.67 ± 0.014 (0.669)
ptsG	1.2 ± 0.019 (1.092)	0.24 ± 0.003 (0.187)
ptsI	1.58 ± 0.017 (1.578)	0.17 ± 0.011 (0.106)
glk	1.47 ± 0.021 (1.765)	1.9 ± 0.006 (1.956)
pgi	1.5 ± 0.025 (1.565)	0.24 ± 0.004 (0.184)
pfkA	2.3 ± 0.011 (2.218)	0.09 ± 0.001 (0.002)
tpiA	2.0 ± 0.011 (1.921)	0.4 ± 0.007 (0.237)
pykA	1.2 ± 0.013 (1.210)	0.8 ± 0.017 (0.834)
pykF	1.0 ± 0.005 (0.789)	0.08 ± 0.001 (0.058)
eno	1.2 ± 0.007 (1.223)	0.03 ± 0.001 (0.021)
ppc	1.2 ± 0.015 (1.197)	1.9 ± 0.042 (2.863)

(*Continued*)

Table 3.15 Growth parameters of *E. coli* BW25113 (parent strain) and *E. coli* JW2711 (*rpoS* mutant) under aerobic growth conditions in LB media (*Continued*)

Genes	Ratio (JW2711/BW25113)	
	Exponential phase	Early stationary phase
Pentose phosphate pathway		
zwf	1.2 ± 0.017 (1.197)	1.2 ± 0.011 (1.230)
gnd	1.4 ± 0.009 (1.368)	1.8 ± 0.003 (1.810)
talA	0.5 ± 0.001 (0.368)	0.08 ± 0.013 (0.086)
talB	2.5 ± 0.025 (1.796)	1.7 ± 0.003 (1.22)
tktA	1.3 ± 0.011 (1.144)	1.4 ± 0.005 (1.366)
tktB	0.3 ± 0.005 (0.315)	0.08 ± 0.001 (0.058)
Gluconeogenic pathway		
sfcA	1.7 ± 0.019 (1.671)	9.25 ± 0.259 (21.597)
maeB	1.6 ± 0.032 (1.38)	0.85 ± 0.035 (0.87)
TCA cycle and glyoxalate shunt		
gltA	4.5 ± 0.057 (4.842)	0.58 ± 0.054 (0.23)
icdA	5.5 ± 0.179 (7.144)	1.2 ± 0.007 (1.210)
mdh	4.0 ± 0.027 (4.296)	0.25 ± 0.004 (0.263)
sdhC	5.3 ± 0.09 (6.421)	0.2 ± 0.09 (0.210)
Glyoxylate shunt		
aceA	0.8 ± 0.015 (0.765)	0.3 ± 0.031 (0.368)
aceB	1.79 ± 0.011 (1.789)	0.09 ± 0.024 (0.034)
Global regulators		
rpoD	1.42 ± 0.007 (1.231)	0.09 ± 0.001 (0.002)
soxR	0.43 ± 0.009 (0.355)	0.32 ± 0.04 (0.280)
soxS	1.5 ± 0.006 (1.236)	0.24 ± 0.07 (0.165)

Values in brackets represent the microarray data of the respective genes

significantly higher at the exponential growth phase. The enzyme activity of NAD^+ specific malic enzyme of the mutant is significantly higher at the early stationary phase. It can be seen that PoxB is active only in the wild-type strain during the early stationary phase of growth, while in the mutant the enzyme activity is not detected in both phases of cell growth.

Table 3.16 Ratio of specific activities of enzymes of *E. coli* BW25113 (parent strain) and *E. coli* JW2711 (*rpoS* mutant) during exponential and early stationary phases of growth

Enzymes	Specific activity at exponential phase (µmole/min/mg of protein)		Specific activity at early stationary phase (µmole/min/mg of protein)	
	BW25113	JWK2711	BW25113	JW2711
Glycolytic pathway				
Pgi	3.52 ± 0.005	4.96 ± 0.09	2.56 ± 0.008	1.1 ± 0.005
Pfk	4.39 ± 0.003	10.91 ± 0.005	0.24 ± 0.07	0.07 ± 0.002
Ppc	0.23 ± 0.01	0.28 ± 0.04	0.11 ± 0.01	0.26 ± 0.003
Pyk	6.28 ± 0.07	7.17 ± 0.1	0.91 ± 0.013	0.38 ± 0.009
Pentose phosphate pathway				
G6PDH	0.15 ± 0.18	0.2 ± 0.1	0.12 ± 0.006	0.2 ± 0.02
PGDH	0.15 ± 0.06	0.21 ± 0.09	0.09 ± 0.001	0.18 ± 0.02
TCA cycle and glyoxylate shunt				
ICDH	3.45 ± 0.09	7.02 ± 0.04	1.77 ± 0.011	2.47 ± 0.09
MDH	0.22 ± .05	0.44 ± 0.09	1.99 ± 0.08	1.54 ± 0.002
ICl	0.048 ± 0.08	0.04 ± 0.05	0.68 ± 0.04	0.19 ± 0.03
Gluconeogenic pathway				
Fbp	0.48 ± 0.011	0.59 ± 0.09	0.63 ± 0.14	1.16 ± 0.13
PckA	0.39 ± 0.08	0.59 ± 0.1	1.46 ± 0.07	0.89 ± 0.15
Mez (NAD+ specific malic enzyme)	0.22 ± 0.13	0.36 ± 0.015	0.03 ± 0.09	0.31 ± 0.011
Malic Enzyme (NADP+ specific malic enzyme)	0.07 ± 0.07	0.10 ± 0.05	0.31 ± 0.06	0.33 ± 0.01
Fermentative pathway				
LDH	1.75 ± 0.11	2.79 ± 0.09	0.82 ± 0.09	0.69 ± 0.14
Ack	2.37 ± 0.09	3.13 ± 0.06	1.42 ± 0.07	0.71 ± 0.09
PoxB	ND	ND	0.11 ± 0.05	ND

ND: not detected

Intracellular metabolite concentrations are also important indicators of carbon flow and metabolic regulation in bacteria. The intracellular metabolite concentrations of four important intermediates of the central metabolism, such as AcCoA, OAA, PYR, and PEP, are shown in Figure 3.35, which indicates that the intracellular AcCoA concentration is very low in the mutant as compared to the wild strain during both phases of growth. The concentrations of three other metabolites in the mutant are lower than the wild-type strain in the exponential phase of growth, but these are increased at the early stationary phase.

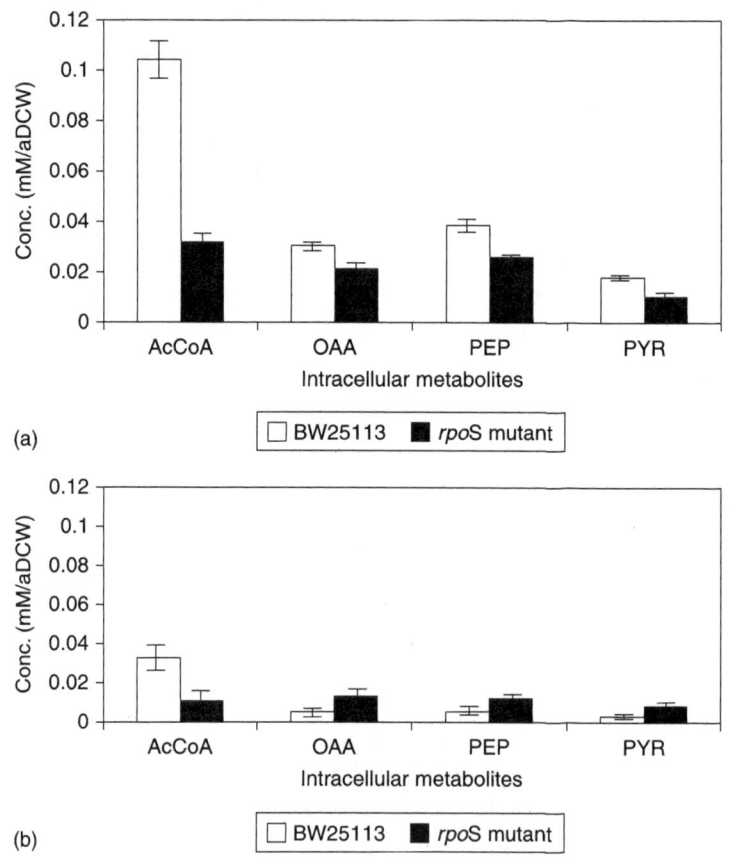

Figure 3.35 Intracellular metabolite concentrations of *E. coli* BW25113 (parent strain) and *E. coli* JWK 2711 (*rpoS* mutant) at: (a) exponential; and (b) early stationary phases of cell growth

When the cell enters into the stationary phase of growth at low glucose concentrations, the *rpoS* gene is known to cause up-regulation of the expression of the carbohydrate metabolism genes of the PTS, such as *crr*, or the glycolytic enzymes encoded by *fbaB* and *pfkB* (Lacour and Landini, 2004). Complying with these observations, several genes, such as *fba*, *pgi*, *pfkB* of the carbohydrate utilization pathway, are down-regulated in the *rpoS* mutant during the early stationary phase of growth. Because of the low level of substrate in the medium at the stationary phase, reduced activity of the glycolytic pathway is expected in the parent strain, and a further decline of this activity in the mutant implies the role of *rpoS* in the regulation of this pathway. The down-regulation of several genes, such as *glgA* and *glgS* involved in glycogen synthesis, is also an indicator of the bacterial adaptive response to carbon-limited conditions in the absence of the *rpoS* background.

Among the PP pathway genes, significant down-regulation is observed for the FAD and FMN synthesizing gene, *ribA*, during the early stationary phases of growth in the mutant, where *ribA* encodes GTP cyclohydrolase II that catalyzes the first committed step in riboflavin biosynthesis from ribulose 5-P and GTP (Zubay, 1993; Eberhardt et al., 2000). Significant down-regulation is also observed for the non-oxidative PP pathway genes. For example, *tktB* and *talA* are significantly down-regulated in the mutant during both phases of growth, although previous studies report the regulation of these two genes by *rpoS* at the stationary phase of growth only (Vijaykumar et al., 2004). Down-regulation of the *tktB* and *talA* genes does not cause any profound change to other pathways that are linked to the non-oxidative PP pathway during the exponential growth phase. However, during the early stationary phase of growth, significant down-regulation is observed among the aromatic amino acid biosynthesis genes, such as *aroM*, *aroL*, and *wrbA*. Among these, the latter two genes are reported to be positively regulated by *rpoS*, whereas *aroM* is known to be co-transcribed with *aroL* (DeFeyter et al., 1986; Lacour and Landini, 2004; Vijaykumar et al., 2004). The reduced activity of the non-oxidative branch of the PP pathway also interrupts the synthesis of a number of vitamins during the early stationary phase of growth. For example, the vitamin B_6 biosynthetic enzyme, such as E4P dehydrogenase (*epd*) that catalyzes the initial step of pyridoxal phosphate synthesis from E4P, is significantly down-regulated (Zubay, 1993; Yang et al., 1998). While the non-oxidative PP pathway branch is affected by the *rpoS* knockout gene during the early stationary phase, some other complementary pathways are induced, which might supply the necessary pentoses for the cell. The expression levels of such genes as *rbsA* and *rbsC*, which encode ribose transport proteins and *rbsK*, and which encode ribokinase are up-regulated. These are known to contribute to the maintenance of the cellular essential level of R5P through increased uptake of D-ribose and subsequent conversion to R5P, respectively (Iida et al., 1984; Sprenger, 1995; Moat et al., 2002). In addition to R5P, which is a precursor of histidine, the histidine biosynthetic pathway genes, such as *hisG*, *hisA*, and *hisB*, are significantly up-regulated.

Knockout of the *rpoS* gene also affects the TCA cycle significantly. The gene expression analysis, enzyme activity data, and intracellular metabolite concentration data demonstrate higher activity of the TCA cycle during the exponential growth phase. Most of the TCA cycle genes, such as *gltA*, *icdA*, *mdh*, *sdhA*, and *sdhC*, are up-regulated, and the concentrations of intracellular metabolites, such as PYR, PEP, OAA, and Ac-CoA, are lower compared to the wild-type strain, indicating rapid consumption of these metabolites by the activated TCA cycle enzymes. However, TCA cycle activity in the stationary phase is considerably lower with a higher accumulation of PYR, PEP, and OAA. The vegetative sigma factor *rpoD* (an ortholog of *rpoS* gene), which contains promoter binding consensus regions for most of the TCA cycle genes, could be involved with these characteristics of the TCA cycle (Farewell et al., 1998; Maeda et al., 2000; Aronis, 2002b).

As mentioned earlier, apart from TCA cycle activity, accumulation of acetate throughout the cultivation period is another notable feature of the *rpoS* mutant. While two genes, *ackA* and *poxB*, involved in acetate production, are down-regulated in the mutant at the early stationary phase, microarray data indicate that acetate production could be stimulated by the up-regulation of L-cysteine biosynthesis genes, such as *cysD*, *cysE*, and *cysK*, catalyzing the reactions that generate acetate as a by-product (Moat et al., 2002). Moreover, enzymes involved in acetate catabolism, such as AcCoA synthetase encoded by *acs*, the glyoxylate shunt enzymes encoded by the *aceBAK* operon, and the TCA cycle genes such as *gltA*, *mdh*, *sdhC*, are down-regulated during the stationary phase (Cronan et al., 1996; Shin et al., 1997; Wei et al., 2000). Of the acetate utilizing pathway genes, *acs* is reported to be regulated by *rpoS* (Shin et al., 1997).

Down-regulation of *acs* during the early stationary phase results in a decrease in the intracellular pool of AcCoA in the mutant compared to the wild-type strain. While the major route for AcCoA formation is less expressed, the other pathways for AcCoA formation rely on fatty acid degradation pathway (Moat et al., 2002). Several genes that participate in β-oxidation of fatty acids, particularly *fadA* and *fadB*, are significantly up-regulated, and the fatty acid biosynthesis genes, such as *accB*, *accC*, *accD*, and *fabF*, are also up-regulated in the mutant. The expression of the fatty acid degradation regulator, *fadR*, is significantly high at the early stationary phase of growth. It is known that *fadR* regulates fatty acid metabolism by binding to the DNA that contains *fadB* promoter binding sites, and in this way *fadR* controls fatty acid metabolism (Marrakchi, 2002; DiRusso, 1993). Down-regulation of *aceA* and *aceB* genes corresponds to the higher expression of *fadR*, where FadR indirectly represses the glyoxylate shunt encoded by *aceBAK* by directly regulating the activation of the glyoxylate shunt repressor, *iclR* (Maloy and Nunn., 1982; Gui et al., 1996). The higher expressions of *fadR* and *iclR* also cause acetate accumulation.

One exception among the TCA cycle genes, at the stationary phase, is the significant up-regulation of a fumarase gene, such as *fumC*, an isoenzyme of *fumA* (Chen et al., 2001). Despite the activation of *fumC*, down-regulation of *mdh* directs the carbon flux toward PYR formation with the up-regulation of the anapleoretic NAD^+ dependent enzyme encoded by *sfcA* (Phue et al., 2005). The accumulation of PYR is also due to down-regulation of the PDH enzyme complex encoded by *aceE*, *aceF*, and *lpdA* (Stephens et al., 1983). Other possible reasons for intracellular pyruvate accumulation could be due to the reduced activity of LDH and PoxB enzyme activities, as observed by enzyme activity measurement and gene expression profiles (Yang et al., 2001; Weber et al., 2005).

Again, as mentioned earlier, OAA accumulates during the early stationary phase of cell growth. Accumulation of intracellular OAA ultimately leads carbon flux toward threonine production from aspartate (Kim et al., 2004). This is indicated by more than 30-fold higher expressions of two major enzymes of threonine biosynthesis pathway encoded by *thrA* and *asd* genes. In addition, the arginine biosynthesis and transport systems are strongly affected in the mutant. Previous works show that some of the arginine transport proteins encoded by *artPM* are under positive control of *rpoS* (Lacour and Landini, 2004). It is demonstrated that the arginine transport protein *artJ*, a periplasmic protein, is significantly up-regulated during the stationary phase, but other proteins of this complex transport system, such as *artQ*, are significantly down-regulated.

3.10 References

Alexeeva, S., Hellingwerf, K.J., and de Mattos, M.J.T. (2003) 'Requirement of ArcA for redox regulation in *Escherichia coli* under microaerobic but not anaerobic or aerobic conditions', *Journal of Bacteriology*, 185: 204–9.

Alexeeva, S., de Kort, B., Sawers, G., Hellingwerf, K.J., and Teixeira de Mattos, M.J. (2000) 'Effects of limited aeration and of the ArcAB system on intermediary pyruvate catabolism in *Escherichia coli*', *Journal of Bacteriology*, 182: 4934–40.

Amabile-Cuevas, C.F. and Demple, B. (1991) 'Molecular characterization of the *soxRS* genes of *Escherichia coli*: two genes control a superoxide stress regulon', *Nucleic Acids Research*, 19: 4479–84.

Amemura, M., Makino, K., Sinagawa, H., and Nakata, A. (1990) 'Cross talk to the phosphate regulon of *Escherichia coli* by PhoM protein: PhoM is a histidine protein kinase and catalyzes phosphorylation of PhoB and PhoM-open reading frame 2', *Journal of Bacteriology*, 172(11): 6300–7.

Arndt, A. and Eikmans, B.J. (2007) 'The alcohol dehydrogenase gene *adhA* in *Corynebacterium glutamicum* is subject to carbon catabolite repression', *Journal of Bacteriology*, 189: 7408–16.

Arnold, C.N., Mcelhanon, J., Lee, A., Leonhard, R., and Siegele, D.A. (2001) 'Global analysis of *Escherichia coli* gene expression during the acetate-induced acid tolerace response', *Journal of Bacteriology*, 183: 2178–86.

Aronis, H.R. (1996) 'Regulation of gene expression during entry into stationary phase', in: Escherichia coli *and* Salmonella typhimurium: *Cellular and Molecular Biology*, 2nd edition, edited by F.C. Neidhardt et al. Washington DC: American Society for Microbiology Press. pp. 1497–512.

Aronis, H.R. (2002a) 'Signal transduction and regulatory mechanisms involved in control of the σ^S (RpoS) subunit of RNA polymerase', *Microbiology and Molecular Microbiology Review*, 66: 373–95.

Aronis, H.R. (2002b) 'Stationary phase gene regulation: what makes an *Escherichia coli* promoter sigmaS-selective?' *Current Opinion in Microbiology*, 5: 591–5.

Arsene, F., Tomoyasu, T., and Bukau, B. (2000) 'The heat shock response of *Escherichia coli*', *International Journal of Food Microbiology*, 55: 3–9.

Atkinson, M.R. and Ninfa, A.J. (1993) 'Mutational analysis of the bacterial signal transducing protein kinase/phosphatase nitrogen regulator II (NRII or NtrB)', *Journal of Bacteriology*, 175: 7016–23.

Atkinson, M.R., Blauwkamp, T.A., and Ninfa, A.J. (2002) 'Context-dependent functions of the PII and GlnK signal transduction proteins in *Escherichia coli*', *Journal of Bacteriology*, 184: 5364–75.

Aulkemeyer, P., Ebner, R., Heilenmann, G., Jahreis, K., Schmid, K. et al. (1991) 'Molecular analysis of two fructokinases involved in sucrose metabolism of enteric bacteria', *Molecular Microbiology*, 2913–22

Ault-Riché, D., Fraley, C.D., Tzeng, C.M., and Kornberg, A. (1998) 'Novel assay reveals multiple pathways regulating stress-induced accumulations of inorganic polyphosphate in *Escherichia coli*', *Journal of Bacteriology*, 180: 1841–7.

Baek, J.H. and Lee, S.Y. (2006) 'Novel gene members in the Pho regulon of *Escherichia coli*', *FEMS Microbiology Letters*, 264: 104–9

Baek, J.H. and Lee, S.Y. (2007) 'Transcriptome analysis of phosphate starvation response in *Escherichia coli*', *Journal of Microbial Biotechnology*, 17(2): 244–52.

Beatty, C.M., Browning, D.F., Busby, S.J.W., and Wolfe, A.J. (2003) 'Cyclic AMP receptor protein-dependent activation of the *Escherichia coli acs* P2 promoter by a synergistic class III mechanism', *Journal of Bacteriology*, 185: 5148–57.

Bell, A.I., Gatson, K.L., Cole, J.A., and Busby, S.J.W. (1989) 'Cloning of binding sequences for *Escherichia coli* transcription activators, FNR and CRP: location of bases involved in discrimination between FNR and CRP', *Nucleic Acids Research*, 17: 3865–74.

Benov, L. and Fridovich, I. (1995) 'Superoxide dismutase protects against aerobic heat shock in *Escherichia coli*', *Journal of Bacteriology*, 177: 3344–6.

Bettenbrock, K., Fischer, S., Kremling, A., Jahreis, K., Sauter, T., and Gilles, E.D. (2006) 'A quantitative approach to catabolite repression in *Escherichia coli*', *Journal of Biological Chemistry*, 281: 2578–84.

Bettenbrock, K., Sauter, T., Jahreis, K., Kremling, A., Lengeler, J.W., and Gilles, E.D. (2007) 'Correlation between growth rates, EIIA^{Crr} phosphorylation, and intracellular cyclic AMP Levels in *Escherichia coli* K-12', *Journal of Bacteriology*, 189: 6891–900.

Blanco, A.G., Sola, M., Gomis-Ruth, F.X., and Coll, M. (2002) 'Tandem DNA recognition by PhoB, a two-component signal transduction transcriptional activator,' *Structure* (Cambridge), 10: 701–13

Blauwkamp, T.A. and Ninfa, A.J. (2002) 'Physiological role of the GlnK signal transduction protein of *Escherichia coli*: Survival of nitrogen starvation', *Molecular Microbiology*, 46: 203–14.

Blencke, H.M. et al. (2003) 'Transcriptional profiling of gene expression in response to glucose in *Bacillus subtilis*: Regulation of the central metabolic pathways', *Metabolic Engineering*, 5: 133–49.

Bo, I., Wing, H., Lee, D., Wu, H.C., and Busby, S. (1998) 'Transcription activation by *Escherichia coli* FNR protein: similarities to, and differences from, the CRP paradigm', *Nucleic Acids Research*; 26: 2057–81.

Booth, I.R. (1985) 'Regulation of cytoplasmic pH in bacteria', *Microbiological Reviews*, 49(4): 329–78.

Bowien, B. and Kusian, B. (2002) 'Genetics and control of CO_2 assimilation in the chemoautotroph *Ralstonia eutropha*', *Archives of Microbiology*, 178: 85–93.

Browning, D.F., Beatty, C.M., Wolfe, A.J., Cole, J.A., and Busby, S.J.W. (2002) 'Independent regulation of the divergent *Escherichia coli* nrfA and acs P1 promoters by a nucleoprotein assembly at a shared regulatory region', *Molecular Microbiology*, 43: 687–701.

Browning, D.F., Beatty, C.M., Sanstad, E.A., Gunn, K.E., Busby, S.J., and Wolfe, A.J. (2004) 'Modulation of CRP-dependent transcription at the *Escherichia coli* acs P2 promoter by nucleoprotein complexes: anti-activation by the nucleoid proteins FIS and IHF', *Molecular Microbiology*, 51: 241–54.

Bunch, P.K., Mat,-Jan, F., Lee, N., and Clark, D.P. (1997) 'The ldhA gene encoding the fermentative lactate dehydrogenase of *Escherichia coli*', *Microbiology*, 143: 187–95.

Camarena, L., Poggio, S., Garcèa, N., and Osorio, A. (1998) 'Transcriptional repression of gdhA in *Escherichia coli* is mediated by the Nac protein', *FEMS Microbiology Letters*, 167: 51–6.

Campbell, J.W. and Cronan, J.E. (2001) '*Escherichia coli* fadR positively regulates transcription of the fabB fatty acid biosynthetic gene', *Journal of Bacteriology*, 183: 5982–90.

Canonaco, F., Hess, T.A., Heri, S., and Wang, T.T. (2001) 'Metabolic flux response to phosphoglucose isomerase knockout in *Escherichia coli* and impact of over-expression of the soluble transhydrogenase *udhA*', *FEMS Microbiology Letters*, 204: 247–52.

Castelli, M.E., Vescovi, E.G., and Soncini, F.C. (2000) 'The phosphatase activity is the target for Mg^{2+} regulation of the sensor protein PhoQ in *Salmonella*', *Journal of Biological Chemistry*, 275: 22948–54.

Castanié-Cornet, M.P. and Foster, J.W. (2001) '*Escherichia coli* acid resistance: cAMP receptor protein and a 20 bp *cis*-acting sequence control pH and stationary phase expression of the *gadA* and *gadBC* glutamate decarboxylase genes', *Microbiology*, 147: 709–15.

Castanié-Cornet, M.P., Penfound, T.A., Smith, D., Elliott, J.F., and Foster, J.W. (1999) 'Control of acid resistance in *Escherichia coli*', *Journal of Bacteriology*, 181: 3525–35.

Castanié-Cornet, M.P., Cam, K, Bénédicte Bastiat, K., Cros A, Bordes P, and Gutierrez C. (2010) 'Acid stress response in *Escherichia coli*: Mechanism of regulation of gadA transcription by RcsB and GadE', *Nucleic Acids Research*, 38(11): 3546–54.

Chao, G., Shen, J., Tseng, C.P., Park, S.J., and Gunsalus, R.P. (1997) 'Aerobic regulation of isocitrate dehydrogenase gene (*icd*) expression in *Escherichia coli* by the *arcA* and *fnr* genes products', *Journal of Bacteriology*, 179: 4299–304.

Chen, D.P., Tseng, C.P., Lin, H.T., and Sun, C.F. (2001) 'Oxygen- and growth rate-dependent regulation of *Escherichia coli* fumarase (FumA, FumB, and FumC) activity', *Journal of Bacteriology*, 183: 461–7.

Chin, A.M., Feldheim, D.A., and Saier, M.H. Jr. (1989) 'Altered transcription patterns affecting several metabolic pathways in strains of *Salmonella typhimurium* which over express the fructose regulon', *Journal of Bacteriology*, 171: 2424–34.

Chirico, W.J., Waters, M.G., and Blobel, G. (1988) '70 K heat shock related proteins stimulate protein translocation into microsomes', *Nature*, 332: 805–10.

Cho, S., Shin, D., Ji, G.E., Heu, S., and Ryu, S. (2005) 'High-level recombinant protein production by overexpression of Mlc in *Escherichia coli*', *Journal of Biotechnology*, 119: 197–203.

Chou, C.H., Bennett, G.N., and San, K.Y. (1994) 'Effect of modulated glucose uptake on high-level recombinant protein production in a dense *Escherichia coli* culture', *Biotechnology Progress*, 10: 644–7.

Chuang, S.E. and Blattner, F.R. (1993) 'Characterization of twenty six new heat shock genes of *Escherichia coli*', *Journal of Bacteriology*, 175: 5242–52.

Claverie-Martin, F. and Magasanik, B. (1991) 'Role of integration host factor in the regulation of the glnHp2 promoter of *Escherichia coli*', *Proceedings of the National Academy of Sciences of USA*, 88: 1631–5.

Collier, D.N., Hager, P.W., and Phibbs, P.V. Jr (1996) 'Catabolite repression control in *Pseudomonads*', Research in Microbiology, 147: 551–61.

Commichau, F.M., Forchhammer, K., and Stülke, J. (2006) 'Regulatory links between carbon and nitrogen metabolism,' *Current Opinion in Microbiology*, 9: 167–72.

Compan, I. and Touati, D. (1994) 'Anaerobic activation of *arcA* transcription in *Escherichia coli*: Roles of Fnr and ArcA', *Molecular Microbiology*; 11: 955–64.

Constantinidou, C., Hobman, J.L., Griffiths, L., Patel, M.D., Penn, C.W. et al. (2006) 'A reassessment of the FNR regulon and transcriptomic analysis of the effects of nitrate, nitrite, NarXL, and NarQP as *Escherichia coli* K12 adapts from aerobic to anaerobic growth', *Journal of Biological Chemistry*, 281: 4802–15.

Cortay, J.C., Negre, D., Galinier, A., Duclos, B., Perriere, G., and Cozzone, A.J. (1991) 'Regulation of the acetate operon in *Escherichia coli*: Purification and functional characterization of the IclR repressor', *European Molecular Biology Organization Journal*, 10: 675–9.

Cotter, P.A. and Gunsalus, R.P. (1992) 'Contribution of the fnr and arcA gene products in coordinate regulation of cytochrome o and d oxidase (cyoABCDE and cydAB) genes in *Escherichia coli*', *FEMS Microbiology Letters*, 70: 31–6.

Cowing, D.W., Bardwell, J.C.A., Craig, E.A., Woolford, C., Hendrix, R.W., and Gross, C.A. (1985) 'Consensus sequence for *Escherichia coli* heat shock gene promoter', *Proceedings of the National Academy of Sciences USA*, 82: 2679–83.

Craig, E.A. and Gross, C.A. (1991) 'Is hsp70 the cellular thermometer?' *Trends in Biochemical Sciences*, 16: 135–40.

Cronan, J.E. Jr and LaPorte, D. (1996) 'Tricarboxylic acid cycle and glyoxylate bypass', in: Escherichia coli *and* Salmonella typhimurium: *Cellular and Molecular Biology*, 2nd edition, vol. 1, edited by F.C. Neidhardt et al. Washington DC: American Society for Microbiology Press. pp. 206–16.

Cronan, J.E. and Subrahmanyam, S. (1998) 'FadR, transcriptional co-ordination of metabolic expediency', *Molecular Microbiology*, 29: 937–44.

Cunningham, L., Gruer, M.J., and Guest, J.R. (1997) 'Transcriptional regulation of the aconitase genes (*acn*A and *acn*B) of *Escherichia coli*', *Microbiology*, 143: 3795–805.

Cunningham, L., Georgellis, D., Green, J., and Guest, J.R. (1998) 'Co-regulation of lipoamide dehydrogenase and 2-oxoglutarate dehydrogenase sysnthesis in *Escherichia coli*-characteristics of ArcA binding site in the *lpd* promoter', *FEMS Microbiology Letters*, 169: 403–8.

Curtis, S.J. and Epstein, W. (1975) 'Phosphorylation of D-glucose in *Escherichia coli* mutants defective in glucosephosphotransferase, mannosephosphotransferase, and glucokinase', *Journal of Bacteriology*, 122: 1189–99.

Curtiss, R. and Kelly, S.M. (1987) '*Salmonella typhimurium* deletion mutants lacking adenylate cyclase and cyclic AMP repressor protein are avirulent and immunogenic', *Infection Immunity*, 55: 3035–43.

D'Alessio, G. and Josse, J. (1971) 'Glyceraldehyde phosphate dehydrogenase, phosphoglycerate kinase, and phosphoglyceromutase of *Escherichia coli*', *Journal of Biological Chemistry*, 246: 4319–25.

Damoglou, A.P. and Dawes, E.A. (1968) 'Studies on the lipid content and phosphate requirement for glucose- and acetate-grown *Escherichia coli*', *Biochemical Journal*, 110: 775–81.

Datta, A.G. and Racker, E. (1961) 'Mechanism of action of transketolase', *Journal of Biological Chemistry*, 236: 617–23.

Death, A. and Ferenci, T. (1994) 'Between feast and famine: Endogenous inducer synthesis in the adaptation of *Escherichia coli* to growth with limiting carbohydrates', *Journal of Bacteriology*, 176: 5101–7.

Death, A., Notley, L., and Ferenci, T. (1993) 'Derepression of LamB protein facilitates outer membrane permeation of carbohydrates into *Escherichia coli* under conditions of nutrient stress', *Journal of Bacteriology*, 175: 1475–83.

De Boris, G. (2008) 'Carbon catabolite repression in bacteria: Many ways to make the most out of nutrients', *Nature Reviews Microbiology*, 6: 613–24.

DeFeyter, R.C., Davidson, B.E., and Pittard, J. (1986) 'Nucleotide sequence of the transcription unit containing the *aro*L and *aro*M genes from *Escherichia coli* K-12', *Journal of Bacteriology*, 165: 233–9.

de Graef, M., Alexeeva, S., Snoepa, J.L., and Teixeira de Mattos, M.J. (1999) 'The steady-state internal redox state (NADH/NAD) reflects the external redox state and is correlated with catabolic adaptation in *Escherichia coli*', *Journal of Bacteriology*, 181: 2351–7.

De Reuse, H. and Danchin, A. (1988) 'The ptsH, ptsI, and crr genes of the *Escherichia coli* phosphoenolpyruvate-dependent phosphotransferase system: a complex operon with several modes of transcription', *Journal of Bacteriology*, 170: 3827–37.

Deutscher, J., Francke, C., and Postma, P.W. (2006) 'How phosphotransferase system-related protein phosphorylation regulates carbohydrate metabolism in bacteria', *Microbiology and Molecular Biology Reviews*, 70: 939–1031.

DiRusso, C.C. and Nystrom, T. (1998) 'The fats of *Escherichia coli* during infancy and old age: regulation by global regulators, alarmones and lipid intermediates', *Molecular Microbiology*, 27: 1–8.

DiRusso, C.C., Heimert, T.L., and Metzger, A.K. (1992) 'Characterization of FadR, a global transcriptional regulator of fatty acid metabolism in *Escherichia coli*', *Journal of Biological Chemistry*, 267: 8685–91.

DiRusso, C.C., Metzger, A.K., and Heimert, T.L. (1993) 'Regulation of transcription of genes required for fatty acid transport and unsaturated fatty acid biosynthesis in *Escherichia coli* by FadR', *Molecular Microbiology*, 7: 311–322.

Djordjevic, G.M., Tchieu, J.H., and Saier, M.H. Jr. (2001) 'Genes involved in control of galactose uptake in *Lactobacillus brevis* and reconstitution of the regulatory system in *Bacillus* subtilis', *Journal of Bacteriology*, 183: 3224–36.

Drapal, N and Sawers, G. (1995) 'Promoter 7 of the *Escherichia coli* pfl operon is a major determinant in the anaerobic regulation of expression by ArcA,' *Journal of Bacteriology*, 177: 5338–41.

Eberhardt, B.S., Fischer, M., Kis, K., and Richter, G. (2000) 'Biosynthesis of vitamin B_2 (riboflavin)', *Annual Review of Nutrition*, 20: 153–67.

Eguchi, Y., Itou, J., Yamane, M., Demizu, R., Yamato, F. et al. (2007) 'B1500, a small membrane protein, connects the two-component systems EvgS/EvgA and PhoQ/PhoP in *Escherichia coli*', *Proceedings of the National Academy of Sciences of USA*, 104: 18712–17.

Farewell, A., Kvint, K., and Nystrom, T. (1998) 'Negative regulation by RpoS: A case of sigma factor competition', *Molecular Microbiology*, 29: 1039–51.

Farewell, A., Diez, A.A., DiRusso, C.C., and Nystrom, T. (1996) 'Role of the *Escherichia coli* FadR regulator in stasis survival and growth phase-dependent expression of the *uspA*, *fad*, and *fab* genes', *Journal of Bacteriology*, 178: 6443–50.

Farmer, W.R. and Liao, J.C. (1997) 'Reduction of aerobic acetate production by *Escherichia coli*', *Applied Environmental Microbiology*, 63: 3205–10.

Fawcett, W.P. and Wolf, R.J. (1995) 'Genetic definition of the *Escherichia coli zwf* "soxbox", the DNA binding site for SoxS-mediated induction of glucose-6-phosphate dehydrogenase in response to superoxide', *Journal of Bacteriology*, 177: 1742–50.

Fischer, E. and Sauer, U. (2003) 'Metabolic flux profiling of *Escherichia coli* mutants in central carbon metabolism using GC-MS', *European Journal of Biochemistry*, 270: 880–91.

Flint, D.H., Tuminello, J.F., and Emptage, M.H. (1993) 'The inactivation of Fe-S cluster containing hydrolyases by superoxide', *Journal of Biological Chemistry*, 268: 22369–76.

Flores, N., Flores, S., Escalante, A., de Anda, R., Leal, L. et al. (2005) 'Adaptation for fast growth on glucose by differential expression of central carbon metabolism and gal regulon genes in an *Escherichia coli* strain lacking the phosphoenolpyruvate: carbohydrate phosphotransferase system', *Metabolic Engineering*, 7: 70–87.

Foster, J.W. (2004) '*Escherichia coli* acid resistance: Tales of an amateur acidophile', *Nature Reviews, Microbiology*, 2: 898–907.

Fraenkel, D.G. and Neidhardt, F. (editors) (1999) Escherichia coli *and* Salmonella: *Cellular and Molecular Biology*. Washington DC: American Society for Microbiology Press.

Frunzke, J., Engels, V., Hasenbein, S., Gätgens, C., and Bott, M. (2008) 'Co-ordinated regulation of gluconate catabolism and glucose uptake in *Corynebacterium glutamicum* by two functionally equivalent transcriptional regulators, GntR1 and GntR2', *Molecular Microbiology*, 67: 305–22.

Gadgil, M., Kapur, V., and Hu, W.S. (2005) 'Transcriptional response of *Escherichia coli* to temperature shift', *Biotechnology Progress*, 21: 689–99.

Galinier, A., Kravanja, M., Engelmann, R., Hengstenberg, W., Kilhoffer, M.C. et al. (1998) 'New protein kinase and protein phosphatase families mediate signal transduction in bacterial catabolite repression', *Proceedings of the National Academy of Sciences of USA*, 95: 1823–8.

Gamer, J., Bujard, H., and Bukau, B. (1992) 'Physical interaction between heat shock proteins DnaK, DnaJ, and GrpE and the bacterial heat shock transcription factor σ^{32}', *Cell*, 69: 833–42.

Gaudu, P. and Weiss, B. (1996) 'SoxR, a [2Fe-2S] transcription factor, is active only in its oxidized form', *Proceedings of the National Academy of Sciences of USA*, 93: 10094–8.

Gaudu, P. and Weiss, B. (2000) 'Flavodoxin mutants of *Escherichia coli* K-12', *Journal of Bacteriology*, 182: 1788–93.

Gentry, D.R., Hernandez, V.J., Nguyen, L.H., Jensen, D.B., and Cashel, M. (1993) 'Synthesis of the stationary-phase sigma factor σ^s is positively regulated by ppGpp', *Journal of Bacteriology*, 175(24): 7982–9.

Georgellis, D., Lynch, A.S., and Lin, E.C.C. (1997) 'In vitro phosphorylation study of the arcA two-component signal transduction system of *Escherichia coli*', *Journal of Bacteriology*, 179: 5429–35.

Georgellis, D., Kwon, O., and Lin, E.C.C. (1999) 'Amplification of signaling activity of the Arc two-component system of *Escherichia coli* by anaerobic metabolites. An *in vitro* study with different protein modules', *Journal of Biological Chemistry*, 274: 35950–4.

Georgellis, D., Kwon, O., and Lin, E.C. (2001) 'Quinones as the redox signal for the arc two-component system of bacteria', *Science*, 292, 552: 2314–16.

Gestmeier, R., Cramer, A., Dangel, P., Schaffer, S., and Eikmanns, B.J. (2004) 'RamB, a novel transcriptional regulator involved in acetate metabolism of *Corynebacterium glutamicum*', *Journal of Bacteriology*, 186: 2798–809.

Gharbi, S., Belaich, A., Murgier, M., and Lazdunski, A. (1985) 'Multiple controls exerted on *in vivo* expression of the pepN Gene in *Escherichia coli*: Studies with *pepN-lacZ* operon and protein fusion strains', *Journal of Bacteriology*, 163: 1191–5.

Goff, S.A., Casson, L.P., and Goldberg, A.L. (1984) 'Heat shock regulatory gene *htpR* influences rates of protein degradation and expression of the *lon* gene in *Escherichia coli*', *Proceedings of the National Academy of Sciences of USA*, 81: 6647–51.

Gong, S., Richard, H., Foster, J.W. (2003) 'YjdE (AdiC) is the arginine: agmatine antiporter essential for arginine-dependent acid resistance in *Escherichia coli*', *Journal of Bacteriology*, 185: 4402–9.

Gorke, B. and Stulke, J. (2008) 'Carbon catabolite repression in bacteria: many ways to make the most out of nutrients', *Nature Reviews, Microbiology*, 6: 613–24.

Gosset, G. (2005) 'Improvement of *Escherichia coli* production strains by modification of the phosphoenolpyruvate:sugar phosphotransferase system', *Microbial Cell Factories*, 4: 14.

Gosset, G., Zhang, Z., Nayyar, S., Cuevas, W.A., and Saier, M.H. Jr. (2004) 'Transcriptome analysis of Crp-dependent catabolite control of gene expression in *Escherichia coli*', *Journal of Bacteriology*, 186: 3516–24.

Green, J, and Guest, J.R. (1993) 'Activation of FNR-dependent transcription by iron: an *in vitro* switch for FNR', *FEMS Microbiology Letters*, 113: 219–22.

Greenberg, J.T. and Demple, B. (1989) 'A global response induced in *Escherichia coli* by redox-cycling agents overlaps with that induced by peroxide stress', *Journal of Bacteriology*, 171: 3933–9.

Greenberg, J.T., Monach, P., Chou, J.H., Josephy, P.D., and Demple, B. (1990) 'Positive control of a global antioxidant defense regulon activated by superoxide-generating agents in *Escherichia coli*', *Proceedings of the National Academy of Sciences of USA*, 87: 6181–5.

Gross, C.A., Straus, D.B., Erickson, J.W., and Yura, T. (1990) 'The function and regulation of heat shock proteins in *Escherichia coli*', in: *Stress Proteins in Biology and Medicine*, edited by R.J. Morimoto et al. New York: Cold Spring Harbor Laboratory Press. pp. 167–89.

Grossman, A.D., Erickson, J.W., and Gross, C.A. (1984) 'The htpR gene product of *E. coli* is a sigma factor for heat-shock promoters', *Cell*, 38: 383–90.

Gruer, M.J. and Guest, J.R. (1994) 'Two genetically-distinct and differentially-regulated aconitases (AcnA and AcnB) in *Escherichia coli*', *Microbiology*, 140: pp. 2531–41.

Guan, C., Wanner, B.L., and Inouye, H. (1983) 'Analysis of regulation of PhoB expression using a *phoB*-cat fusion', *Journal of Bacteriology*, 156: 710–17

Gui, L., Sunnarborg, A., and LaPorte, D.C. (1996) 'Regulated expression of a repressor prote: FadR activates *iclR*', *Journal of Bacteriology*, 178: 4704–9.

Gunasekera, T.S., Csonka, L.N., and Paliy, O. (2008) 'Genome wide transcriptional responses of *Escherichia coli* K-12 to continuous osmotic and heat stresses', *Journal of Bacteriology*, 190: 3712–20.

Gunnewijk, M. and Poolman, B. (2000) 'Phosphorylation state of HPr determines the level of expression and the extent of phosphorylation of the lactose transport protein of *Streptococcus thermophilus*', *Journal of Biological Chemistry*, 275: 34073–9.

Gunnewijk, M., van den Bogaard, P.T., Veenhoff, L.M., Heuberger, E.H., de Vos, W.M. et al. (2001) 'Hierarchical control versus autoregulation of carbohydrate utilization in bacteria', *Journal of Molecular Microbiology and Biotechnology*, 3: 401–13.

Gunsalus, R.P. (1992) 'Control of electron flow in *Escherichia coli*: co-ordinated transcription of respiratory pathway genes', *Journal of Bacteriology*, 174: 7069–74.

Halbedel, S. et al. (2007) 'Transcription in *Mycoplasma pneumoniae*: Analysis of the promotersof the ackA and ldh genes', *Journal of Molecular Biology*, 371: 596–607.

Hall, M.N. and Silhavy, T.J. (1981) 'The ompB locus and the regulation of the major outer membrane porin proteins of *Escherichia coli* K12', *Journal of Molecular Biology*, 146: 23–43.

Hanson, R.L. and Rose, C. (1980) 'Effects of insertion mutation in a locus affecting pyridine nucleotide transhydrogenase (*pnt* Tn5) on the growth of *Escherichia coli*', *Journal of Bacteriology*, 141: 401–4.

Harris, R.M., Webb, D.C., Howitt, S.M., and Cox, G.B. (2001) 'Characterization of PitA and PitB from *Escherichia coli*', *Journal of Molecular Biology*, 183: 5008–501.

Hasan, C.M. and Shimizu, K. (2008) 'Effect of temperature up-shift on fermentation and metabolic characteristics in view of gene expressions in *Escherichia coli*', *Microbial Cell Factories*, 7: 35.

Helling, R.B. (1998) 'Pathway choice in glutamate synthesis in *Escherichia coli*', *Journal of Molecular Biology*, 180: 4571–5.

Henkin, T.M., Grundy, F.J., Nicholson, W.L., and Chambliss, G.H. (1991) 'Catabolite repression of α-amylase gene expression in *Bacillus subtilis* involves a trans-acting gene product homologous to the *Escherichia coli lacI* and *galR* repressors', *Molecular Microbiology*, 5: 575–84.

Hidalgo, E. and Demple, B., Lin, F.C.C., and Lynch, A.S. (editors) (1996) *Regulation of Gene Expression in Escherichia coli*. Austin, TX: Landes. pp. 435–52.

Hockney, R.C. (1994) 'Recent developments in heterologous protein production in *Escherichia coli*', *Trends in Biotechnology*, 12: 456–63.

Hoffmann, F. and Rinas, U. (2001) 'Plasmid amplification in *Escherichia coli* after temperature up-shift is impaired by induction of recombinant protein synthesis', *Biotechnology Letters*, 23: 1819–25.

Hoffmann, F., Weber, J., and Rinas, U. (2002) 'Metabolic adaptation of *Escherichia coli* during temperature-induced recombinant protein production: 1. Readjustment of metabolic enzyme synthesis', *Biotechnology and Bioengeering*, 80: 313–19.

Hogema, B.M., et al. (1998) 'Inducer exclusion in *Escherichia coli* non-PTS substrates: the role of the PEP to pyruvate ratio in determining the phosphorylation state of enzyme II A^{Glc},' *Molecular Microbiology*, 30: 487–98.

Hogema, B.M., Arents, J.C., Bader, R., and Postma, P.W. (1999) 'Autoregulation of lactose uptake through the LacY permease by enzyme II A^{Glc} of the PTS in *Escherichia coli* K-12', *Molecular Microbiology*, 31: 1825–33.

Hosono, K., Kakuda, H., and Ichihara, S. (1995) 'Decreasing accumulation of acetate in a rich medium by *Escherichia coli* on introduction of genes on a multicopy plasmid', *Bioscience, Biotechnology and Biochemistry*, 59: 256–61.

Hsieh, Y.J. and Wanner, B.L. (2010) 'Global regulation by the seven-component P_i signaling system', *Current Opinion in Microbiology*, 13(2): 198–203.

Hua, Q., Yang, C., Baba, T., Mori, H., and Shimizu, K. (2003) 'Response of the central metabolism in *Escherichia coli* to phosphoglucose isomerase and glucose-6-phosphate dehydrogenase knockouts', *Journal of Bacteriology*, 185: 7053–67.

Hua, Q., Yang, C., Oshima, T., Mori, H., and Shimizu, K. (2004) 'Analysis of gene expression in *Escherichia coli* in response to changes of growth-limiting nutrient in chemostat cultures', *Applied Environmental Microbiology*, 70: 2354–66.

Iida, A., Harayama, S., Iino, T., and Hazelbauer, G.L. (1984) 'Molecular cloning and characterization of genes required for ribose transport and utilization in *Escherichia coli* K-12', *Journal of Bacteriology*, 158: 674–82.

Inada, T., Kimata, K., and Aiba, H. (1996a) 'Mechanism responsible for glucose-lactose diauxie in *Escherichia coli*: challenge to the cAMP model', *Genes Cells*, 1: 293–301.

Inada, T., Takahashi, H., Mizuno, T., and Aiba, H. (1996b) 'Down-regulation of cAMP production by cAMP receptor protein in *Escherichia coli*: An assessment of the contributions of transcriptional and posttranscriptional control of adenylate cyclase', *Molecular and General Genetics*, 253: 198–204.

Ishige, K., Nagasawa, S., Tokishita, S.I., and Mizuno, T. (1994) 'A novel device of bacterial signal transducers', *European Molecular Biology Organization Journal*, 13: 5195–202.

Ishige, T., Krause, M., Bott, M., Wendisch, V.F., and Sahm, H. (2003) 'The phosphate starvation stimulon of *Corynebacterium glutamicum* determined by DNA microarray analyses', *Journal of Bacteriology*, 185: 4519–29.

Iuchi, S. (1993) 'Phosphorylation/dephosphorylation of the receiver module at the conserved aspartate residue controls transphosphorylation activity of histidine kinase sensor protein ArcB of *Escherichia coli*', *Journal of Biological Chemistry*, 268: 23972–80.

Iuchi, S. and Lin, E.C.C. (1988) 'arcA (dye), a global regulatory gene in *Escherichia coli* mediating repression of enzyme in aerobic pathways', *Proceedings of the National Academy of Sciences of USA*, 85: 888–1892.

Iuchi, S. and Lin, E.C.C. (1992) 'Mutational analysis of signal transduction by arcB: A membrane sensor protein for anaerobic expression of operons involved in the central aerobic pathways in *Escherichia coli*', *Journal of Bacteriology*, 174: 3972–80.

Iuchi, S., Cameron, D.C., and Lin, E.C.C. (1989) 'A second global regulator gene (arcB) mediating repression of enzymes in aerobic pathways of *Escherichia coli*', *Journal of Bacteriology*, 171: 868–73.

Iuchi, S., Chepuri, V., Fu, H.A., Gennis, R.B., and Lin, E.C.C. (1990) 'Requirement for terminal cytochromes in generation of the aerobic signal for the arc regulatory system in *Escherichia coli*: Study utilizing deletions and lac fusions of cyo and cyd', *Journal of Bacteriology*, 172: 6020–5.

Iyer, R., Williams, C., and Miller, C. (2003) 'Arginine-agmatine antiporter in extreme acid resistance in *Escherichia coli*', *Journal of Bacteriology*, 185: 6556–61.

Jaggi, R., van Heeswijk, W.C., Westerhoff, H.V., Ollis, D.L., and Vasudevan, S.G. (1997) 'The two opposing activities of adenylyltransferase reside in distinct homologous domains, with intramolecular signal transduction', *European Molecular Biology Organization Journal*, 16: 5562–71.

Jair, K.W., Martin, R.G., Rosner, J.L., Fujita, N., Isihhama, A., and Wolf, R.E. (1995) 'Purification and regulatory properties of MarA protein, a transcriptional activator of Escherichia coli multiple antibiotic and superoxide resistance promoters', *Journal of Bacteriology*, 177: 7100–4.

Jiang, P. and Ninfa, A.J. (2007) '*Escherichia coli* PII signal transduction protein controlling nitrogen assimilation acts as a sensor of adenylate energy charge *in vitro*', *Biochemistry*, 46: 12979–96.

Jiang, P., Peliska, J.A., and Ninfa, A.J. (1998) 'The regulation of *Escherichia coli* glutamine synthetase revisited: Role of 2-ketoglutarate in the regulation of glutamine synthetase adenylylation state', *Biochemistry*, 37: 12802–10.

Jault, J.M., Fieulaine, S., Nessler, S., Gonzalo, P., Di Pietro, A. et al. (2000) 'The HPr kinase from Bacillus subtilis is a homo-oligomeric enzyme which exhibits strong positive cooperativity for nucleotide and fructose 1,6-bisphosphate binding', *Journal of Biological Chemistry*, 275: 1773–80.

Jordan, P.A., Tang, Y., Bradbury, A.J., Thomson, A.J., and Guest, J.R. (1999) 'Biochemical and spectroscopic characterization of *Escherichia coli* aconitases (AcnA and AcnB)', *Biochemical Journal*, 344: 739–46.

Jung, I.L., Phyo, K.H., and Kim, I.G. (2005) 'RpoS-mediated growth-dependent expression of the *Escherichia coli* tkt genes encoding transketolases isoenzymes', *Current Micobiology*, 50: 314–18.

Kabir, M.M. and Shimizu, K. (2006) 'Investigation into the effect of *soxR* and *soxS* genes deletion on the central metabolism of *Escherichia coli* based on gene expressions and enzyme activities', *Biochemical Engineering Journal*, 30: 39–47.

Kang, Y., Weber, K.D., Qin, Y., Kiley, P.J., Blattner, F.R. (2005) 'Genome-wide expression analysis indicates that FNR of *Escherichia coli* K-12 regulates a large number of genes of unknown function', *Journal of Bacteriology*, 187: 1135–60.

Kao, C.M. (1999) 'Functional genomics technologies: creating new paradigms for fundamental and applied biology', *Biotechnology Progress*, 15: 304–11.

Kato, N., Mizuno, T. Shimizu, T., and Hakoshima, T. (1997) 'Insights into multi-step phosphorelay from the crystal structure of the C-terminal HP-1 domain of arcB', *Cell*, 88: 717–23.

Kessler, D. and Knappe, J. (1996) 'Anaerobic dissimilation of pyruvate', in: Escherichia coli *and* Salmonella typhimurium: *Cellular and Molecular Biology*, vol. 1, 2nd edition, edited by F.C. Neidhardt et al. Washington DC: American Society for Microbiology Press. pp. 199–205.

Kiley, J.P. and Reznikoff, W. (1991) '*fnr* mutants that activate gene expression in the presence of oxygen', *Journal of Molecular Biology*, 173: 16–22.

Kim, S.Y., Nam, T.W., Shin, D., Koo, B.M., Seok, Y.J., and Ryu, S. (1999) 'Purification of Mlc and analysis of its effects on the *pts* expression in *Escherichia coli*', *Journal of Biological Chemistry*, 274: 25398–402.

Kim, Y.H., Park, J.S., Cho, J.Y., Cho, K.M., Park, Y.H., and Lee, J. (2004) 'Proteomic response analysis of a threonine-overproducing mutant of *Escherichia coli*', *Biochemical Journal*, 381: 823–9.

Kimata, K., Takahashi, H., Inada, T., Postma, P., and Aiba, H. (1997) 'cAMP receptor protein – cAMP plays a crucial role in glucose-lactose diauxie by activating the glucose transporter gene in *Escherichia coli*', *Proceedings of the National Academy of Sciences of USA*, 94: 12914–19.

Kimata, K., Inada, T., Tagami, H., and Aiba, H. (1998) 'A global repressor (Mlc) is involved in glucose induction of the *ptsG* gene encoding major glucose transporter in *Escherichia coli*', *Molecular Microbiology*, 29: 1509–19.

Kimura, S., Makino, K., Shinagwa, H., Amemura, M., and Nakata, A. (1989) 'Regulation of the phosphate regulon of *Escherichia coli*: characterization of the promoter of the *pstS* gene,' *Molecular and General Genetics*, 215: 374–80.

Kirkpatrick, C., Maurer, M., Oyelakin, N., Yoncheva, Y.N., Maurer, R., and Slonczewski, J.L. (2001) 'Acetate and formate stress: opposite responses in the proteome of *Escherichia coli*', *Journal of Bacteriology*, 183: 6466–77.

Kitagawa, M., Miyakawa, M., Matsumura, Y., and Tsuchido, T. (2002) '*E. coli* small heat shock proteins, IbpA and IbpB, protect enzymes from inactivation by heat and oxidants', *European Journal of Biochemistry*, 269: 2907–17.

Koebmann, B.J., Westerhoff, H.V., Snoep, J.L., Nilsson, D., and Jensen, P.R. (2002) 'The glycolytic flux in *Escherichia coli* is controlled by the demand for ATP', *Journal of Bacteriology*, 184: 3909–16.

Kornberg, H.L. (2001) 'Routes for fructose utilization by *Escherichia coli*', *Journal of Molecular Microbiology and Biotechnology*, 3: 355–9.

Krapp, A.R., Rodriguez, R.E., Poli, H.O., Paladini, D.H., Palatnik, J.F., and Carrillo, N. (2002) 'The flavoenzyme ferredoxin (flavodoxin)-NADP(H) reductase modulates NADP(H) homeostasis during the *sox*RS response of *Escherichia coli*', *Journal of Bacteriology*, 184: 1474–80.

Kumar, R. and Shimizu, K. (2010) 'Metabolic regulation of *Escherichia coli* and its *gdhA, glnL, gltB, D* mutants under different carbon and nitrogen limitations in the continuous culture', *Microbial Cell Factories*, 9: 8.

Kumari, S., Beatty, C.M., Browning, D.F., Busby, S.J.W., Simel, E.J. et al. (2000) 'Regulation of Acetyl Coenzyme A Synthetase in *Escherichia coli*', *Journal of Bacteriology*, 182: 4173–9.

Kusano, S., Ding, Q., Fujita, N., and Ishihama, A. (1996) 'Promoter selectivity of *Escherichia coli* RNA polymerase Eσ^{70} and Eσ^{38} holoenzymes: Effect of DNA supercoiling', *Journal of Biological Chemistry*, 271: 1998–2004.

Kustu, S., Santero, E., Keener, J., Popham, D., and Weiss, D. (1989) 'Expression of sigma 54 (*ntrA*)-dependent genes is probably united by a common mechanism', *Microbiology Reviews*, 53: 367–76.

Lacour, S. and Landini, P. (2004) 'SigmaS-dependent gene expression at the onset of stationary phase in *Escherichia coli*: function of σ^S-dependent genes and identification of their promoter sequences', *Journal of Bacteriology*, 186: 7186–95.

Lamarche, M.G., Wanner, B.L., Sebastien, C., and Harel, J. (2008) 'The phosphate regulon and bacterial virulence: a regulatory network connecting phosphate homeostasis and pathogenesis', *FEMS Microbiology Reviews*, 32(3): 461–73.

Lazzaroni, J.C. and Portailer, R.C. (1985) 'Regulation of the lkyB gene expression in *Escherichia coli* K-12 strains carrying an *lkyB-lacZ* gene fusion', *Molecular and General Genetics*, 201: 323–8.

Lee, S.J., Boos, W., Bouche, J.P., and Plumbridge, J. (2000) 'Signal transduction between a membrane-bound transporter, PtsG, and a soluble transcription factor, Mlc, of *Escherichia coli*', *European Molecular Biology Organization Journal*, 19: 5353–61.

Lee, J.H., Lee, D.E., Lee, B.U., and Kim, H.S. (2003) 'Global analyses of transcriptomes and proteomes of a parent strain and an l-threonine-over-producing mutant strain', *Journal of Bacteriology*, 185: 5442–51.

Li, Z. and Demple, B. (1994) 'SoxS, an activator of superoxide stress genes in *Escherichia coli*: Purification and interaction with DNA', *Journal of Biological Chemistry*, 269: 18371–7.

Liang, A. and Houghton, R.L. (1981) 'Coregulation of oxidized nicotinamide adenine dinucleotide (phosphate) transhydrogenase and glutamate dehydrogenase activities in enteric bacteria during nitrogen limitation', *Journal of Bacteriology*, 146: 997–1002.

Liberek, K., Galitski, T.P., Zylicz, M., and Georgopoulos, C. (1992) 'The DnaK chaperone modulates the heat-shock response of *Escherichia coli* by binding to the σ^{32} transcription factor', *Proceedings of the National Academy of Science USA*, 89: 3516–20.

Lin, H., Bennett, G.N., and San, K.Y. (2005) 'Chemostat culture characterization of *Escherichia coli* mutant strains metabolically engineered for aerobic succinate production: A study of the modified metabolic network based on metabolite profile, enzyme activity, and gene expression profile', *Metabolic Engineering*, 7: 337–52.

Lin, J., Lee, I.S., Frey, J., Slonczewski, J.L., and Foster, J.W. (1995) 'Comparative analysis of extreme acid survival in *Salmonella typhimurium, Shigella flexneri,* and *Escherichia coli*', *Journal of Bacteriology*, 177: 4097–04.

Liochev, S.I. and Fridovich, I. (1992) 'Fumarase C, the stable fumarase of *Escherichia coli*, is controlled by the *soxRS* regulon', *Proceedings of the National Academy of Science USA*, 89: 5892–6.

Liochev, S.I. and Fridovich, I. (1994) 'The role of O_2^- in the production of OH: *in vitro* and *in vivo*', *Free Radical Biology and Medicine*, 16: 29–33.

Liochev, S.I., Hausladen, A., and Fridovich, I. (1999) 'Nitroreductase A is regulated as a member of the *soxRS* regulon of *Escherichia coli*', *Proceedings of the National Academy of Science USA*, 96: 3537–9.

Liu, M., Durfee, T., Cabrera, J.E., Zhao, K., Jin, D.J., and Blattner, F.R. (2005) 'Global transcriptional programs reveal a carbon source foraging strategy by *Escherichia coli*', *Journal of Biological Chemistry*, 280: 15921–7.

Lowen, P.C., Ossowski, I.V., Switala, J., and Mulvey, M.R. (1993) 'KatF (σ_s) synthesis in *Escherichia coli* is subject to post-transcriptional regulation', *Journal of Bacteriology*, 175: 2150–3.

Lu, C., Bentley, W.E., and Rao, G. (2003) 'Comparison of oxidative stress response genes in aerobic *Escherichia coli* fermentations,' *Biotechnology and Bioengeering*, 83: 864–70.

Lugtenberg, B., Peters, R., Bernheimer, H., and Berendsen, W. (1976) 'Influence of cultural conditions and mutations on the composition of the outer membrane proteins of *Escherichia coli*', *Molecular and General Genetics*, 147: 251–62.

Lunin, V.V., Li, Y., Schrag, J.D., Iannuzzi, P., Cygler, M., and Matte, A. (2004) 'Crystal structures of *Escherichia coli* ATP-dependent glucokinase and its complex with glucose', *Journal of Bacteriology*, 186: 6915–27.

Lynch, A.S. and Lin, E.C.C. (1996a) 'Transcriptional control mediated by the ArcA two-component response regulator protein of *Escherichia coli*: characterization of DNA binding at target promoters', *Journal of Bacteriology*, 178: 6238–49.

Lynch, A.S. and Lin, E.C.C. (1996b) 'Responses to molecular oxygen', in: Escherichia coli *and* Salmonella typhimurium: *Cellular and Molecular Biology*, 2nd edition, edited by F.C. Neidhardt et al. Washington DC: American Society for Microbiology Press. pp. 1526–38.

Ma, Z., Gong, S., Richard, H., Tucker, D.L., Conway, T., and Foster, J.W. (2003) 'GadE (YhiE) activates glutamate decarboxylase-dependent acid resistance in *Escherichia coli* K-12', *Molecular Microbiology*, 49: 1309–20

Maeda, H., Fujita, N., and Ishihama, A. (2000) 'Competition among seven *Escherichia coli* sigma subunits: relative binding affinities to the core RNA polymerase', *Nucleic Acids Research*, 28: 3497–503

Magasanik, B. (1961) *Catabolite Repression*. Cold Spring Harbor Symposia on Quantitative Biology, 26: 249–56.

Magasanik, B. (1996) 'Regulation of nitrogen utilization', in Escherichia coli *and* Salmonella typhimurium: *Cellular and Molecular Biology*, edited by F.C. Neidhardt et al. Washington, DC: American Society for Microbiology Press. pp. 1344–56.

Maheswaran, M. and Forchhammer, K. (2003) 'Carbon-source-dependent nitrogen regulation in *Escherichia coli* is mediated through glutamine-dependent GlnB signaling', *Microbiology*, 149: 2163–72.

Makino, K., Shinagawa, H. and Nakata, A. (1985) 'Regulation of the phosphate regulon of *Escherichia coli* K-12: Regulation and role of the regulatory gene *phoR*', *Journal of Molecular Biology*, 184: 231–40.

Makino, K., Shinagawa, H., Amemura, M., Kimura, S., Nakata, A., and Ishihama, A. (1988) 'Regulation of the phosphate regulon of *Escherichia coli*: activation of *pstS* transcription by PhoB protein *in vitro*', *Journal of Molecular Biology*, 203: 85–95.

Makino, K., Shinagawa, H., Amemura, M., Kawamoto, T., Yamada, M., and Nakata, A. (1989) 'Signal transduction in the phosphate regulon of *Escherichia coli* involves phosphotransfer between PhoR and PhoB proteins', *Journal of Molecular Biology*, 210: 551–9.

Makino, K., Amemura, M., Kim, S.K., Nakata, A., and Shinagawa, H. (1993) 'Role of the σ70 subunit of RNA polymerase in transcriptional activation by activator *Escherichia coli*', *Genes Development*, 17(1): 149–60.

Maloy, S.R. and Nunn, W.D. (1982) 'Genetic regulation of the glyoxylate shunt in *Escherichia coli* K-12', *Journal of Bacteriology*, 149: 173–80.

Malpica, R., Franco, B., Rodriguez, C., Malpica, R., Franco, B. et al. (2004) 'Identification of a quinone-sensitive redox switch in the ArcB sensor kinase', *Proceedings of the National Academy of Sciences of the USA*, 101(36): 13318–23.

Mao, X.J., Huo, Y.X., Buck, M., Kolb, A., and Wang, Y.P. (2007) 'Interplay between CRP-cAMP and PII-Ntr systems forms novel regulatory network between carbon metabolism and nitrogen assimilation in *Escherichia coli*', *Nucleic Acids Research*, 35: 1432–40.

Marrakchi, H., Zhang, Y.M., and Rock, C.O. (2002) 'Mechanistic diversity and regulation of Type II fatty acid synthesis', *Biochemical Society Transaction*, 30: 1050–5.

Martin-Galiano, A.J., Ferrandiz, M.J., and de La Campa, A.G. (2001) 'The promoter of the operon encoding the F_0F_1 ATPase of *Streptococcus pneumoniae* is inducible by pH', *Molecular Microbiology*, 41: 1327–38.

Marzan, L.W., Siddiquee, K.A.Z., and Shimizu, K. (2011) 'Metabolic regulation of *fnr* gene knockout *Escherichia coli* under oxygen limitation', *Bioengineered Bugs*, 2(6): November/December.

Masuda, N. and Church, G.M. (2003) 'Regulatory network of acid resistance genes in *Escherichia coli*', *Molecular Microbiology*, 48(3): 699–712.

Matsushita, A. and Mizuno, T. (1998) 'A dual-signalling mechanism mediated by the ArcB hybrid sensor kinase containing the histidine-containing phosphotransfer domain in *Escherichia coli*', *Journal of Bacteriology*, 180: 3973–37.

Merrick, M.J. and Edwards, R.A. (1995) 'Nitrogen control in bacteria', *Microbiology Reviews*, 4: 604–22.

Mijakovic, I., Poncet, S., Galinier, A., Monedero, V., Fieulaine, S. et al. (2002) 'Pyrophosphate-producing protein dephosphorylation by HPr kinase/phosphorylase: A relic of early life?' *Proceedings of the National Academy of Science, USA*, 99: 13442–7.

Mikulskis, A., Aristarkhov, A., and Lin, E.C.C. (1997) 'Regulation of expression of ethanol dehydrogenase gene (*adhE*) in *Escherichia coli* by catabolite repressor activator protein Cra', *Journal of Bacteriology*, 179: 7129–34.

Misko, T.P., Mitchell, W.J., Meadow, N.D., and Roseman, S. (1987) 'Sugar transport by the bacterial phosphotransferase system. Reconstitution of inducer exclusion in *Salmonella typhimurium* membrane vesicles', *Journal of Biological Chemistry*, 262: 16261–6.

Moat, A.G., Foster, J.W., and Spector, M.P. (2002) *Microbiology*, 4th edition, New York: Wiley-Liss, Inc.

Mogk, A., Schlieker, C., Friedrich, K.L., Schonfeld, H.J., Vierling, E., and Bukau, B. (2003) 'Refolding of substrates bound to small Hsps relies on a diaggregation reaction mediated most efficiently by ClpB/DnaK', *Journal of Biological Chemistry*, 278: 31033–42.

Monod, J. (1942) 'Recherches sur la Croissance de cultures Bacteriennes', PhD Thesis, Hermann et Cie, Paris.

Moreno, R. and Rojo, F. (2008) 'The target for the *Psedomonas putida* Crc global regulator at the benzoate degradation pathway is the BenR transcriptional regulator', *Journal of Bacteriology*, 190: 1539–45.

Moreno, M.S., Schneider, B.J., Maile, R.R., Weyler, W., and Saier, M.H. Jr. (2001) 'Catabolite repression mediated by CcpA protein in *Bacillus subtilis*: novel modes of regulation revealed by whole-genome analyses', *Molecular Microbiology*, 39: 1366–81.

Moreno, R., Ruiz-Manzano, A., Yuste, L., and Rojo, F. (2007) 'The *Pseudomonas putida* Crc global regulator is an RNA binding protein that inhibits translation of the AlkS transcriptional regulator', *Molecular Microbiology*, 64: 665–75.

Morita, T., Waleed, E., Yuya, T., Toshifumi, I., and Hiroji, A. (2003) 'Accumulation of glucose 6-phospate or fructose 6-phosphate is responsible for destabilization of glucose transporter mRNA in *Escherichia coli*', *Journal of Biological Chemistry*, 278: 15608–14.

Muller, C., Petruschka, L., Cuypers, H., Burchhardt, C., and Herrmann, H. (1996) 'Carbon catabolite repression of phenol degradation in *Pseudomonas putida* is mediated by the inhibition of the activator protein PhlR', *Journal of Bacteriology*, 178: 2030–6.

Murai, T., Tokushige, M., Nagai, J., and Katsuki, H. (1971) 'Physiological functions of NAD^+- and $NADP^+$-linked malic enzymes in *E. coli*', *Biochemical and Biophysical Research Communications*, 43: 75–881.

Nakano, M.M., Zuber, P., and Sonenshein, A. (1998) 'Anaerobic regulation of *Bacillus subtilis* krebs cycle gene', *Journal of Bacteriology*, 180: 3304–11.

Nakata, A., Amemura, M., and Shinagawa, H. (1984) 'Regulation of the phosphate regulon in *Escherichia coli* K-12: Regulation of the negative regulatory gene *phoU* and identification of the gene product', *Journal of Bacteriology*, 159(3): 979–85.

Nam, T., Cho, S., Shin, D., Kim, J., Jeong, J. et al. (2001) 'The *Escherichia coli* glucose transporter enzyme IICBGlc recruits the global repressor Mlc', *European Molecular Biology Organization Journal*, 120: 491–8.

Nanchen, A., Schicker, A., Revelles, O., and Sauer, U. (2008) 'Cyclic AMP-dependent catabolite repression is dominant control mechanism of metabolic fluxes under glucose limitation', *Journal of Bacteriology*, 190: 2323–30.

Nelson, S.O., Wright, J.K., and Postma P.W. (1983) 'The mechanism inducer expression: Direct interaction between purified III^{Glc} of the phosphoenolpyruvate: Sugar phosphotransferase system and the lactose carrier of *Escherichia coli*', *European Molecular Biology Organization Journal*, 2: 715–20.

Nesmeianova, M.A., Gonina, S.A., and Kulaev, I.S. (1975) 'Biosynthesis of *Escherichia coli* polyphosphatases under control of the regulatory genes usual with alkaline phosphatase', *Dokl Akad Nauk SSSR*, 224(3): 710–12.

Nicholson, T.L., Chie, K., and Stephens, R.S. (2004) 'Chlamydia trachomatis lacks an adaptive response to changes in carbon source availability', *Infection Immunity*, 72: 4286–9.

Nikaido, H. (1996) *Outer Membrane*, edited by F.C. Neidhardt. Washington, DC: American Society for Microbiology Press. pp. 1325–43.

Nikaido, H. and Nakae, T. (1979) 'The outer membrane of Gram-negative bacteria', *Advanced Microbology and Physiology*, 20: 163–250.

Nikaido, H. and Rosenberg, E.Y. (1983) 'Porin channels in *Escherichia coli*: studies with liposomes reconstituted from purified proteins', *Journal of Bacteriology*, 153: 241–52.

Nikaido, H. and Vaara, M. (1985) 'Molecular basis of bacterial outer membrane permeability', *Microbiology Review*, 49: 1–32.

Ninfa, A.J. and Atkinson, M.R. (2000) 'PII signal transduction proteins', *Trends in Microbiology*, 8: 172–9.

Ninfa, A.J. and Jiang, P. (2005) 'PII signal transduction proteins: sensors of α-ketoglutarate that regulate nitrogen metabolism', *Current Opinion in Microbiology*, 8: 168–73.

Ninfa, A.J., Jiang, P., Atkinson, M.R., and Peliska, J.A. (2000) 'Integration of antagonistic signals in the regulation of nitrogen assimilation in *Escherichia coli*', *Current Topics in Cellular Regulation*, 36: 31–75.

Nizam, S.A. and Shimizu, K. (2008) 'Effects of *arc*A and *arc*B genes knockout on the metabolism in *Escherichia coli* under anaerobic and microaerobic conditions', *Biochemical Engineering Journal*, 42; 229–36.

Nizam, S.A., Ehu, J., and Shimizu, K. (2009) 'Effects of *arc*A and *arc*B genes knockout on the metabolism in *Escherichia coli* under aerobic condition', *Biochemical Engineering Journal*, 44: 240–50.

Oarche, S. et al. (2006) 'Lactose-over-glucose preference in *Bifidobacterium longum* NCC2705: glcP, encoding a glucose transporter, is subject to lactose repression', *Journal of Bacteriology*, 188: 1260–5.

Oganesyan, V., Oganesyan, N., Adams, P.D., Jancarik, J., Yokota, H.A. et al. (2005) 'Crystal structure of the "PhoU-like" phosphate uptake regulator from *Aquifex aeolicus*', *Journal of Bacteriology*, 187: 4238–44.

Pahel, G., Rothstein, D.M., and Magasanik, B. (1982) Complex *glnA-glnL-glnG* operon of *Escherichia coli*', *Journal of Bacteriology*, 150: 202–13.

Park, S.J. and Gunsalus R.P. (1995a) 'Oxygen, iron, carbon, and super-oxide control of the fumarase *fum*A and *fum*C genes of *Escherichia coli*: role of the *arc*A, *fnr*, and *sox*R gene products', *Journal of Bacteriology*, 177: 6255–62.

Park, S.J., Tseng, C.P., and Gunsalus, R.P. (1995) 'Regulation of succinate dehydrogenase (sdhCDAB) operon expression in Escherichia coli in response to carbon supply and anaerobiosis: role of ArcA and Fnr', *Molecular Microbiology*, 15: 473–82.

Park, S.J., Chao. G., and Gunsalus, R.P. (1997) 'Aerobic regulation of the sucABCD genes of Escherichia coli, which encode α-ketoglutarate dehydrogenase and succinyl coenzyme A synthetase: roles of ArcA, Fnr, and the upstream sdhCDAB promoter', *Journal of Bacteriology*, 179: 4138–42.

Park, S.J., McCabe, J., Turna, J., and Gunsalus, R.P. (1994) 'Regulation of the citrate synthase (gltA) gene of Escherichia coli in response to anaerobiosis and carbon supply: Role of the arcA gene product' *Journal of Bacteriology*, 176: 5086–92.

Park, Y.H., Lee, B.R., Seok, Y.J., and Peterkofsky, A. (2006) 'In vitro reconstitution of catabolite repression in Escherichia coli', Journal of Biological Chemistry, 281: 6448–54.

Parkinson, J.S. (1993) 'Signal transduction schemes of bacteria', Cell, 73: 857–71.

Paul, L., Mishra, P.K., Blumenthal, R.M., and Matthews, R.G. (2007) 'Integration of regulatory signals through involvement of multiple global regulators: control of the Escherichia coli gltBDF operon by Lrp, IHF, Crp, and ArgR', BMC Microbiology, 7: 2.

Peng, L. and Shimizu, K. (2006) 'Effect of fadR gene knockout on the metabolism of Escherichia coli based on analyses of protein expressions, enzyme activities and intracellular metabolite concentrations', Enzyme and Microbial Technology, 38: 512–20.

Perrenoud, A. and Sauer, U. (2005) 'Impact of global transcriptional regulation by ArcA, ArcB, Cra, Crp, Cya, Fnr, and Mlc on glucose catabolism in Escherichia coli', Journal of Bacteriology, 187: 3171–9.

Petersen, S. and Young, G.M. (2002) 'Essential role for cyclic AMP and its receptor protein in Yersinia enterocolitica virulence', Infection Immunuity, 70: 3665–72.

Philippidis, G.P., Smith, T.K., and Wyman, C.E. (2004) 'Study of the enzymatic hydrolysis of cellulose for production of fuel ethanol by the simultaneous saccharification and fermentation process', Biotechnology and Bioengineering, 41(9): 846–53.

Phue, J.N., Noronha, S.B., Hattacharyya, R., Wolfe, A.J., and Shiloach, J. (2005) 'Glucose metabolism at high density growth of E. coliB and E. coliK: differences in metabolic pathways are responsible for efficient glucose utilization in E. coliB as determined by microarrays and Northern Blot Analysis', Biotechnology and Bioengineering, 90: 805–20.

Plumbridge, J. (1998a) 'Expression of ptsG, the gene for the major glucose PTS transporter in Escherichia coli, is repressed by Mlc and induced by growth on glucose', Molecular Microbiology, 29: 1053–63.

Plumbridge, J. (1998b) 'Control of the expression of the manXYZ operon in Escherichia coli: Mlc is a negative regulator of the mannose PTS', Molecular Microbiology, 27: 369–80.

Pomposiello, P.J. and Demple, B. (2001) 'Redox-operated genetic switches: the soxR and OxyR transcription factors', Trends in Biotechnology, 19: 109–14.

Poolman, B. and Konings, W.N. (1993) Biochemisty and Biophysics ACTA, 1183: 5–39.

Poolman, B., Knol, J., Mollet, B., Nieuwenhuis, B., and Sulter, G. (1995) 'Regulation of bacterial sugar-H^+ symport by phosphoeneolpyruvate-dependent enzyme I/HPr-mediated phosphorylation', Proceedings of the National Academy of Science USA, 92: 778–82.

Postma, P.W., Lengeler, J.W., and Jacobson, G.R. (1996) 'Phosphoenolpyruvate: Carbohydrate phosphotransferase systems', in: Escherichia coli and Salmonella typhimurium: Cellular and Molecular Biology, edited by F.C. Neidhardt et al. Washington, DC: American Society for Microbiology Press. pp. 1149–74.

Pratt, C. and Torriani, A. (1977) 'Complementation test between alkaline phosphate regulatory mutations phoB and phoRc in Escherichia coli', Genetics, 85: 203–8.

Pratt, L.A. and Silhavy, T.J. (1996) 'The response regulator SprE controls the stability of RpoS', *Proceedings of the National Academy of Science USA*, 93: 2488–92.

Privalle, C.T. and Fridovich, I. (1987) 'Induction of superoxide dismutase in *Escherichia coli* by heat shock', *Proceedings of the National Academy of Science USA*, 84: 2723–6.

Prost, J.F., Negre, D., Oudot, C., Murakami, K., Ishihama, A. et al. (1999) 'Cra-dependent transcriptional activation of the icd gene of *Escherichia coli*', *Journal of Bacteriology*, 181: 893–8.

Rahman, M. and Shimizu, K. (2004) 'Altered acetate metabolism and biomass production in several *Escherichia coli* mutants lacking *rpoS*-dependent metabolic pathway genes', *Molecular Biosystems*, 4: 160–9.

Rahman, M., Hasa, M.M., and Shimizu, K. (2008) 'Growth phase-dependent changes in the expression of global regulatory genes and associated metabolic pathways in *Esherichia coli*', *Biotechnology Letters*, 30: 853–60.

Rahman, M., Hasan, M.R., Oba, T., and Shimizu, K. (2006) 'Effect of *rpo*S gene knockout on the metabolism of *Escherichia coli* during exponential growth phase and early stationary phase based on gene expressions, enzyme activities and intracellular metabolite concentrations', *Biotechnology and Bioengineering*, 94: 585–95.

Ramseier, T.M., Chien, S.Y., Saier, M.H. Jr. (1996) 'Cooperative interaction between Cra and Fnr in the regulation of the cydAB operon of *Escherichia coli*', *Current Microbiology*, 33: 270–4.

Ramseier, T.M., Bledig, S., Michotey, V., Fegahali, R., Saier, M.H. Jr. (1995) 'The global regulatory protein FruR modulates the direction of carbon flow in *Escherichia coli*', *Molecular Microbiology*, 16: 1157–69.

Ramström, H., Bourotte, M., Philippe, C., Schmitt, M., Haiech, J., and Bourguignon, J.J. (2004) 'Heterocyclic bis-cations as starting hits for design of inhibitors of the bifunctional enzyme histidine-containing protein kinase/phosphatase from *Bacillus subtilis*', *Journal of Biological Chemistry*, 47: 2264–75.

Reitzer, L. (2003) 'Nitrogen assimilation and global regulation in *Escherichia coli*', *Annual Reviews in Microbiology*, 57: 155–76.

Reitzer, L.J. and Magasanik, B. (1985) 'Expression of *glnA* in *Escherichia coli* is regulated at tandem promoters', *Proceedings of the National Academy of Science USA*, 82: 1979–83.

Reizer, J., Hoischen, C., Titgemeyer, F., Rivolta, C., Rabus, R. et al. (1998) 'A novel protein kinase that controls carbon catabolite repression in bacteria', *Molecular Microbiology*, 27: 1157–69.

Rektorschek, M., Buhmann, A., Weeks, D., Schwan, D., Bensch, K.W. et al. (2000) 'Acid resistance of *Helicobacter pylori* depends on the UreI membrane protein and an inner membrane proton barrier', *Molecular Microbiology*, 36(1): 141–52.

Riba, L., Becerril, B., Serven-Gonzaèlez, L., Valle, F., and Bolivar, F. (1988) 'Identification of a functional promoter for the *Escherichia coli gdhA* gene and its regulation', *Gene*, 71: 233–46.

Richard, H.T. and Foster, J.W. (2003) 'Acid resistance in *Escherichia coli*', *Advanced Applied Microbiology*, 52: 167–86.

Richard, H.T. and Foster, J.W. (2004) '*Escherichia coli* glutamate- and arginine-dependent acid resistance systems increase internal pH and reverse transmembrane potential', *Journal of Bacteriology*, 86(18): 6032–41.

Richmond, C.S., Glasner, D.J., Mau, R., Jin, H., and Blattner, F.R. (1999) 'Genome-wide expression profiling in *Escherichia coli*', *Nucleic Acids Research*, 27: 3821–35.

Rosen, R. and Ron, E.Z. (2002) 'Proteome analysis in the study of the Bacterial Heat-shock response,' *Mass Spectrometer Reviews*, 21: 244–65.

Ruiz, N. and Silhavy, T.J. (2003) 'Constitutive activation of the *Escherichia coli* Pho regulon up-regulates *rpoS* translation in an Hfq dependent fashion', *Journal of Bacteriology*, 185(20): 5984–92.

Ryu, S., Ramseier, T., Michotey, V., Saier, M.H. Jr., and Garges, S. (1995) 'Effects of the FruR regulator on transcription of the pts operon of *Escherichia coli*', *Journal of Biological Chemistry*, 270: 2489–96.

Saier, M.H. (2000) 'Vectorial metabolism and the evolution of transport systems', *Journal of Bacteriology*, 182: 5029–35.

Saier, M.H., Jr. and Ramseier, T.M. (1996) 'The catabolite repressor/activator (Cra) protein of enteric bacteria', *Journal of Bacteriology*, 178: 3411–17.

Saier, M.H. Jr., Ramseier, T.M., and Reizer, J. (1997) 'Regulation of carbon utilization', in Escherichia coli *and* Salmonella typhimurium: *Cellular and Molecular Biology*, edited by F.C. Neidhardt et al. Washington, DC: American Society for Microbiology Press. pp. 1325–43.

Salmon, K., Hung, S., Mekjian, K., Baldi, P., Hatfield, G.W., and Gunsalus, R.P. (2003) 'Global gene expression profiling in *Escherichia coli* K12: the effect of oxygen availability and FNR', *Journal of Biological Chemistry*, 278: 29837–55.

Sarkar, D. and Shimizu, K. (2008) 'Effect of *cra* gene knockout together with other genes knockouts on the improvement of substrate consumption rate in *Escherichia coli* under microaerobic conditions', *Biochemical Engineering Journal*, 42: 224–8.

Sarkar, D., Siddiquee, K.A.Z., Arauzo-Bravo, M.J., Oba, T., Shimizu, K. (2008) 'Effect of *cra* gene knockout together with *edd* and *iclR* genes knockout on the metabolism in *Escherichia coli*', *Archives of Microbiology*, 190: 559–71.

Sato, M., Machida, K., Arikado, E., Saito, H., Kakegawa, T., and Kobayashi, H. (2000) 'Expression of outer membrane proteins in *Escherichia coli* growing at acid pH', *Applied and Environmental Microbiology*, 66(3): 943–7.

Sauer, U., and Eikmanns, B.J. (2005) 'The PEP-pyruvate-oxaloacetate node as the switch point for carbon flux distribution in bacteria', *FEMS Microbiology Reviews*, 29: 765–94.

Sawers, G. and Suppmann, B. (1992) 'Anaerobic induction of pyruvate formate-lyase gene expression is mediated by the ArcA and FNR proteins', *Journal of Bacteriology*, 174: 3474–8.

Sawers, G. and Watson, G. (1998) 'A glycyl radical solution: oxygen-dependent interconversion of pyruvate formate-lyase', *Molecular Microbiology*, 29: 945–54.

Schlieker, C., Bukau, B., and Mogk, A. (2002) 'Prevention and reversion of protein aggregation by molecular chaperones in the *E. coli* cytosol: implications for their applicability in biotechnology', *Journal of Biotechnology*, 96: 13–21.

Schuetz, R., Kuepfer, L., and Sauer, U. (2006) 'Systematic evaluation of objective functions for predicting intracellular fluxes in *Escherichia coli*', *Molecular Systems Biology*, 3: 119.

Schumacher, M.A., Seidel, G., Hillen, W., and Brennan, R.G. (2007) 'Structual mechanism for the fine-tuning of CcpA function by the small molecule effectors glucose 6-phosphate and fructose 1,6-bisphosphate', *Journal of Molecular Biology*, 368: 1042–50.

Schurdell, M.S., Woodbury, G.M., and McCleary, W.R. (2007) 'Genetic evidence suggests that the intergenic region between *pstA* and *pstB* plays a role in the regulation of *rpoS* translation during phosphate limitation', *Journal of Bacteriology*, 189(3): 1150–3.

Seidel, G., Diel, M., Fuchsbauer, N., and Hillen, W. (2005) 'Quantitative interdependence of coeffectors, CcpA and cre in carbon catabolite regulation of *Bacillus subtilis*', *Federation of European Biochemical Societies Journal*, 272: 2566–77.

Shalel-Levanon, S., San, K.Y., and Bennett, G.N. (2005a) 'Effect of oxygen on the *Escherichia coli* ArcA and FNR regulation systems and metabolic responses', *Biotechnology and Bioengeering*, 89: 556–64.

Shalel-Levanon, S, San, K.Y., and Bennett, G.N. (2005b) 'Effect of oxygen, and ArcA and FNR regulators on the expression of genes related to the electron transfer chain and the TCA cycle in *Escherichia coli*', *Metabolic Engineering*, 7: 364–74.

Shapiro, B.M., and Stadtman, E.R. (1968) 'Glutamine synthetase deadenylylating enzyme', *Biochemical and Biophysical Research Communications*, 30: 32–7.

Shin, P.K and Seo, J.H. (1990) 'Analysis of *E. coli phoA-lacZ* fusion gene expression inserted into a multicopy plasmid and host cell's chromosome', *Biotechnology and Bioengeering*, 36(11): 1097–104.

Shin, S., Song, S.G., Lee, D.S., Pan, J.G., and Park, C. (1997) 'Involvement of *iclR* and *rpoS* in the induction of *acs*, the gene for acetyl coenzyme A synthetase of *Escherichia coli* K-12', *FEMS Microbiology Letters*, 146: 103–8.

Shin, D., Lim, S., Seok, Y., and Ryu, S. (2001) 'Heat shock RNA polymerase (Eσ^{32}) is involved in the transcription of *mlc* and crucial for induction of the Mlc regulon by Glucose in *Escherichia coli*', *Journal of Biological Chemistry*, 276: 25871–5.

Shinagawa, H., Makino, K., Amemura, M., and Nakata, A. (1987) 'Structure and function of the regulatory gene for the phosphate regulon in *Escherichia coli*', in: *Phosphate Metabolism and Cellular Regulation in Microorganisms*, edited by A. Torriani-Gorini, F.G. Rothman, S. Silver, A. Wright, and E. Yagil. Washington, DC: American Society for Microbiology Press. pp. 20–5.

Sipro, S., Gatson, K.L., Bell, A.I., Roberts, R.E., Busby, S.J.W., and Guest, J.R. (1990) 'Interconversion of the DNA-binding specificity of two related transcription regulators, CRP and FNR', *Molecular Microbiology*, 4: 1831–8.

Skelly, S., Colemant, T., Fu, C.F., Brot, N., and Weissbach, H. (1987) 'Correlation between the 32-kDa σ factor levels and *in vitro* expression of *Escherichia coli* heat shock genes', *Proceedings of the National Academy of Science USA*, 84: 8365–9.

Smirnova, I. et al. (2007) 'Sugar binding induces an outward facing conformation of LacY', *Proceedings of the National Academy of Sciences of USA*, 104: 16504–9.

Smith, J.L. (2003) 'The role of gastric acid in preventing food-borne disease and how bacteria overcome acid conditions', *Journal of Food Protection*, 66: 1292–303.

Smith, M.W. and Payne, J.W. (1992) 'Expression of periplasmic binding proteins for peptide transport is subject to negative regulation by phosphate limitation in *Escherichia coli*', *FEMS Microbiology Letters*, 79: 183–90.

Soini, J., Falschlehner, C., Mayer, C., Böhm, D., Weinel, S. et al. (2005) 'Transient increase of ATP as a response to temperature up-shift in *Escherichia coli*', *Microbial Cell Factories*, 4: 9.

Sorensen, H.P. and Mortensen, K.K. (2005) 'Soluble expression of recombinant proteins in the cytoplasm of *Escherichia coli*', *Microbial Cell Factories*, 4: 1.

Spira, B. and Yagil, E. (1999) 'The integration host factor (IHF) affects the expression of the phosphate-binding protein and of alkaline phosphatase in *Escherichia coli*', *Current Microbiology*, 38: 80–5.

Spira, B., Silberstein, N., and Yagil, E. (1995) 'Guanosine 39, 59- bispyrophosphate (ppGpp) synthesis in cells of *Escherichia coli* starved for P_i', *Journal of Bacteriology*, 177: 4053–8.

Sprenger, G.A., Schorken, U., Sprenger, G., and Sahm, H. (1995) 'Transaldolase B of *Escrichia coli* K-12: cloning of its gene, *tal*B, and characterization of the enzyme from recombinant strains', *Journal of Bacteriology*, 177: 5930–6.

Stadtman, E.R. (1990) 'Discovery of glutamine synthetase cascade,' *Methods in Enzymology*, 182: 793–809.

Stephen, P.E., Darlison, M.G., Lewis, H.M., and Guest, J.R. (1983) 'The pyruvate dehydrogenase complex of *Escherichia coli* K12. Nucleotide sequence encoding the pyruvate dehydrogenase component', *European Journal of Biochemistry*, 133: 155–62.

Sternberg, N. (1973) 'Properties of a mutant of *Escherichia coli* in bacteriophage λ head formation (groE), II the propagation of phase λ', *Journal of Molecular Biology*, 76: 25–44.

Stincone, A., Daudi, N., Rahman, A.S., Antczak, P., Henderson, I. et al. (2011) 'A systems biology approach sheds new light on *Escherichia coli* acid resistance', *Nucleic Acids Research*, 39(17): 7512–28.

Stock, J.B., Ninfa, A., and Stock, A.M. (1989) 'Protein phosphorylation and regulation of adaptive responses in bacteria', *Microbiology Review*, 53: 450–90.

Storz, G., Tartaglia, L.A., and Ames, B.N. (1990) 'Transcriptional regulator of oxidative stress-inducible genes: direct activation by oxidation', *Science*, 248: 189–94.

Straus, D.B., Walter, W.A., and Gross, C.A. (1987) 'The heat shock response of *Escherichia coli* is regulated by changes in the concentration of σ^{32}', *Nature*, 329: 348–51.

Stryer, L. (1988) *Biochemistry*. New York: W.H. Freeman and Company. pp. 511–12.

Suppmann, B. and Sawers, G. (1994) 'Isolation and characterization of hypophosphite-resistant mutants of *Escherichia coli*: Identificationof the FocA protein, encoded by the pfl operon, as a putative formate transporter', *Molecular Microbiology*, 11: 965–82.

Suziedeliene, E., Suziedelis, K., Garbenciute, V., and Normak, S. (1999) 'The acid-inducible asr gene in *Escherichia coli*: Transcriptional control by the PhoBR operon', *Journal of Bacteriology*, 181(7): 2084–93.

Tanaka, Y., Kimata, K. and Aiba, H. (2000) 'A novel regulatory role of glucose transporter of *Escherichia coli*: Membrane sequestration of a global repressor Mlc', *European Molecular Biology Organization Journal*, 19: 5344–52.

Tanaka, Y., Kimata, K., Inada, T., Tagami, H., and Aiba, H. (1999) 'Negative regulation of the *pts* operon by Mlc: Mechanism underlying glucose induction in *Escherichia coli*', *Genes Cells*, 4: 391–9.

Tang, Y., Quail, M.A., Artymiuk, P.J., Guest, J.R., and Green, J. (2002) '*Escherichia coli* aconitases and oxidative stress: post-transcriptional regulation of *sod*A expression', *Microbiology*, 148: 1027–37.

Taschner, N.P., Yagil, E., and Spira, B. (2004) 'A differential effect of σ^S on the expression of the PHO regulon genes of *Escherichia coli*', *Microbiology*, 150: 2985–92.

Taschner, N.P., Yagil, E., and Spira, B. (2006) 'The effect of IHF on σ^S selectivity of the *phoA* and *pst* promoters of *Escherichia coli*', *Archives of Microbiology*, 185(3): 234–7.

Taylor, W.E., Straus, D.B., Grossman, A.D., Burton, Z.F., Gross, C.A., and Burgess, R.R. (1984) 'Transcription from a heat-inducible promoter causes heat shock regulation of the sigma subunit of *E. coli* RNA polymerase', *Cell*, 1984, 38: 371–81.

Tchieu, J.H., Norris, V., Edwards, J.S., and Saier, M.H. Jr. (2001) 'The complete phosphotranferase system in *Escherichia coli*', *Journal of Molecular Microbiology and Biotechnology*, 3: 329–46.

Tilly, K., Spence, J., and Georgopoulos, C. (1983) 'Modulation of stability of the *Escherichia coli* heat shock regulatory factor σ^{32}', *Journal of Bacteriology*, 171(3): 1585–9.

Tilly, K., McKittrick, N., Zylicz, M., and Georgopoulos, C. (1983) 'The dnaK protein modulates the heat-shock response of *Escherichia coli*', *Cell*, 34: 641–6.

Tilly, K., Erickson, J., Sharma, S., and Georgopoulos, C. (1986) 'Heat shock regulatory gene *rpoH* mRNA level increases after heat shock in *Escherichia coli*', *Journal of Bacteriology*, 168: 1155–8.

Titgemeyer, F. and Hillem, W. (2002) 'Global control of sugar metabolism: a Gram-positive solution', *Antonie van Leeuwenhoek*, 82: 59–71.

Titgemeyer, F., Mason, R.E., and Saier, M.H. Jr. (1994) 'Regulation of the raffinose permease of *Escherichia coli* by the glucose-specific enzyme A of the phosphoenolpyruvate: Sugar phosphotransferase system', *Journal of Bacteriology*, 176: 543–6.

Torriani, A. and Ludke, D.N. (1985) 'The pho regulon of *Escherichia coil*', in: *The Molecular Biology of Bacterial Growth*, edited by M. Schaechter, F.C. Neidhardt, J. Ingraham, and N.O. Kjeldgaard. Boston, MA: Jones and Bartlett.

Tsai, S.P. and Lee, Y.H. (1988) 'Application of metabolic pathways stoichiometry to statistical analysis of bioreactor measurement data', *Biotechnology and Bioengeering*, 32: 713–15.

Tsaneva, I.R. and Weiss, B. (1990) '*soxR*, a locus governing a superoxide response regulon in *Escherichia coli* K-12', *Journal of Bacteriology*, 172: 4197–205.

Tseng, C.P., Albrecht, J., and Gunsalus, R.P. (1996) 'Effect of microaerophilic cell growth conditions on expression of the aerobic (cyoABCDE and cydAB) and anaerobic (narGHJI, frdABCD, and dmsABC) respiratory pathway genes in *Escherichia coli*', *Journal of Bacteriology*, 178: 1094–8.

Tsfasman, I.M. and Nesmeianova, M.A. (1981) 'Membrane proteins in *Escherichia coli*: effect of orthophosphate and mutation on regulatory genes of secreted alkaline phosphate', *Molecular Biology*, 15(2): 298–309.

Tsuzuki, M., Ishege, K., and Mizuno, T. (1995) 'Phosphotransfer circuitry of the putative multi-signal transducer, ArcB, of *Escherichia coli: in vitro* studies with mutants', *Molecular Microbiology*, 18: 953–62.

Tucker, D.L., Tucker, N., Ma, Z., Foster, J.W., Miranda, R.L. et al. (2003) 'Genes of GadX-GadW regulon in *Escherichia coli*', *Journal of Bacteriology*, 185: 3190–201.

Unden, G., Achebach, S., Holighaus, G., Tran, H.G., Wackwitz, B., and Zeuner, Y. (2002) 'Control of Fnr function of *Escherichia coli* by O_2 and reducing conditions', *Journal of Molecular Microbiology and Biotechnology*, 4: 263–8.

van Aalten, D.M., DiRusso, C.C., and Knudsen, J. (2001) 'The structural basis of acyl coenzyme A-dependent regulation of the transcription factor FadR', *European Molecular Biology Organization Journal*, 20: 2041–50.

van de Walle, M. and Shiloach, J. (1998) 'Proposed mechanism of acetate accumulation in two recombinant *Escherichia coli* strains during high density fermentation', *Biotechnology and Bioengeering*, 57: 71–8.

Van Dien, S.J. and Keasling, J.D. (1998) 'A dynamic model of the *Escherichia coli* phosphate-starvation response', *Journal of Theoretical Biology*, 190: 37–49.

van Heeswijk, W.C., Hoving, S., Molenaar, D., Stegeman, B., Kahn, D., Westerhoff, H.V. (1996) 'An alternative PII protein in the regulation of glutamine synthetase in *Escherichia coli*', *Molecular Microbiology*, 21: 133–46.

Van Wezel, G.P. et al. (2007) 'A new piece of an old jigsaw: glucose kinase is activated posttranslationally in a glucose transport-dependent manner in *Streptomyces coelicolor* A3(2)', *Journal of Molecular Microbiology and Biotechnology*, 12: 67–74.

Vanbogelen, R.A., Vaughn, V., and Neidhardt, F.C. (1983) 'Gene for heat-inducible lysyl tRNA synthetase (*lys U*) maps near cadA in *Escherichia coli*', *Journal of Bacteriology*, 153(2): 1066–8.

Varghese, S., Tang, Y., and Imlay, J.A. (2003) 'Contrasting sensitivities of *Escherichia coli* aconitases A and B to oxidation and iron depletion', *Journal of Bacteriology*, 185: 221–30.

Vemuri, G.N., Eiteman, M.A., and Altman, E. (2005) 'Increased recombinant protein production in Escherichia coli strains with over expressed water-forming NADH oxidase and a deleted ArcA regulatory protein', *Biotechnology and Bioengeering*, 90: 64–76.

Vemuri, G.N., Eiteman, M.A., and Altman, E. (2006a) 'Increased recombinant protein production in *Escherichia coli* strains with over-expressed water-forming NADH oxidase and a deleted ArcA regulatory protein', *Biotechnology and Bioengineering*, 94: 538–42.

Vemuri, G.N., Altman, E., Sangurdekar, D.P., Khodursky, A.B., and Eitman, M.A. (2006b) 'Overflow metabolism in *Escherichia coli* during steady-state growth: transcriptional regulation and effect of the redox ratio', *Applied Environmental Microbiology*, 72: 3653–61.

von Meyenburg, K. and Nikaido, H. (1977) 'Outer membrane of gram-negative bacteria. XVII. Secificity of transport process catalyzed by the lambda-receptor protein in *Escherichia coli*', *Biochemical and Biophysical Research Communications*, 78: 1100–7.

Vijaykumar, S.R.V., Kirchhof, M.G., Patten, C.L., and Schellhorn, H.E. (2004) 'RpoS-regulated genes of *Escherichia coli* identified by random *lacZ* fusion mutagenesis', *Journal of Bacteriology*, 186: 8499–507.

Wagner, A.F.V., Schulz, S., Bomke, J., Pils, T., Lehmann, W.D., and Knappe, J. (2001) 'YfiD of *E. coli* and Y061 of bacteriophage T4 as autonomous glycyl radical cofactors reconstituting the catalytic center of oxygen-fragmented pyruvate formate-lyase, *Biochemical and Biophysical Research Communications*, 285: 456–62.

Wanner, B.L. (1987) 'Phosphate regulon of gene expression in *Escherichia coli*', in: Escherichia coli *and* Salmonella typhimurium: *Cellular and Molecular Biology*, edited by F.C. Neidhardt et al. Washington DC: American Society for Microbiology Press. pp. 1326–33.

Wanner, B.L. (1993) 'Gene regulation by phosphate in enteric bacteria', *Journal of Cellular Biochemistry*, 51: 47–54

Wanner, B.L (1996) 'Phosphorus assimilation and control of the phosphate regulon', in: Escherichia coli *and* Salmonella typhimurium: *Cellular and Molecular Biology*, edited by F.C. Neidhardt et al. Washington DC: American Society for Microbiology Press. pp. 1357–81.

Wanner, B.L. (1997) 'Phosphate signaling and the control of gene expression in *Escherichia coli*', in: *Metal Ions in Gene Regulation*, edited by S. Silver and W. William. New York: Chapman and Hall. pp. 104–28.

Wanner, B.L., Wilmes, M.R., and Young, D.C. (1988) 'Control of bacterial alkaline phosphatase synthesis and variation in an *Escherichia coli* K-12 *phoR* mutant by adenyl cyclase, the cyclic amp receptor protein, and the *phoM* operon', *Journal of Bacteriology*, 170(3): 1092–102

Warner, J.B. and Lolkema, J.S. (2003) 'A Crh-specific function in carbon catabolite repression in *Bacillus subtilis*', *FEMS Microbiology Letters*, 220: 277–80.

Weber, J., Hoffmann, F., and Rinas, U. (2002) 'Metabolic adaptation of *Escherichia coli* during temperature-induced recombinant protein production: 2. Redirection of metabolic fluxes', *Biotechnology and Bioengeering*, 80: 320–30.

Weber, H., Polen, T., Heuveling, J., Wendisch, V.F., and Hengge, R. (2005) 'Genome-wide analysis of the general stress response network in *Escherichia coli*: σ^S-dependent genes, promoters, and sigma factor selectivity', *Journal of Bacteriology*, 187: 1591–603.

Wei, B., Shin, S., LaPorte, D., Wolfe, A.J., and Romeo, T. (2000) 'Global regulatory mutations in *csrA* and *rpoS* cause severe central carbon stress in *Escherichia coli* in the presence of acetate', *Journal of Bacteriology*, 182: 1632–40.

Wendisch, V.F. (2006) 'Genetic regulation of *Corynebacterium glutamicum* metabolism', *Journal of Microbiology and Biotechnology*, 16: 999–1009

Wendische, V.F., de Graaf, A.A., Sahm, H., and Eikmans, B.J. (2000) 'Quantitative determination of metabolic fluxes during cultivation of two carbon source: comparative analyses with *Corynebacterium glutamicum* during growth on acetate and/or glucose', *Journal of Bacteriology*, 182: 3088–96.

Williams, R., Bell, A., Sims, G., and Busby, S. (1991) 'The role of two surface exposed loops in transcription activation by the *Escherichia coli* CRP and FNR proteins', *Nucleic Acids Research*, 24: 6705–12.

Willis, R.C., Iwata, K.K., and Furlong, C.E. (1975) 'Regulation of glutamine transport in *Escherichia coli*', *Journal of Bacteriology*, 122: 1032–7.

Winkler, H.H. and Wilson, T.H. (1967) 'Inhibition of β-galactoside transport by substrates of the glucose transport system in *Escherichia coli*', *Biochemisty and Biophysics ACTA*, 135: 1030–51.

Wittmann, C., Weber, J., Betiku, E., Krömera, J., Böhmb, D., and Rinas, U. (2007) 'Response of fluxome and metabolome to temperature-induced recombinant protein synthesis in *Escherichia coli*', *Journal of Biotechnology*, 132: 375–84.

Wolf, R.E., Prather, D.M., and Shea, F.M. (1979) 'Growth-rate-dependent alteration of 6-phosphogluconate dehydrogenase and glucose 6-phosphate dehydrogenase levels in *Escherichia coli* K-12', *Journal of Bacteriology*, 139, pp. 1093–96.

Yamada, M., Makino, K., Amemura, M., Shinagawa, H., and Nakata, A. (1989) 'Regulation of the phosphate regulon of *Escherichia coli*: Analysis of mutant *phoB* and *phoR* genes causing different phenotypes', *Journal of Bacteriology*, 171: 5601–6

Yan, D. (2007) 'Protection of the glutamate pool concentration in enteric bacteria,' *Proceedings of the National Academy Science USA*, 104: 9475–80.

Yang, Y.T., Bennet, G.N., and San, K.Y. (2001) 'The effects of feed and intracellular pyruvate levels on the redistribution of metabolic fluxes in *Escherichia coli*', *Metabolic Engineering*, 3: 115–23.

Yang, Y., Zhao, G., Man, T.K., and Winkler, M.E. (1998) 'Involvement of the *gap*A- and *epd* (*gap*B)-encoded dehydrogenases in pyridoxal 5′-phosphate coenzyme biosynthesis in *Escherichia coli* K-12', *Journal of Bacteriology*, 180: 4294–9.

Yao, R., Hirose, Y., Sarkar, D., Nakahigashi, K., Ye, Q., and Shimizu, K. (2011) 'Catabolic regulation analysis of *Escherichia coli* and its *crp, mlc, mgsA, pgi* and *ptsG* mutants', *Microbial Cell Factories*, 10: 67.

Yoon, S.H., Han, M., Lee, S.Y., Jeong, K.J., and Yoo, J.S. (2003) 'Combined transcriptome and proteome analysis of *Escherichia coli* during high cell density culture', *Biotechnology and Bioengeering*, 81: 753–66.

Yura, T., Nagai, H., and Mori, H. (1993) 'Regulation of the heat shock response in bacteria', *Annual Reviews in Microbiology*, 47: 321–50.

Yura, T. and Nakahigashi, K. (1999) 'Regulation of heat shock response', *Current Opinion in Microbiology*, 1999, 2: 153–8.

Zhao, G. and Winklerm, M.E. (1994) 'An *Escherichia coli* K-12 tktA tktB mutant deficient in transketolase activity requires pyridoxine (vitamin B_6) as well as the aromatic amino acids and vitamins for growth', *Journal of Bacteriology*, 176: 6134–8.

Zhang, Z., Gosset, G., Barabote, R., Gonzalez, C.S., Cuevas, and W.A., Saier, M.H. Jr. (2005) 'Functional interactions between the carbon and iron utilization regulators, Crp and Fur, in *Escherichia coli*', *Journal of Bacteriology*, 187: 980–90.

Zhang, X., Jantama, K., Moore, J.C., Shanmugam, K.T., and Ingram, L.O. (2007) 'Production of L-alanine by metabolically engineered *Escherichia coli*', *Applied Microbiology and Biotechnology*, 77: 355–66.

Zhu, J. and Shimizu, K. (2004) 'The effect of *pfl* genes knockout on the metabolism for optically pure D-lactate production by *Escherichia coli*', *Applied Microbiology and Biotechnology*, 64: 367–75.

Zhu, J. and Shimizu, K. (2005) 'Effect of a single-gene knockout on the metabolic regulation in *E.coli* for D-lactate production under microaerobic condition', *Metabolic Engineering*, 7: 104–15.

Zhu, J., Shalel-Levanon, S., Bennett, G., and San, K.Y. (2006) 'Effect of the global redox sensing/regulation networks on *Escherichia coli* and metabolic flux distribution based on C-13 labeling experiments', *Metabolic Engineering*, 8: 619–27.

Ziegelhoffer, E.C. and Kiley, P.J. (1995) '*In vitro* analysis of a constitutive active mutant form of the *Escherichia coli* global transcription factor FNR', *Journal of Molecular Biology*, 245: 351–61.

Zubay, G. (1993) *Biochemistry*. Iowa: Wm. C. Brown Communications, Inc.

Zuckier, G., Ingenito, E., and Torriani, A. (1980) 'Pleiotropic effects of alkaline phosphate regulatory mutations *phoB* and *phoT* on anaerobic growth of and polyphosphate synthesis in *Escherichia coli*', *Journal of Bacteriology*, 143: 934–41.

4

Conventional flux balance analysis and its applications

Abstract: The conventional flux balance analysis method is briefly explained. Its application to the metabolic flux analysis of the *Chlorella* cell, under autotrophic, heterotrophic, and mixotrophic conditions, focuses on energy generation. Other applications to such single-gene knockout *Escherichia coli* strains, such as *pflA*, *pta*, *ppc*, *adhE*, and *pykF* mutants cultivated under anaerobic conditions, are also explained, as is metabolic regulation analysis based on metabolic fluxes, enzyme activities, and intracellular metabolite concentrations.

Key words: flux balance analysis; *Chlorella* cell; *pflA* mutant; *pta* mutant; *adhE* mutant; *ppc* mutant; *pykF* mutant; anaerobic conditions; photosynthetic bacteria.

4.1 Introduction

It is important to clarify metabolic fluxes in practice, where a set of metabolic fluxes describe the cell physiology. The information on metabolic flux distribution is useful in metabolic engineering (Bailey, 1991; Stephanopoulos and Vallino, 1991; Stephanopoulos, 1999). In principle, metabolic flux analysis is based on mass conservation of key metabolites. Intracellular fluxes can be calculated from the measured specific rates by applying mass balances to these intracellular metabolites together with stoichiometric equations. The number of measurable extracellular fluxes is limited in practice, and the stoichiometric constraints often lead to an under-determined algebraic system. Therefore, cofactor balances sometimes need to be introduced into the stoichiometric model, or the appropriate objective functions have to be introduced for optimization to determine the fluxes.

The central metabolic pathway has both anabolic and catabolic functions, as it provides cofactors and building blocks for macromolecular synthesis (anabolism), as well as energy (ATP) generation (catabolism). Optimization may be made in terms of catabolism and anabolism, or both, under the constraints of the stoichiometric equations. Flux balance analysis (FBA) has been extensively used to predict steady-state metabolic fluxes in order to maximize cell growth rates (Schilling and Palsson, 1998; Edwards and Palsson, 2000; Price and Papin, 2003). This method has been significantly extended to genome-scale models which include thousands of metabolic reactions (Palsson, 2009). *E. coli* utilizes carbon almost optimally, and this is a special feature of this organism (Edwards et al., 2001; Ibarra et al., 2002; Yuan et al., 2006).

However, the accuracy of the flux calculation depends on the validity of the cofactor assumptions and an appropriate choice of the objective function(s) employed (Schuetz et al., 2007). The presence of unknown reactions that generate or consume the cofactors may invalidate the assumption that their concentrations remain in balance, and the selected objective functions may not be appropriate, or their validity may be limited to certain states of the cell. Note that this approach cannot essentially compute such fluxes as: i) recycled fluxes; ii) bidirectional fluxes; and iii) parallel fluxes, due to the singularity of the stoichiometric matrix. These fluxes can be computed based on ^{13}C-matabolic flux analysis, which is explained in Chapter 5. Note, however, that the conventional FBA may be reasonably applied for cases without recycles, such as anaerobic or microaerobic cultivation or by lumping several pathways together. Here we consider metabolic flux analysis based on mass balances with stoichiometric equations and show its several applications.

4.2 Basis for metabolic flux analysis

As shown by Figure 1.1, when using glucose as a carbon source, the overall main metabolic reactions under aerobic condition may be expressed as:

$$\text{Substrate + nitrogen source} + O_2 \rightarrow \text{biomass} + \text{metabolites} + CO_2 + H_2O \quad (4.1)$$

Let r_i be the specific consumption or production rate of the ith substance, and let E be the stoichiometric matrix. Then the following mass balance holds at steady state, where the specific rate is defined as the consumption or production rate of the substance of interest per cell per hour:

$$E \cdot r = 0 \quad (4.2)$$

Conventional flux balance

where r is the vector containing r_i. Let r be subdivided into measured specific rate vector r_m and the other rate vector as r_c. Then Equation 4.2 can be expressed as:

$$Er = E_m r_m + E_c r_c = 0 \qquad (4.3)$$

where E_m and E_c are the sub-matrices of E corresponding to r_m and r_c, respectively. From Equation 4.3, r_c may be expressed as:

$$r_c = -E_c^{-1} E_m r_m \qquad (4.4)$$

where E_c is assumed to be the square and non-singular. For example, consider the cultivation of yeast using glucose as a carbon source. The stoichiometric equation may be expressed as:

$$C_6H_{12}O_6 + O_2 + NH_3 \rightarrow CH_{1.8}O_{0.5}N_{0.15} + C_2H_5OH + CO_2 + H_2O \qquad (4.5)$$

Glucose biomass ethanol

where the stoichiometric balance with respect to C, H, O, and N may be expressed as:

$$\begin{array}{c} C \\ H \\ O \\ N \end{array} \begin{bmatrix} 6 & 0 & 0 & 1 & 1 & 2 & 0 \\ 12 & 0 & 3 & 1.8 & 0 & 6 & 2 \\ 6 & 2 & 0 & 0.5 & 2 & 1 & 1 \\ 0 & 0 & 1 & 0.15 & 0 & 0 & 0 \end{bmatrix} \begin{bmatrix} r_S \\ r_{O_2} \\ r_n \\ r_X \\ r_{CO_2} \\ r_e \\ r_w \end{bmatrix} = 0 \qquad (4.6)$$

where the specific glucose consumption rate r_s, the specific oxygen consumption rate r_{O_2}, and the specific ammonia consumption rate r_n are negative, while the specific growth rate r_x, the specific CO_2 production rate r_{CO_2}, the specific ethanol production rate r_e, and the specific water formation rate r_w, are positive. Suppose that r_s, r_x, and r_{CO_2} can be measured, then the other specific rates can be estimated from Equation 4.4. Let E be subdivided and re-express the above equation as:

$$E_m r_m + E_c r_c = \begin{bmatrix} 6 & 1 & 1 \\ 12 & 0 & 1.8 \\ 6 & 2 & 0.5 \\ 0 & 0 & 0.15 \end{bmatrix} \begin{bmatrix} r_s \\ r_{CO_2} \\ r_x \end{bmatrix} + \begin{bmatrix} 0 & 0 & 2 & 0 \\ 0 & 3 & 6 & 2 \\ 2 & 0 & 1 & 1 \\ 0 & 1 & 0 & 0 \end{bmatrix} \begin{bmatrix} r_{O_2} \\ r_n \\ r_e \\ r_w \end{bmatrix} = 0 \qquad (4.7)$$

Thus r_s, r_x, and r_{CO_2} can be used to estimate the other parameters by the following equation:

$$\begin{bmatrix} \hat{r}_{O_2} \\ \hat{r}_n \\ \hat{r}_e \\ \hat{r}_w \end{bmatrix} = - \begin{bmatrix} 3 & 1.5 & 0.4125 \\ 0 & 0 & 0.15 \\ 6 & 1 & 0.5 \\ -3 & -1.5 & -0.15 \end{bmatrix} \begin{bmatrix} r_s \\ r_{CO_2} \\ r_x \end{bmatrix} \tag{4.8}$$

Note that this equation is valid where ethanol (and CO_2) is assumed to be the sole metabolic product. When other metabolites, such as acetate, etc. are formed, the accuracy of the estimation based on the above equation will degrade.

Let us consider next the metabolic pathway equations. Let A be the metabolic stoichiometric coefficient matrix, r be the intracellular flux vector, and q be the extracellular measured specific rate vector, which include the specific substrate consumption rate, and the specific metabolite production rate, etc. Then the mass balance equation becomes:

$$Ar = q \tag{4.9}$$

where A is known by assuming the active metabolic network. If q is also known by measurement of all the specific rates, then the problem is to estimate r from the above equation. Since the A matrix is not necessarily square, multiply A^T from the left, and then multiply $(A^T A)^{-1}$ from the left, and finally we have:

$$r = (A^T A)^{-1} A^T q \tag{4.10}$$

In practice, the following least square method is used:

$$\hat{r} = (A^T \Sigma A)^{-1} A^T \Sigma q \tag{4.11}$$

where Σ is the variance-covariance matrix for the measurement error.

Consider another way of estimating the metabolic fluxes from the measured specific rates (Tsai and Lee, 1988). Let Equation 4.9 be expressed as:

$$\begin{bmatrix} q_m \\ q_c \end{bmatrix} = \begin{bmatrix} A_{11} & A_{12} \\ A_{21} & A_{22} \end{bmatrix} \begin{bmatrix} r_1 \\ r_2 \end{bmatrix} \tag{4.12}$$

where q_m is the measured specific rate, and q_c is the intracellular specific rate. Suppose A_{22} is the square matrix and non-singular. In general, we can set q_c to zero, and the above equation may be expressed as:

Conventional flux balance

$$q_m = A_{11}r_1 + A_{12}r_2 \qquad (4.13a)$$

$$0 = A_{21}r_1 + A_{22}r_2 \qquad (4.13b)$$

The following equation may be derived from Equation 4.13b:

$$r_2 = -A_{22}^{-1} A_{21} r_1 \qquad (4.14)$$

Substituting this into Equation 4.13a gives:

$$q_m = Br_1 \qquad (4.15)$$

where

$$B \equiv A_{11} - A_{22}^{-1} A_{21} A_{12} \qquad (4.16)$$

B is not necessarily the square matrix, and thus the least square estimate for r_1 may be obtained from Equation 4.15 with Equation 4.16 as:

$$\hat{r}_1 = (B^T B)^{-1} B^T q_m \qquad (4.17)$$

If the covariance matrix for measurement Σ is given, the least square estimate for r_1 becomes:

$$\hat{r}_1 = [B^T \Sigma^{-1} B]^{-1} B^T \Sigma^{-1} q_m \qquad (4.18)$$

Note that the following equation holds from Equation 4.14:

$$\hat{r}_2 = -A_{22}^{-1} A_{21} \hat{r}_1 \qquad (4.19)$$

In the following, let us consider the metabolic flux analysis based on the above equations.

4.3 Application to photosynthetic bacteria

4.3.1 Background for microalgae cell

Microalgal cultures have received much attention in recent years because of their potential application for industrial CO_2 removal, production of many valuable metabolites, and life support in space (Borowitzke, 1988; Glombitza and Koch, 1989; Kurano and Miyachi, 1995). There have been several studies on the effects of medium composition, illumination intensity, and various photobio-reactors on the growth and photosynthetic rates of microalgae (Lee and Palsson, 1994; Mandanam and Palsson, 1998). In order to improve the performance of microalgal cultures, a deep understanding of the carbon and energy metabolism in microalgal cells is needed. In addition, an essential understanding of the metabolic mechanism of microalgal cells may help to investigate the metabolism inside the plant cells, because there are many similarities between microalgae and plant cells. In fact, there are many distinct features about microalgal cell metabolism compared with other microorganisms. Microalgae can perform oxygenic photosynthesis and fix CO_2 through the Calvin cycle, like plant cells. That is, microalgal cells can trap light energy as the energy source and assimilate CO_2 as the carbon source. Moreover, organic substrates can also be utilized as carbon and energy sources by many microalgae (Droop, 1974). Therefore, by varying the nature of carbon and energy sources, the different underlying metabolic status of cells, especially the influence of light on carbon and energy metabolism, can be elucidated.

Here, metabolic flux analysis is considered to quantitatively assess intracellular fluxes through the main metabolic pathways of microalgal cells (Vallino and Stephanopulos, 1993; Bonarius et al., 1996). The performance of microalgal culture systems can be evaluated and compared through the efficiency with which the supplied energy to the culture can be utilized for biomass production (Tredici and Zittelli, 1998). In order to improve the energy utilization efficiency of the culture, we need a fundamental understanding of the energy conversion from the supplied energy to biomass formation, i.e. how the microalgal cells harvest light or chemical energy from the environment and then convert them into ATP, the universal currency of free energy for cell growth. The elucidation of carbon and energy metabolism of microalgae from metabolic flux distributions can provide a basis for the investigation of the energy utilization efficiency when cells are grown on different energy sources (Yang et al., 2000).

4.3.2 Metabolism of Chlorella cell

Consider the cultivation of *Chlorella pyrenoidosa* C-212, cultivated under autotrophic, mixotrophic, and cyclic light-autotrophic/dark-heterotrophic conditions (Yang et al., 2000). Metabolic flux analysis is used to clarify the metabolism of microalgae grown on different carbon and energy sources. Of particular interest is the effect of light on carbon and energy metabolism. Figure 4.1 shows the metabolic networks for autotrophic, heterotrophic, and mixotrophic cultures. The corresponding metabolic pathway reactions are listed in Tables 4.1 and 4.2 (Yang et al., 2000).

Algae cells can use light as their energy source. Light quanta absorbed by pigments drive the photosynthetic electron transport, which results in the reduction of $NADP^+$ and couples to the formation of ATP. It has been

Conventional flux balance

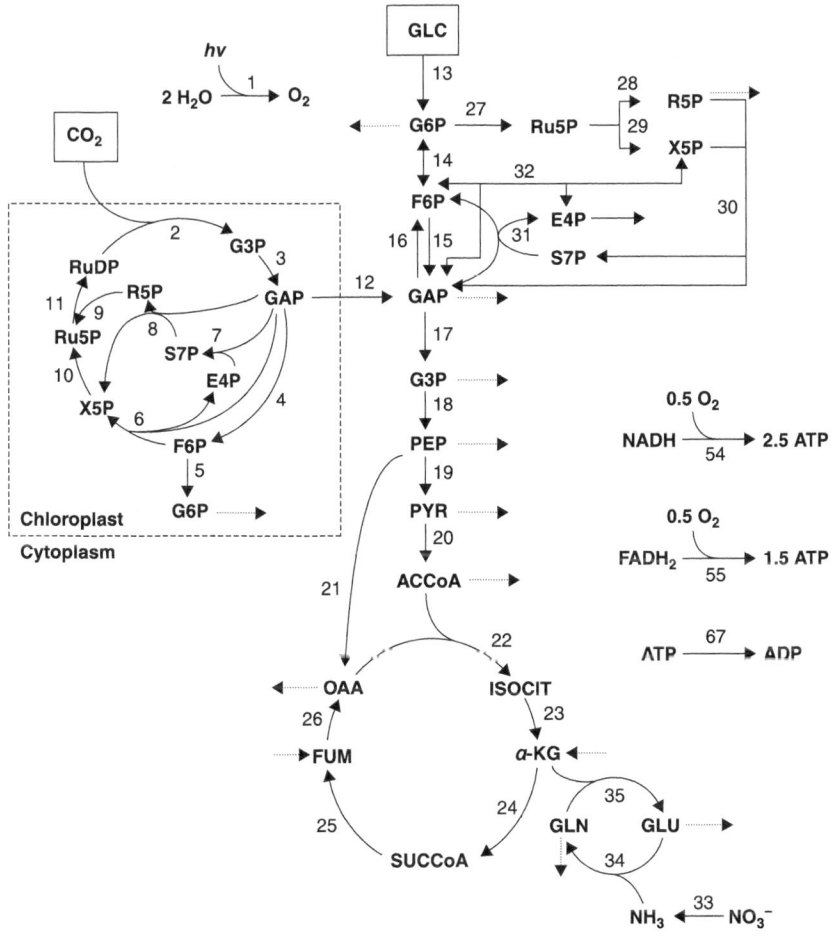

Figure 4.1 Central reaction network for microalgae metabolism. Numbers correspond to the reactions shown in Tables 4.1 and 4.2. The dotted lines represent fluxes for cell synthesis

reported that the P/2e⁻ ratio (the number of ATP molecules formed per pair of electrons moving through the photosynthetic electron transport chain) is about 1.3 (Avron, 1989). According to the widely accepted two-step model of photosynthesis, 8 mol quanta of light are required to evolve 1 mol O_2. Therefore, the photosynthetic O_2 evolution rate can be determined from the photon flux absorbed. NADPH and ATP formed by the action of light reduce CO_2 by a series of dark reactions. The first step in the Calvin cycle, photosynthetic CO_2 fixation, is catalyzed by ribulose

1,5-bisphosphate carboxylase. This enzyme is also an oxygenase, which can react with O_2 and lead to a different pathway called photorespiration. Algae have the photorespiration pathway, and photosynthesis is inhibited by high O_2 concentration. However, under normal conditions, CO_2 loss by algal photorespiration is minimal (Tolbert, 1980). Here, the photorespiration pathway is not included in the metabolic network.

Photosynthesis reactions, including light reactions, the Calvin cycle, and starch synthesis, are located in chloroplasts. Glyceraldehyde 3-phosphate (GAP) is withdrawn from the Calvin cycle and exported to the cytoplasm for consumption. Although algae and plant cells have a high degree of sub-cellular compartmentation of metabolism, compartmentation of most metabolites between chloroplast, mitochondria, and cytoplasm is not considered because the extent to which biosynthetic reactions are localized in chloroplasts in algae cells is not understood or not yet fully established (Heldt and Flugge, 1987; Rees, 1987).

After the export of GAP from chloroplast to cytoplasm, the flow of carbon is divided into the synthesis of sugars or oxidation through the glycolytic pathway to pyruvate. Sugars, including sucrose, are the major storage products in the cytoplasm of plant cells. In addition, the structural carbohydrates, such as cell wall components, are considered to be synthesized in the cytoplasm. Some researchers found that the alga cell cultivated with [1-^{14}C] glucose yields starch in which glucose is still predominantly labeled in position 1 and the amount of glucose that remains unchanged is about 70% (Akazawa and Okakoto, 1980). This suggests that glucose can be directly converted to starch without prior conversion to GAP and then taken up by the chloroplast. Therefore, in the network of mixotrophic metabolism, one part of the exogenous glucose is directly converted to starch, and the remainder is oxidized through the glycolytic pathway.

In plant cells, replenishment of carbon to maintain the operation of the TCA cycle is achieved by anaplerotic reactions involving CO_2 fixation by Ppc (Wiskich, 1980). The pentose phosphate (PP) pathway has been reported to operate in the cytoplasm at the same time that the Calvin cycle is functioning in the chloroplast (Lloyde, 1974; Kelly and Latzko, 1980). In metabolic networks for autotrophic and mixotrophic cultures, only the PP pathway is considered to supply pentose phosphate for nucleic acid synthesis and produce E4P for the synthesis of shikimi acid, because the production of aromatic amino acids occurs in the cytoplasm and the transport of pentose phosphate out of the chloroplast is not possible. However, there is evidence that the PP pathway may function in both cytoplasm and chloroplast in plant cells under dark condition (Graham, 1980). Therefore, both pathways located in different compartments are combined in the network of heterotrophic metabolism.

Nitrate is the predominant form of nitrogen available to most plants. Nitrate is reduced to ammonia before becoming available for assimilation. The reduction is divided into two steps: nitrate is reduced by a cytoplasmic NADH-dependent nitrate reductase to nitrite, which is further reduced to ammonia by a chloroplast-located NADPH-linked nitrite reductase (Morris, 1974). Glutamate dehydrogenase (GDH) and glutamine synthetase (GS) are considered as the important entries of ammonia into organic form. If both reactions are included in the metabolic network, a singularity arises. However, labeling experiments show the predominant operation of GS in algae cells (Miflin and Lea, 1980). Hence, only GS is included, and the glutamine formed provides the nitrogen donor to α-KG by the action of glutamate synthetase. GS/glutamate synthetase is considered in the reactions of ammonia assimilation in metabolic networks.

There are three different mechanisms of electron transport in the plant respiratory chain (Douce et al., 1987). De Gucht and van der Plas (1995) determined the activities of these pathways and calculated the P/O values for ATP production, which are about 2.5. Therefore, here it is assumed that the P/O ratios are 2.5 and 1.5 for NADH and $FADH_2$, respectively. It is not clear whether transhydrogenase, which catalyzes the reversible transfer of the hydride ion between NAD and NADP, is present in plant cells. Some experiments demonstrate the presence of this enzyme but with a very low activity, so this reaction can be disregarded in metabolic networks (Storey, 1980).

Of all the pigments, chlorophyll a takes a major fraction, thus only chlorophyll a formation is considered here. δ-Aminolevulinic acid (δ-ALA) is the key chlorophyll precursor molecule, which may be formed from different routes. The classical succinate-glycine pathway is the condensation of glycine and succinyl-CoA catalyzed by δ-ALA synthetase. In addition, 5-carbon compounds, glutamate, and α-KG are found to be incorporated into δ-ALA much more efficiently than are glycine and succinate in many green cells. Since the incorporation of all the synthesis reactions of δ-ALA in the metabolic network leads to a singularity and the C_5 route is a major contribution, this route can be considered to be the only contributor for δ-ALA formation (Castelfranco and Beale, 1981).

Eukaryotic algae and higher plants have a variety of lipids. Since these cells contain chloroplasts and have the biological function of photosynthesis, they have some unique lipids that are responsible for the characteristic features of chloroplast membrane (Nichols, 1965). For metabolic flux analysis, all lipids but pigments may be lumped into diacylglycerol (DG), which is the key precursor in the synthesis of triglyceride, phospholipids, galactolipids, and perhaps sulfolipids. Although most of the synthesis reactions of fatty acids occur in the chloroplast, the source of AcCoA derives from its synthesis in the mitochondria. It should be pointed out that the fatty acid composition of the lipids isolated from *Chlorella* cells grown under different conditions varies considerably, particularly for the α-linolenic acid (C18:3) content (Wood, 1974). This leads to the difference in the synthetic reactions of DG in the three metabolic networks (Tables 4.1 and 4.2).

The protein composition may be obtained from Yanagi et al. (1995), and the composition of RNA and DNA is given in Wanka et al. (1970). The amino acid compositions of protein are assumed to be constant under different cultivation conditions, since the reported composition values for different cultivation conditions are similar (Fowden, 1962). The nucleotide compositions of DNA and RNA are also assumed to be maintained under different growth conditions. Three matrixes of stoichiometric coefficients are constructed using the reactions listed in Tables 4.1 and 4.2, and applied to calculate the metabolic fluxes under different culture conditions (Yang et al., 2000).

4.3.3 Cultivation of Clorella under different conditions

Figures 4.2–4.4 show the cell growth, the cellular composition, the consumption of glucose and nitrate, and the CO_2 production rate and O_2 uptake rates in autotrophic, mixotrophic, and cyclic light-autotrophic/

Table 4.1 Reactions in the networks of three different types of metabolism of *Chlorella* cell

---

Autotrophic network (CO_2 as a carbon source):

1, 2–11, 12, 14, 16–26, 27–32, 33–35, 36–53, 54–55, 56, 57, 60–62, 63, 66, 67

Heterotrophic network (glucose as a carbon source):

13–15, 17–26, 27–32, 33–35, 36–53, 54, 55, 57, 60–62, 64, 66, 67

Mixotrophic network (glucose and CO_2 as a carbon source):

1, 2–11, 12, 13–15, 17–26, 27–32, 33–35, 36–53, 54, 55, 57–62, 65–67

Table 4.2 Biochemical reactions for *Chlorella* cell

Light reactions

(4.1) $2H_2O+2NADP+2ADP+2Pi+0.125APF \Rightarrow 2NADPH+2H+2.6ATP+O_2$

Calvin cycle

(4.2) $H_2O+CO_2+RuDP \Rightarrow G3Pchl$

(4.3) $G3Pchl+ATP+NADPH+H \Rightarrow GAPchl+ADP+NADP+Pi$

(4.4) $2GAPchl+H_2O \Rightarrow F6Pchl+Pi$

(4.5) $F6Pchl \Leftrightarrow G6Pchl$

(4.6) $F6Pchl+GAPchl \Rightarrow X5Pchl+E4Pchl$

(4.7) $E4Pchl+GAPchl+H2O \Rightarrow S7Pchl+P_i$

(4.8) $S7Pchl+GAPchl \Rightarrow R5Pchl+X5Pchl$

(4.9) $R5Pchl \Rightarrow Ru5Pchl$

(4.10) $X5Pchl \Rightarrow Ru5Pchl$

(4.11) $Ru5Pchl+ATP \Rightarrow RuDP+ADP$

Transport of triose phosphate from chloroplast to cytoplasm

(4.12) $GAPchl \Rightarrow GAP$

Glycolytic pathway and tricarboxylic acid cycle

(4.13) $GLC+ATP \rightarrow G6P+ADP+H$

(4.14) $G6P \Leftrightarrow F6P$

(4.15) $F6P+ATP \rightarrow 2GAP+ADP+H$

(4.16) $2GAP+H_2O \Rightarrow F6P+P_i$

(4.17) $GAP+NAD+P_i+ADP \Leftrightarrow G3P+ATP+NADH+H$

(4.18) $G3P \Leftrightarrow PEP+H_2O$

(4.19) $PEP+ADP \Rightarrow PYR+ATP$

(4.20) $PYR+NAD+CoA \Rightarrow AcCoA+NADH+CO_2+H$

(4.21) $PEP+CO_2+ADP \Rightarrow OAA+ATP$

(4.22) $OAA+AcCoA+H2O \Leftrightarrow ISOCIT+CoA+H$

(4.23) $ISOCIT+NAD \Leftrightarrow AKG+NADH+CO_2$

(4.24) $AKG+CoA+NAD \Rightarrow SUCCoA+NADH+CO_2+H$

(4.25) $SUCCoA+ADP+Pi+FAD \Leftrightarrow FUM+FADH_2+ATP+CoA$

(4.26) $FUM+NAD+H_2O \Leftrightarrow OAA+NADH+H$

Pentose phosphate pathway

(4.27) $G6P+2NADP+H_2O \Rightarrow Ru5P+CO_2+2NADPH+2H$

(4.28) $Ru5P \Leftrightarrow R5P$

(Continued)

Table 4.2 Biochemical reactions for *Chlorella* cell (*Continued*)

(4.29) $Ru5P \Leftrightarrow X5P$

(4.30) $R5P+X5P \Leftrightarrow S7P+GAP$

(4.31) $S7P+GAP \Leftrightarrow F6P+E4P$

(4.32) $X5P+E4P \Leftrightarrow F6P+GAP$

Assimilation of nitrate

(4.33) $NO3+NADH+3NADPH+5H \Rightarrow NH3+3NADP+NAD+3H_2O$

(4.34) $GLU+NH3+ATP \Rightarrow GLN+ADP+Pi$

(4.35) $AKG+NADPH+GLN \Rightarrow 2GLU+NADP$

Amino acid synthesis

(4.36) $GLU+ATP+2NADPH+2H \Rightarrow PRO+ADP+Pi+H_2O+2NADP$

(4.37) $GLU+AcCoA+ASP+GLN+CO_2+NADPH+5ATP+3H2O \Rightarrow$
ARG+AKG+CoA+AC+5ADP+FUM+5P_i+NADP+6H

(4.38) $ASP+PYR+2NADPH+SUCCoA+GLU+ATP+2H \Rightarrow$
LYS+SUC+AKG+CO_2+2NADP+CoA+ADP+Pi

(4.39) $G3P+GLU+NAD+H_2O \Rightarrow SER+AKG+P_i+H+NADH$

(4.40) $SER+THF \rightarrow GLY+METHF+H_2O$

(4.41) $SER+AcCoA+SO_4+4NADPH+4H+ATP \Rightarrow$
CYS+AC+CoA+4NADP+ADP+3H_2O+P_i

(4.42) $OAA+GLU \Rightarrow ASP+\alpha KG$

(4.43) $ASP+GLN+2ATP+H_2O \Rightarrow ASN+GLU+2ADP+2P_i$

(4.44) $ASP+2ATP+2NADPH+H+H2O \rightarrow THR+2ADP+2P_i+2NADP$

(4.45) $ASP+2NADPH+SUCCoA+CYS+MYTHF+ATP \Rightarrow$
MET+CoA+SUC+PYR+NH3+ADP+Pi+THF+2NADP

(4.46) $THR+PYR+NADPH+GLU+2H \Rightarrow ILE+NH_3+NADP+H_2O+CO_2+AKG$

(4.47) $PYR+GLU \rightarrow ALA+AKG$

(4.48) $2PYR+NADPH+2H+GLU \Rightarrow VAL+AKG+CO_2+NADP+H_2O$

(4.49) $2PYR+NADPH+AcCoA+GLU+NAD+H_2O \Rightarrow$
LEU+AKG+CoA+2CO_2+NADP+NADH

(4.50) $2PEP+E4P+NADPH+ATP+GLU+H \Rightarrow PHE+AKG+CO_2+H_2O+ADP+4Pi+NADP$

(4.51) $2PEP+E4P+NADPH+ATP+GLU+NAD \Rightarrow$
TYR+AKG+CO_2+NADH+ADP+4Pi+NADP

(4.52) $2PEP+E4P+NADPH+GLN+R5Pcyt+3ATP+SER \Rightarrow$
TRP+6P_i+CO_2+GAP+GLU+2H+PYR+H_2O+3ADP+NADP

(4.53) $R5P+6ATP+GLN+2NAD+ASP+FTHF \Rightarrow$
HIS+AKG+FUM+2NADH+6ADP+7P_i+THF

Oxidative phosphorylation

(4.54) $NADH+0.5O_2+2.5ADP+2.5P_i+3.5H \Rightarrow 3.5H_2O+NAD+2.5ATP$

(4.55) $FADH_2+0.5O_2+1.5ADP+1.5P_i+2.5H \Rightarrow 2.5H_2O+FAD+1.5ATP$

Biosynthesis of macromolecules

(4.56) $G6P_{chl}+2ATP \Rightarrow STA$

(4.57) $G6P+2ATP \Rightarrow CAR$

(4.58) $G6P_{chl}+2ATP \Rightarrow 0.3STA$

(4.59) $G6P+2ATP \Rightarrow 0.7STA$

(4.60) $R5P+1.2ASP+2.1GLN+0.54GLY+1.1FTHF+0.54CO_2+8.2ATP+0.79NAD+2.2H_2O \Rightarrow RNA+2.1GLU+0.75FUM+1.1THF+8.2ADP+8.2P_i+0.79NADH+9.3H$

(4.61) $R5P+1.2ASP+0.25SER+2GLN+0.25GLY+FTHF+0.5CO_2+2.3H_2O+NADPH+0.76NAD+8ATP \Rightarrow DNA+0.75FUM+2GLU+THF+8ADP+8P_i+0.76NADH+NADP+9H$

(4.62) $0.09959ALA+0.0342ARG+0.02058ASN+0.04756ASP+0.01046CYS+0.05354GLU+0.02322GLN+0.08406GLY+0.0274HIS+0.02988ILE+0.07003LEU+0.05309LYS+0.01555MET+0.03176PHE+0.04268PRO+0.0193SER+0.04024THR+0.009932TRP+0.02583TYR+0.05044VAL+4ATP \Rightarrow PROT$

(4.63) $GAP+NADH+17AcCoA+33NADPH+34H+15ATP+3O_2 \Rightarrow DG+NAD+33NADP+15ADP+15P_i+17CoA+4H2O$

(4.64) $GAP+NADH+17AcCoA+32.4NADPH+34H+15ATP+2.2O_2 \Rightarrow DG+NAD+32NADP+15ADP+15Pi+17CoA+2.4H_2O$

(4.65) $GAP+NADH+17AcCoA+32.6NADPH+34H+15ATP+2.5O_2 \Rightarrow DG+NAD+33NADP+15ADP+15P_i+17CoA+2.7H2O$

(4.66) $8GLU+12AcCoA+21ATP+24NADPH+Mg_2{++}MYTHF+3O_2 \Rightarrow CHLO+4NH_3+10CO_2+24NADP+1HF+21ADP$

Miscellaneous reaction

(4.67) $ATP \Rightarrow ADP$

dark-heterotrophic cultures. It is assumed that the contents of RNA and DNA are maintained at 5.5 and 0.4% for all cultures (Nishimura et al., 1988). The experimental results for these cultures are summarized in Table 4.3. The autotrophic culture converts about 0.15 mol of CO_2 into biomass and evolves as much as 0.21 mol of O_2. Nevertheless, the increase in biomass results in a rapid reduction of the cell growth rate due to light limitation, hence the final cell concentration achieved by autotrophic growth is very low. With the increase in cell density, the contents of protein, lipids, and chlorophylls increase up to maximum values, then

Bacterial cellular metabolic systems

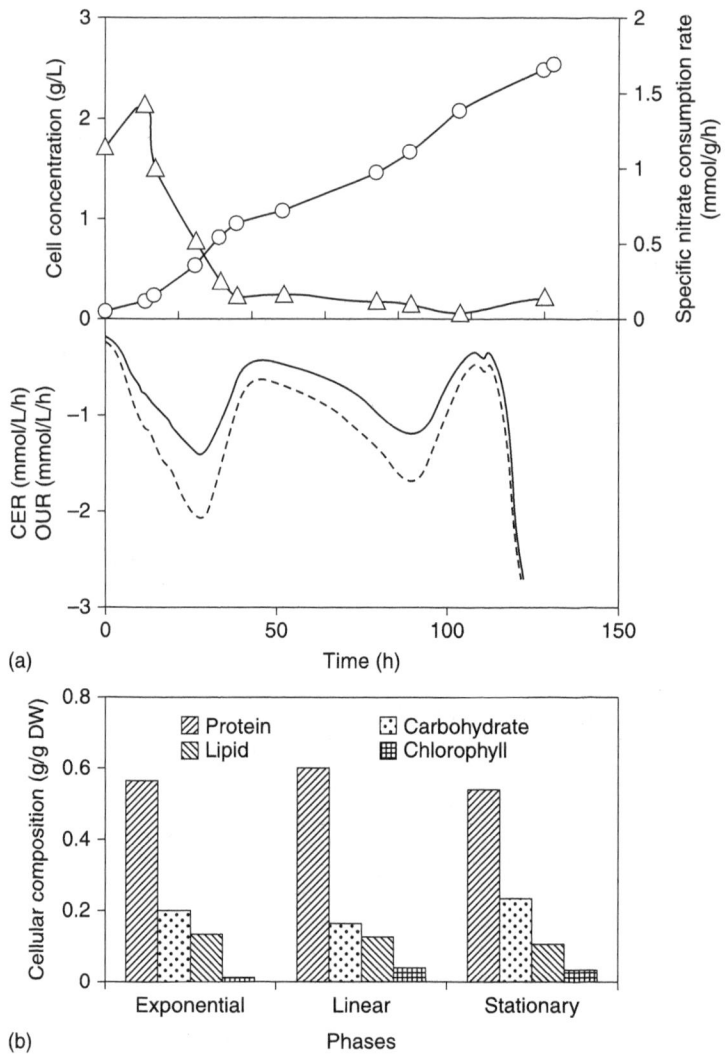

Figure 4.2 Cultivation results of *C. pyrenoidosa* under autotrophic conditions: (a) Time profiles of cell growth (○), consumption of nitrate (△), rates of CO$_2$ production (solid line), and O$_2$ uptake (dotted line); (b) Contents of carbohydrates, protein, lipids, and chlorophylls in cells during the various growth phases. The incident light intensity was 500 μmol/m^2/s. The various growth phases were identified from the curve of cell growth: exponential phase (0–48 h), linear phase (48–120 h) and stationary phase (120 h–end)

Published by Woodhead Publishing Limited, 2013

Conventional flux balance

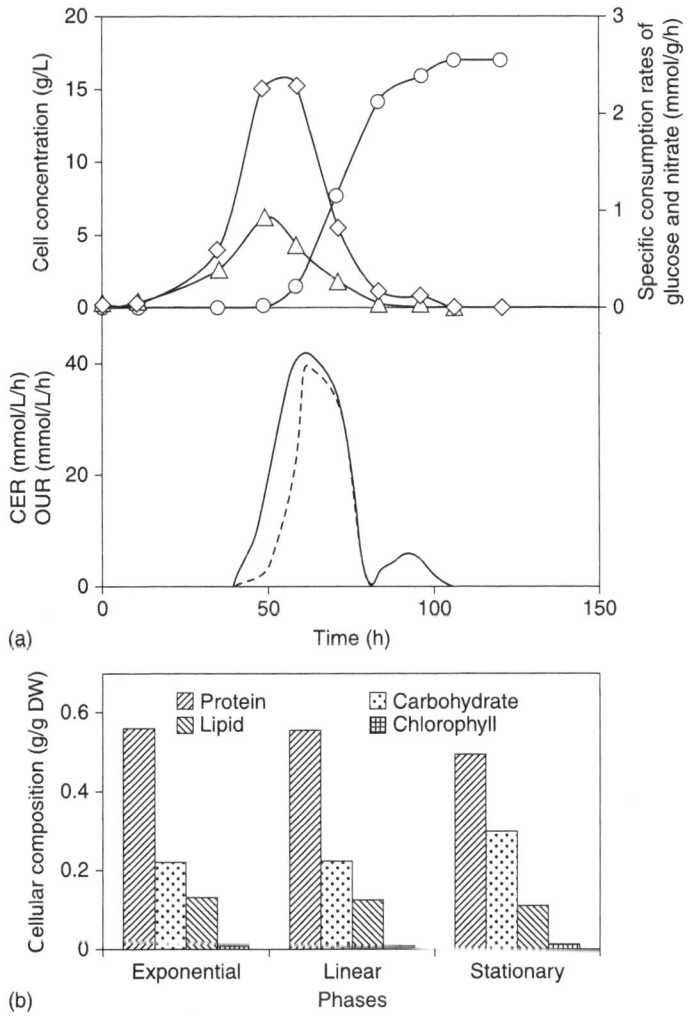

Figure 4.3 Cultivation results of C. pyrenoidosa under mixotrophic conditions: (a) Time profiles of cell growth (○), consumption of glucose (◇) and nitrate (△), and the rates of CO_2 production (solid line) and O_2 uptake (dotted line); (b) Contents of carbohydrates, protein, lipids, and chlorophylls in cells during the various growth phases. The light intensity was 500 µmol/m²/s. Glucose was added to the culture and glucose concentration in the medium was controlled at 5 g/l. The various growth phases were identified from the curve of cell growth: exponential phase (0–58 h), linear phase (58–83 h), and stationary phase (83 h–end)

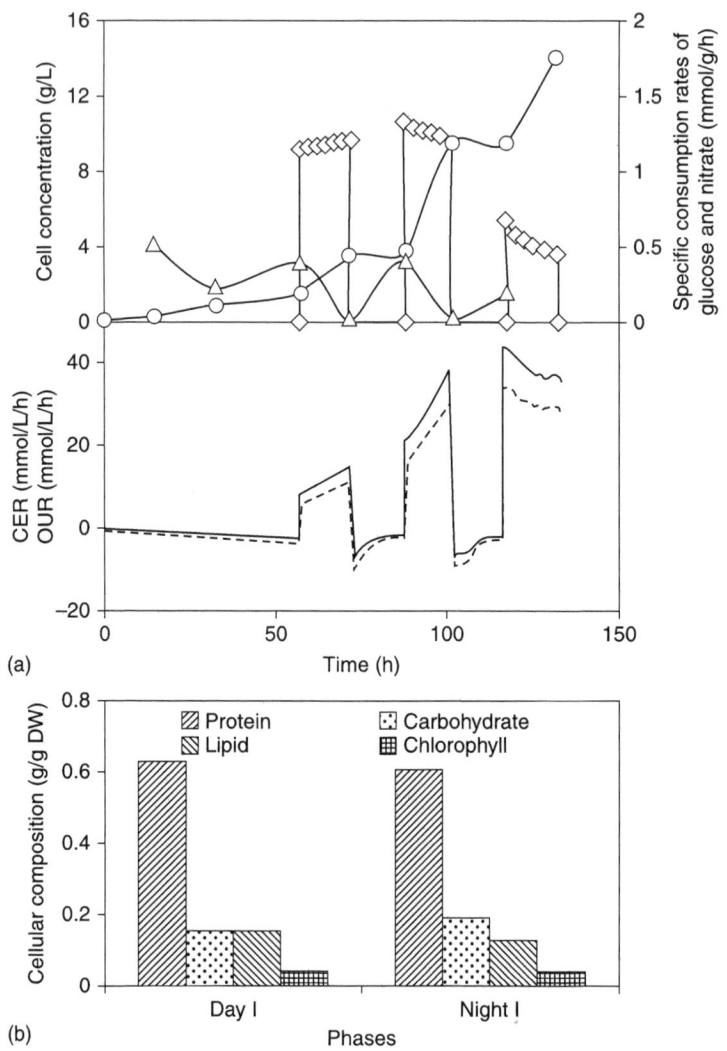

Figure 4.4 Cultivation results of C. pyrenoidosa under cyclic light-autotrophic/dark-heterotrophic conditions: (a) Time profiles of cell growth (○), consumption of glucose (◇) and nitrate (△), and the rates of CO_2 production (solid line) and O_2 uptake (dotted line); (b) Contents of carbohydrates, protein, lipids, and chlorophylls in cells during the first light/dark cycle. Cells were cultivated autotrophically for 24 h and then subjected to 12 h-light/12 h-dark cycles. The light intensity during the light period was 500 μmol/m²/s. Glucose was added during the dark period and glucose concentration in the medium was controlled at 5 g/l

Conventional flux balance

Table 4.3 The consumption of glucose, CO_2 production, and O_2 uptake of *C. pyrenoidosa* under different cultivation conditions

Experiment	$Y_{GLC/X}$ (mol/g)	$Y_{CO_2/X}$ (mol/g)	$Y_{O_2/X}$ (mol/g)
Autotrophic cultivation	–	–0.0398	–0.0556
Mixotrophic cultivation	0.0179	0.0653	0.0478
Cyclic cultivation	0.0170	0.0585	0.0384

decline gradually in the stationary growth phase, while the amount of carbohydrates decreases to the minimum value, and then accumulates to a high level. This result is in agreement with the variation of cellular ultra-structure, as reported by Hu et al. (1998), who suggest that this modification is characteristic of photo-adaption, so as to optimize light harvesting and light utilization.

As expected for mixotrophic and cyclic cultures, a significant improvement is observed in the biomass concentration, because the ability of *Chlorella* to grow on organic carbon sources is exploited in both cultures. However, the uptake of glucose also results in the consumption of O_2 and evolution of CO_2. The growth yields on glucose in the mixotrophic and cyclic cultures are close to each other. With light and organic carbon provided simultaneously as energy sources, the mixotrophic culture reaches the maximum final cell concentration, but forms a much smaller content of chlorophyll in comparison to the autotrophic culture. From the time profiles of the cyclic culture, it is found that the subsequent light autotrophic cell growth is not adversely affected by the carbon source addition during the night, showing that *Chlorella* cells can swiftly switch from autotrophic to heterotrophic metabolism and vice versa (Yang et al., 2000).

4.3.4 Metabolic flux distribution for different culture condition

By cultivating cells under autotrophic, mixotrophic, and cyclic autotrophic-heterotrophic cultures, it is possible to create different conditions for analyzing cell metabolism. Let us consider metabolic flux analysis to elucidate three different metabolisms. Before flux analysis, the data can be analyzed for the presence of measurement errors using elemental balances (Wang and Stephanopoulos, 1983). It can be shown

Bacterial cellular metabolic systems

that the elemental balances for C and N can be closed and that no gross measurement errors are present. Therefore, a flux analysis can be carried out using the data given in Figures 4.2–4.4. Figure 4.5 shows the estimated fluxes for autotrophic, heterotrophic, and mixotrophic cultures (Yang et al., 2000). These values represent the flux distributions of the exponential growth phase during autotrophic and mixotrophic cultivations and the first dark period during the cyclic autotrophic/heterotrophic cultivation. The specific growth rates are almost the same for all three systems. The flux values are expressed in mmol produced metabolites per gram cell per unit time (mmol/g/h).

4.3.5 Effect of light on respiratory metabolism

The cells obtain substrates, reducing power, and ATP for biosynthesis through respiratory pathways, i.e. glycolysis, the PP pathway, and the TCA cycle linked to mitochondrial oxidative electron transport pathways. However, since the light reactions of photosynthesis can provide much reducing power and ATP, respiratory metabolism is unlikely to be essential for these normal functions. It has been proposed by some researchers that the respiratory activity of plants is inhibited in the light, while another indicates little or no effect of light on the respiratory rate (Kelly and Latzko, 1980). Thus the physiological evidence for the effect of light on respiratory metabolism in algae and plant cells is conflicting. Here the autotrophic flux distribution estimated from metabolic flux analysis can be assessed for the respiratory activity of microalgae in the light.

As shown in Figure 4.5, the autotrophic culture shows comparable activities of the glycolytic pathway and the TCA cycle with the heterotrophic culture. This result is consistent with the labeling experiments by Gibbs' laboratory using the alga *Scenedesmus*, which shows that the rates of equilibration of ^{14}C through the intermediates of the TCA cycle are essentially equivalent in light and dark (Kelly and Latzko, 1980). In addition, a significant flux through Ppc is observed for all cultures, suggesting the important role of this enzyme in maintaining the operation of the TCA cycle.

It is found in different plant cells that the percentage of glucose metabolism via the PP pathway is relatively small, i.e. between 5% and 15%, as compared to the total glycolytic flux (Rees, 1980). However, using mass balance to determine the flux distribution of algal cells in heterotrophic culture, the result indicates much higher activity in the PP

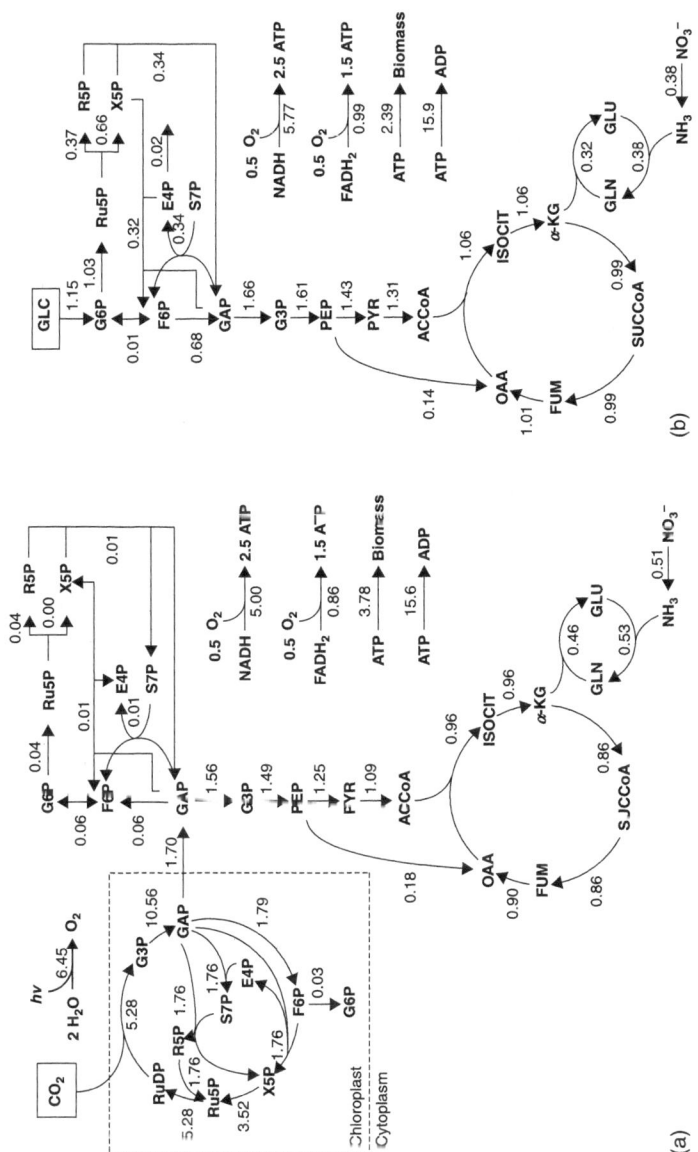

Figure 4.5 Metabolic flux distribution of *Chlorella* cells in: (a) autotrophic; (b) heterotrophic; and

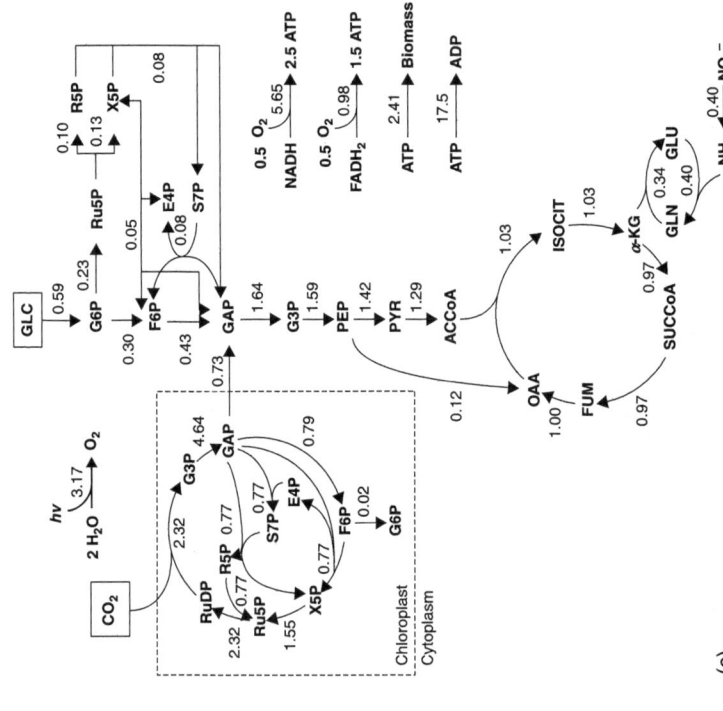

Figure 4.5 (*Continued*) (c) mixotrophic cultures. The flux values are expressed in mmol/g/h. These values represent the flux distributions of the exponential growth phase during the autotrophic (11 h) and mixotrophic (35 h) cultivations and the first dark period during the cyclic autotrophic/heterotrophic cultivation (48 h). The specific growth rates for all three systems are approximately 0.066 h^{-1}

Published by Woodhead Publishing Limited, 2013

pathway than found by others: about 90% of the glucose proceeds via G6PDH.

In the flux distribution under autotrophic condition, the flux through the PP pathway is very small. Since the function of NADPH synthesis is provided by photosynthetic electron transport, the main synthetic functions of the PP pathway in the light appear to be only the supply of precursors for synthesis of nucleic acid and amino acids. G6PDH is known as the major site for regulation in the PP pathway, and its activity is strongly inhibited by NADPH (Turner et al., 1980). It has already been proposed that the light-modulated regulation of G6PDH may be due to a change in the ratio of NADPH/NADP$^+$ (Copeland and Turner, 1987). The low activity of the PP pathway in the autotrophic culture is therefore probably a result of light-mediated control on G6PDH.

It has been suggested that the mitochondrial electron transport will be light-inhibited because of the high ATP/ADP ratios in the light (Kelly and Latzko, 1980). However, other data indicate that any increase in the cytoplasmic ATP/ADP ratios observed in photosynthetic cells upon light to dark transitions is only transient and that steady-state cytoplasmic ATP/ADP ratios are similar under both light and dark conditions (Raymond et al., 1987). Furthermore, it has been shown that light has no direct effect on the activity of the respiratory chain in green algae and plant cells (Dry et al., 1987). It seems that the function of mitochondrial electron transport may have been underestimated and that oxidative phosphorylation can provide a significant fraction of energy for cell growth.

4.3.6 Energy metabolism of microalgae under different culture condition

From the estimated flux distribution, the fluxes involved in the generation and utilization of ATP can be obtained (Table 4.4). Under autotrophic condition, a significant fraction (40%) of ATP is formed from mitochondrial oxidative phosphorylation, suggesting its important role in ATP production. The Calvin cycle is the main ATP sink in the autotrophic culture, and the ATP demand for the assimilation of CO_2 accounts for about 77% of the total. From Table 4.4, the theoretical yields of biomass on ATP for three different cultures can be calculated (Table 4.5), where these values are comparable to the yield found in plant and microbial systems (de Gucht and van der Plas, 1995; Aristidou et al., 1999). Obviously, the ATP yield decreases in the following order:

Table 4.4 The generation and utilization of ATP in the autotrophic, heterotrophic, and mixotrophic cultures

Autotrophic culture		Heterotrophic culture		Mixotrophic culture	
ATP production					
Direct ATP	3.83	Direct ATP	3.54	Direct ATP	3.70
Oxidative phosphorylation	13.8	Oxidative phosphorylation	15.9	Oxidative phosphorylation	15.5
Photo-phosphorylation	16.8			Photo-phosphorylation	8.24
ATP consumption					
Calvin cycle	15.8	Glucose uptake	1.15	Glucose uptake	0.59
Synthesis of cell mass	3.78	Synthesis of cell mass	2.39	Synthesis of cell mass	2.41
				Calvin cycle	6.96

heterotroph > mixotroph > autotroph. Since the Calvin cycle requires a large amount of ATP, the difference in the contribution of the Calvin cycle to total carbon metabolism leads to the different ATP yields. From the flux of excess ATP, the growth-related maintenance ATP requirements can be estimated (Table 4.5). It can be seen that a significant amount of available ATP is required for maintenance. Maintenance processes take up as much as 45–82% of the total ATP produced. This result is consistent with the data reported by de Gucht and van der Plas (1995), who found that maintenance processes require 50–75% of the available ATP in the continuous cultures of plant cells. This maintenance energy includes ATP requirements for transport, translocation, futile cycles, and so on. Especially for algae and plant cells, which have a high degree of subcellular

Table 4.5 Theoretical yields of biomass on ATP and ATP maintenance requirements in the autotrophic, heterotrophic, and mixotrophic cultures

Culture	Theoretical ATP yield (g/mol)	ATP maintenance demand (mmol/g/h)
Autotrophic culture	3.11	15.6
Heterotrophic culture	19.3	15.9
Mixotrophic culture	6.64	17.5

compartmentation of metabolism, various transport reactions are involved in the metabolic pathways (Rees, 1987). These transport processes may consume a large amount of energy.

In autotrophic and heterotrophic cultures, there are only two energy contributors: light or glucose. However, both light and glucose are sources for ATP production in the mixotrophic culture. Since GAP generated by photosynthesis partly enters the glycolytic pathway and the TCA cycle, both light and glucose account for the production of NADH and $FADH_2$ in the TCA cycle. Therefore, ATP production from light includes not only ATP produced from photophosphorylation but also ATP provided by the oxidative phosphorylation of NADH and $FADH_2$ derived from photosynthesis. The contribution of light and glucose to NADH and $FADH_2$ can be calculated using the fractional contribution model (Xie and Wang, 1996). With this information, the contribution of light and glucose to ATP production can be obtained (Table 4.6). The amount of ATP produced from photosynthesis is about 63%. Hence, light is the major source for ATP production in the early phase of mixotrophic cultivation.

4.3.7 Energy conversion under different culture condition

In all three cases, the energy provided by light and/or glucose is absorbed by the microalgal cells, then transformed into ATP for various energy demands inside the cells. Thus the conversion of energy involves three energy forms, the energy supplied to the culture, the energy absorbed by the cells, and the high free energy stored in the phosphoanhydride bonds of ATP. Figure 4.6 shows the conversion efficiencies among the three energy forms in autotrophic, heterotrophic, and mixotrophic cultures. Since the concentration of the organic substrate is maintained as constant, the energy provided by the addition of glucose is completely utilized by

Table 4.6 Contributions of light energy and glucose to ATP production in the exponential phase of mixotrophic cultures

Energy source	ATP production	
	mmol/g/h	%
Light	17.3	63.1
Glucose	10.1	36.9

Bacterial cellular metabolic systems

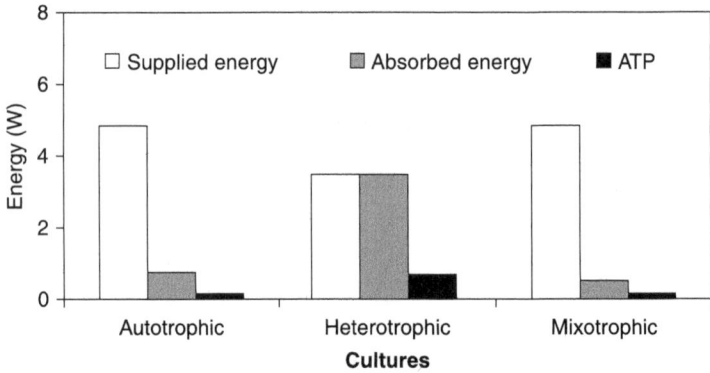

Figure 4.6 Energy conversion efficiency between the energy supplied to the culture, the energy absorbed by the cells, and the high free energy stored in ATP. These values represent the energy conversion efficiency of the exponential growth phase during the autotrophic and mixotrophic cultivations and the first dark period during the cyclic autotrophic/heterotrophic cultivation. The supplied and absorbed light energy were calculated from the appropriate equations (Grima et al., 1997), and the estimation of the total ATP produced was based on the flux distribution shown in Figure 4.5. Assuming that the wavelength of fluorescent light is 600 nm, 1 mol of photons has an energy content of 200.8 kJ. The supplied glucose energy was calculated by multiplying the glucose consumption rate by the free energy change in the reaction of glucose oxidation (2868.852 J/mol). The free energy stored in 1 mol ATP is 30.5 kJ

the cells under heterotrophic and mixotrophic conditions. The capture of light energy in the autotrophic culture is found to be as low as 14% of the total supplied light energy, while in the mixotrophic culture, for which both light and glucose provide energy, the conversion efficiency of the supplied energy to the absorbed energy is even lower than that of the autotrophic culture. The low energy availability in the mixotrophic culture is the result of a lower pigment content in the cells.

From Figure 4.6, microalgal cells transfer 10, 18, and 12% of the absorbed energy into ATP in the autotrophic, heterotrophic, and mixotrophic cultures, respectively. The maximum thermodynamic efficiency of ATP formation from the absorbed energy can be calculated from the fluxes through the relevant metabolic networks at zero growth rates (Figure 4.7). This figure shows that the theoretical yield of ATP on the absorbed energy, Y_{ATP}/AE_{max}, is highest in the heterotrophic culture, while maximum ATP production in the autotrophic culture is only 16% of the harvested light energy. In the mixotrophic culture, Y_{ATP}/AE_{max} is a linear function of the fraction of absorbed light energy of the total. Since the mitochondrial oxidative phosphorylation is a more efficient energy-producing pathway than the photophosphorylation, the conversion efficiency of ATP from the absorbed energy depends on the contributions of both phosphorylation systems to the total ATP production (Lee and Erickson, 1987). If the energy conversion efficiency through the photosynthetic electron transport can be improved, a higher availability of ATP from the absorbed energy will be expected. It has been reported that an algae that lacks photosystem I requires a single photon rather than two in the process of photosynthesis (Greenbaum et al., 1995). Therefore, the conversion efficiency of light energy into chemical energy can be potentially doubled (Yang et al., 2000).

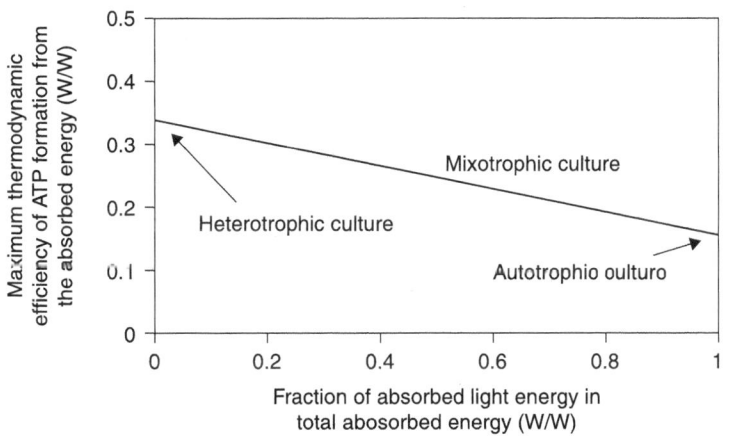

Figure 4.7 Theoretical thermodynamic efficiency of ATP formation from the absorbed energy (Y_{ATP}/AE_{max}) as a linear function of the fraction of the absorbed light energy of the total. Y_{ATP}/AE_{max} was calculated from the fluxes through the relevant metabolic network at zero growth rate. For the fraction of the light energy of the total absorbed energy, 0 represents heterotrophic culture, and 1 represents autotrophic culture

From the above analysis, about 18% of the supplied energy is transformed to ATP in the heterotrophic culture, while the percentages decrease to 1.5% and 1.1% in the autotrophic and mixotrophic cultures, respectively. Apparently, the difference is caused by the different energy sources supplied to the cultures. It seems that compared with the organic substrate, light energy is difficult to trap and convert to ATP by the cells.

4.3.8 Energy economy under different culture condition

The energy economy of microalgae cultures can be evaluated through the efficiency of energy utilization. Table 4.7 shows the overall yield of biomass on the supplied energy for the autotrophic, mixotrophic, and cyclic autotrophic/heterotrophic cultivations. Not surprisingly, the energetic growth yield in the autotrophic culture is lowest due to the inefficient conversion of light energy into biomass, as discussed above. For mixotrophic and cyclic autotrophic/heterotrophic cultures, both light and glucose are the energy sources. However, the difference in the energy supply results in different bioenergetic yields. The biomass energetic yield in the mixotrophic culture is lower than that in the cyclic culture.

The conversion efficiencies of total energy and light energy can be calculated for different growth phases during mixotrophic cultivation (Figure 4.8). Total ATP production and ATP derived from light can be estimated from metabolic fluxes. Figure 4.8 shows that in the exponential phases, where light plays a major role in ATP production, only 0.73% of the supplied light energy is transferred into ATP. This value is much lower than that in the autotrophic culture, because of pigment in the mixotrophic cells. The formation of photosynthetic apparatus is disturbed due to the

Table 4.7 Biomass yields on the supplied energy ($Y_{X/SE}$) in the autotrophic, mixotrophic, and cyclic autotrophic/heterotrophic culture experiments

Experiment	Cell produced (g)	Glucose supplied (kJ)	Light supplied (kJ)	$Y_{X/SE}$ (g/kJ)
Autotrophic culture	3.78	–	2144.6	0.00177
Mixotrophic culture	25.5	1307.6	2099.4	0.00749
Cyclic culture	21.1	1041.1	1238.6	0.00924

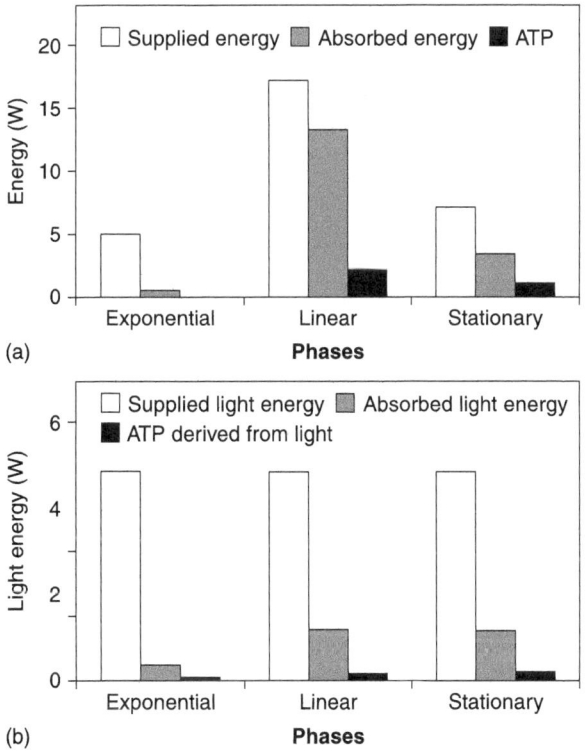

Figure 4.8 Conversion efficiency of: (a) total energy; and (b) light energy during the various growth phases of the mixotrophic cultivation. The total ATP production and ATP derived from light were estimated from the results of metabolic flux analysis

presence of the organic substrate. With the age of the culture and the increase in cell concentration, the contribution of light energy to ATP production is decreased according to flux analysis. In the linear phase, nearly all the ATP produced originated from glucose, and the light energy supplied to the culture cannot be efficiently utilized for cell growth due to high cell density. Therefore, energy is not utilized efficiently in the mixotrophic cultivation. This result is unexpected, since the mixotrophic culture is often applied for commercial algal production. It is expected that the two processes of photosynthesis and glucose catabolism proceed independently and do not interact with each other (Ogawa and Aiba, 1981). However, from the above analysis, it seems that the photosynthetic capacity of microalgal cells is reduced significantly, due to the uptake of glucose (Yang et al., 2000).

Figure 4.9 shows the energy conversion efficiency between the three energy forms during the first light/dark cycle of the cyclic autotrophic/heterotrophic culture. In the cyclic culture, cells are cultivated autotrophically for the first two days, and then subjected to dark/light cycles. From Figure 4.9, during the first two days of autotrophic growth, the light energy absorbed by the cells has reached the maximum value, indicating that autotrophic growth had undergone a rapid growth phase and the availability of light to the culture has started to be limited. In the subsequent dark period, the addition of glucose to the culture enhanced cell growth and results in a significant increase in cell concentration. Therefore, the advantage of the cyclic culture is that autotrophic cell growth and photosynthetic processes are not adversely affected by the addition of organic substrate during the night. In the late light period of the cyclic culture, although the growth rates are very low, autotrophic growth can improve the contents of some important components, such as pigments and linolenic acid. Thus the nutritional value and product quality of the microalgae will be better than those in the simple heterotrophic culture. More importantly, the cyclic culture can be easily employed for the utilization of solar energy. As is well known, utilization of solar energy is very desirable, but the solar light supply is not continuous due to diurnal and seasonal changes (Ogbonna and Tanaka,

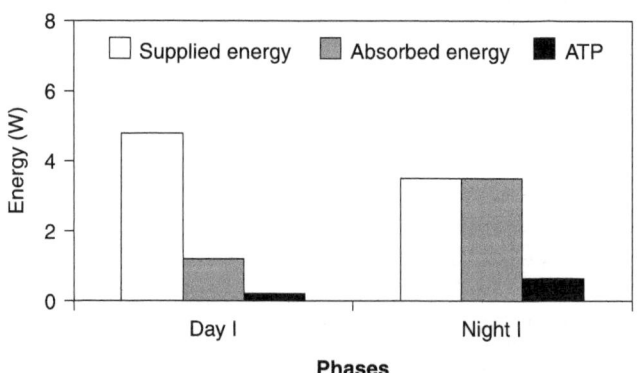

Figure 4.9 Energy conversion efficiency between the three energy forms during the first light/dark cycle of the cyclic autotrophic/heterotrophic cultivation. The total ATP production and ATP derived from light were estimated from the results of metabolic flux analysis

1996). For autotrophic and mixotrophic cultures, which require continuous illumination, it is necessary to capture and concentrate the solar energy for the light supply during the night, while for the illumination of the cyclic culture, solar light energy can be used directly. Therefore, from the viewpoint of economy, the cyclic light-autotrophic/dark-heterotrophic culture can be employed for efficient production of microalgae.

In summary, the metabolic flux distribution of microalgal cells for autotrophic, heterotrophic, and mixotrophic cultures, and the subsequent analysis indicate that the glycolytic pathway, the TCA cycle, and mitochondrial oxidative phosphorylation maintain high activity during illumination, implying little effect of light on these pathways. However, the flux through the PP pathway during illumination is very small due to light-mediated regulation. The theoretical yields of biomass on ATP and the maintenance ATP requirements can be estimated, and the results show that the difference in the contribution of the Calvin cycle to total carbon metabolism leads to different ATP yields of the three cultures, and a significant amount of the available ATP is required for maintenance processes in microalgal cells. The energy conversion efficiency between the energy supplied to culture, the energy absorbed by cells, and the free energy conserved in ATP can be calculated, and the heterotrophic culture generates more ATP from the supplied energy than the autotrophic and mixotrophic cultures. The maximum thermodynamic efficiency of ATP production from the absorbed energy (Y_{ATP}/AE_{max}), which is calculated from the metabolic fluxes at zero growth rate, is the highest in the heterotrophic culture and as low as 16% in the autotrophic culture. In the mixotrophic culture, Y_{ATP}/AE_{max} is a linear function of the fraction of absorbed light energy of the total. The biomass yield on the supplied energy is lowest in the autotrophic cultivation, and the cyclic culture displays the most efficient utilization of energy for biomass production. The analysis of the energy utilization efficiency may be useful for providing information concerning cell energetics and guidance to improve microalgal cell culture performance (Yang et al., 2000).

4.4 Metabolic flux analysis of a single gene knockout *E. coli* under anaerobic conditions

Although the conventional metabolic flux analysis has some limitations, as stated in the introduction, this may be applied to anaerobic culture, since the metabolic pathway network does not include cyclic pathways or closed loops, and the reaction mainly occurs unidirectionally. Here we consider the metabolic flux analysis for several single-gene knockout *E. coli* cultivated under anaerobic condition. Before metabolic flux analysis, let us consider the fermentation characteristics (Table 4.8) and some enzyme activities for *pflA*, *pta*, *ppc*, *adhE*, and *pykF* mutants, as compared to the wild type BW25113 (Table 4.9) (Zhu and Shimizu, 2005). The *pflA* mutant shows much higher activity of GAPDH, LDH, Ppc, and Ack, as compared to the parent strain. The up-regulation of GAPDH and LDH implies the coupling between the NADH production and consumption between the two corresponding reactions. The specific Pyk activity of the *pflA* mutant is higher than that of the parent strain. Since ATP can only be produced through substrate level phosphorylation in glycolysis and the Pta–Ack pathway, the up-regulation of the enzyme activity of Pyk and Ack implies ATP requirement for the cell. Similarly to the *pflA* mutant, the simultaneous up-regulation of GAPDH and LDH is also observed in the *pta* mutant (Table 4.9). Interestingly, Pfl is inactivated in this mutant, which implies a common regulatory mechanism that controls the expressions of *pta* and *pfl* genes. The significant difference between *pta* and *pflA* mutants is the specific activity of Ppc. In the *pta* mutant, this enzyme activity changes little compared to the parent strain. However, the *pta* mutant produces more succinate than the *pflA* mutant, despite lower Ppc activity (Table 4.9), which implies that the intracellular pool size of PEP plays another important role in the metabolic regulation.

The enzyme activity of Ack and ADH, both of which are the AcCoA assimilation pathway enzymes, increases in the *ppc* mutant. However, the enzyme activity of Pfl, which supplies AcCoA for these two reactions, is much lower than that of the parent strain (Table 4.9). Less acetate and ethanol are produced in the *ppc* mutant than those of the parent strain. Although the activity of LDH is lower, lactate production is higher in the *ppc* mutant compared to that of the parent strain. The activity of Pyk in the *ppc* mutant is about one-quarter of that in the parent strain. Pyk activity in the *adhE* mutant is also down-regulated. The activity of Ack is higher in these two strains and the *pykF* mutant, which may be due to ATP generation under anaerobic condition. The activities of LDH and Pfl are both down-regulated in the *adhE* mutant, which is consistent with relatively lower lactate and formate yield in this strain (Table 4.8). The *pykF* mutant shows higher Ppc activity, while the activities of LDH and ADH are both down-regulated as compared to the parent strain (Zhu and Shimizu, 2005).

Table 4.8 Comparison of fermentation results

Strain	Specific growth rate μ (h^{-1})	Yield on glucose (g/g)%						
		Biomass	Lactate	Acetate	Formate	Succinate	Ethanol	Pyruvate
Wild	0.85±0.08	30.1±0.	1.2±0.0	24.3±0.1	30.0±0.2	5.0±0.2	14.7±0.3	0.2±0.0
ΔPflA	0.18±0.04	8.1±0.1	72.5±0.0	1.1±0.1	0.0±0.2	0.0±0.0	1.0±0.0	1.1±0.0
Δpta	0.24±0.04	6.9±0.1	51.3±0.0	1.6±0.1	2.1±0.2	4.5±0.1	0.0±0.0	0.0±0.0
Δppc	0.01±0.02	4.6±0.1	32.2±0.0	13.5±0.1	9.4±0.2	0.3±0.0	6.2±0.1	4.6±0.0
ΔadhE	0.03±0.02	11.4±0.1	9.7±0.0	42.4±0.1	12.1±0.2	3.2±0.1	0.2±0.0	0.0±0.0
ΔpykF	0.48±0.05	13.0±0.1	2.1±0.0	34.0±0.1	22.9±0.2	6.9±0.2	12.0±0.2	0.0±0.0

Table 4.9 Comparison of enzyme activities

	Enzyme activities (nmol/min mg protein)									
	G6PDH	6PGDH	GAPDH	Pyk	_DH	Ppc	Pfl	Ack	ADH
wild	0.152±0.023	0.18±0.002	0.008±0.001	0.657±0.102	0.5C3±0.120	0.008±0.003	0.071±0.012	0.038±0.006	0.006±0.000
ΔpflA	0.252±0.061	0.079±0.010	0.076±0.011	0.726±0.170	2.0C3±0.512	0.343±0.122	0.004±0.004	4.088±0.200	0.011±0.002
Δpta	0.137±0.027	0.001±0.001	0.060±0.008	0.512±0.006	1.621±0.306	0.005±0.001	0.013±0.001	3.465±0.195	0.005±0.001
Δppc	0.017±0.002	0.000±0.000	0.014±0.001	0.183±0.060	0.164±0.020	0.003±0.002	0.025±0.006	0.629±0.035	0.040±0.009
ΔadhE	0.013±0.002	0.004±0.001	0.008±0.003	0.108±0.021	0.0C0±0.000	0.005±0.001	0.043±0.001	0.536±0.035	0.000±0.000
ΔpykF	0.052±0.006	0.017±0.003	0.020±0.005	0.009±0.003	0.069±0.030	0.013±0.001	0.050±0.004	0.293±0.016	0.001±0.001

Let us consider how the intracellular metabolite concentrations change in response to specific-gene knockout (Table 4.10). The pool sizes of G6P, FDP, and PYR increase 3.5 to 20 times in the *pflA* mutant, as compared to the parent strain. The PEP concentration is lower in the *pflA* mutant. The low PEP concentration causes less succinate production, even with high Ppc activity. The metabolite concentrations in the *pta* mutant are similar to those in the *pflA* mutant, except for significantly accumulated PEP. In the case of the *pykF* mutant, all intracellular metabolites in glycolysis are significantly accumulated, as compared to the parent strain and other mutants. Two of the lactate producing strains, such as *pflA* and *pta* mutants, show a higher NADH/NAD$^+$ ratio as compared to the parent strain. Therefore, it is reasonable to consider that lactate production is promoted by intracellular redox balance pressure in these two strains. The value of ATP /AXP (ATP+ADP+AMP) is less in the mutants as compared to the parent strain, and this ratio is low in *pta* and *pykF* mutants (Zhu and Shimizu, 2005).

Consider the intracellular metabolic flux distribution estimated, as stated in Section 4.2, for each strain using the fermentation data during the exponential growth phase. Based on the biochemistry framework (Figure 4.10), the constraints imposed by the stoichiometry and the measured specific rates during the exponential phase can be used to obtain the flux distribution. The flux values are shown in Table 4.11. Although one of the catabolic pathways is blocked by gene knockout, the higher glucose uptake rate (v_1), followed by higher glycolytic flux, may be seen in such strains as *pflA*, *pta*, and *pykF* mutants, as compared to the parent strain. In particular, the glucose uptake rate is about 40% higher in the *pykF* mutant than that in *E. coli* BW25113. This is consistent with the significantly higher intracellular concentration levels of the glycolytic metabolites in the *pykF* mutant. The flux through GAPDH (v_3) is shown to be regulated by the intracellular NADH/NAD$^+$ ratio (Garrigues et al., 1997; De Graef et al., 1999). It has been shown that NADH competitively combines with GAPDH and inhibits the reaction through this enzyme. The GAPDH activity in the mutant strains, such as *pflA*, *pta*, and *pykF* mutants, are 9.5-fold, 7.5-fold, and 2.5-fold of that in the parent strain, respectively (Table 4.9). The intracellular NADH/NAD$^+$ ratio is significantly lower in the *pykF* mutant (Table 4.10) and, therefore, the higher glycolysis flux is not restricted by the competitive inhibition of NADH on the reaction through relatively high levels of GAPDH. While the intracellular NADH/NAD$^+$ ratio is significantly higher in *pflA* and *pta* mutants than the other strains, lactate is produced in these two mutants to regenerate NAD$^+$ needed for

Table 4.10 Comparison of intracellular metabolite concentrations in *E. coli* mutants

	Metabolite concentration (mM/g DCW)											
	G6P	FDP	PEP	PYR	AcCoA	ATP	ADP	AMP	ATP/AXP	NADH	NAD	NADH/NAD
BW25113	0.05±0.01	4.59±0.02	0.32±0.12	8.21±0.01	0.07±0.00	3.06±0.17	0.38±0.01	0.21±0.01	0.84	0.018±0.002	0.143±0.001	0.126
ΔpflA	1.09±0.01	16.41±0.11	0.12±0.04	25.49±0.01	0.05±0.00	1.69±0.11	0.20±0.01	0.20±0.01	0.81	0.056±0.001	0.060±0.002	0.933
Δpta	0.54±0.03	15.08±0.03	1.27±0.20	21.83±0.55	0.01±0.00	1.52±0.16	1.17±0.08	1.12±0.17	0.44	0.047±0.001	0.129±0.007	0.680
Δppc	0.84±0.03	24.46±0.04	0.96±0.07	38.72±2.26	0.05±0.02	4.19±0.07	3.71±1.14	1.16±0.43	0.53	0.018±0.004	0.170±0.010	0.106
ΔadhE	0.32±0.01	11.41±0.03	0.46±0.08	18.40±0.25	0.01±0.00	1.38±0.04	0.69±0.02	0.46±0.04	0.55	0.010±0.006	0.058±0.009	0.167
ΔpykF	1.15±0.02	30.90±0.90	3.67±0.06	46.96±1.98	0.08±0.02	2.35±0.10	2.27±0.05	1.01±0.08	0.47	0.002±0.005	0.177±0.008	0.010

Bacterial cellular metabolic systems

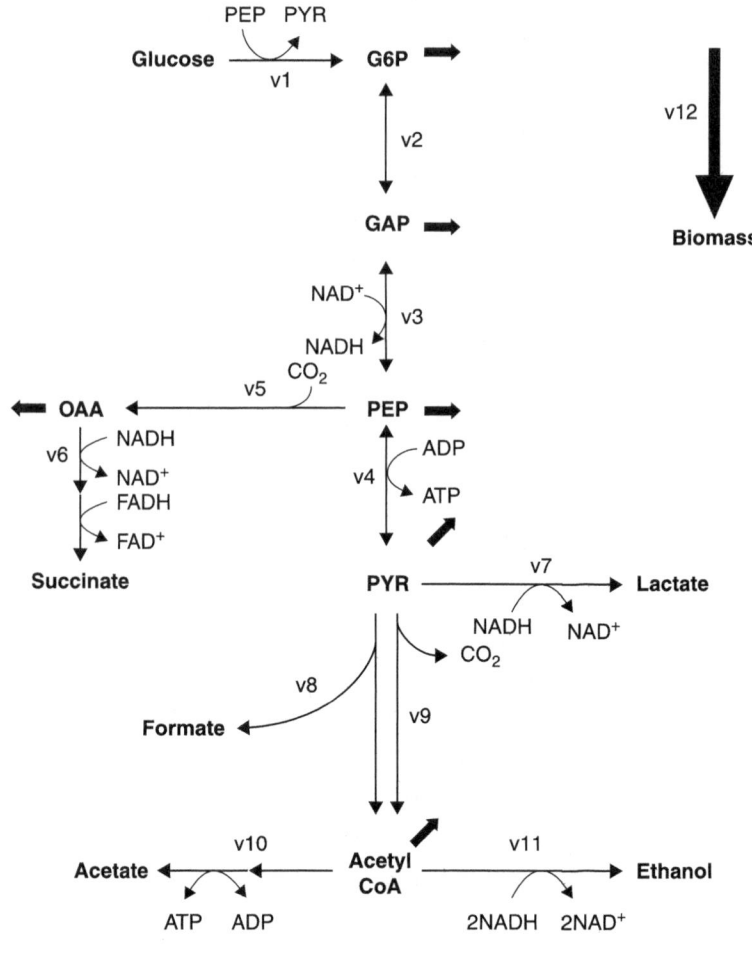

Figure 4.10 General metabolic pathway of *E. coli* under oxygen-limited conditions. The fluxes through each pathway are designated v_1–v_{12}

glycolysis. Correspondingly, the GAPDH expression level is also significantly high in these two strains. These results indicate that the demand for glycolytic flux dominates over the intracellular redox balance. The high glycolytic flux results in a high intracellular NADH/NAD⁺ ratio, and subsequently lactate is produced to regenerate NAD⁺ so that the glycolysis continues to work (Zhu and Shimizu, 2005).

The lactate producing flux (v_7) in the *ppc* mutant is lower than those of the other two lactate producing mutants, such as *pflA* and *pta* mutants.

Table 4.11 Effect of a single-gene knockout on the flux distribution

	\multicolumn{12}{c}{Fluxes (mmol/g DCW/h)}											
	v_1	v_2	v_3	v_4	v_5	v_6	v_7	v_8	v_9	v_{10}	v_{11}	v_{12}
Wild	6.30±0.06	5.90±0.07	11.73±0.10	3.85±0.09	1.39±0.09	0.42±0.06	0.09±0.06	7.28±0.35	1.97±0.23	4.47±0.12	3.52±0.11	3.94±0.10
ΔpflA	8.15±0.06	7.58±0.07	14.93±0.10	5.72±0.10	0.78±0.10	0.18±0.11	11.69±0.7	0.00±0.00	1.64±0.11	0.46±0.2	0.45±0.10	1.62±0.11
Δpta	6.51±0.06	5.79±0.06	11.27±0.10	3.03±0.10	1.37±0.10	0.77±0.10	7.36±0.83	0.70±0.06	0.92±0.10	0.61±0.3	0.31±0.12	1.26±0.12
Δppc	1.59±0.02	1.48±0.02	2.90±0.03	1.12±0.02	0.14±0.02	0.00±0.03	1.13±0.1	0.62±0.07	0.66±0.03	0.72±0.08	0.45±0.06	0.20±0.02
ΔadhE	1.10±0.02	1.04±0.02	2.05±0.03	0.76±0.04	0.15±0.03	0.07±0.02	0.23±0.03	0.53±0.06	1.03±0.05	1.45±0.05	0.00±0.02	0.27±0.04
ΔpykF	9.02±0.07	7.05±0.07	13.21±0.07	0.00±0.00	3.21±0.33	1.54±0.23	0.13±0.2	7.82±0.29	2.56±0.43	7.75±0.4	2.93±0.31	2.48±0.23

It is consistent with the lower glycolytic flux in the *ppc* mutant. As compared to the parent strain, the lactate production rate in the *ppc* mutant is about 7 times higher. However, the lactate yield is about 30-fold higher as compared to that of the parent strain (Table 4.8). Comparing the glucose uptake rates (v_1) in these two strains, the glycolytic flux in the *ppc* mutant is about 17% of that in the wild type. Table 4.12 shows that the percentage of the flux partitioned at the pyruvate node to lactate is significantly higher in the *ppc* mutant, as compared to the parent strain. The ratio is even higher than the other lactate producing strains, such as *pflA* and *pta* mutants. It is known that the reactions through LDH and Pfl are competing with each other (Böck and Sawers, 1996). The high intracellular NADH/NAD⁺ ratio is a plausible reason for the significantly up-regulated LDH activity and down-regulated Pfl activity in the *pflA* and *pta* mutants.

The flux through Ppc (v_5) in the *pykF* mutant is about 2.6-fold that in the parent strain. From enzyme activity data, Ppc activity in the *pykF* mutant is up-regulated by about 60% as compared to the parent strain. The reason for the high Ppc flux may be due to high PEP concentration and synergistic activation by FDP (Table 4.10) (Smith et al., 1980; McAlister et al., 1981). The Ppc flux is low in *ppc* and *adhE* mutants, as compared to the parent strain, which is consistent with the higher relative flux going toward pyruvate at the PEP node (Table 4.12).

Table 4.12 Effect of a single-gene knockout on flux partitions

	Percentage partitioned at PEP node to pyruvate $\left(\dfrac{v_1 + v_4}{v_1 + v_4 + v_4}\right) \times 100$	Percentage partitioned at pyruvate node to lactate $\left(\dfrac{v_7}{v_7 + v_8 + v_9}\right) \times 100$	Percentage partitioned at AcCoA node to acetate $\left(\dfrac{v_{10}}{v_{10} + v_{11}}\right) \times 100$
Wild type	88.0	1.0	55.9
ΔpflA	94.7	87.7	50.5
Δpta	87.4	82.0	66.3
Δppc	95.1	46.9	61.5
ΔadhE	92.5	12.8	100.0
ΔpykF	73.8	0.0	72.6

Since the product of the reaction through Ppc is OAA, which is the precursor of some amino acids such as aspartate and asparagine, the low Ppc flux is consistent with the low biomass synthesis flux (v_{12}) in these two mutants. As a consequence, the glycolytic flux, which is used to supply energy for cell growth, is low in *ppc* and *adhE* mutants (Zhu and Shimizu, 2005).

Table 4.12 shows that the flux partitioning at the AcCoA node is significantly different in *adhE* and *pykF* mutants, as compared to the others. The reaction through ADH to ethanol is blocked by the *adhE* knockout gene, and therefore AcCoA can only be used for cell growth and acetate production. In the *pykF* mutant, the flux distribution at the AcCoA node is determined by the expression of Ack and ADH. The Ack expression in *pflA* and *pta* mutants is significantly up-regulated. However, the ratios of the flux partitioned to acetate are 50.5% and 66.3% in *pflA* and *pta* mutants, respectively, which are little changed by high Ack activity in these two strains, as compared to the ratio of 55.9% in the parent strain (Zhu and Shimizu, 2005).

Pyruvate is competed for by the reactions through Pfl and LDH. In wild-type *E. coli*, the LDH reaction is not as competitive as the reaction through Pfl and, therefore, acetate and formate are the main metabolites instead of lactate (Kessler et al., 1992; Kessler and Knappe, 1996). However, the knockouts of *pflA*, *pta*, and *ppc* genes significantly change enzyme expression and intracellular states, resulting in over-production of lactate. The knockout of the *pflA* gene blocks the pyruvate assimilation through the Pta–Ack and ADH pathways, which are commonly used for ATP production and NADH re-oxidation, respectively, in the wild-type *E. coli*. Since the glycolytic flux is promoted by anaerobiosis with a higher ATP requirement, lactate is produced to satisfy both the stoichiometric and intracellular redox balances (Zhu and Shimizu, 2004). Looking at the enzyme activity results (Table 4.9), the LDH activity is up-regulated by about 3-fold, as compared to the parent strain. Similar phenomena can be seen in the *pta* mutant. The Pfl activity in the *pta* mutant is about one-fifth of that in the parent strain, and LDH activity is up-regulated by more than 2-fold. As a result, the flux through Pfl (v_8) is about one-tenth of that in the parent strain (Table 4.11), and the ratio of flux partitioned at the pyruvate node to lactate is comparable to that in the *pflA* mutant (Table 4.12).

The competition for PYR by reactions through Pfl and LDH is shown in Figure 4.11, by plotting the LDH fluxes against Pfl and LDH activity

Bacterial cellular metabolic systems

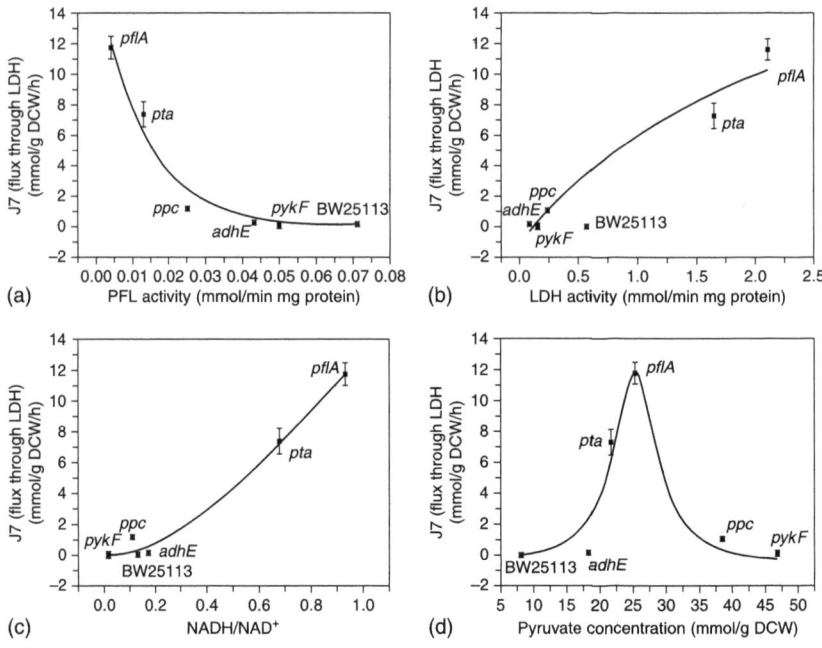

Figure 4.11 Factors influencing lactate producing flux. The effects of Pfl, LDH activities, NADH/NAD$^+$ ratio, and intracellular pyruvate concentration on flux toward lactate generation are represented in (a), (b), (c), and (d), respectively. Error bars indicate deviation of fluxes

in different strains. The LDH flux increases as the Pfl activity becomes low. The change in LDH activity also contributes to the LDH flux. The deviation index of the specific flux J with respect to the enzyme E, defined as $(\Delta J/\Delta E)(E^r/J^r)$ (Small and Kacser, 1993), may be used to compare the effects of LDH and Pfl activity on the lactate production. The deviation index is related to the control coefficient, and can be used as a measure of the effects of large changes in enzyme activities or other effectors on the flux (Stephanopoulos et al., 1998; Yang and San, 1999). The superscript r refers to the reference value at the new perturbed state. Since the enzyme activity varies largely to the wild-type state, the deviation indexes may be evaluated using the neighboring points as the original state (Table 4.11). Table 4.13 shows that LDH controls lactate production in the *pflA* mutant, where the deviation index for LDH flux with respect to Pfl is low due to the low Pfl activity value. Pfl significantly dominates over LDH in the cases of *ppc* and *adhE* mutants. This may explain why there is still high lactate production in *ppc* and *adhE* mutants, although LDH activity is lower, as compared to the parent strain. The large value of the deviation index of the *adhE* mutant for the LDH flux with respect to Pfl is due to the low LDH flux value in this mutant. This result is in contrast with the previous work on *Lactococcus lactis* (Andersen et al., 2001), which shows that the control by lactate dehydrogenase on lactate production is close to zero. In their study, LDH activity ranges from 1% to 133% of that in the wild type. Here, similar results may be seen when LDH activity is comparable or lower than that in the wild type. However, further amplification of LDH activity to 3–5-fold higher by corresponding gene modification shows that the amplified LDH regained control on the LDH flux. Similar results may be seen in Yang and San (1999), who show that the over-expression of LDH from 1.3 to 15.3 units significantly increases lactate production and the deviation index is about 0.57 and 1.16, respectively, according to different original states. Here, the simultaneous decrease in the activity of Pfl also contributes to the increase in LDH flux. Table 4.13 shows that LDH and Pfl activity have approximately equal control on lactate production in the *pta* mutant.

Looking at the relationship between the flux through LDH (v_7) and the intracellular NADH/NAD$^+$ ratio (Figure 4.11C), it is clear that the high flux toward lactate production or the high LDH activity might be induced by the high intracellular redox balance pressure. Similar to the deviation index, the sensitivity index, $(\Delta J/\Delta P)(Pr/Jr)$, may be defined (Small and

Table 4.13 Deviation index for LDH flux

New state	Original state	Deviation index for LDH flux	
		$\left(\dfrac{\Delta J_7}{\Delta LDH}\right)\left(\dfrac{LDH^r}{J_7^r}\right)$	$\left(\dfrac{\Delta J_7}{\Delta Pfl}\right)\left(\dfrac{Pfl^r}{J_7^r}\right)$
ΔpflA	Δpta	1.64	−0.16
Δpta	Δppc	0.94	−0.92
Δppc	ΔadhF	0.80	−1.11
ΔadhE	Δppc	0.00	−9.35
ΔpykF	Wild type	−0.05	−0.73

Kacser, 1993; Yang et al., 2001), where ΔP is the perturbation of the intracellular metabolite concentration. In *pflA* and *pta* mutants, the sensitivity indexes for LDH flux with respect to NADH/NAD$^+$ ratio are 1.37 and 1.28, respectively, which indicate that the higher intracellular redox balance pressure causes more lactate formation. The effect of intracellular pyruvate concentration on LDH flux (v_7) is shown in Figure 4.11d. Although the flux toward lactate production increases as the pyruvate concentration increases, further increase in pyruvate concentration reduces the LDH flux in *ppc* and *pykF* mutants. The pyruvate is the substrate for LDH reaction. However, the high lactate production in *pflA* and *pta* mutants cannot be simply explained by the high intracellular pyruvate concentration. The results indicate that the high LDH activity and NADH/NAD$^+$ ratios contribute more than intracellular pyruvate to the high LDH flux in these two strains. It should be noted that the high pyruvate concentration in the mutants is the result of several factors such as glucose uptake rate, Pfl, and LDH activity, etc.

In the parent strain and its several mutants, such as *pflA*, *pta*, and *ppc* knockout strains, the percentage of the flux partitioned at the AcCoA node to acetate shows a small change, which varies (Table 4.12) (AcCoA used for biomass synthesis is not considered to calculate the flux partition at the AcCoA node). Studies of strictly anaerobic cultivation have shown that the ratio of acetate production and ethanol production is about 1:1 in the wild-type *E. coli* (Alexeeva et al., 2000; Vemuri et al., 2002). This ratio may be affected by the available AcCoA and the intracellular redox balance caused by the residual activity of PDH (Alexeeva et al., 2000; De Graef et al., 1999). The flux partition at the AcCoA node is different in *pykF* and *adhE* mutants (Table 4.11). Most AcCoA is forced by the Pta–Ack pathway to form acetate in the *adhE* mutant, since the ethanol formation pathway is blocked. Note the higher acetate production in the *pykF* mutant as compared to the parent strain. It has been shown that the higher intracellular pyruvate concentration favors acetate production (Yang et al., 2001). The feed and intracellular pyruvate levels effect the redistribution of metabolic fluxes in *E. coli* (Yang et al. 2001). However, this may not be the case in the present example. No AcCoA accumulation occurs, while it is expected in the mutants such as *pta* and *adhE* strains. The relationship between the Pta–Ack flux and the intracellular AcCoA concentration can be seen in Figure 4.12a. The sensitivity indexes for the Pta–Ack flux with respect to AcCoA concentration in the *pykF* mutant and the parent strain are 0.93 and 0.79, respectively.

Conventional flux balance

This indicates that AcCoA availability is a limiting factor for relatively high flux toward acetate formation. Figure 4.12b shows that the Pta–Ack flux increases as Pfl activity increases. The control of Pfl on the Pta–Ack flux is not significant in *pflA* and *pta* mutants, which show low acetate production. The deviation indexes for the Pta–Ack flux with respect to Pfl activity are 0.15 and 0.36 for *pflA* and *pta* mutants, respectively. However, it is shown that the LDH flux is significantly controlled by LDH in these two strains. Therefore, the

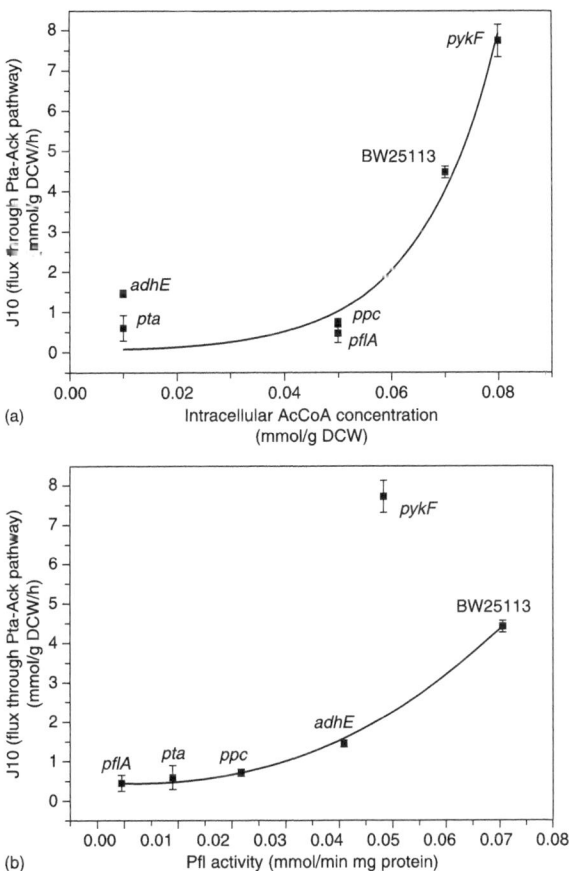

Figure 4.12 Factors influencing flux through Pta–Ack pathway. The effects of intracellular AcCoA concentration and Pfl activity are shown in a and b, respectively. Error bars indicate deviation of fluxes

flux through Pfl, the competitive process of LDH, is low due to high LDH activity in *pflA* and *pta* mutants. Therefore, the Pta–Ack flux depends on the flux through Pfl for AcCoA supply (Zhu and Shimizu, 2005).

The reaction from PEP to OAA through Ppc is the main pathway to replenish OAA in *E. coli* under oxygen-limited conditions (Vemuri et al., 2002). Succinate is derived from OAA, and two equivalents of NADH are required for 1 mole of succinate production. In PTS, PEP is used to transport phosphate for glucose utilization. The reaction through Pyk also uses PEP to produce PYR and ATP. It has been reported that glycolysis is down-regulated in the *pykF* mutant *E. coli* under aerobic condition (Emmerling et al., 2002; Siddiquee et al., 2004a,b). However, when the *pykF* mutant is cultivated under anaerobic condition, the specific glucose uptake rate increases by about 50%, as compared to the parent strain. Flux through glycolysis is reported to be controlled by the ATP requirement in *E. coli*. By optimizing additional ATP hydrolysis, the glycolytic flux increases significantly (Koebmann et al., 2002). Under anaerobic conditions, ATP is considered to be produced by the reactions through the Pyk and Ack pathway due to ATP balance. It can be seen that the flux through the Ack reaction is controlled by the flux through Pfl instead of Ack activity. The changes in the glycolytic flux and the flux through Pyk with respect to Pyk activity are shown in Figure 4.13. The deviation index for the Pyk flux with respect to Pyk activity is 1 for all the strains, which indicates that this flux is controlled by the Pyk enzyme. The Pyk enzyme also shows significant control on the glycolytic flux (Figure 4.13b). The deviation indexes for glycolytic flux (v_3) with respect to Pyk activity vary from 0.60 to 1.16 in the strains considered, except for the *pykF* mutant, in which the glycolytic flux is obviously independent of Pyk activity (Zhu and Shimizu, 2007).

Since OAA is one of the important precursors for the cell synthesis, the flux through Ppc is considered to affect cell growth. This indicates that the single gene knockout significantly reduce the growth. This can also be shown by the low biomass yields in the mutants (Table 4.8). Ppc activity in *pflA* is high compared to other strains (Table 4.9). This may be due to intracellular redox balance pressure. Although Ppc activity is high, the intracellular PEP concentration is low in the *pflA* mutant as compared to other strains. Therefore, the flux through Ppc (v_5) is relatively low in this mutant.

Conventional flux balance

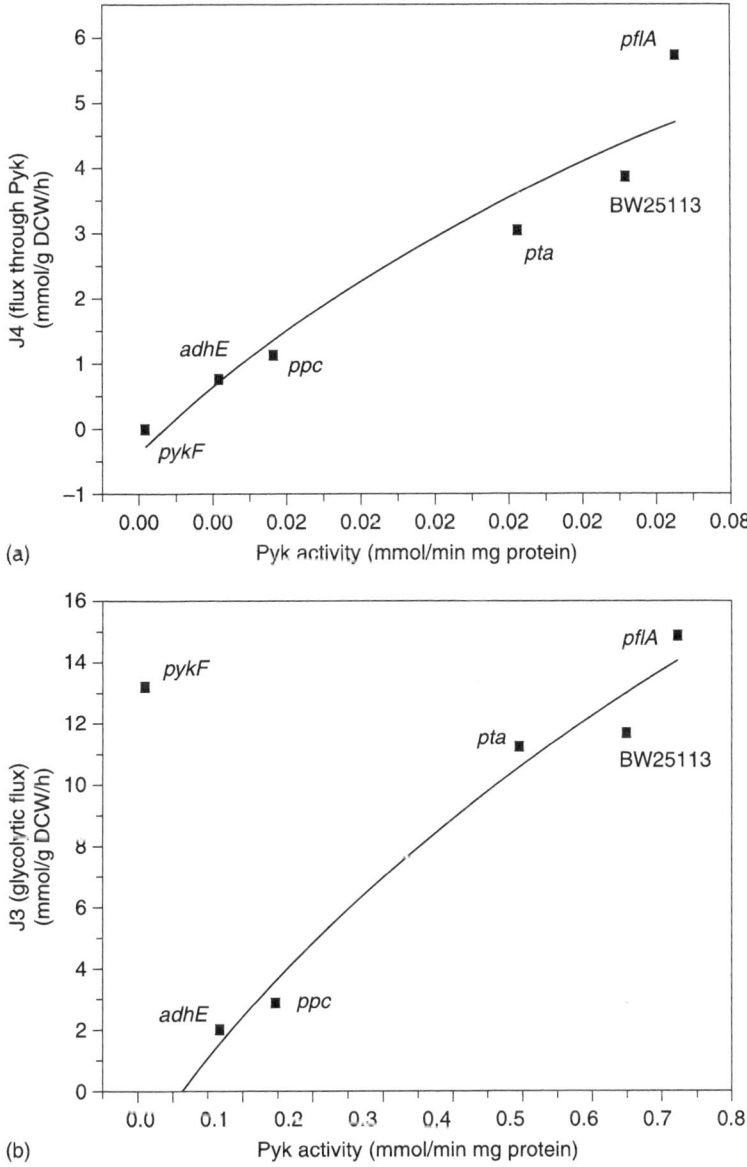

Figure 4.13 The effect of Pyk activity on: (a) Pyk flux; and (b) glycolytic flux. Error bars indicate deviation of fluxes

4.5 References

Akazawa, T. and Okamoto, K (1980) 'Biosynthesis of sucrose', in: *The Biochemistry of Plants*, Vol. 3, edited by J. Preiss. New York: Academic Press. pp. 199–218.

Alexeeva, S., de Kort, B., Sawers, G., Hellingwerem, K.J., and de Mattos, M.J.T. (2000) 'Effects of limited aeration and of the ArcAB system on intermediary pyruvate catabolism in *Escherichia coli*', *Journal of Bacteriology*, 182: 4934–40.

Andersen, H.W., Pedersen, M.B., Hammer, K., and Jensen, P.R. (2001) 'Lactate dehydrogenase has no control on lactate production but has a strong negative control on formate production in *Lactococcus lactis*', *European Journal of Biochemistry*, 268: 6379–89.

Aristidou, A.A., San, K.Y., and Bennett, G.N. (1999) 'Metabolic flux analysis of *Escherichia coli* expressing the *Bacillus subtilis* acetolactate synthase in batch and continuous cultures', *Biotechnology and Bioengineering*, 63: 737–49.

Avron, M. (1989) 'Photosynthetic electron transport and photophosphorylation', in: *The Biochemistry of Plants*, vol. 8, edited by M.D. Hatch and N.K. Boardman. New York: Academic Press. pp. 164–89.

Bailey, J.E. (1991) 'Toward a science of metabolic engineering', *Science*, 252: 1668–75.

Böck, A. and Sawers, G. (1996) 'Fermentation', In: Escherichia coli *and* Salmonella. *Cellular and Molecular Biology*, 2nd edition, edited by F.C. Neidhardt et al. Washington DC: American Society for Microbiology Press.

Bonarius, H.P.J., Hatzimanikatis, V., Meesters, H.P.H, de Gooijer, C.D., Schmid, G., and Tramper, J. (1996) 'Metabolic flux analysis of hybridoma cells in different culture media using mass balances', *Biotechnology and Bioengineering*, 50: 299–318.

Borowitzka, MA. (1988) 'Vitamins and fine chemicals from microalgae', in: *Microalgal Biotechnology*, edited by M.A. Borowitzka and L.J. Borowitzka. Cambridge: Cambridge University Press. pp. 153–96.

Castelfranco, P.A. and Beale, S.I. (1981) 'Chlorophyll biosynthesis', in: *The Biochemistry of Plants*, vol. 8, edited by M.D. Hatch and N.K. Boardman. New York: Academic Press. pp. 376–414.

Copeland, L. and Turner, J.F. (1987) 'The regulation of glycolysis and the pentose phosphate pathway,' in: *The Biochemistry of Plants*, vol. 11, edited by D.D. Davies. New York: Academic Press. pp. 107–25.

de Graef, M.R., Alexeeva, S., Snoep, J.L., and de Mattos, M.J.T. (1999) 'Steady-state internal redox state (NADH/NAD) reflects the external redox state and is correlated with catabolic adaptation in *Escherichia coli*', *Journal of Bacteriology*, 181: 2351–7.

de Gucht, L.P.E. and van der Plas, L.H.W. (1995) 'Growth kinetics of glucose-limited *Petunia hybrida* cells in chemostat cultures: Determination of experimental values for growth and maintenance parameters', *Biotechnology and Bioengineering*, 47: 42–52.

Douce, R., Brouquisse, R., and Journet, E.P. (1987) 'Electron transfer and oxidative phosphorylation in plant mitochondria', in: *The Biochemistry of Plants*, vol. 11, edited by D.D. Davies. New York: Academic Press. pp. 177–207.

Droop, M.R. (1974) 'Heterotrophy of carbon', in: *Algal Physiology and Biochemistry*, edited by W.D.P. Stewart. University of California Press. pp. 530–60.

Dry, I.B., Bryce, J.H., and Wiskich, J.T. (1987) 'Regulation of mitochondrial respiration', in: *The Biochemistry of Plants*, vol. 11, edited by D.D. Davies. New York: Academic Press. pp. 214–7.

Edwards, J.S. and Palsson, B.O. (2000) 'The *Escherichia coli* MG1655 *in silico* metabolic genotype: Its definition, characteristics, and capabilities', *Proceedings of the National Academy of Sciences of USA*, 97: 5528–33.

Edwards, J.S., Ibarra, R.U., and Palsson, B.O. (2001) '*In silico* predictions of *Escherichia coli* metabolic capabilities are consistent with experimental data', *Nature Biotechnology*, 19: 125–30.

Emmerling, M., Dauner, M., Ponti, A., Fiaux, J., Hochuli, M. et al. (2002) 'Metabolic flux responses to pyruvate kinase knockout in *Eshcherichia coli*', *Journal of Bacteriology*, 184: 152–64.

Fowden, L. (1962) 'Amino acids and proteins', in: *Physiology and Biochemistry of Algae*, edited by R.A. Lewin. New York: Academic Press. pp. 189–206.

Garrigues, C., Loubiere, P., Lindle, N.D., and Cocaign-bousquet, M. (1997) 'Control of the shift from homolactic acid to mixed-acid fermentation in *lactococcus lactis*: predominant role of the NADH/NAD+ ratio', *Journal of Bacteriology*, 179: 5282–7.

Glombitza, K.W. and Koch, M. (1989) 'Secondary metabolites of pharmaceutical potential', in: *Algal and Cyanobacterial Biotechnology*, edited by R.C. Cresswell, T.A.V. Rees, and H. Shah. New York: Longman. pp. 161–219.

Graham, D. (1980) 'Effects of light on dark respiration', in: *The Biochemistry of Plants*, vol. 2, edited by D.D. Davies. New York: Academic Press. pp. 526–75.

Greenbaum, E., Lee, J.W., Tevault, C.V., Blankinshi, S.L., and Mets, L.J. (1995) 'CO_2 fixation and photoevolution of H_2 and O_2 in a mutant of Chlamydomonas lacking photosystem I', *Nature*, 376: 438–41.

Grima, E.M., Camacho, F.G., Perez, J.A.S, Fernandez, F.G.A., and Sevilla, J.F. (1997) 'Evaluation of photosynthetic efficiency in microalgal cultures using averaged irradiance', *Enzyme and Microbial Technology*, 21: 375–81.

Heldt, H.W. and Flugge, U.I. (1987) 'Subcellular transport of metabolites in plant cells', in: *The Biochemistry of Plants*, vol. 12, edited by D.D. Davies. New York: Academic Press. pp. 50–80.

Hu, Q., Kurano, N., Kawachi, M., Iwasaki, I., and Miyachi, S. (1998) 'Ultrahigh-cell-density culture of a marine green alga *Chlorococcum littorale* in a flat-plate photobioreactor', *Applied Microbiology and Biotechnology*, 49: 652–5.

Ibarra, R.U., Edwards, J.S., and Palsson, B.O. (2002) '*Escherichia coli* K-12 undergoes adaptive evolution to achieve in silico predicted optimal growth', *Nature*, 420: 186–9.

Kelly, G.J. and Latzko, E. (1980) 'The cytosol', in: *The Biochemistry of Plants*, vol. 1, edited by N.E. Tolbert. New York: Academic Press. pp. 183–205.

Kessler, D. and Knappe, J. (1996) 'Anaerobic dissimilation of pyruvate', In: Escherichia coli *and* Salmonella. *Cellular and Molecular Biology*, 2nd edition, edited by F.C. Neidhardt et al., Washington DC: American Society of Microbiology Press.

Kessler, D., Herth, W., and Knappe, J. (1992) 'Ultrastructure and pyruvate formate-lyase radical quenching property of the multi-enzymatic adhE protein of *Escherichia coli*', *Journal of Biological Chemistry*, 267: pp. 18073–9.

Koebmann, B.J., Westerhoff, H.V., Snoep, J.L., Nilsson, D., and Jensen, P.R. (2002) 'The glycolytic flux in *Escherichia coli* is controlled by the demand for ATP', *Journal of Bacteriology*, 184: 3909–16.

Kurano, N, and Miyachi, S. (1995) 'Fixation and utilization of carbon dioxide by microalgal photosynthesis', *Biotechnology and Bioengineering*, 36: 689–92.

Lee, H.Y. and Erickson, L.E. (1987) 'Theoretical and experimental yields for photo-autotrophic mixotrophic and photo-heterotrophic growth', *Biotechnology and Bioengineering*, 29: 476–81.

Lee, C.G. and Palsson, B.O. (1994) 'High-density algal photo-bioreactors using light-emitting diodes', *Biotechnology and Bioengineering*, 44: 1161–7.

Lloyd, D. (1974) 'Dark respiration', in: *Algal Physiology and Biochemistry*, edited by W.D.P. Stewart. University of California Press. pp. 505–30.

Mandalam, R.K. and Palsson, B.O. (1998) 'Elemental balancing of biomass and medium composition enhances growth capacity in high-density *Chlorella vulgaris* cultures', *Biotechnology and Bioengineering*, 59: pp. 605–11.

McAlister, L.E., Evans, E.L., and Smith, T.E. (1981) 'Properties of a mutant *Escherichia coli* phosphoenolpyruvate carboxylase deficient in co-regulation by intermediary metabolites', *Journal of Bacteriology*, 146: 200–8.

Miflin, B.J. and Lea, P.J. (1980) 'Ammonia assimilation', in: *The Biochemistry of Plants*, vol. 5, edited by B.J. Miflin. New York: Academic Press. pp. 169–99.

Morris, I. (1974) 'Nitrogen assimilation and protein synthesis', in: *Algal Physiology and Biochemistry*, edited by W.D.P. Stewart. University of California Press. pp. 583–610.

Nichols, B.W. (1965) 'Light induced changes in the lipids of *Chlorella vulgaris*', *Biochemistry and Biophysics ACTA*, 106: 274–9.

Nishimura, T., Pachpande, R.R., and T. Iwamura, T. (1988) 'A heterotrophic synchronous culture of *Chlorella*', *Cell Structure Function*, 13: 207–15.

Ogawa, T, and Aiba, S. (1981) 'Bioenergetic analysis of mixotrophic growth in *Chlorella vulgaris* and *Scenedesmus acutus*', *Biotechnology and Bioengineering*, 23: 1121–32.

Ogbonna, J.C. and Tanaka, H. (1996) 'Night biomass loss and changes in biochemical composition of cells during light/dark cyclic culture of *Chlorella pyrenoidosa*', *Journal of Fermentation and Bioengineering*, 82: 558–64.

Palsson, B. (2009) 'Metabolic systems biology', FEBS Letters, 583: 3900–3904.

Price, N.D., Papin, J.A., Schilling, C.H., and Palsson, B,O. (2003) 'Genome-scale microbial in silico models: the constraints-based approach', *Trends in Biotechnology*, 21: 162–9.

Raymond, P., Gidrol, X., Salon, C., and Pradet, A. (1987) 'Control involving adenine and pyridine nucleotides', in: *The Biochemistry of Plants*, vol. 11, edited by D.D. Davies. New York: Academic Press. pp. 130–68.

Rees, T.A. (1980) 'Assessment of the contribution of metabolic pathways to plant respiration', in: *The Biochemistry of Plants*, vol. 2, edited by D.D. Davies. New York: Academic Press. pp. 1–27.

Rees, T.A. (1987) 'Compartmentation of plant metabolism', in: *The Biochemistry of Plants*, vol. 12, edited by D.D. Davies. New York: Academic Press. pp. 87–113.

Schilling, C.H. and Palsson, B.O. (1998) 'The underlying pathway structure of biochemical reaction networks', *Proceedings of the National Academy of Sciences of USA*, 95: 4193–8.

Schuetz, R., Kuepfer, L., and Sauer, U. (2007) 'Systematic evaluation of objective functions for predicting intracellular fluxes in *Escherichia coli*', *Molcular Systems Biology*, 3: 1–14.

Siddiquee, K.A.Z., Arauzo-Bravo, M., and Shimizu, K. (2004a) 'Metabolic Flux Analysis of *pykF* gene knockout *Escherichia coli* based on ^{13}C-labeled experiment together with measurements of enzyme activities and intracellular metabolite concentrations', *Applied Microbiology and Biotechnology*, 63: 407–17.

Siddiquee, K.A.Z., Arauzo-Bravo, M., and Shimizu, K. (2004b) 'Effect of pyruvate kinase (*pykF* gene) knockout mutation on the control of gene expression and metabolic fluxes in *Escherichia coli*', *FEMS Microbiology Letters*, 235: 25–33.

Small, J.R. and Kacser, H. (1993) 'Response of metabolic systems to large changes in enzyme activities and effectors: 1. The linear treatment of unbranched chains', *European Journal of Biochemistry*, 213: 613–24.

Smith, T.E., Balasubramanian, K.A., and Beezley, A. (1980) '*Escherichia coli* phoshpoenol -pyruvate carboxylase: Studies on the mechanism of synergistic activation by nucleotides', *Journal of Biological Chemistry*, 255: 1635–42.

Stephanopoulos, G.N. (1999) 'Metabolic fluxes and metabolic engineering', *Metabolic Engineering*, 1: 1–11.

Stephanopoulos, G.. and Vallino, J.J. (1991) 'Network rigidity and metabolic engineering in metabolite overproduction', *Science*, 252: 1675–81.

Stephanopoulos, G.N., Aristidou, A.A., and Nielsen, J. (1988) 'Metabolic control analysis', in: *Metabolic Engineering. Principles and Methodologies*, edited by G.N. Stephanopoulos, A.A. Aristidou, and J. Nielsen. New York: Academic Press. pp. 461–533.

Storey, B.T. (1980) 'Electron transport and energy coupling in plant mitochondria', in: *The Biochemistry of Plants*, vol. 2, edited by D.D. Davies. New York: Academic Press. pp. 125–87.

Tolbert, N.E. (1980) 'Photorespiration' *The Biochemistry of Plants*, vol. 2, edited by D.D. Davies. New York: Academic Press. pp. 488–521.

Tredici, M.R. and Zittelli, G.C. (1998) 'Efficiency of sunlight utilization: Tubular versus flat photo-bioreactors', *Biotechnology and Bioengineering*, 57: 187–97.

Tsai, S.P. and Lee, Y.H. (1988) 'Application of metabolic pathway stoichiometry to statistical analysis of bioreactor measurement data', *Biotechnology and Bioengineering*, 32: 713–16.

Turner, J.F. and Turner, D.H. (1980) 'The regulation of glycolysis and the pentose phosphate pathway', in: *The Biochemistry of Plants*, vol. 2, edited by D.D. Davies. New York: Academic Press. pp. 279–312.

Vallino, J.J. and Stephanopoulos, G.N. (1993) 'Metabolic flux distributions in *Corynebacterium glutamicum* during growth and lysine over-production', *Biotechnology and Bioengineering*, 41: 633–46.

Vemuri, G.N., Eiteman, M.A., and Altman, E. (2002) 'Effects of growth mode and pyruvate carboxylase on succinic acid production by metabolically engineered strains of *Escherichia coli*', *Applied Environmental Microbiology*, 68: 1715–27.

Wang, N.S. and Stephanopoulos, G..N. (1983) 'Application of macroscopic balances to the identification of gross measurement errors', *Biotechnology and Bioengineering*, 25: 2177–208.

Wanka, F., Joosten, H.F.P., and de Grip, W.J. (1970) 'Composition and synthesis of DNA in synchronously growing cells of *Chlorella pyrenoidosa*', *Archives of Microbiology*, 75: 25–36.

Wiskich, T. (1980) 'Control of the Kerbs cycle', in: *The Biochemistry of Plants*, vol. 2, edited by D.D. Davies. New York: Academic Press. pp. 244–75.

Wood, B.J.B. (1974) 'Fatty acids and saponifiable lipids', in: *Algal Physiology and Biochemistry*, edited by W.D.P. Stewart. University of California Press. pp. 236–66.

Xie, L. and Wang, D.I.C. (1996) 'Energy metabolism and ATP balances in animal cell cultivation using a stoichiometrically based reaction network', *Biotechnology and Bioengineering*, 52: 591–601.

Yanagi, M., Watanabe. Y., and Saiki, H. (1995) 'CO_2 fixation by *Chlorella* sp. HA-1 and its utilization', *Biotechnology and Bioengineering*, 36: 713–16.

Yang, Y. and San, K.Y. (1999) 'Redistribution of metabolic fluxes in *Escherichia coli* with fermentative lactate dehydrogenase over-expression and deletion', *Metabolic Engineering*, 1: 141–52.

Yang, C., Hua, Q., and Shimizu, K. (2000) 'Energetics and carbon metabolism during growth of microalgal cells under photoautotrophic, mixotrophic, and cyclic light-autotrophic/dark-heterotrophic conditions', *Biochemical Engineering Journal*, 6: 87–102.

Yang, Y.T., Bennett, G.N., and San, K.Y. (2001) 'The effects of feed and intracellular pyruvate levels on the redistribution of metabolic fluxes in *Escherichia coli*', *Metabolic Engineering*, 3: 115–23.

Yuan, J., Fowler, W.U., Kimball, E., Lu, W., and Rabinowitz, J.D. (2006) 'Kinetic flux profiling of nitrogen assimilation in *Escherichia coli*', *Nature Chemical Biology*, 2: 529–30.

Zhu, J. and Shimizu, K. (2004) 'The effect of *pfl* genes knockout on the metabolism for optically pure D-lactate production by *Escherichia coli*', *Applied Microbiology and Biotechnology*, 64: 367–75.

Zhu, J. and Shimizu, K. (2005) 'Effect of a single-gene knockout on the metabolic regulation in *E. coli* for D-lactate production under microaerobic conditions', *Metabolic Engineering*, 7: 104–15.

5

^{13}C-metabolic flux analysis and its applications

Abstract: The basic idea and the algorithm for ^{13}C metabolic flux analysis are explained. Its application to wild-type *Escherichia coli* using glucose or acetate as a carbon source is explained based on isotopomer data obtained by GC-MS. Other applications are explained for metabolic flux analysis of *pckA, pgi,* and *zwf* gene knockout *E. coli* strains based on the isotopomer data obtained by NMR. By combining the information available from NMR, GC-MS, and metabolite balances, intracellular flux distributions in *Synechocystis* grown under the heterotrophic and mixotrophic conditions can be quantified, where the pentose phosphate pathway is the major pathway of glucose catabolism in the heterotrophic culture, which is switched to the Calvin cycle when light is supplied to cyanobacterial cells. The operation of a C_4 pathway in cyanobacterial cells can be identified, based on labeling data.

Key words: ^{13}C-Metabolic flux analysis; GC-MS; NMR; wild-type *Escherichia coli*; cyanobacteria; *pckA* mutant *E. coli*; *pgi* mutant *E. coli*; *zwf* mutant *E. coli*.

5.1 Introduction

As stated in Chapter 4, metabolic fluxes can be computed based on measured specific rates with stoichiometric constraints. However, this approach cannot compute such fluxes as: i) recycled fluxes; ii) bidirectional fluxes; and iii) parallel fluxes, etc., due to the singularity of the stoichiometric matrix. This problem can be overcome by using ^{13}C-labeled flux analysis by adding more information on isotopomer distribution.

Therefore, the potential of using a ^{13}C-labeled substrate has been recognized to clarify metabolic fluxes. Early efforts have been made to utilize NMR spectra for estimating metabolic fluxes (Walsh and Koshland, 1984; Jeffrey et al., 1991; Kelleher, 2001).

Determination of flux distribution in a metabolic network has been significantly facilitated by the introduction of labeling experiments (Noronha *et al.*, 2000). In such an experiment, cells are fed with a ^{13}C-labeled substrate, and the labeling patterns of certain intracellular metabolites are determined. Since the measurements of labeling patterns of the intracellular metabolites are usually difficult to perform due to their small pool sizes, analysis of the amino acids has been frequently used for clarification of the labeling states in the central metabolism. These labeling data provide additional and yet independent constraints on intracellular fluxes and thus determine the fluxes that are unobservable by conventional flux analysis using only metabolite balances.

A powerful approach for accurately quantifying intracellular fluxes in a complex metabolic network is based on ^{13}C-labeling experiments followed by measurement of isotopomers using either nuclear magnetic resonance (NMR) or mass spectrometry (Szyperski, 1995; Marx et al., 1996). These data may be used to distinguish metabolic fluxes between alternative pathways. One particular type of labeling experiment is based on growing cells on a mixture of uniformly ^{13}C-labeled [U-^{13}C] and unlabeled glucose. The resulting ^{13}C-labeling patterns of intracellular metabolites are analyzed by 2D NMR spectroscopy. The observed multiplet intensities are transformed into the relative abundance of intact carbon fragments that originate from a single source of glucose molecule. Since alternative pathways leading to the same metabolites yield different intact fragments, this method identifies active pathways in a bioreaction network and the analysis of the ratios of intracellular fluxes. In addition to direct interpretation of 2D NMR data for flux ratio analysis, isotopomer measurement data in combination with biomass composition and all extracellular flux data can also be used for quantification of intracellular fluxes in a metabolic network (Schmidt et al., 1997; Dauner et al., 2001). Based on the balance of metabolites and isotopomers, a mathematical framework relating to metabolic fluxes with isotopomer measurement data is constructed. The intracellular flux distribution is then estimated by finding a best fit to all available data in an iterative fitting procedure. This method has been used successfully to estimate intracellular fluxes in complex reaction networks (Sauer et al., 1997).

^{13}C-metabolic flux analysis

The detailed analysis of NMR spectra has proved useful in investigating metabolic regulation of *E. coli* (Szyperski, 1995). The problem of using NMR is that it requires a relatively large amount of samples. However, the amount of samples required for GC-MS (gas chromatography mass spectrometry) is less, and therefore, GC-MS is becoming popular (Sauer, 2006; Wittman, 2007). Moreover, CE-MS (capillary electrophoresis mass spectrometry) and CE-TOF (time of flight)/MS may become popular due to their high sensitivity and high throughput performance, where these can detect the mass isotopomer distribution of intracellular metabolites (Toya et al., 2007; 2010).

Here, the basic idea and the principles of ^{13}C-metabolic flux analysis are briefly explained, and its practical applications are illustrated for cases using GC-MS, NMR, or a combination of both.

5.2 Basic principle for flux calculation based on ^{13}C-labeling experiment

The isotopomer distribution of a molecule may be expressed in several ways. First is the positional representation, which shows enrichment of ^{13}C for each carbon with an n-carbon molecule (positional representation) (Zupke and Stephanopoulos, 1994; Marx et al., 1996; Wiechert and de Graaf, 1997). Second is the isotopomer representation, which shows the distribution of 2^n isotopomers (isotopomer representation) (Schmidt et al., 1997, 1999b; Christensen and Nielsen, 1999; Mollney et al., 1999; Wiechert et al., 1999).

In isotopic tracer experiments, cells are cultivated with the substrate labeled at the specific C atom. Isotopomer balance is made for the flux estimation. In order to express the labeling state, the specific activity may be expressed as the fractional enrichment of a specific atom within the molecule of interest. Let us introduce a metabolite activity vector (MAV) (Zupke and Stephanopoulos, 1994), where the ith entry of MAV is the fractional enrichment of the ith carbon of the molecule. For example, MAV is expressed as follows, for the case of a molecule with two carbons:

$$A = \begin{bmatrix} A(1) \\ A(2) \end{bmatrix} \qquad (5.1)$$

where A(1) and A(2) represent the ^{13}C fractional enrichment of the first and the second carbons, respectively.

The next step is to construct a matrix to map or describe the transfer of atoms from substrates (reactants) to products, where such matrices are called atom mapping matrices (AMMs). Consider the simple example in Figure 5.1a, where AMMs are expressed as:

$$AMM_{S>A} = \begin{bmatrix} 1 & 0 \\ 0 & 1 \end{bmatrix} \quad AMM_{A>B} = \begin{bmatrix} 1 & 1 \end{bmatrix} \quad AMM_{A>C} = \begin{bmatrix} 0 & 1 \\ 1 & 0 \end{bmatrix}$$

$$AMM_{C>A} = \begin{bmatrix} 1 & 0 \\ 0 & 1 \end{bmatrix} \quad AMM_{B>C} = \begin{bmatrix} 1 \\ 1 \end{bmatrix} \quad AMM_{C>P} = \begin{bmatrix} 1 & 0 \\ 0 & 1 \end{bmatrix} \quad (5.2)$$

where $AMM_{i>j}$ denotes the atom mapping matrix from i to j. For example, $AMM_{S>A}$ is the atom mapping matrix from S to A, where it indicates that the first carbon of S is converted to the first carbon of A, and the second carbon of S is converted to the second carbon of A, and so on. The ^{13}C balance for A, B, and C for the example shown in Figure 5.1a may be expressed as:

A:

$$\vec{v_1} AMM_{S>A} \begin{bmatrix} S(1) \\ S(2) \end{bmatrix} + \overleftarrow{v_2} AMM_{C>A} \begin{bmatrix} C(1) \\ C(2) \end{bmatrix}$$

$$= (\vec{v_2} AMM_{A>C} + \vec{v_3} AMM_{A>C} + \vec{v_4} AMM_{A>B}) \begin{bmatrix} A(1) \\ A(2) \end{bmatrix}$$

B: $\vec{v_4} AMM_{A>B} \begin{bmatrix} A(1) \\ A(2) \end{bmatrix} = \vec{v_5} B(1)$ (5.3)

C:

$$\vec{v_2} AMM_{A>C} \begin{bmatrix} A(1) \\ A(2) \end{bmatrix} = \vec{v_3} AMM_{A>C} \begin{bmatrix} A(1) \\ A(2) \end{bmatrix} + \vec{v_5} AMM_{B>C} B(1)$$

$$= \vec{v_6} AMM_{C>P} \begin{bmatrix} C(1) \\ C(2) \end{bmatrix}$$

where the fluxes may be obtained by these equations, and the way of calculating fluxes will be explained later in this section.

Consider the following reaction:

$$A + B \xrightarrow{E} C + D$$

where enzyme E catalyzes the reaction from A and B to C and D. In this case, four AMMs are defined for A→C, A→D, B→C, and B→D, where the MAV of C and D are expressed as functions of A and B as:

^{13}C-metabolic flux analysis

$$AMM_{A>C}A + AMM_{B>C}B = C, \quad AMM_{A>D}A + AMM_{B>D}B = D \quad (5.4)$$

Consider the example of the pyruvate dehydrogenase complex (PDHc) reaction from PYR to AcCoA. Let PYR(i) (I = 1,2,3) be the fractional enrichment of the *i*th carbon of PYR, and AcCoA(i) (i = 1,2) be defined in the same way. In this case, the first carbon of PYR becomes CO_2, and thus the second and third carbons of PYR become the first and the second carbons of AcCoA, respectively, and the PDHc reaction can be expressed as:

$$AMM_{PYR>AcCoA} PYR = \begin{bmatrix} 0 & 1 & 0 \\ 0 & 0 & 1 \end{bmatrix} \begin{bmatrix} PYR(1) \\ PYR(2) \\ PYR(3) \end{bmatrix} = \begin{bmatrix} PYR(2) \\ PYR(3) \end{bmatrix}$$

$$= \begin{bmatrix} AcCoA(1) \\ AcCoA(2) \end{bmatrix} = AcCoA \quad (5.5)$$

Although it is easy to understand the notion of fractional enrichment or metabolite activation vectors, it may not be useful for practical cases of utilizing NMR and MS data. In order to express isotopomer distribution of a metabolite, Schmidt et al. (1997) introduced the isotopomer distribution vector (IDV). Since the carbon atom of a molecule is either ^{13}C-labeled or unlabeled (^{12}C), this can be expressed by a binary digit of 0 or 1, and thus the number of components of the isotopomer distribution

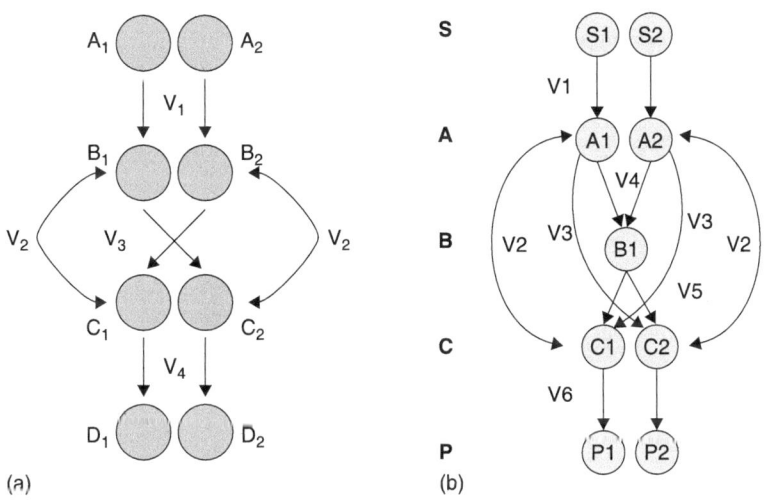

Figure 5.1 Simple examples of flux calculation

for a molecule with n carbons is 2^n. For example, the isotopomer distribution of A with two carbons is expressed as:

$$IDV_A = \begin{bmatrix} I_A(0) \\ I_A(1) \\ I_A(2) \\ I_A(3) \end{bmatrix} = \begin{bmatrix} I_A(00_{bin}) \\ I_A(01_{bin}) \\ I_A(10_{bin}) \\ I_A(11_{bin}) \end{bmatrix} \quad (5.6)$$

where

$$\sum_{i=0}^{3} I_A(i) = 1 \quad (5.7)$$

The subscript '$_{bin}$' denotes binary expression. Namely, $I_A(0)$ denotes the fraction of a molecule with only ^{12}C, $I_A(1)$ the fraction of a molecule with the first carbon labeled by ^{13}C, $I_A(2)$ the fraction of a molecule with the second carbon labeled by ^{13}C, and $I_A(3)$ denotes the fraction of a molecule with all carbons labeled by ^{13}C. Figure 5.2 illustrates the difference between MAV and IDV for all three carbons. Note that there exists 2^3 isotopomers, where the components of IDV are summed up to 1, whereas the components of MAV, as shown to the right, do not necessarily become 1 by their summation.

In the same way as AMM relates to the positional enrichment, Schmidt et al. (1997) considered isotopomer mapping matrix (IMM) for the isotopomer distribution vector. Like AMM, IMM is defined for one substrate and one product, where the row of IMM is equal to the number of the components of IDV of the product, while the column is equal to the number of components of IDV of the substrate. Namely, the first row of $IMM_{A>B}$ corresponds to the unlabeled molecule A (00_{bin}), and the second row corresponds to the molecule A with only the first carbon labeled by ^{13}C (01_{bin}).

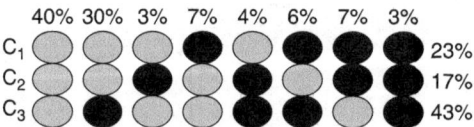

Figure 5.2 IDV and MAV: ●^{13}C, ○^{12}C

For example, consider the reaction $A + B \rightarrow C$, where it is assumed that A has two carbons, B has one carbon, and C has three carbons. Let AMM be given by:

$$\text{AMM}_{A>C} = \begin{bmatrix} 0 & 1 \\ 1 & 0 \\ 0 & 0 \end{bmatrix}, \quad \text{AMM}_{B>C} = \begin{bmatrix} 0 \\ 0 \\ 1 \end{bmatrix} \quad (5.8)$$

Then the IDVs are expressed as:

$$IDV_A = \begin{bmatrix} I_A(00) \\ I_A(01) \\ I_A(10) \\ I_A(11) \end{bmatrix}, \quad IDV_B = \begin{bmatrix} I_B(0) \\ I_B(1) \end{bmatrix}, \quad DV_C = \begin{bmatrix} I_C(000) \\ I_C(001) \\ I_C(010) \\ I_C(011) \\ I_C(100) \\ I_C(101) \\ I_C(110) \\ I_C(111) \end{bmatrix} \quad (5.9)$$

and the IMMs are expressed as:

$$\text{IMM}_{A>C} = \begin{bmatrix} 1 & 0 & 0 & 0 \\ 1 & 0 & 0 & 0 \\ 0 & 0 & 1 & 0 \\ 0 & 0 & 1 & 0 \\ 0 & 1 & 0 & 0 \\ 0 & 1 & 0 & 0 \\ 0 & 0 & 0 & 1 \\ 0 & 0 & 0 & 1 \end{bmatrix}, \quad \text{IMM}_{B>C} = \begin{bmatrix} 1 & 0 \\ 0 & 1 \\ 1 & 0 \\ 0 & 1 \\ 1 & 0 \\ 0 & 1 \\ 1 & 0 \\ 0 & 1 \end{bmatrix} \quad (5.10)$$

The IDVc can be expressed using IMMs as:

$$IDV_C = (IMM_{A>C} \cdot IDV_A) \otimes (IMM_{B>C} \cdot IDV_B) \quad (5.11)$$

where the operator $\otimes$ ($R^8 \times R^8 \rightarrow R^8$) denotes element-wise multiplication for the two vectors.

Consider the following reaction:

$$A + B \rightarrow C + D$$

where A, B, C, and D are the MAVs for A, B, C, and D molecules. The mass balance at the steady state can be expressed as:

$$A + B - C - D = 0 \quad (5.12)$$

Let $AMM_{A>C}$, $AMM_{B>C}$, $AMM_{A>D}$, and $AMM_{B>D}$ be the AMMs from A and B to C and D, then the isotopomer balance is expressed as shown above. In a similar way, let I_A, I_B, I_C, and I_D be the isotopomer distribution vectors of A, B, C, and D, respectively. Let $IMM_{A>C}$, $IMM_{B>C}$, $IMM_{A>D}$, and $IMM_{B>D}$ be the isotopomer mapping matrices from A and B to C and D. Then the following equation holds from the isotopomer balance:

$$(IMM_{A \to C} \cdot I_A) \otimes (IMM_{B \to C} \cdot I_B) = I_C,$$
$$(IMM_{A \to D} \cdot I_A) \otimes (IMM_{B \to D} \cdot I_B) = I_D \quad (5.13)$$

Note that the sizes of AMM and IMM are different. For example, suppose A has n carbons and C has m carbons. Then the dimension of $AMM_{A>C}$ is n × m, whereas the dimension of $IMM_{A>C}$ becomes $2^n \times 2^m$, and the latter becomes exponentially increased as n and m increase.

5.3 Simple example for flux calculation based on ^{13}C-labeling experiment

The number of isotopomers for the molecule with n carbon atoms is 2^n, and the isotopomer distribution can be detected by NMR and/or MS with some limitation (Figure 5.3), for the case of n = 3 (Yang et al., 2002a).

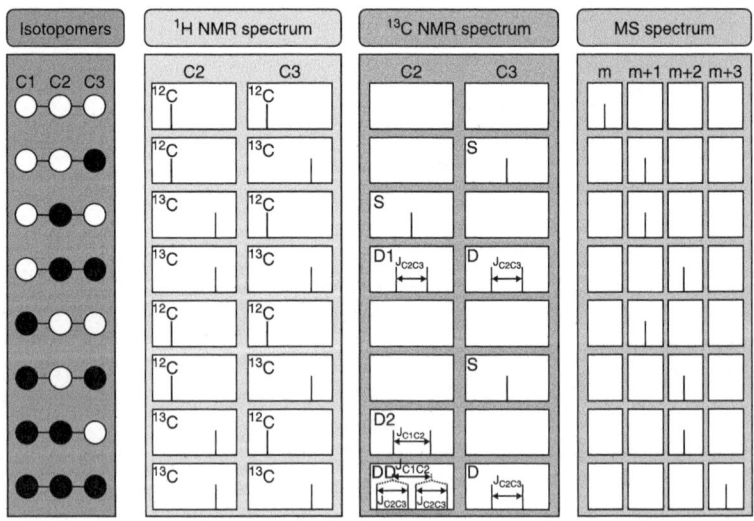

Figure 5.3 Schematic illustration of the data NMR and MS, compared to isotopomer distribution

The problem is how to quantify the fluxes based on limited information on the isotopomer distribution. Let us consider the simple example in Figure 5.1b to understand the basic principles for flux computation based on isotopomer labeling patterns together with measured specific rates. Suppose that the metabolic system consists of an input substrate S with a known isotopomer pattern, intracellular metabolites such as A, B, and C, and an output extracellular metabolite P, where B is assumed to have one carbon atom, and the others are assumed to have two carbons. In Figure 5.1b, V_1 is the system input flux, and V_6 is the system output flux, where these are assumed to be measurable, and the remaining fluxes V_2, V_3, V_4, and V_5 are the intracellular fluxes. V_2 and V_3 keep the metabolite A together, but with different fates for the carbon atoms, while V_4 splits A into two molecules of B, and V_5 reunites the atoms. V_2 is assumed to take place in both directions, and the other fluxes are unidirectional (i.e. $\overleftarrow{v}_1 = \overleftarrow{v}_3 = \overleftarrow{v}_4 = \overleftarrow{v}_5 = \overleftarrow{v}_6 = 0$). The mass balances for A, B, and C are (Yang et al., 2002c):

$$A: \quad \vec{v}_1 + \vec{v}_2 = \overleftarrow{v}_2 + \vec{v}_3 + \vec{v}_4$$
$$B: \quad \vec{v}_4 = \vec{v}_5 \tag{5.14}$$
$$C: \quad \overleftarrow{v}_2 + \vec{v}_3 + \vec{v}_5 = \vec{v}_2 + \vec{v}_6$$

Let $\vec{v}_2$, $\vec{v}_3$, $\vec{v}_4$ be the free fluxes (the fluxes to be determined), then the remaining fluxes are expressed as:

$$\vec{v}_6 = \vec{v}_1$$
$$\vec{v}_5 = \vec{v}_4 \tag{5.15}$$
$$\overleftarrow{v}_2 = \vec{v}_2 + \vec{v}_3 + \vec{v}_4 - \vec{v}_1$$

where $\vec{v}_1$ is known as the measured specific substrate consumption rate. Then the isotopomer balances are expressed by assuming isotopic steady state as:

$$A_{01}: \quad (\vec{v}_2 + \vec{v}_3 + \vec{v}_4)a_{01} = \overleftarrow{v}_2 c_{01} + \vec{v}_1 s_{01}$$
$$A_{10}: \quad (\vec{v}_2 + \vec{v}_3 + \vec{v}_4)a_{10} = \overleftarrow{v}_2 c_{10} + \vec{v}_1 s_{10}$$
$$A_{11}: \quad (\vec{v}_2 + \vec{v}_3 + \vec{v}_4)a_{11} = \overleftarrow{v}_2 c_{11} + \vec{v}_1 s_{11}$$
$$B_1: \quad 2\vec{v}_5 b_1 = \vec{v}_4 a_{01} + \vec{v}_4 a_{10} + 2\vec{v}_4 a_{11} \tag{5.16}$$
$$C_{01}: \quad (\vec{v}_2 + \vec{v}_6)c_{01} = \vec{v}_2 a_{01} + \vec{v}_3 a_{10} + \vec{v}_5 (1-b_1)b_1$$
$$C_{10}: \quad (\vec{v}_2 + \vec{v}_6)c_{10} = \vec{v}_2 a_{10} + \vec{v}_3 a_{01} + \vec{v}_5 (1-b_1)b_1$$
$$C_{11}: \quad (\vec{v}_2 + \vec{v}_6)c_{11} = \vec{v}_2 a_{11} + \vec{v}_3 a_{11} + v_5 b_1^2$$

where a_{ij}, b_i, c_{ij}, and s_{ij} (ij = 0,1) are the isotopomer fractions of A, B, C, and S, respectively. The '$_0$' in the suffixes corresponds to unlabeled carbon (^{12}C), while '$_1$' represents labeled carbon (^{13}C). For example, a_{01} is the mole fraction of A with the second carbon labeled. The isotopomer distribution of C is the same as that of P. Note that the unlabeled molecules such as A_{00}, B_0, and C_{00} are not included in the above equations, since the sum of all isotopomer fractions for the corresponding molecule must be equal to 1, and thus unlabeled molecule can be computed from the fractions of the labeled molecules. For example, $a_{00} = 1 - (a_{01} + a_{10} + a_{11})$.

From Equation 5.12, we have the following relationships:

$$a_{01} + a_{10} + 2a_{11} = 2b_1 = c_{01} + c_{10} + 2c_{11} = s_{01} + s_{10} + 2s_{11} \quad (5.17)$$

The first and the third relationships of this equation represent the sum of the positional labeling data (i.e. ^{1}H-NMR measurements) for the two carbons of metabolites A and C, respectively, which means that the ^{13}C enrichment of the first carbon atom is redundant with that of the second carbon atom. This indicates that it is not possible to determine the free fluxes from positional enrichment data of A and B. However, the isotopomer distributions contain additional information on the fluxes, and thus all free fluxes can be determined. From Equation 5.16, we have (Yang et al., 2002a):

$$a_{01} - a_{10} = (\vec{v}_1\vec{v}_2 + \vec{v}_1\vec{v}_3 + \vec{v}_1\vec{v}_4) \cdot (s_{01} - s_{10}) / \delta$$
$$c_{01} - c_{10} = (\vec{v}_1\vec{v}_2 - \vec{v}_1\vec{v}_3) \cdot (s_{01} - s_{10}) / \delta$$
$$a_{11} = [4(\vec{v}_1\vec{v}_2 + \vec{v}_1\vec{v}_3 + \vec{v}_1\vec{v}_4)s_{11} + (\vec{v}_2\vec{v}_4 + \vec{v}_3\vec{v}_4 + \vec{v}_4^2 - \vec{v}_1\vec{v}_4) \quad (5.18)$$
$$(s_{01} + s_{10} + 2s_{11})^2] / \zeta$$
$$c_{11} = [4(\vec{v}_1\vec{v}_2 + \vec{v}_1\vec{v}_3)s_{11} + (\vec{v}_2\vec{v}_4 + \vec{v}_3\vec{v}_4 + \vec{v}_4^2)(s_{01} + s_{10} + 2s_{11})^2] / \zeta$$

where δ and ζ are defined as:

$$\delta \equiv 2\vec{v}_3^2 + \vec{v}_4^2 + 2\vec{v}_2\vec{v}_3 + \vec{v}_2\vec{v}_4 + 3\vec{v}_3\vec{v}_4 + \vec{v}_1\vec{v}_2 - \vec{v}_1\vec{v}_3$$
$$\zeta \equiv 4(\vec{v}_4^2 + \vec{v}_2\vec{v}_4 + \vec{v}_3\vec{v}_4 + \vec{v}_1\vec{v}_2 + \vec{v}_1\vec{v}_3)$$

From Equations 5.17 and 5.18, all the isotopomer distributions of A and B can be expressed as the functions of the fluxes as well as the known isotopomer pattern (distribution) of the substrate.

Consider the case of using GC-MS data. Let A_m be the fraction of molecules with all the carbons as ^{12}C, let A_{m+1} be the fraction of molecules with only one carbon labeled by ^{13}C, and let A_{m+2} be the case of two carbons labeled by ^{13}C. Then the following equation holds:

$$\begin{bmatrix} A_{m+1} \\ A_{m+2} \end{bmatrix} = \begin{bmatrix} 1 & 1 & 0 \\ 0 & 0 & 1 \end{bmatrix} \begin{bmatrix} A_{01} \\ A_{10} \\ A_{11} \end{bmatrix} \quad (5.19)$$

^{13}C-metabolic flux analysis

where A_{ij} represents the components of isotopomer distribution. The left-hand side of the above equation is known as the mass distribution vector (MDV) (Schmidt et al., 1997). Note that A_m and A_{00} are not included in the above equation, since these can be obtained for IDV, so that the summation of the components must be equal to 1. Consider the case of using NMR data. As shown in Figure 5.3, the isotopomer distribution can be converted to the signal of NMR for the case of three carbons as:

$$\begin{bmatrix} 0 & 0 & 0 & 0 & 1 & 1 & 0 & 0 \\ 0 & 0 & 1 & 0 & 0 & 0 & 0 & 0 \\ 0 & 1 & 0 & 0 & 0 & 1 & 0 & 0 \end{bmatrix} I_A = \begin{bmatrix} S'_{C1} \\ S'_{C2} \\ S'_{C3} \end{bmatrix} \quad (5.20)$$

where $I_A = \begin{bmatrix} 0 & 0 & 0 \\ 0 & 0 & 1 \\ 0 & 1 & 0 \\ 0 & 1 & 1 \\ 1 & 0 & 0 \\ 1 & 0 & 1 \\ 1 & 1 & 0 \\ 1 & 1 & 1 \end{bmatrix}$

Note that the fifth and sixth columns of the first row of the conversion matrix appear to be 1, which means that the fifth and the sixth entries of the isotopomer vector contribute to the singlet of C_1. In the same way, the doublet (D_1, D_2) and the doublet of doublet (DD) can also be converted as follows, while the explanation on the singlet, doublet, and doublet of doublet will follow later in this chapter:

$$\begin{bmatrix} 0 & 0 & 0 & 0 & 0 & 0 & 0 & 0 \\ 0 & 0 & 0 & 1 & 0 & 0 & 0 & 0 \\ 0 & 0 & 0 & 1 & 0 & 0 & 0 & 1 \end{bmatrix} I_A = \begin{bmatrix} D1'_{C1} \\ D1'_{C2} \\ D1'_{C3} \end{bmatrix} \quad (5.21a)$$

$$\begin{bmatrix} 0 & 0 & 0 & 0 & 0 & 0 & 1 & 1 \\ 0 & 0 & 0 & 0 & 0 & 0 & 1 & 0 \\ 0 & 0 & 0 & 0 & 0 & 0 & 0 & 0 \end{bmatrix} I_A = \begin{bmatrix} D2'_{C1} \\ D2'_{C2} \\ D2'_{C3} \end{bmatrix} \quad (5.21b)$$

$$\begin{bmatrix} 0 & 0 & 0 & 0 & 0 & 0 & 0 & 0 \\ 0 & 0 & 0 & 0 & 0 & 0 & 0 & 1 \\ 0 & 0 & 0 & 0 & 0 & 0 & 0 & 0 \end{bmatrix} I_A = \begin{bmatrix} DD'_{C1} \\ DD'_{C2} \\ DD'_{C3} \end{bmatrix} \quad (5.21c)$$

In practice, the signal values are normalized, as shown below for the case of a singlet at C_2:

$$S_{C2} = \frac{S'_{C2}}{S'_{C2} + D1'_{C2} + D2'_{C2} + DD'_{C2}} \qquad (5.22)$$

As stated above, the signals obtained from MS and NMR are the subset of IDV or a linear combination of its components. Let A (m_0^a, m_1^a, m_2^a) and C(m_0^C, m_1^C, m_2^C) be the mass distribution obtained by GC-MS, and let A (s_{C2}^a, d_{C2}^a) and C (s_{C2}^C, d_{C2}^C) be the multiplet patterns for the second carbons of A and C measured by NMR. Then the relationships between the components of IDV and the measured signals are expressed as (Yang et al., 2002a):

$$\begin{aligned} a_{01} + a_{10} &= m_1^a \\ a_{11} &= m_2^a \\ c_{01} + c_{10} &= m_1^c \\ c_{11} &= m_2^c \\ \frac{a_{01}}{a_{01} + a_{11}} &= s_{C2}^a \\ \frac{c_{01}}{c_{01} + c_{11}} &= s_{C2}^c \end{aligned} \qquad (5.23)$$

As stated above, the following functional relationship exists:

$$\gamma : (\vec{v}_2, \vec{v}_3, \vec{v}_4) \rightarrow (m_1^a, m_2^a, m_1^c, m_2^c, s_{C2}^a, s_{C2}^c) \qquad (5.24)$$

For example, m_1^a can be expressed as:

$$\begin{aligned} m_1^a = s_{01} + s_{10} + 2s_{11} - \\ [8(\vec{v}_1\vec{v}_2 + \vec{v}_1\vec{v}_3 + \vec{v}_1\vec{v}_4)s_{11} + 2(\vec{v}_2\vec{v}_4 + \vec{v}_3\vec{v}_4 + \vec{v}_4^2 - \vec{v}_1\vec{v}_4) \\ (s_{01} + s_{10} + 2s_{11})^2]/\zeta \end{aligned} \qquad (5.25)$$

Since the isotopomer distribution of the substrate (s_{01}, s_{10}, s_{11}) is known, the MS and NMR signals can be calculated by the above equation if the metabolic flux values are given. Then the next step is to find the fluxes that give the minimum between the measured MS and NMR signals and the corresponding estimated signals. As implied by Equation 5.25, the functional relationship of Equation 5.24 is nonlinear and requires nonlinear optimization to search for the best fluxes.

The algorisms for flux computation based on ^{13}C-labeled experiment have been developed by several researchers (Zupke and Stephanopoulos, 1994; Marx et al., 1996; Wiechert et al., 1997; Schmidt et al., 1997: Arauzo-Bravo and Shimizu, 2003; Zhao et al., 2003). For these, the atom mapping matrices (AMMs) that track the transfer of carbon atoms from substrates to products, and the isotopomer mapping matrices (IMMs), are used as described above. Moreover, the cumomer concept (Wiechert et al., 1999), theoretical bondomer concept (van Winden et al., 2002), isotopomer path tracing concept (Forbes et al., 2001), and the elemental metabolic unit (EMU) framework (Antoniewics et al., 2007) have been proposed.

The problem for computer-based flux analysis, which will be explained later, is that it is difficult to see the relationship between the fluxes and the resulting isotopomer distribution due to its black box nature, and thus the problem formulation is not transparent. It is, therefore, not easy to obtain confident results without experiences and trial and error manipulations based on statistical analysis of the estimation of confidence intervals in practice. Let us consider next the relationship between the fluxes and the isotopomer patterns before going into detailed flux analysis.

5.4 Analytical approach for flux computation

Although the application may be limited in practice, it is useful to consider an analytical approach. Klapa et al. (1999) traced the labeled carbon atoms through the TCA cycle and derived several useful expressions, which are verified by experiments (Park et al., 1999). Noronha et al. (2000) considered the steady state isotopomer balance to analyze acetate production from glucose by assuming the labeling pattern of PEP. Sherry et al. (2004) also derived analytical expressions for the relationship between the isotopomer distribution and the fluxes around the TCA cycle, in its application to rat heart. As stated above, it is important to grasp how the metabolic fluxes affect the isotopomer distribution of metabolites. Let us consider the relationships between the fluxes and the isotopomer distribution by mathematical formulation, and make the flux analysis simpler and gain a deeper insight into the flux analysis for the labeling pattern of the substrate such as glucose, acetate, and pyruvate. Let us subdivide the main metabolic pathways into two, namely the upper part for glycolysis together with the pentose phosphate (PP) pathway and the lower part for glycolysis with the TCA cycle (Matsuoka and Shimizu, 2010).

5.4.1 Case of using glucose as a carbon source

Consider the metabolic pathways in Figure 5.4, where Figure 5.4a shows the use of glucose as a carbon source, while Figure 5.4b shows the use of either pyruvate or acetate as a carbon source.

Bacterial cellular metabolic systems

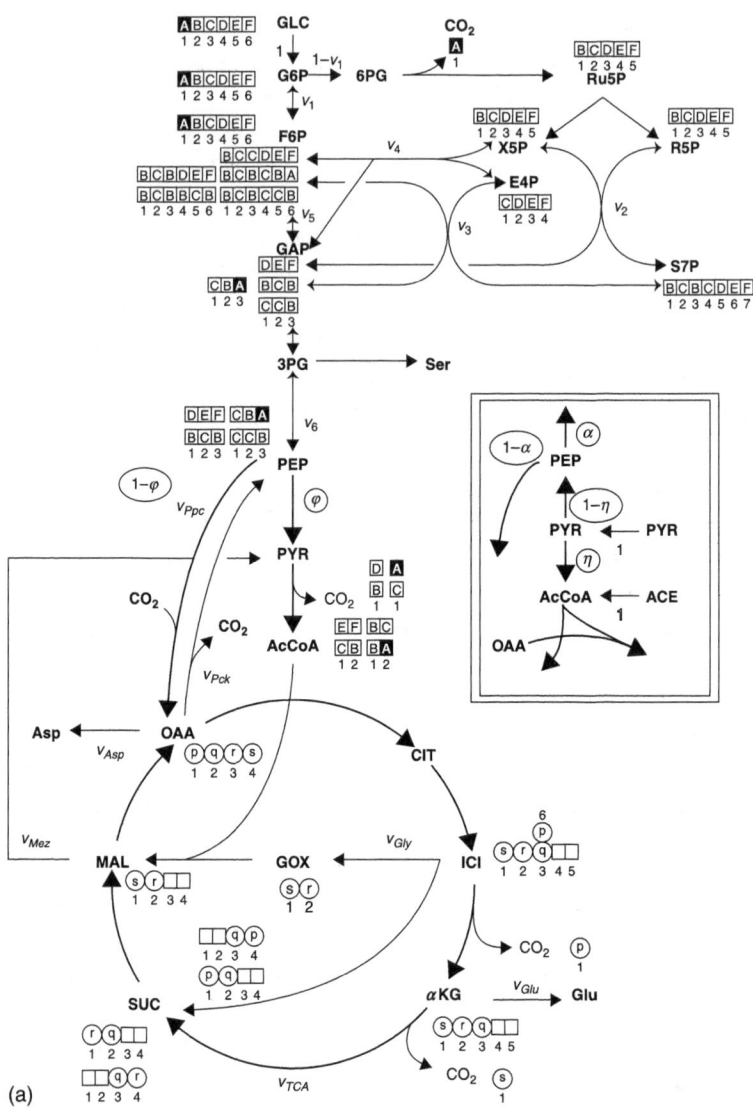

Figure 5.4 The fate of labeled carbons when using: (a) glucose; and (b) acetate or pyruvate, as a carbon source

Let us consider the mass balances and the isotopomer balances along the glycolysis together with the PP pathway. Referring to Figure 5.4a, let v_1 be the normal flux toward glycolysis at G6P and let $1 - v_1$ be the flux toward the oxidative PP pathway, where the input flux to the system is normalized to be 1. Let us lump Ru5P, X5P, and R5P into P5P for

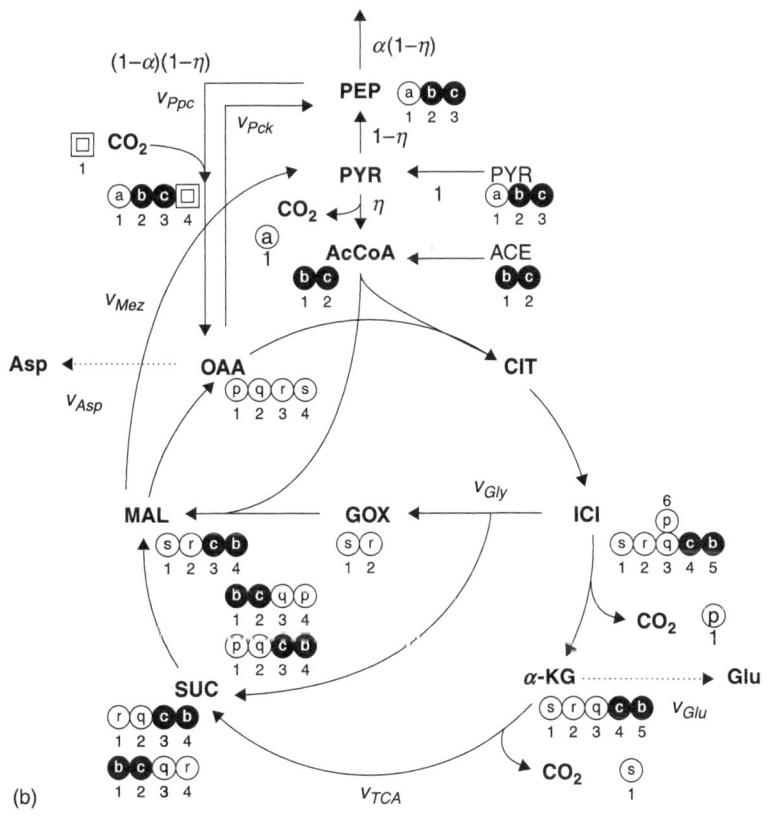

Figure 5.4 Continued

simplicity. Then the following equations may be obtained from mass balances for the metabolites at steady state:

P5P: $1 - v_1 = 2v_2 + v_4$ (5.26a)

S7P: $v_2 = v_3$ (5.26b)

F4P: $v_3 = v_4$ (5.26c)

F6P: $v_1 + v_3 + v_4 = v_5$ (5.26d)

GAP: $v_2 + v_4 + 2v_5 = v_3 + v_6$ (5.26e)

from which we have the following relationships:

$$v_2 = v_3 = v_4 = \frac{1-v_1}{3} \tag{5.27a}$$

$$v_5 = \frac{v_1+2}{3} \tag{5.27b}$$

$$v_6 = \frac{v_1+5}{3} \tag{5.27c}$$

Note that here we considered only the net fluxes.

Let us now consider the isotopomer balances. Since glucose has 6 carbons, the number of its isotopomers is $2^6 = 64$, so let these be expressed as $G_0, G_1, \ldots, G_{123456}$, where G_0 denotes the unlabeled glucose, let G_1 be the [1-^{13}C] glucose, and let G_{123456} be the [U-^{13}C] glucose. The isotopomer balance equations for the kth metabolite may be expressed as:

$$I_k \sum_{i=1} v_i = \sum_{j=1} v_j I_j^{v_j} \quad (k=1,2,\cdots) \tag{5.28a}$$

where I_k is the isotopomer distribution vector (IDV) for the kth metabolite.

The left-hand side of Equation 5.28a is the output of the kth metabolite, and the right-hand side is the input to the metabolite. $I_j^{v_j}$ in Equation 5.28a may be expressed as:

$$I_j^{v_j} = \begin{cases} IMM_{j \to k} \cdot I_j & \text{(uni molecular reaction)} \\ IMM_{j_1 \to k} \cdot I_{k_1} \otimes IMM_{j_2 \to k} \cdot I_{k_2} & \text{(bi molecular reaction)} \end{cases} \tag{5.28b}$$

where $\otimes$ is the elementwise multiplication operation, as explained previously. The number of isotopomer balance equations is $2^6 = 64$ for G6P, $2^5 = 32$ for P5P, $2^7 = 128$ for S7P, $2^4 = 16$ for E4P, $2^6 = 64$ for F6P, and $2^3 = 8$ for GAP. The same number of isotopomers are present and, therefore, the isotopomer distribution for each metabolite can be obtained uniquely as a function of the labeling pattern of glucose on condition that the fluxes are given. Since P5P, E4P, and GAP (or 3PG) are the precursors for biosynthesis, it is preferred to express these isotopomer distributions with respect to the labeling pattern of glucose. For instance, the isotopomer balance for GAP_1 can be expressed after several manipulations as:

$$\begin{aligned} & v_1 [G_3 + G_4 + G_{14} + G_{24} + 2G_{34} + G_{35} + G_{36} + G_{124} \\ & + G_{134} + G_{234} + G_{345} + G_{346} + G_{356} + G_{1234} + G_{3456}] \\ & + v_2 [G_4 + G_{14} + G_{24} + G_{124} + G_{34} + G_{134} + G_{234} + G_{1234}] \\ & + v_3 \, GAP_1 + v_4 \, [2G_4 + 2G_{14} + 2G_{24} + 2G_{124} + 2G_{34} \\ & + 2G_{134} + 2G_{234} + 2G_{1234}] = (2v_1 + 3v_4 + v_2 + 2v_3) GAP_1 \end{aligned} \tag{5.29a}$$

where GAP_1 is the fraction of the first carbon labeled GAP. Note that GAP_i is the isotopomer representation, and not the positional representation. Equation 5.29a can be solved for GAP_1, and the following equation can be derived using the relationships of Equations 5.27a–c:

$$GAP_1 = \frac{3}{5+v_1}\{v_1(G_3 + G_{34} + G_{35} + G_{36} + G_{345} + G_{346} + G_{356} + G_{3456}) \\ + G_4 + G_{14} + G_{24} + G_{124} + G_{34} + G_{134} + G_{234} + G_{1234}\} \quad (5.29b)$$

The fraction of the other isotopomers can be derived in a similar way, and the overall equations may be expressed as:

$$I_{GAP} = \Lambda_{Glc} I_{Glc} \quad (5.30)$$

where I_{GAP} ($\in R^8$) and I_{Glc} ($\in R^{64}$) are the IDV of GAP and glucose, respectively, and Λ_{Glc} ($\in R^8 \times R^{64}$) is a function of fluxes (Matsuoka and Shimizu, 2010). A similar equation may be obtained for P5P and E4P. For the special case of using a mixture of unlabeled glucose (G_0), [1-^{13}C] glucose (G_1), and [U-^{13}C] glucose (G_{123456}), which are often used for the labeling experiments, Equation 5.30 reduces to:

$$\begin{bmatrix} GAP_0 \\ GAP_3 \\ GAP_{123} \end{bmatrix} = \frac{1}{5+v_1} \begin{bmatrix} 5+v_1 & 5-2v_1 & 0 \\ 0 & 3v_1 & 0 \\ 0 & 0 & 5+v_1 \end{bmatrix} \begin{bmatrix} G_0 \\ G_1 \\ G_{123456} \end{bmatrix} \quad (5.31)$$

Note that although G_6 may sometimes be used, this may not be as useful as G_4 and G_5 in identifying v_1, as can be seen from Λ_{Glc}.

Let us consider the lower part of glycolysis and the TCA cycle (Figure 5.4a). Here the labeling pattern of 3PG (or GAP) is obtained from Equation 5.30, and the flux from 3PG is normalized to 1 for formulation purposes without loss of generality. In other words, we can connect the upper part of glycolysis plus the PP pathway with the lower part of glycolysis plus the TCA cycle. Let φ be the flux from PEP to PYR through Pyk, and let $1 - \varphi$ be the rest of the flux that flows from PEP to OAA by the anaplerotic pathway of Ppc (Figure 5.4a). v_{TCA} is the normalized flux that flows from α-ketoglutarate (αKG) to succinate (SUC), v_{Gly} the flux from isocitrate (ICI) through the glyoxylate pathway, and v_{Asp} and v_{Glu} are the fluxes for cell synthesis via aspartate and glutamate formations, respectively. From the mass balance around OAA, the following equation may be obtained (Noronha et al., 2000):

$$v_{TCA} + 2v_{Gly} + v_{Ppc} = v_{TCA} + v_{Gly} + v_{Glu} + v_{Asp} \quad (5.32a)$$

or

$$v_{Gly} + v_{Ppc} = v_{Glu} + v_{Asp} \quad (5.32b)$$

Note that $v_{Ppc} = 1 - \varphi$ and $v_{TCA} = \varphi - 2v_{Gly} - v_{Glu}$. Similar to the upper part of glycolysis and the PP pathway, let us consider the isotopomer balances for OAA. The number of OAA isotopomers is $2^4 = 16$, and let these isotopomers be expressed as $O_0, O_1, O_2, \ldots, O_{1234}$. Then the isotopomer balances can be described for each isotopomer. For example, the isotopomer balance for O_1 is expressed as:

$$\left(\frac{v_{TCA}}{2}\right)[(O_3 + O_{13} + O_{34} + O_{134})(P_0 + P_1) + (O_0 + O_1 + O_4 + O_{14})(P_2 + P_{12})]$$
$$+ \left(\frac{v_{Gly}}{2}\right)[2(O_0 + O_1 + O_2 + O_{12})(P_2 + P_{12})$$
$$+ (O_1 + O_{13} + O_{14} + O_{134})(P_0 + P_1) + (O_0 + O_3 + O_4 + O_{34})(P_2 + P_{12})]$$
$$+ v_{Ppc} P_1 C_0 = (v_{TCA} + 2v_{Gly} + v_{Ppc}) O_1 \quad (5.33a)$$

where the left-hand side of Equation 5.33a is the formation rate of O_1, and the right-hand side is the consumption rate of O_1. P_i and C_i are the mole fractions of the isotopomers labeled at the ith positions of PEP and CO_2, respectively, where the first carbon of PEP is lost as CO_2 through PDHc reaction. Dividing Equation 5.33a by v_{TCA}, the following equation can be obtained:

$$\frac{1}{2}[(O_3 + O_{13} + O_{34} + O_{134})(P_0 + P_1) + (O_0 + O_1 + O_4 + O_{14})(P_2 + P_{12})]$$
$$+ \frac{z}{2}[2(O_0 + O_1 + O_2 + O_{12})(P_2 + P_{12}) + (O_1 + O_{13} + O_{14} + O_{134})(P_0 + P_1)$$
$$+ (O_0 + O_3 + O_4 + O_{34})(P_2 + P_{12})]$$
$$+ yP_1 C_0 = (1 + 2z + y) O_1 \quad (5.33b)$$

where $y \equiv v_{Ppc}/v_{TCA}$ and $z \equiv v_{Gly}/v_{TCA}$. Then the overall equation can be expressed as:

$$\Lambda_{PEP} I_{OAA} = b_{PEP} \quad (5.34)$$

where $I_{OAA} (\in R^{16})$ is the isotopomer vector of OAA, $\Lambda_{PEP} (\in R^{16} \times R^{16})$ is the coefficient matrix, and $b_{PEP} (\in R^{16})$ is the residual vector, where Λ_{PEP} and b_{PEP} are given elsewhere (Matsuoka and Shimizu, 2010). Equation 5.34 may be solved if Λ_{PEP} is non-singular as:

$$I_{OAA} = \Lambda_{PEP}^{-1} b_{PEP} \quad (5.35)$$

^{13}C-metabolic flux analysis

This can be connected to Equation 5.30 by letting $I_{PEP} = I_{GAP}$ to express O_i as a function of G_i, where I_{PEP} is the isotopomer vector of PEP.

5.4.2 Case of acetate or pyruvate as a carbon source

Referring to Figure 5.4b, let η be the flux toward AcCoA from PYR, and let $(1 - \eta)$ be the other flux toward PEP. Moreover, let α be the fraction toward the gluconeogenic pathway from PEP, and let $(1 - \alpha)$ be the other fraction toward the anaplerotic pathway through Ppc to OAA. From the mass balance for OAA, the normalized molar formation rate of OAA is expressed as:

$$v^{+}_{OAA} = v_{TCA} + 2v_{Gly} + v_{Ppc} \tag{5.36}$$

and the molar consumption rate of OAA is expressed as:

$$v^{-}_{OAA} = v_{TCA} + v_{Gly} + v_{Glu} + v_{Asp} + v_{Mez} + v_{Pck} \tag{5.37}$$

Since $v^{+}_{OAA} = v^{-}_{OAA}$ holds at steady state, the following expression holds:

$$v_{Gly} + v_{Ppc} = v_{Glu} + v_{Asp} + v_{Mez} + v_{Pck} \tag{5.38a}$$

Since $v_{Ppc} = (1 - \alpha)(1 - \eta)$, the following expression holds:

$$(1 - \eta)\alpha = 1 - \eta + v_{Gly} - (v_{Glu} + v_{Asp}) - (v_{Mez} + v_{Pck}) \tag{5.38b}$$

or

$$\alpha = 1 + \frac{v_{Gly} - (v_{Glu} + v_{Asp}) - (v_{Mez} + v_{Pck})}{1 - \eta} \tag{5.38c}$$

Moreover, since $v_{TCA} = \eta - (v_{Glu} + 2v_{Gly}) \geq 0$, and from Equation 5.38c, the following constraints must be satisfied for η:

$$v_{Glu} + 2v_{Gly} \leq \eta \leq 1 + v_{Gly} - (v_{Glu} + v_{Asp}) - (v_{Mez} + v_{Pck}) \tag{5.39}$$

Then the isotopomer balance for O_1 may be expressed as:

$$\left(\frac{v_{TCA}}{2}\right)[(O_3 + O_{13} + O_{34} + O_{134})A_0 + (O_0 + O_1 + O_4 + O_{14})A_1]$$
$$+ \left(\frac{v_{Gly}}{2}\right)[2(O_0 + O_1 + O_2 + O_{12})A_1 + (O_1 + O_{13} + O_{14} + O_{134})A_0 \tag{5.40a}$$
$$+ (O_0 + O_3 + O_4 + O_{34})A_1]$$
$$+ v_{Ppc}P_1C_0 = (v_{TCA} + 2v_{Gly} + v_{Ppc})O_1$$

where the left-hand side of Equation 5.40a is the formation rate of O_1, and the right-hand side is the consumption rate of O_1. In Equation 5.40a, A_i and P_i are the mole fractions of acetate and PYR isotopomers labeled at ith carbons, respectively, and C_i ($i = 0,1$) is the mole fraction of labeled CO_2 through the Ppc reaction, where C_0 is the fraction of unlabeled CO_2. Dividing Equation 5.40a by v_{TCA}, we have:

$$\frac{1}{2}[(O_3 + O_{13} + O_{34} + O_{134})A_0 + (O_0 + O_1 + O_4 + O_{14})A_1]$$
$$+ \frac{z}{2}[2(O_0 + O_1 + O_2 + O_{12})A_1 + (O_1 + O_{13} + O_{14} + O_{134})A_0 \quad (5.40b)$$
$$+ (O_0 + O_3 + O_4 + O_{34})A_1]$$
$$+ yP_1C_0 = (1 + 2z + y)O_1$$

The similar equations can be derived for all the other isotopomers, and the set of equations can be expressed by matrix and vector representation as (Matsuoka and Shimizu, 2010):

$$\Lambda_{AP} I_{OAA} = b_{AP} \quad (5.41)$$

where Λ_{AP} ($\in R^{16} \times R^{16}$) is the coefficient matrix, and b_{AP} ($\in R^{16}$) is the residual vector, where those are given in Appendix 5A in the present chapter. Equation 5.41 may be solved as (Matsuoka and Shimizu, 2010):

$$I_{OAA} = \Lambda_{AP}^{-1} b_{AP} \quad (5.42)$$

if Λ_{AP} is non-singular. However, Λ_{AP} turns out to be singular, and Equation 5.41 cannot be solved by Equation 5.42 directly. Recall that the isotopomer fractions must be summed to 1, so the following equation must be satisfied:

$$\sum_i O_i = 1 \quad (5.43)$$

Therefore, Equation 5.16 can be solved by replacing one of the equations in Equation 5.41 by this equation.

5.4.3 Special cases of using acetate as a carbon source

Let us consider the case of using acetate as a carbon source by letting $v_{Ppc} = 0$ in Figure 5.4b or $y = 0$ in Equations 5.40b or 5.41. If the mixture of $[1-^{13}C]$ acetate (A_1) and the unlabeled acetate is used, only 4 isotopomers

(O_0, O_1, O_4, O_{14}) are present, and if $A_1 = 1$, these can be expressed as (Matsuoka and Shimizu, 2010):

$$\begin{bmatrix} O_1 \\ O_4 \\ O_{14} \end{bmatrix} = \begin{bmatrix} \dfrac{1+2z}{(1+z)(2+3z)} \\ \dfrac{1}{2+3z} \\ \dfrac{z}{1+z} \end{bmatrix} \quad (5.44)$$

If [2-^{13}C] acetate (A_2) and the unlabeled acetate are used as a carbon source, 12 isotopomers are present, and these reduce to 3 isotopomers when $A_2 = 1$, as follows (Matsuoka and Shimizu, 2010):

$$\begin{bmatrix} O_{23} \\ O_{123} \\ O_{234} \end{bmatrix} = \begin{bmatrix} \dfrac{z}{1+z} \\ \dfrac{1}{2+3z} \\ \dfrac{1+2z}{(1+z)(2+3z)} \end{bmatrix} \quad (5.45)$$

5.4.4 Special cases of using pyruvate as a carbon source

Consider where pyruvate is used as a carbon source. In this case, the fraction η goes through the PDHc reaction, while the remainder goes through Pps to PEP, and part of it continues to OAA through the Ppc reaction, while the other continues via the gluconeogenic pathway (Figure 5.4b). Here we set $v_{Mez} = 0$ and $v_{Pck} = 0$ for simplicity. Let us consider the special case where 100% [1-^{13}C] pyruvate is used as a carbon source. Then Equation 5.42 gives the following equation (Matsuoka and Shimizu, 2010):

$$\begin{bmatrix} O_0 \\ O_1 \\ O_4 \end{bmatrix} = \begin{bmatrix} \dfrac{1}{1+y} \\ \dfrac{y}{1+y} \\ 0 \end{bmatrix} \quad (5.46)$$

In the case where 100% [2-^{13}C] pyruvate is used, the following equations can be derived (Matsuoka and Shimizu, 2010):

Bacterial cellular metabolic systems

$$O_1 = O_4 = \frac{1}{2(1+y)^2(1+2y)} \tag{5.47a}$$

$$O_2 = \frac{y}{1+y} \tag{5.47b}$$

$$O_{13} = O_{24} = \frac{y}{(1+y)(1+2y)} \tag{5.47c}$$

$$O_{14} = \frac{y}{(1+y)^2(1+2y)} \tag{5.47d}$$

where $P_3 = 1$, [3-^{13}C] pyruvate is transferred to the second carbon of AcCoA, and the following expressions can be derived (Matsuoka and Shimizu, 2010):

$$O_2 = 0 \tag{5.48a}$$

$$O_3 = \frac{y}{1+y} \tag{5.48b}$$

$$O_{13} = O_{24} = \frac{y}{(1+y)(1+2y)} \tag{5.48c}$$

$$\begin{bmatrix} O_{23} \\ O_{123} \\ O_{234} \end{bmatrix} = \begin{bmatrix} \dfrac{y}{(1+y)^2(1+2y)} \\ \dfrac{1}{2(1+y)^2(1+2y)} \\ \dfrac{1}{2(1+y)^2(1+2y)} \end{bmatrix} \tag{5.48d}$$

So far, we have considered how isotopomer distribution is expressed with respect to the flux. In practice, the fluxes must be found from the measured signals for the isotopomer distribution. Let these signals be obtained by either NMR or MS, and let these be $s_i (i = 1,2,...)$. Then the fluxes may be obtained by minimizing the following objective function:

$$J = (\hat{s} - s)^T \Sigma_s^{-1} (\hat{s} - s) \tag{5.49}$$

where s represents the measured signal vector containing s_i, $\hat{s}$ is its estimated vector, and Σ_s is the variance-covariance matrix associated with the measured errors. Let us now consider the metabolic flux analysis (MFA) based on ^{13}C-labeling experiment in practice.

5.5 ^{13}C metabolic flux analysis based on GC-MS

Here, we consider the metabolic flux analysis based on GC-MS (Schmidt et al., 1997, Zhao et al., 2003). Compared to NMR, GC-MS analysis is sensitive and may thus provide labeling measurements with higher precision. This technique has been applied successfully to estimate the intracellular flux distribution for a variety of organisms (Wittman, 2007).

Consider the mass distribution vectors (MDV) and fragment mass distribution vectors (FMDV) for dealing with mass isotopomer distribution in metabolic flux analysis (Wittmann and Heinzle, 1999). The transformation of IDVs into MDVs and FMDVs is shown in Figure 5.5, by grouping isotopomers with the same number of labeled carbon atoms together, where FMDV are caused by the fragment mass distribution analysis of the compound of interest by EI mass spectrometry (Zhao et al., 2003). Such fragments can give valuable information for the estimation of metabolic fluxes in that they can increase the resolution of the labeling pattern.

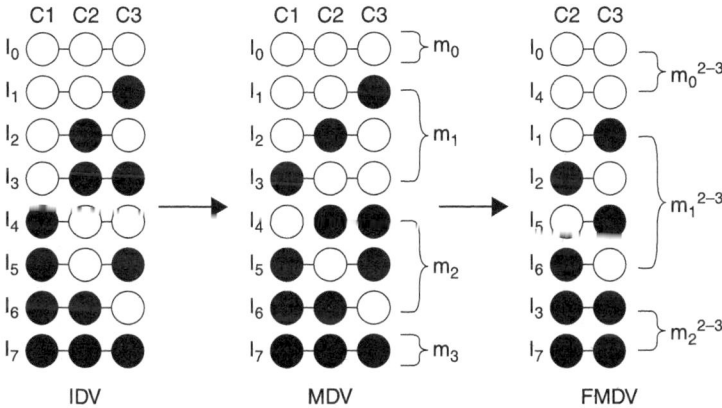

Figure 5.5 Transformation of IDVs into MDVs and FMDVs for a C_3 molecule. The individual isotopomers in IDVs are expressed as mole fractions and are enumerated from 0 to 7, according to the binary number formed using white circles (^{12}C) as 0's and the black circles (^{13}C) as 1's. MDVs and FMDVs are formed by grouping isotopomers with the same number of labeled carbon atoms together

Incorporation of FMDVs into an appropriate model for metabolic flux analysis can provide a highly accurate methodology to determine metabolic flux distribution (Möllney et al., 1999). The mathematical framework is based on an iterative scheme and error minimization procedure. For every bidirectional reaction step, a concept of exchange flux v_i^{exch} can be used to represent a forward and backward reaction $\vec{v}_i$ and $\overleftarrow{v}_i$ by net flux v_i^{net} and v_i^{exch} (Wiechert et al. 1997; Schimidt et al., 1999a) (Figure 5.6). To circumvent the difficulties involved in the numerical treatment of high exchange fluxes, non-linear mapping between v_i^{exch} and exchange coefficient $v_i^{exch[0,1]}$ is employed (Schmidt et al., 1999b). Nonlinear mapping of the exchange fluxes v^{exch} to exchange coefficients $v^{exch[0,1]}$ (Wiechert et al., 1997), as given in Equation 5.51, may be used to overcome the numerical problems arising from large parameter values. Here, β is a constant in the order of magnitude of the net fluxes, where either $\beta = 1.0$ or $100(\%)$ is often used to normalize the fluxes with respect to the substrate consumption rate:

$$v^{exch} = \min(\vec{v}, \overleftarrow{v}) \quad (5.50)$$

$$v^{exch[0,1]} = \frac{v^{exch}}{\beta + v^{exch}} \quad (5.51)$$

Let us consider the MFA in practice for wild-type *E. coli* by using acetate as a carbon source, where the stoichiometric reactions are given in Table 5.1, and the stoichiometric reactions for using glucose as a carbon source are given in Table 5.2. The reaction network for acetate metabolism consists of 25 intracellular metabolites and 26 unknown forward fluxes plus 24 unknown backward fluxes, where the flux via *gnd* (here refers to a grouped reaction catalyzed by multi-enzyme system including genes of *zwf*, *pgl*, and *gnd*) and *acs* were considered to be irreversible. Through the sensitivity analysis by computer simulation with isotopomer balance models, as well as by referring to Gibbs free

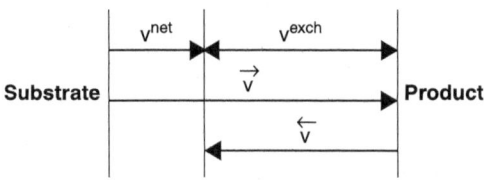

Figure 5.6 Schematic illustration of net flux and exchange flux

Table 5.1 Metabolic reactions for acetate metabolism

	Reaction		Gene coding for enzymes	
1.	(Acetate)$_{out}$	$\Rightarrow$	(Acetate)$_{in}$	
2.	(Acetate)$_{in}$	$\Rightarrow$	AcA	acs
3.	AcA + OAA	$\Leftrightarrow$	CIT	gltA
4.	AcA + Glyoxy	$\Leftrightarrow$	MAL	aceB
5.	CIT	$\Leftrightarrow$	ICIT	acnA
6.	ICIT	$\Leftrightarrow$	SUC + Glyoxy	aceA
7.	ICIT	$\Leftrightarrow$	OGA + CO_2	icdA
8.	OGA	$\Leftrightarrow$	SUC + CO_2	SucAB
9.	SUC	$\Leftrightarrow$	FUM	sdhABCD
10.	FUM	$\Leftrightarrow$	MAL	fum
11.	MAL	$\Leftrightarrow$	OAA	mdh
12.	OAA	$\Leftrightarrow$	PEP + CO_2	pckA
13.	PEP	$\Leftrightarrow$	PGA	eno
14.	PGA	$\Leftrightarrow$	T3P (GAP)	pgk
15.	T3P1(GAP) + T3P2(DHAP)	$\Leftrightarrow$	F6P	fbp
16.	F6P	$\Leftrightarrow$	G6P	pgi
17.	G6P	$\Rightarrow$	Ru5P + CO_2	gnd
18.	Ru5P	$\Leftrightarrow$	R5P	rpi
19.	Ru5P	$\Leftrightarrow$	X5P	rpe
20.	R5P + X5P	$\Leftrightarrow$	S7P + T3P	tk1
21.	X5P + E4P	$\Leftrightarrow$	F6P + T3P	tk2
22.	S7P + T3P	$\Leftrightarrow$	E4P + F6P	tal
23.	MAL	$\Leftrightarrow$	PYR + CO_2	maeB
24.	PYR	$\Leftrightarrow$	AcA + CO_2	aceEF
25.	PGA	$\Leftrightarrow$	SER	ser
26.	SER	$\Leftrightarrow$	GLY + C1	glyA
27.	GLY	$\Leftrightarrow$	C1 + CO_2	gcvHTP
28.	(CO_2)in	$\Rightarrow$	$(CO_2)_{out}$	
29.	Precursors	$\Rightarrow$	Biomass	

Table 5.2 Metabolic reactions for glucose metabolism

	Reaction		Gene coding for enzymes	
1.	GLC	⇒	G6P	glk
2.	g6p	⇔	F6P	pgi
3.	F6P	⇔	T3P1 + T3P2	pfk
4.	T3P	⇔	PGA	pgk
5.	PGA	⇔	PEP	eno
6.	PEP	⇔	PYR	pyk
7.	PYR	⇔	ACA + CO_2	aceEF
8.	ACA	⇔	(Acetate)$_{in}$	acs
9.	(Acetate)$_{in}$	⇔	(Acetate)$_{out}$	
10.	G6P	⇒	Ru5P + CO_2	gnd
11.	Ru5P	⇔	X5P	rpe
12.	Ru5P	⇔	R5P	rpi
13.	R5P + X5P	⇔	S7P + T3P (GAP)	tk1
14.	S7P + T3P(GAP)	⇔	E4P + F6P	tal
15.	X5P + E4P	⇔	F6P + T3P(GAP)	tk2
16.	PGA	⇔	SER	ser
17.	SER	⇔	GLY + C1	glyA
18.	GLY	⇔	C1 + CO_2	gcvHTP
19.	AcA + OAA	⇔	CIT	gltA
20.	CIT	⇔	ICIT	acnA
21.	ICIT	⇔	OGA + CO_2	icdA
22.	OGA	⇔	SUC + CO_2	SucAB
23.	SUC	⇔	FUM	sdhABCD
24.	FUM	⇔	MAL	fum
25.	MAL	⇔	OAA	mdh
26.	PEP + CO_2	⇔	OAA	ppc-pckA
27.	$(CO2)_{in}$	⇒	$(CO_2)_{out}$	
28.	Precursors	⇒	Biomass	

energy changes, only the backward fluxes of *aceB*, *aceA*, *SucAB*, *pckA*, and *aceEF* have a significant effect on the mass distribution of proteinogenic amino acids. Thus, other backward fluxes may be negligible in general.

^{13}C-metabolic flux analysis

Thus, the problem of determining metabolic fluxes is to estimate 26 net fluxes and 5 exchange coefficients. As for the 25 intracellular metabolite pools, the quasi-steady is assumed such that:

$$A \, v^{net} = 0 \tag{5.52}$$

where v^{net} is the steady state vector for the net fluxes. A is the stoichiometric matrix, which is composed of 25 species linked by 29 reactions. The rank of A is 25 and the number of degrees of freedom is 4. That is, four independent fluxes should be specified to determine the rest. The values of these four free net fluxes are then varied within the simulation runs or obtained from extracellular measurements. The aim of the labeling data obtained from mass isotopomer analysis is to supply the information for determination of the free net fluxes and the other five exchange coefficients. The algorithm for metabolic flux calculation based on isotope distribution of proteinogenic amino acids is as follows:

Skewing effects of natural isotopes (^{13}C, ^{2}H, ^{17}O, ^{18}O, ^{15}N, ^{29}Si, ^{30}Si) existing in MTBSTFA-derived amino acids must be corrected (Paul Lee, 1991; 1992; Wittman and Heinzel, 1999). Three net fluxes, such as v_6^{net}, v_{12}^{net}, and v_{17}^{net} are randomly chosen, and these values as well as those for five exchange coefficients, are first given with arbitrary values. By rearranging and partitioning A and v, such that v_1^{net}, v_6^{net}, v_{12}^{net} and v_{17}^{net} are at the bottom of v, all the net fluxes are determined by:

$$v_u = -A_u^{-1} A_m v_m \tag{5.53}$$

where A_u and A_m are 25 × 25 and 25 × 4 matrices, and v_u and v_m are 25 × 1 and 4 × 1 matrices, respectively. All the forward fluxes $\vec{v}$ and backward fluxes $\overleftarrow{v}$ are expressed by the net fluxes v^{net} and the exchange coefficients. By using the concept of isotopomer mapping matrices, and the assumed values of v^{net} and exchange coefficient, the corresponding steady-state isotopomer distributions in the intracellular metabolite pools are obtained by an iterative scheme. The steady-state isotopomer distributions estimated are transformed into simulated fragment mass isotopomer distribution based on precursor-amino acid relationships, where Figure 5.7 shows the precursor-amino acid relationships.

The simulated fragment mass isotopomer distributions are compared to the measured data. The optimization may be made by the hybrid algorithm by combining genetic algorithm (GA) (Davis, 1991) with Levenberg-Marquardt algorithm (LMA) (Moré, 1977) to find a global

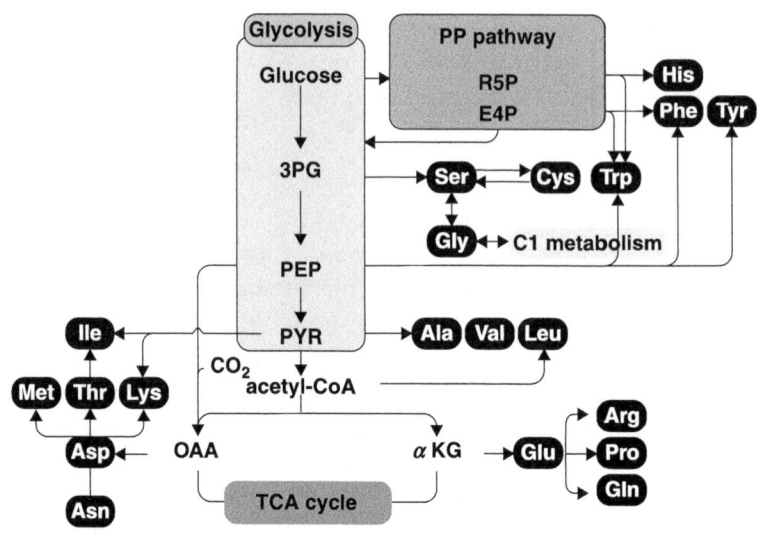

Figure 5.7 The relationship between the precursor and amino acid

minimum among numerous local ones (Zhao and Shimizu, 2003). The objective function to be minimized by GA is defined as:

$$\varepsilon(v, I_{ace}) = \sum_{i=1}^{n} \left(\frac{M_{i,meas} - M_{i,calc}(v, I_{ace})}{\delta_i} \right)^2 + \sum_{j=1}^{m} \left(\frac{S_{j,meas} - S_{j,calc}}{\delta_j} \right)^2 \quad (5.54)$$

where v is the vector containing free net fluxes and exchange coefficients to be optimized, and I_{ace} is the isotopomer distribution vector of the input (sodium) acetate. $M_{i,meas}$ are the n individual labeling measurements obtained from mass isotopomer analysis and $M_{i,calc}$ their corresponding simulated values computed, based on the assumed values in v. $S_{j,meas}$ is a vector containing the measured values of m extracellular fluxes and $S_{j,calc}$ the vector containing the simulated values of extracellular fluxes. δ_i and δ_j are the corresponding measurement errors. If the sum of two errors in Equation 5.54 is smaller than the pre-specified criterion or if the algorithm reaches a specified maximum number of iterations, the calculated fluxes are considered to be a suitable solution for further optimization.

A statistical analysis must be made to check the reliability of flux estimates and investigate the sensitivity of the estimated values to the measurement inaccuracies. One approach is to generate several hundreds of signal datasets of mass distribution by addition of normally distributed measurement noise to the simulated measurement dataset corresponding to the best fit flux distribution. The same optimization is made to estimate the flux distribution from such generated datasets. Then, from the probability distribution of these flux distributions, confidence limits can be obtained for the estimated parameters.

The application of MTBSTFA for deriving amino acids yields the corresponding TBDMS derivatives in good characteristics for GC-MS measurement. The typical mass fragmentation pattern of the derivatives ($[M-57]^+$ and $[M-159]^+$) makes them useful for isotopomer analysis (Figure 5.8).

Consider the sensitivity of mass distribution upon changes in fluxes and exchange coefficients in order to check the reliability of the metabolic network assumed. Consider glutamate as an example for the case of using acetate as a carbon source, where the effect of altering the flux Icl (*aceA*) on the mass distribution can be determined. The result, as shown in Table 5.3, indicates that the mass distribution is sensitive to the flux change of Icl (*aceA*). The fractional labeling of the fragment of glutamate ($[M-159]^+$) increases with the decrease in the flux *aceA*, thus proving that mass distribution can be used as a sensitive indicator for flux analysis. Similarly, the sensitivity of mass distribution with respect to the changes in exchange coefficients can be evaluated, and the results suggest that certain exchange coefficients significantly affect mass distribution, whereas other exchange coefficients show minor effects. For example, the change in the reversibility of Pck can alter the mass distribution of glutamate significantly, as shown in Table 5.4, while almost no change can be found in the mass distribution if different reversibility of Tkt is applied to the network system. If mass distribution for all amino acids is insensitive to one exchange coefficient, then perhaps the reversibility of this reaction contributes little to the isotopomer balance. Through this

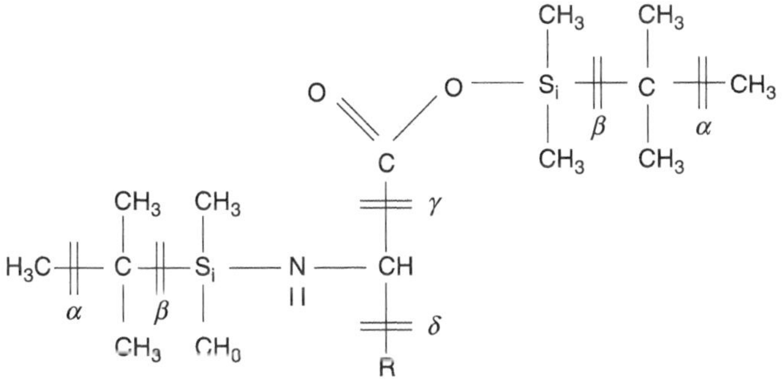

Figure 5.8 Derivatization to generate M-57^+ and M-159^+

Table 5.3 Sensitivity of mass distribution (fragment [M-159]⁺ of glutamate) upon changes in fluxes of Icl (aceA)

Flux of Icl (aceA)	m0	m1	m2	m3	m4
5	0.6793	0.2763	0.0416	0.0027	0.0001
10	0.6937	0.2667	0.0373	0.0022	0.0000
20	0.7110	0.2550	0.0324	0.0016	0.0000
40	0.7218	0.2473	0.0295	0.0014	0.0000

Table 5.4 Sensitivity of mass distribution (fragment [M-159]⁺ of glutamate) upon changes in exchange coefficients of Pck (pckA)

$exch_{pckA}^{[0,1]}$	m0	m1	m2	m3	m4
0.05	0.6935	0.2667	0.0373	0.0022	0.0000
0.95	0.6961	0.2650	0.0367	0.0022	0.0000
0.99	0.7011	0.2615	0.0353	0.0020	0.0000

sensitivity analysis, five exchange coefficients for MS(aceB), Icl (aceA), KGDH (sucAB,lpdA), Pck (pckA), and PDHc (aceEF,lpdA) are more effective in isotopomer distribution than others. Thus, these five exchange coefficients must be incorporated into the isotopomer balance model for flux calculation when using acetate as a carbon source.

Simulated mass distributions obtained with the best-fit flux distribution are compared with experimental data with correction for naturally labeled elemental isotopes. Considering the standard deviation (in the range from 0.002 to 0.006) for mass distribution, only mass signals of m, $m+1$, and $m+2$ are compared. A good agreement between simulated data and experimental measurements may be obtained (Table 5.5), except for two amino acids in glucose metabolism, where a relatively larger difference may be seen for m_0 of Phe and m_1 of Gly.

Analysis of 90% confidence limits is made for the estimated net fluxes and exchange coefficients to evaluate the quality of the estimates. A total of 500 simulated measurement datasets (including mass distribution data and extracellular flux data) are generated by addition of normally distributed measurement errors to an ideal simulated dataset that corresponds to the best fit flux distribution. The standard deviation for

Table 5.5 Experimental determined (exp) and calculated (cal) fragment mass distribution of TBDMS-derived amino acids from *E. coli* K12 hydrolysates (chemostat culture by using acetate and glucose as the carbon source; D = 0.22 h^{-1})

Amino acids	fragment	Origin	Acetate metabolism			Glucose metabolism		
			m0	m1	m2	m0	m1	m2
Gly	[M-57]$^+$	Exp:	0.9207	0.0781	0.0012	0.9202	0.0154	0.0644
		Cal:	0.9219	0.0770	0.0011	0.9176	0.0274	0.0550
Val	[M-57]$^+$	Exp:	0.7950	0.1869	0.0175	0.7646	0.0944	0.1140
		Cal:	0.7981	0.1846	0.0166	0.7602	0.0952	0.1152
Leu	[M-159]$^+$	Exp:	0.7716	0.2057	0.0198	0.7487	0.1097	0.1103
		Cal:	0.7736	0.2042	0.0212	0.7422	0.1112	0.1121
Ile	[M-159]$^+$	Exp:	0.8002	0.1832	0.0130	0.7689	0.0964	0.1116
		Cal:	0.8002	0.1831	0.0160	0.7632	0.0977	0.1132
Ser	[M-159]$^+$	Exp:	0.9061	0.0918	0.0017	0.9044	0.0260	0.0696
		Cal:	0.9071	0.0908	0.0021	0.8990	0.0274	0.0737
Thr	[M-57]$^+$	Exp:	0.8552	0.1373	0.0065	0.8267	0.0788	0.0565
		Cal:	0.8554	0.1372	0.0072	0.8237	0.0797	0.0573
Phe	[M-159]$^+$	Exp:	0.6986	0.2589	0.0330	0.7683	0.0466	0.1058
		Cal:	0.7001	0.2563	0.0400	0.7553	0.0470	0.1092
Glu	[M-159]$^+$	Exp:	0.8271	0.1616	0.0103	0.7789	0.1112	0.0930
		Cal:	0.8267	0.1619	0.0110	0.7733	0.1129	0.0950

mass distribution measurement is 0.006, which is computed from multiple GC-MS analysis, and the values in the measurement of extracellular fluxes are assumed to be 5%. From these 500 simulated datasets, the 90% confidence limits for the estimates can be obtained and are given in Tables 5.6 and 5.7 as the lower and upper bounds. Small

Table 5.6 90% confidence limits for estimated net fluxes and exchange coefficients in acetate metabolism

	$D = 0.22\ h^{-1}$		$D = 0.11\ h^{-1}$	
	Optimal estimates*	90% confidence region	Optimal estimates*	90% confidence region
aceA	21.80	[21.00, 22.60]	20.60	[19.87, 21.55]
pckA	10.00	[9.90, 10.20]	9.81	[9.66, 9.96]
gnd	1.57	[1.51, 1.63]	1.70	[1.61, 1.79]
$Exch_{aceB}^{[0,1]}$	0.963	[0.961, 0.965]	0.920	[0.915, 0.924]
$Exch_{aceA}^{[0,1]}$	0.118	[0.115, 0.121]	0.130	[0.126, 0.135]
$Exch_{sucAB}^{[0,1]}$	0.054	[0.051, 0.056]	0.048	[0.044, 0.053]
$Exch_{pckA}^{[0,1]}$	0.967	[0.964, 0.969]	0.978	[0.975, 0.981]
$Exch_{aceEF}^{[0,1]}$	0.078	[0.076, 0.080]	0.082	[0.079, 0.084]

* Net flux rates are given relative to the molar acetate uptake rate and exchange coefficient defined in the the model were limited to values between 0.00 and 0.99

Table 5.7 90% confidence limits for estimated net fluxes and exchange coefficients in glucose metabolism

	$D = 0.22\ h^{-1}$		$D = 0.11\ h^{-1}$	
	Optimal estimates*	90% confidence region	Optimal estimates*	90% confidence region
tk1	13.00	[12.99, 13.50]	8.00	[7.90, 8.45]
$Exch_{pgi}^{[0,1]}$	0.761	[0.753, 0.768]	0.919	[0.910, 0.925]
$Exch_{eno}^{[0,1]}$	0.402	[0.050, 0.920]	0.251	[0.050, 0.950]
$Exch_{tk1}^{[0,1]}$	0.443	[0.140, 0.520]	0.503	[0.250, 0.630]
$Exch_{tal}^{[0,1]}$	0.125	[0.110, 0.139]	0.100	[0.080, 0.115]
$Exch_{tk2}^{[0,1]}$	0.021	[0.020, 0.023]	0.018	[0.016, 0.020]
$Exch_{ppc-pckA}^{[0,1]}$	0.138	[0.130, 0.146]	0.204	[0.195, 0.208]
$Exch_{mdh}^{[0,1]}$	0.019	[0.015, 0.230]	0.002	[0.001, 0.190]

* Net flux rates are given relative to the molar glucose uptake rate and exchange coefficient defined in the the model were limited to values between 0.00 and 0.95

^{13}C-metabolic flux analysis

confidence limits are found for the exchange coefficients in acetate metabolism, whereas relatively larger confidence intervals are estimated for some coefficients in glucose metabolism. It may be seen that the net fluxes are reliable, because of the small confidence limits in general.

The best-fit flux distributions in acetate-limited and glucose-limited chemostat cultures at the dilution rates of 0.11 h^{-1} and 0.22 h^{-1} are given in Figures 5.9 and 5.10, and the measured specific rates are shown in Tables 5.8 and 5.9. Although the experimental data of the specific acetate uptake rates (q_{ace}), the oxygen consumption rates (q_{O2}), and the CO_2 evolution rates (q_{CO2}) increase in proportion to the cell growth rate (Table 5.8), the flux distribution shows little change between the two different growth rates. For example, the flux through ICDH is always regulated to approximately 2.45 times the throughput of Icl, and the flux ratio of CS to Pck is always maintained at around 7.38. The pentose phosphate (PP) pathway, although an important route for glucose metabolism, appears to contribute little when using acetate as a carbon source. A PP pathway flux is estimated to be below 2% of the acetate uptake rate, and the cell growth rate has little influence on the relative

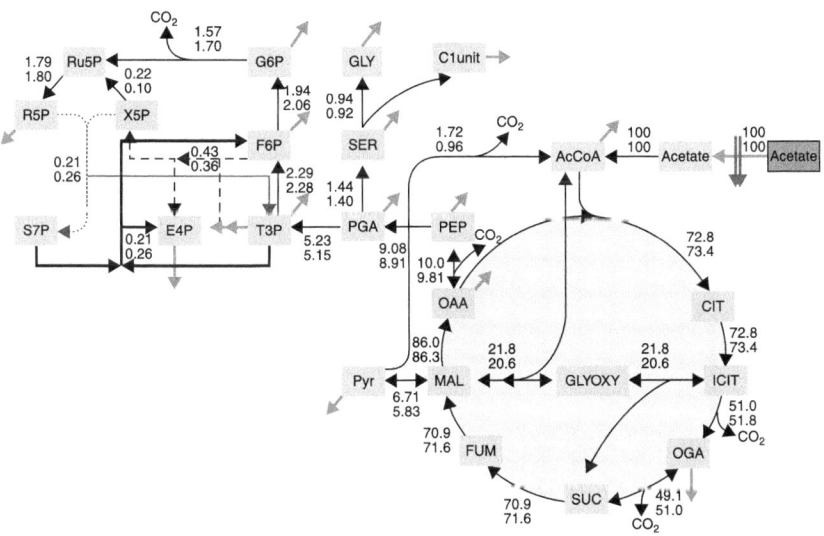

Figure 5.9 Net flux distribution in acetate metabolism of *E. coli* K12 in chemostat cultures at D of 0.11 and 0.22 h^{-1}, flux values at D of 0.22 h^{-1} are at upper values, and those for 0.11 h^{-1} are at lower values

Bacterial cellular metabolic systems

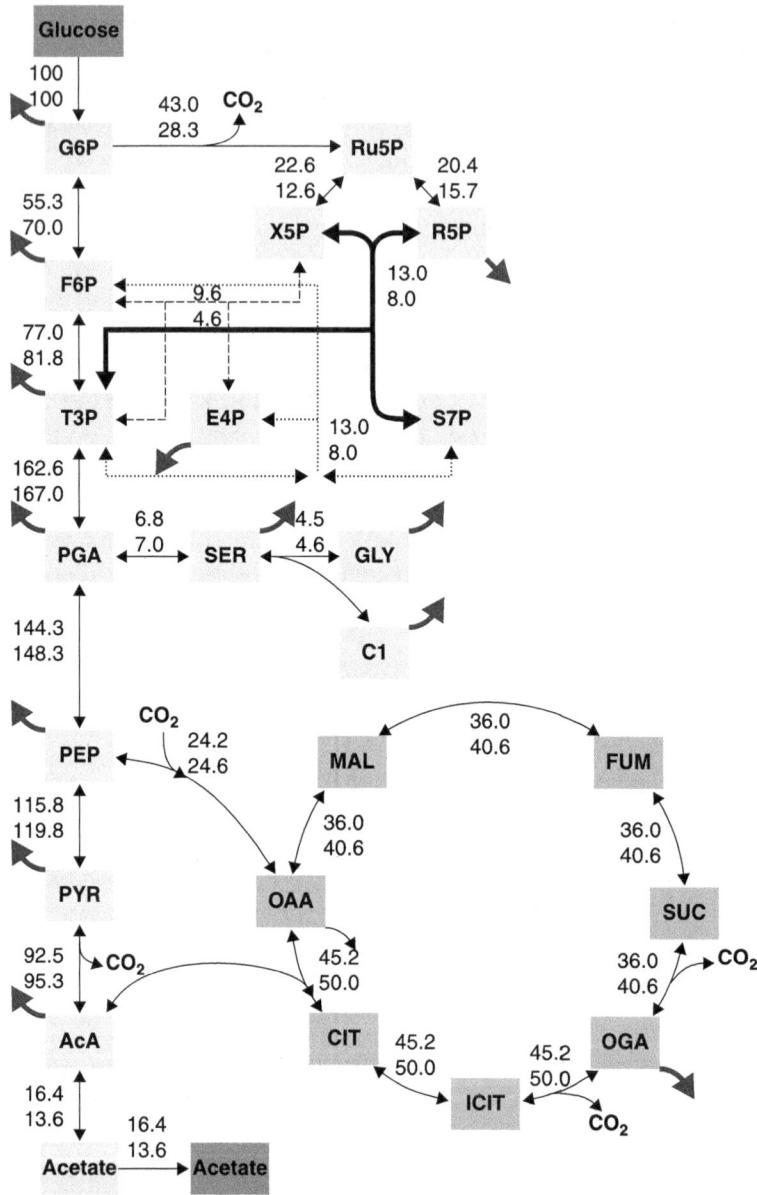

Figure 5.10 Net flux distribution in glucose metabolism of *E. coli* K12 in chemostat cultures at D of 0.11 and 0.22 h^{-1}, flux values at D of 0.22 h^{-1} are at upper values and those for 0.11 h^{-1} are at lower values

Table 5.8 Growth parameters of *E. coli* K12 at a D of 0.11 and 0.22 h^{-1}, where acetate is used as the sole carbon source

Dilution rate (h^{-1})	Acetate consumption rate ($mmol\,g^{-1}h^{-1}$)	CO_2 formation rate ($mmol\,g^{-1}h^{-1}$)
0.11	7.18	8.76
0.22	14.43	17.32

Table 5.9 Growth parameters of *E. coli* K12 at a D of 0.11 and 0.22 h^{-1}, where glucose is used as the sole carbon source

Dilution rate (h^{-1})	Glucose consumption rate ($mmol\,g^{-1}h^{-1}$)	CO_2 formation rate ($mmol\,g^{-1}h^{-1}$)	Acetate formation rate ($mmol\,g^{-1}h^{-1}$)
0.11	2.78	5.25	0.38
0.22	5.92	11.37	0.97

PP pathway flux. However, in glucose metabolism, a significant change in flux distribution can be seen when the specific growth rate increases (Figure 5.10). It can be observed that the relative flux for the PP pathway increases, while that of the TCA cycle decreases as the cell growth rate increases. A similar change in flux pattern of *Bacillus subtilis* has also been observed (Sauer, 1996).

The network model is an important factor affecting the reliability of flux analysis, as illustrated by the case of using acetate as a carbon source, where the stoichiometric matrix is assumed to be composed of 25 species linked by 29 reactions. The rank of A is 25 and the number of degrees of freedom is 4. Among the three extracellular fluxes, such as the specific acetate consumption rate, the specific CO_2 evolution rate, and the specific growth rate, only two are independent, because of an overall carbon balance. Therefore, it is impossible to determine all the internal rates solely based on the external measurements and metabolic flux balancing. With the ^{13}C tracer experiment, not only the full distribution of the net fluxes in the entire network, but also certain exchange fluxes that also contribute to the isotopomer distribution, can be quantified. In theory, most of the metabolic reactions are bidirectional, but some have only an insignificant influence on the value of the isotopomer distribution with respect to the specified network structure.

Different from the glucose metabolism, the central metabolism of *E. coli* K12 growing on acetate is characterized by high activity of the TCA cycle, high activity of the glyoxylate cycle as the anaplerotic reaction, and PEP/PYR formation from OAA/MAL. During metabolism, acetate is fluxed via central metabolic pathways to the precursors required for the synthesis of biomass and also to generate the reducing power and ATP required for the cell growth. Two important enzymes, Icl encoded by *aceA* and ICDH encoded by *icdA*, lead the carbon flux to two different cycles, which are responsible for these two functions. From the mass isotopomer analysis of the entire cellular network, a subtle regulation mechanism exists in certain key junctions of the network system. At the steady state, the flux through ICDH is always regulated to be approximately 2.45 times that of the flux of Icl. In general, the flux through Icl generates the precursors used for biosynthesis, while the flux through ICDH is dedicated to the supply of reducing power and ATP for biosynthesis. The effect of the regulation system is to maintain the flux ratio of ICDH to Icl, so that the rate of supplying NADPH and ATP is equal to the demands for biosynthesis. Thus, when the growth rate increases from $0.11\ h^{-1}$ to $0.22\ h^{-1}$, with the increase in absolute flux through Icl to generate more precursors for biosynthesis, the flux through ICDH also increases in proportion to produce more energy.

From the flux distribution in the entire network, it can be seen that ICDH is the major source of NAD(P)H in acetate metabolism, causing the flux to oxidative PP pathway to be of minor importance. This is the main reason why a slight change in flux distribution is observed for acetate metabolism at different growth rates. The fact that almost no change in the flux through the PP pathway, when the cell growth rate increased from $0.11\ h^{-1}$ to $0.22\ h^{-1}$, reveals that the main role of the PP pathway in acetate metabolism is to provide E4P and pentose phosphates for biosynthesis of nucleotides and aromatic amino acids, and its function as the source of NADPH is negligible.

In the case where glucose is used as a carbon source, it may be seen that the lower activity of the TCA cycle, the absence of glyoxylate cycle, and anaplerosis is made by carboxylation of PEP. From mass isotopomer analysis, it may be concluded that when the specific growth rate increases, a significantly increased PP pathway flux is observed for *E. coli* at the expense of the TCA cycle. It is obvious that when growing on preferred carbon sources such as glucose, ICDH cannot fulfill the NADPH demand for the cell growth by itself, causing the oxidative PP pathway to be utilized to a larger extent to complement the NADPH demand in addition to its normal function for pentose formation.

5.6 ^{13}C-MFA using NMR

5.6.1 Measurement of isotopomer pattern by NMR

Consider next the metabolic flux analysis based on nuclear magnetic resonance (NMR) spectroscopy (Sauer et al., 1999). The use of ^{1}H-NMR for measuring the positional ^{13}C enrichment of purified amino acids and the application of these measurements for determining the intracellular fluxes in *Corynebacterium glutamicum* have been reported (Marx et al., 1996). The 2D ^{1}H-^{13}C NMR spectroscopy has also been used for analyzing the labeling states of proteinogenic amino acids in crude biomass hydrolysates (Szyperski, 1995). This technique, which allows the direct monitoring of the breakage of carbon bonds, has been used to estimate the flux distribution in *E. coli* (Schmidt et al., 1999a), *B. subtilis* (Sauer et al., 1997), and *Aspergillus niger* (Schmidt et al., 1999b).

Isotope labeling experiments are usually applied for continuous culture and are initiated after the culture reaches a steady state, which may be inferred from the stable oxygen and carbon dioxide concentrations in the off-gas and stable optical density in the effluent medium for at least twice or three times as long as the residence time. The feed medium with unlabeled glucose is then replaced by an identical medium containing, for example, a mixture of uniformly labeled glucose [U-^{13}C], the first carbon labeled glucose [1-^{13}C], and naturally labeled glucose. Another combination of labeling substrate may be used, but its determination should be made based on the available measurement apparatus and the cost of labeling substrate. Note that the accuracy of the metabolic flux distribution depends on the mixture of labeled substrate and the measurement apparatus.

At the end of the labeling experiment, the samples are harvested followed by centrifugation, and then the cell is hydrolyzed. In the resulting hydrolysate, 16 proteinogenic amino acids are present, since cysteine and trytophan are oxidized, while asparagine and glutamine are deaminated during the acid hydrolysis. The labeling patterns of amino acids in the hydrolysates can be determined by NMR spectroscopy. Two-dimensional proton-detected ^{1}H-^{13}C heteronuclear multiple-quantum correlation (HMQC) spectra may be recorded.

5.6.2 ^{13}C resonance fine structure in fractional labeled substrates

In general, 2^n isotopomers exist for a molecule with n carbon atoms, where NMR can detect ^{13}C isotope at any specified position. The corresponding scalar coupling fine structure observed in the NMR spectrum is a superposition of the fine structures of these isotopomers weighted by their relative abundance (Szyperski, 1995). Moreover, only 1-bond scalar coupling constants $^1J_{cc}$ are sufficiently large to be resolved in the ^{13}C dimension in the [^{1}H, ^{13}C]-COSY spectrum, and therefore the ^{13}C-fine structure is solely determined by the ^{13}C-labeling pattern of the directly attached carbon atoms. Figure 5.11 shows the relevant isotopometric labeling patterns and the corresponding multiplets, as observed in ^{13}C-NMR spectra (Szyperski, 1995).

For a peripheral carbon, only two isotopomer patterns are observed, where one of them exhibits a doublet 'd' split by $^1J_{cc}$, and the other a singlet 's' fine structure (Figure 5.11a) (Szyperski, 1995). Let I_s and I_d be the relative intensities for a weighted superposition of the singlet and doublet with $I_s + I_d = 1$, thereby defining the vector $I_{term} = [I_s, I_d]$. If the two scalar coupling constants with the control central carbon position are different, such as $^1J_{cc}$ and $^1J_{cc}*$ (Figure 5.11b), all four isotopomers give rise to different scalar coupling fine structures. Hence, a maximal number of nine resonance peaks is expected in a weighted superposition of all possible multiplets (Szypersli, 1995), where these arise from a singlet 's', a doublet split by a small coupling constant 'da', a doublet split by a larger coupling constant 'db', and a doublet of doublets 'dd'. The

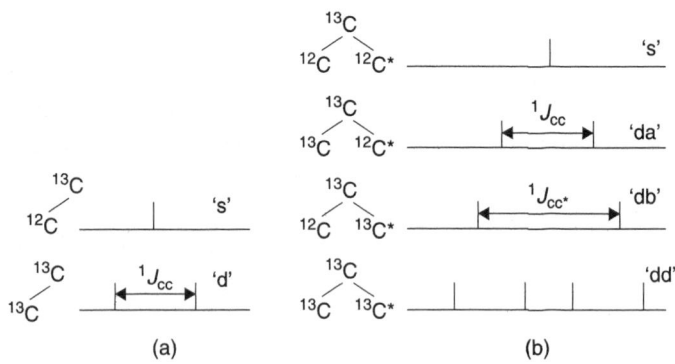

Figure 5.11 ^{13}C isotopomer patterns and NMR multiple spectral patterns

^{13}C-metabolic flux analysis

corresponding relative intensities are denoted as I_s, I_{da}, and I_{db}, I_{dd} with $I_s + I_{da} + I_{db} + I_{dd} = 1$ and define $I_{central-1} = [I_s, I_{da}, I_{db},$ and $I_{dd}]$. If two coupling constants $^1J_{cc}$ and $^1J_{cc}*$ are identical, then at most five resonance peaks are expected where (Figure 5.11c), these correspond to a weighted superposition of a singlet 's', doublet 'd', and triplet 't' with I_s, I_d, and I_t as their relative intensities with $I_s + I_d + I_t = 1$, and define $I_{central-2} = [I_s, I_d, I_t]$.

The contiguous carbon fragment originating from a single source molecule of glucose is incorporated via precursor metabolites in the main metabolic pathways into the proteinogenic amino acids during biosynthesis (Figure 5.12) (Szyperski, 1995). To elucidate the carbon flux from labeled substrate to metabolite precursors, the quantitative analysis of ^{13}C fine structures must unravel to which extent a certain carbon atom possesses neighboring carbons originating from the same source molecule of the substrate. Hence, the above defined 'I' have to be related to the relative abundances of such intact carbon fragments (Szyperski, 1995). Three basic probabilities must be considered here. Let P_n be the degree of ^{13}C labeling at natural abundance where $P_n = 0.01107$, P_1 be the overall degree of ^{13}C labeling, and P_f be the fraction of ^{13}C glucose relative to the total amount of glucose. Then P_1 may be expressed as:

$$P_1 = P_f + (1-P_f)P_n \qquad (5.55)$$

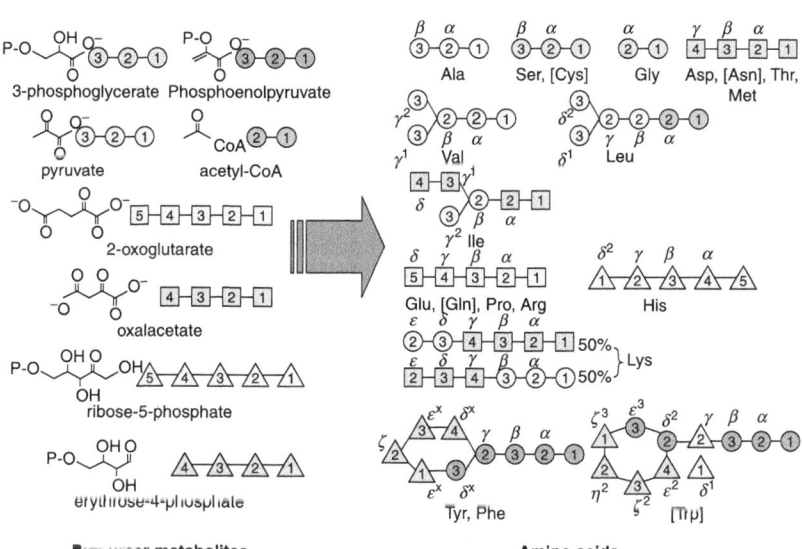

Figure 5.12 The fate of carbons of precursor metabolites to amino acids

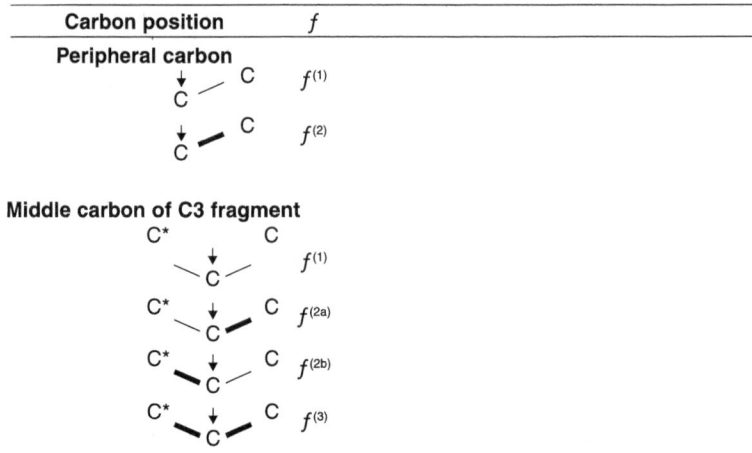

Figure 5.13 Carbon position and f values

For a peripheral carbon atom, only two situations have to be considered, where it may be attached to a carbon arising from the same source molecule, or it may arise from a different source molecule. The fraction of molecules for the former case is denoted as $f^{(2)}$, where the superscript indicates that the peripheral carbon is located in an intact C_2 fragment originating from one source molecule, while the fraction for the latter case is denoted as $f^{(1)}$ (Szyperski, 1995). Considering that the intact C_2 fragments are either fully ^{13}C labeled or contain ^{13}C at natural abundances, the multiplet intensities arising from both fractions can be calculated as shown in Figure 5.13, where, for example, the fraction of molecules carrying the intact C_2 fragments give rise to a relative intensity of $K_d^{(2)} = [P_f + P_n^2(1 - P_f)]/P_1$ for the doublet, $K_s^{(2)} = P_n(1 - P_n)(1 - P_f)/P_1$ for the singlet, and $K^{(2)} = [K_s^{(2)}, K_d^{(2)}]$. In the similar way, $K^{(1)} = [K_s^{(1)}, K_d^{(1)}]$ can be also defined, and $I_{term} = f^{(1)}K^{(1)} + f^{(2)}K^{(2)}$ yields the desired fractions $f^{(1)}$ and $f^{(2)}$ (Szyperski, 1995).

In the same way, for a carbon atom that is centrally embedded in a C_3 fragment and exhibits different scalar coupling constants to the attached carbons, the following equation can be derived:

$$I_{cent-1} = f^{(1)}K^{(1)} + f^{(2a)}K^{(2a)} + f^{(2b)}K^{(2b)} + f^{(3)}K^{(3)} \tag{5.56a}$$

In the same way, where the carbon atom is centrally located within a C_3 fragment in which the scalar coupling constants to both attached carbons are identical, the following equation can be derived:

$$I_{cent-2} = f^{(1)}K^{(1)} + f^{(2)}K^{(2)} + f^{(3)}K^{(3)} \tag{5.56b}$$

The relative multiplet intensities (I values) can be used to calculate the relative abundances of intact carbon fragments (f values) (Szyperski, 1995). These f values can be calculated from the observed relative multiplet intensities, I, by:

$$f = K^{-1} I \tag{5.57}$$

The transformation matrix, K, is composed of the probabilistic expressions that are functions of the degree of ^{13}C labeling at natural abundance ($P_n = 0.01107$), the overall degree of ^{13}C labeling, P_1, and the fraction of [U-^{13}C] glucose to the total amount of glucose, P_f, where we consider the use of a mixture of [U-^{13}C] glucose and unlabeled glucose as a carbon source for ^{13}C-labeling experiment.

Consider the ^{13}C-labeling experiment in the continuous culture. Assuming that the unlabeled biomass in the bioreactor follows first-order washout kinetics during the labeling experiment, P_1 can be calculated using the following equation:

$$P_1 = (1 - e^{-\mu t})[P_s + (1 - P_s)P_n] + e^{-\mu t}P_n \tag{5.58}$$

where t is the time elapsed between the initiation of labeling experiments and biomass harvesting for NMR measurements, and P_s the fraction of [U-^{13}C] glucose in the feed. The fraction of labeled glucose to the total glucose, P_f, can be calculated by the following equation, by considering the presence of unlabeled biomass in the bioreactor:

$$P_f = (1 - e^{-\mu t})P_s \tag{5.59}$$

The probabilistic expressions are used to calculate the components in the matrix K, which denote the relative multiplet intensities arising from different intact carbon fragments. The matrix K is given in Table 5.10 (Yang et al., 2003).

The relative abundances of intact carbon fragments in the amino acids (and glycerol) can be used to derive information on the origin of the precursor intermediates such as E4P, Tpi, 3PG, PEP, PYR, AceCoA, αKG, and OAA. Here, the equations that are required to investigate the anaplerotic metabolism are shown below.

5.6.3 Flux ratio analysis of pckA mutant E. coli

Flux ratio analysis can be used to identify the metabolic network structure of a cell grown under various conditions, and the flux distribution in the

Bacterial cellular metabolic systems

Table 5.10 Transformation matrix K for calculating f values

Carbon position	K
Terminal	$K_{term} = \begin{bmatrix} K_s^{(1)} & K_s^{(2)} \\ K_d^{(1)} & K_d^{(2)} \end{bmatrix} = \begin{bmatrix} 0.904 & 0.104 \\ 0.096 & 0.896 \end{bmatrix}$
Central in C_3 fragment:	$K_{central} = \begin{bmatrix} K_s^{(1)} & K_s^{(2a)} & K_s^{(2b)} & K_s^{(3)} \\ K_{da}^{(1)} & K_{da}^{(1a)} & K_{da}^{(2b)} & K_{da}^{(3)} \\ K_{db}^{(1)} & K_{db}^{(2a)} & K_{db}^{(2b)} & K_{db}^{(3)} \\ K_{dd}^{(1)} & K_{dd}^{(2a)} & K_{dd}^{(2b)} & K_{dd}^{(3)} \end{bmatrix} = \begin{bmatrix} 0.819 & 0.094 & 0.094 & 0.102 \\ 0.086 & 0.811 & 0.010 & 0.001 \\ 0.086 & 0.010 & 0.811 & 0.001 \\ 0.009 & 0.085 & 0.085 & 0.896 \end{bmatrix}$

identified bioreaction network can be quantified by metabolic flux analysis. The derivation for the fraction of OAA synthesized via the glyoxylate shunt is given as follows:

The equations for calculation of the flux ratios involved in anaplerotic metabolism can be expressed as (Yang et al., 2003):

OAA from PEP:

$$X^{ppc} = X(OAA \leftarrow PEP) = \frac{f^{(2a)}\{Asp-\alpha\} + f^{(3)}\{Asp-\alpha\}}{f^{(2a)}\{Phe-\alpha, Tyr-\alpha + f^{(3)}\{APhe-\alpha, Tyr-\alpha\}}$$

$$= \frac{f^{(2a)}\{Asp-\beta\} + f^{(3)}\{Asp-\beta\}}{f^{(2a)}\{Phe-\alpha, Tyr-\alpha\} + f^{(3)}\{Phe-\alpha, Tyr-\alpha\}}$$

(5.60)

PEP from OAA:

$$X(PEP \leftarrow OAA) = \frac{f^{(2b)}\{Phe-\alpha, Tyr-\alpha\}}{f^{(2b)}\{Asp-\alpha\}}$$

(5.61)

PYR from MAL:

$$X^{1b}(PYP \leftarrow MAL) = \frac{f^{(2b)}\{Ala-\alpha\} - f^{(2b)}\{Phe-\alpha\}}{1 - f^{(2b)}\{Phe-\alpha\}}$$

(5.62)

$$X^{ub}(PYP \leftarrow MAL) = \frac{f^{(2b)}\{Ala-\alpha\} - f^{(2b)}\{Phe-\alpha\}}{f^{(2b)}\{Asp-\alpha\} - f^{(2b)}\{Phe-\alpha\}}$$

(5.63)

OAA via glyoxylate shunt:

$$f^{(2b)}\{Asp-\alpha\} - X^{ppc} \cdot (1 - 0.5 \cdot X^{exch}) \cdot \{f^{(2b)}\{Phe-\alpha\} - (1 - X^{ppc}) \cdot D\, X^{glo} = 0.5(A + B) - D$$

or

$$f^{(2b)}\{Asp-\beta\} - X^{ppc} \cdot 0.5 \cdot X^{exch}\, f^{(2b)}\{Phe-\alpha\} - (1 - X^{ppc}) \cdot D\, X^{glo} = 0.5\,(A + C) - D \qquad (5.64)$$

where

$$A = 0.5 \cdot (f^{(2b)}\{Leu-\alpha\} + f^{(2b)}\{Asp-\alpha\} + f^{(3)}\{Asp-\alpha\})$$
$$B = (1 - 0.5 \cdot X^{exch}) \cdot (f^{(2b)}\{Asp-\beta\} + f^{(3)}\{Asp-\beta\}) + 0.5 \cdot X^{exch} \cdot f^{(2b)}\{Leu-\alpha\}$$
$$C = 0.5 \cdot X^{exch} \cdot (f^{(2b)}\{Asp-\beta\} + f^{(3)}\{Asp-\beta\}) + (1 - 0.5 \cdot X^{exch}) \cdot f^{(2b)}\{Leu-\alpha\}$$
$$D = 0.5 \cdot (f^{(2)}\{Glu-\beta\} + f^{(2b)}\{Glu-\gamma\})$$

where Equation 5.60 is derived based on the fact that only the reaction catalyzed by Ppc leads to the introduction of intact C_2–C_3 connections in OAA. Equation 5.61 is derived based on the fact that the intact C_1–C_2 fragments in the PEP pool originate from OAA via the Pck. Here Pps is assumed to be inactive. The *lb* denotes the lower bound, and *ub* denotes the upper bound. The conversion of MAL to PYR via the Mez leads to an excess intact C_1–C_2 fragment in the PYR pool when compared to the PEP pool, and the upper bound is derived under the assumption that MAL is entirely synthesized from OAA.

If the glyoxylate shunt is inactive, the intact C_1–C_2 and C_3–C_4 fragments in the OAA pool originate from αKG and PEP, yielding:

$$f^{(2b)}\{Asp-\alpha\} = X^{ppc}\, f^{(2b)}\{Phe-\alpha\} + (1 - X^{ppc}) \cdot 0.5\,(f^{(2)}\{Glu-\beta\} + f(2b)\{Glu-\gamma\}) \qquad (5.65a)$$

$$f^{(2b)}\{Asp-\beta\} = (1 - X^{ppc}) \cdot 0.5 \cdot \{f^{(2)}\{Glu-\beta\} + f^{(2b)}\{Glu-\gamma\} \qquad (5.65b)$$

In the above equation, X^{ppc}, the fraction of OAA from PEP, can be calculated using Equation 5.60, and the $(1 - X^{ppc})$ term is obtained by assuming full symmetry of ^{13}C labeling patterns due to the symmetry of succinate or fumarate. However, the symmetry of ^{13}C labeling patterns arising from reversible interconversion from OAA to FUM may not be considered. The fraction of OAA molecules that are at least once reversibly interconverted to FUM, denoted by $X^{exch} = X(OAA \leftrightarrow FUM)$, is given by the following equation, which is derived based on this reaction leading to the introduction of intact C_4–C_3–C_2 fragments into OAA molecules, due to the symmetry of fumarate:

$$X^{exch} = X(OAA \longleftrightarrow FUM) = \frac{2 \cdot f^{(3)}\{Asp-\beta\}}{f^{(3)}\{Asp-\alpha\}+f^{(3)}\{Asp-\beta\}} \quad (5.66)$$

Thus, the symmetry of ^{13}C labeling patterns arising from reversible interconversion from OAA to FUM is considered, yielding:

$$f^{(2b)}\{Asp-\alpha\} = X^{ppc} \cdot (1 - 0.5\, X_{exch}) \cdot f^{(2b)}\{Phe-\alpha\}+(1 - X^{ppc})$$
$$\cdot 0.5 \cdot (f^{(2)}\{Glu-\beta\}+ f^{(2b)}\{Glu-\gamma\} \quad (5.67a)$$

$$f^{(2b)}\{Asp-\beta\} = X^{ppc} \cdot 0.5\, X^{exch}\{f^{(2b)}\{Phe-\alpha\}+(1 - X^{ppc})$$
$$\cdot 0.5 \cdot (f^{(2)}\{Glu-\beta\}+ f^{(2b)}\{Glu-\gamma\}) \quad (5.67b)$$

Equation 5.67 is used to identify the activity of the glyoxylate shunt. Namely, if the glyoxylate shunt is inactive, Equation 5.67 will be satisfied within experimental error, while the active glyoxylate shunt will result in obviously higher values of $f^{(2b)}\{Asp\text{-}\alpha\}$ and $f^{(2b)}\{Asp\text{-}\beta\}$ than the values calculated from Equation 5.67.

If the glyoxylate shunt is active, excess intact C_1–C_2 and C_3–C_4 connectivities in OAA are introduced via the glyoxylate shunt, yielding:

$$f^{(2b)}\{Asp-\alpha\} = X^{ppc} \cdot (1 - 0.5\, X^{exch}) \cdot f^{(2b)}\{Phe-\alpha\}+ X^{glo}$$
$$\cdot 0.5 \cdot (A + B) + (1 - X^{ppc} - X^{glo}) \cdot D \quad (5.68a)$$

$$f^{(2b)}\{Asp-\beta\} = X^{ppc}(1 - 0.5\, X^{exch}) \cdot f^{(2b)}\{Phe-\alpha\}+ X^{glo}$$
$$\cdot 0.5 \cdot (A + C) + (1 - X^{ppc} - X^{glo}) \cdot D \quad (5.68b)$$

where

$A = 0.5\, (\, f^{(2b)}\{Leu-\alpha\} + f^{(2b)}\{Asp-\alpha\} + f^{(3)}\{Asp-\alpha\}$
$B = (1 - 0.5\, X^{exch}) \cdot (\, f^{(2b)}\{Asp-\beta\} + f^{(3)}\{Asp-\beta\}+ 0.5\, X^{exch} \cdot f^{(2)}\{Leu-\alpha\}$
$C = 0.5 \cdot X^{exch} \cdot (f^{(2)}\{Asp-\beta\} + f^{(3)}\{Asp-\beta\})+ (1 - 0.5\, X^{exch}) \cdot f^{(2b)}\{Leu-\alpha\}$
$D = 0.5 \cdot (\, f^{(2)}\{Glu-\beta\} + f^{(2b)}\{Glu-\gamma\})$

In Equations 5.64 and 5.68, A, B, and C represent the intact C_1–C_2 or C_3–C_4 fragments introduced via the glyoxylate shunt into OAA, and the symmetry of ^{13}C labeling patterns due to reversible interconversion from MAL to FUM is also considered. Thus, the fraction of OAA synthesized via the glyoxylate shunt, X^{glo}, can be obtained using Equation 5.64. The ratios of other metabolic fluxes can be assessed in a similar way (Szyperski,

1995). All results of flux ratios contain the standard error introduced by experimental error, which is estimated from the analysis of redundant ^{13}C scalar coupling fine structures and the signal-to-noise ratio of the 2D NMR spectra.

Consider the continuous cultivation of *E. coli* wild-type W3110 and its *pck* knockout mutant JWK3366 in glucose-limited chemostats (Yang et al., 2003). The growth parameters are summarized in Table 5.11.

Abolition of Pck activity by deletion of the corresponding gene *pck* results in a significant increase in the biomass yield on glucose and a drastic reduction in CO_2 production rates. The macromolecular composition of biomass is known to change with environmental conditions and growth rates. Note that the relative fraction of the major biomass components of *E. coli* is as follows: protein, RNA, and glycogen, where glycogen is negligible (~1.4%), and the protein content is 70% ± 7%, 68% ± 8%, and 65% ± 5%, and the RNA content is 7% ± 1%, 8% ± 1%, and 12% ± 1% in the wild-type chemostat cultures at dilution rates of 0.10, 0.32, and 0.55 h^{-1}, respectively. Note that the RNA content increases with the growth rate. The contents of protein and RNA in the *pck* mutant are 68% ± 8% and 7% ± 1%, respectively. Hence, deletion of *pck* gives no significant impact on the macromolecular biomass composition. The remaining fraction of biomass is given in Dauner and Sauer (2001).

Table 5.11 Growth parameters of chemostat cultures of *E. coli* wild-type W3110 and *pck* mutant (JWK3366)

	E. coli strain (growth rate)			
Growth parameters	W3110 (0.10 h^{-1})	W3110 (0.32 h^{-1})	W3110 (0.55 h^{-1})	JWK3366 (0.10 h^{-1})
Biomass yield (g g–1)	0.40 ± 0.02	0.44 ± 0.02	0.48 ± 0.03	0.46 ± 0.02
Glucose uptake rate (mmol g–1 h–1)	1.4 ± 0.1	4.0 ± 0.2	6.4 ± 0.3	1.2 ± 0.1
O_2 uptake rate (mmol g–1 h–1)	4.0 ± 0.7	10.7 ± 1.6	16.3 ± 2.4	2.7 ± 0.5
CO_2 evolution rate (mmol g–1 h–1)	4.2 ± 0.4	11.1 ± 1.2	16.6 ± 1.8	2.9 ± 0.3
C balance (%)	99 ± 7	103 ± 8	104 ± 8	97 ± 7

5.6.4 Network identification by flux ratio analysis of pckA mutant E. coli

The results of flux ratio analysis allows identification of the network of active reactions. The flux ratio analysis of wild-type *E. coli* W3110 shows the activities of Pck and Mez, which are generally considered to be inactive in *E. coli* grown on glucose (Table 5.12). The gluconeogenic Pck is active, even at high growth rates (0.55 h^{-1}). The glyoxylate shunt consisting of Icl and MS is inactive in the wild type, because Equation 5.67 is satisfied within experimental error. The anaplerotic Ppc is responsible for about half of the OAA molecules synthesized in the wild type (Table 5.12). Comparing the results of flux ratio analysis of the wild type at different growth rates, similar flux ratio patterns may be seen. The only exception is the reduction in both the fraction of PEP molecules originating from OAA and the fraction of PYR arising from MAL, caused by an increase in growth rates (Table 5.12). The results in the table show the absence of PEP molecules originating from OAA in the *pck* mutant. This information is obtained from the probabilistic interpretation of the

Table 5.12 Origins of metabolic intermediates in chemostat cultures of *E. coli* W3110 and *pck* mutant JWK3366 determined by flux ratio analysis

Metabolite	% of Total pool in *E. coli* strain at growth rate			
	W3110 (0.10 h^{-1})	W3110 (0.32 h^{-1})	W3110 (0.55 h^{-1})	Δ *pckA* (0.10 h^{-1})
P5P from G6P (lb)	17 ± 2	23 ± 2	21 ± 2	13 ± 2
P5P from T3P + S7P	82 ± 2	76 ± 2	78 ± 2	87 ± 2
P5P from E4P	35 ± 2	27 ± 2	19 ± 2	31 ± 2
OAA from PEP	56 ± 2	53 ± 3	48 ± 2	18 ± 3
PEP from OAA	28 ± 3	19 ± 3	11 ± 2	0 ± 2
PYR from MAL (lb)	2 ± 2	0 ± 2	0 ± 2	7 ± 3
PYR from MAL (ub)	8 ± 4	0 ± 2	0 ± 2	14 ± 3
ACoA from PYR	>98	>99	>98	>98
AKG from OAA + ACoA	>97	>98	>98	>97
OAA exchanged to FUM	76 ± 10	65 ± 10	72 ± 8	94 ± 14
OAA via glyoxylate shunt	–	–	–	38 ± 12

lb: lower bound; ub: upper bound

^{13}C-metabolic flux analysis

$^{13}C-^{13}C$ scalar coupling fine structure of Phe-α in the [^{13}C, ^{1}H]–COSY spectra.

As can be seen in Figure 5.14, the scalar coupling multiplets of Phe-α, which represent the labeling in PEP, show absence of the db component for the *pck* deletion mutant, indicating the lack of intact C_1–C_2 fragments in PEP due to the abolishment of Pck activity. This result also excludes the possibility of an active Pps in the *pck* mutant and confirms the activity of Pck in wild-type *E. coli* grown on glucose. A decrease in the fraction

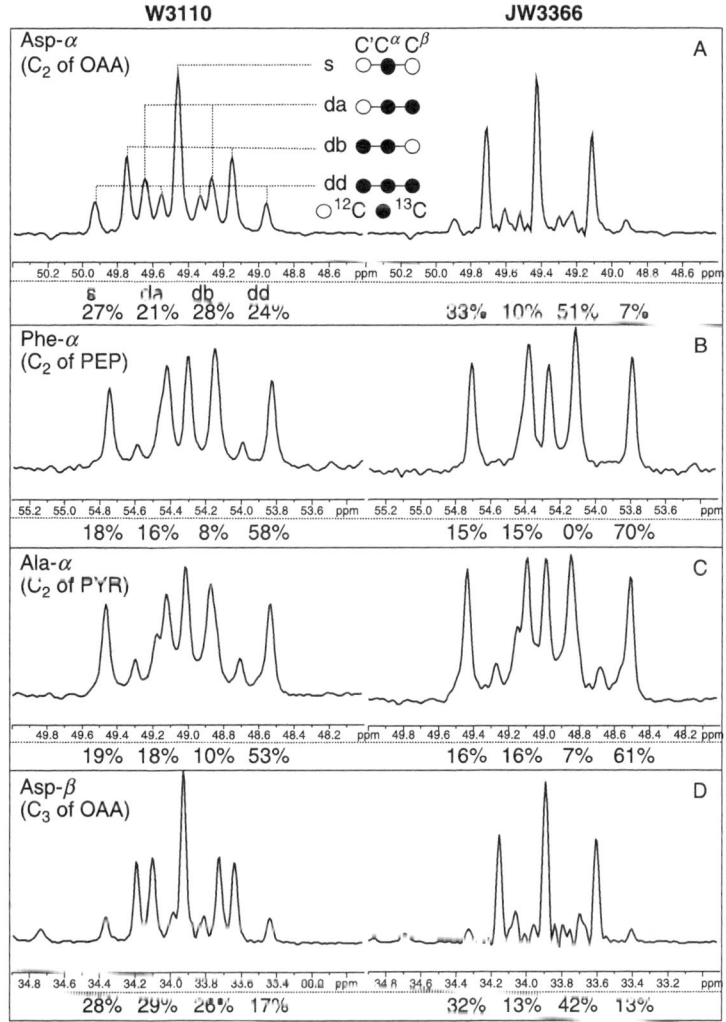

Figure 5.14 $^{13}C-^{13}C$ scalar coupling for W3110 and JW3366 (Δ *pckA*)

of OAA molecules arising from PEP is observed for the *pck* deletion mutant when compared to the wild-type strain (Table 5.12). This can also be qualitatively assessed from direct inspection of the $^{13}C-^{13}C$ scalar coupling fine structure of Asp-α (Figure 5.14). The abundance of the da and dd components in the multiplets of Asp-α relate to the OAA molecules (the direct precursor of Asp) that possess intact C_2–C_3 connectivities. Since the intact C_2–C_3 fragments in OAA can be introduced only by the anaplerotic reaction of PEP carboxylation, the abundance of the da and dd components reflects the *in vivo* activity of the Ppc. The abundances of the da and dd components are significantly lower for the *pck* mutant than those for the wild-type *E. coli* (Figure 5.14), indicating a smaller fraction of OAA synthesized from the anaplerotic reaction of PEP carboxylation in the mutant. Finally, the activity of the glyoxylate shunt in the *pck* mutant is identified by the flux ratio analysis. Visual inspection of the ^{13}C multiplets of Asp-α and Asp-β reveals a high abundance of the db component for the *pck* deletion mutant (Figure 5.14). Using flux ratio analysis, it is found that the intact C_1–C_2 and C_3–C_4 fragments in OAA cannot be derived entirely from the TCA cycle, and excess intact C_1–C_2 and C_3–C_4 connectivities are introduced via the glyoxylate shunt in the *pck* mutant. That is, in this case, the value of the left-hand side of Equation 5.67 is significantly higher than the value of the right-hand side. Hence, the flux ratio analysis shows evidence of the *in vivo* activity of the glyoxylate shunt, which is normally required for growth on carbon sources such as acetate or fatty acids and generally considered to be repressed in *E. coli* grown on glucose. From Equation 5.64, it can be seen that more than one-third of OAA molecules are synthesized via the glyoxylate shunt in the Pck-deficient mutant (Table 5.12). Consequently, the results of flux ratio analysis reveal that abolishment of Pck activity leads to reduced *in vivo* activity of anaplerotic Ppc and the activation of the glyoxylate shunt.

5.6.5 Metabolic flux distribution of pckA mutant E. coli

To obtain higher resolution on the intracellular fluxes, the relative abundances of $^{13}C-^{13}C$ scalar coupling multiplets, the biomass composition, and the extracellular flux data are combined for flux quantification. Based on the results of flux ratio analysis, the reactions catalyzed by Ppc, Pck, and Mez are included in the bioreaction network for metabolic flux analysis. Moreover, since the flux ratio analysis

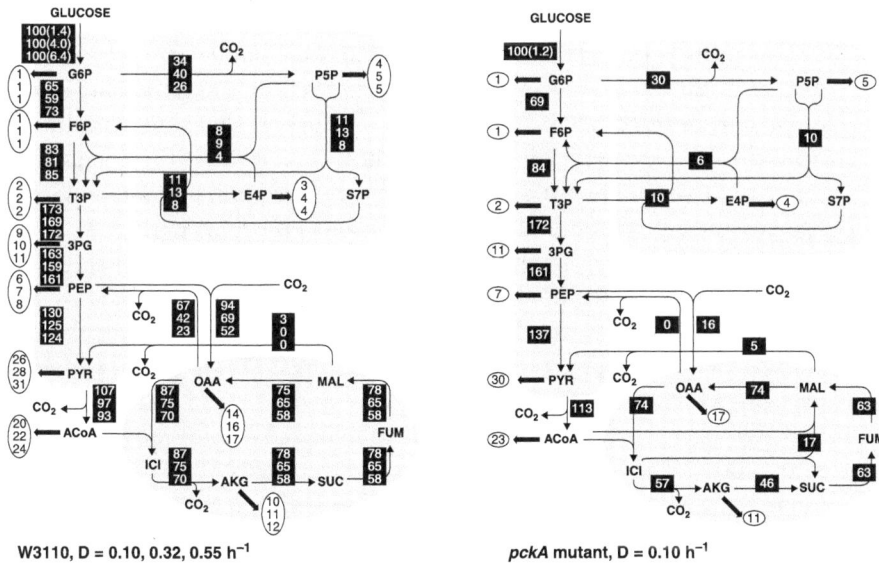

Figure 5.15 Metabolic flux distribution of the wild type and *pckA* mutant

provides direct evidence for the activity of the glyoxylate shunt in the *pck* mutant, this pathway has to be considered for quantification of the intracellular fluxes in the mutant.

Figure 5.15 shows the intracellular flux distributions in *E. coli* wild-type W3110 grown at different dilution rates. It can be seen that in wild-type *E. coli*, there is a significant interchange between PEP and OAA. The anaplerotic reaction catalyzed by Ppc replenishes the TCA cycle with a carbon flux of 94%, while the backflow through the Pck is 67% at the dilution rate of 0.10 h^{-1}. The *in vivo* activity of the gluconeogenic Pck is high, and the backward flux from the TCA cycle to glycolysis carried by Pck is 23% of the total glucose consumed, even at the dilution rate of 0.55 h^{-1}. As can be seen in Figure 5.16, different growth rates have little influence on the overall flux distribution. The only differences are the reduced fluxes through the TCA cycle and the simultaneous decrease in the fluxes through the Ppc and Pck at the higher growth rates. In addition, the malic enzyme flux is 3% at the low dilution rate (0.10 h^{-1}) and almost negligible at higher dilution rates.

The flux ratio analysis indicates the *pck* knockout results in reduced activity of Ppc and the activation of the glyoxylate shunt. As can be seen in Figure 5.15, the carbon flux through Ppc decreases to 16% in the *pck* mutant. Furthermore, when the glyoxylate shunt is activated, 23% of the isocitrate is funneled into the glyoxylate shunt, and 77% is converted further in the TCA cycle. The MAL synthesized by the glyoxylate shunt (34% carbon flux) serves to replenish the carbon skeletons withdrawn from the TCA cycle for biosynthesis. These indicate that the Ppc cannot solely fulfill the anaplerotic function in the *pck* deletion mutant, and both Ppc and the glyoxylate shunt serve to replenish the TCA cycle.

The estimated flux distributions allow the calculation of the balance of NADPH, which is required for biomass formation and produced by the oxidative PP pathway and ICDH reaction. According to this calculation, an NADPH inbalance is observed for all cultures. This is probably due to the interconversion between NADPH and NADH catalyzed by transhydrogenase, which has been shown to be active in *E. coli* (Canonaco et al., 2001). The transhydrogenase flux converting NADPH to NADH is found to be 21%, 7%, and −41% for wild-type *E. coli* at the dilution rates of 0.10, 0.32, and 0.55 h^{-1}, and −39% for the *pck* mutant, respectively.

The flux solutions with the lowest χ^2 values are shown in Figure 5.15. These χ^2 values are 69, 68, and 49 for the wild type at the dilution rates of 0.10, 0.32, and 0.55 h^{-1}, and 92 for the *pck* mutant, respectively. The low χ^2 values imply that the estimated fluxes fit well to the experimentally determined data and thus provide a reliable description of the metabolic processes in *E. coli*. The flux solutions are also subjected to statistical error analysis based on Monte Carlo simulations (Schmidt et al., 1999a; Yang et al., 2002b). For most fluxes, we can obtain 90% confidence intervals that are less than 8% of the estimated flux. One exception is the oxidative PP pathway flux, for which 90% confidence intervals are less than 25%. Let us compare the flux estimates to the flux ratios. For example, from the estimated flux distribution in the *pck* deletion mutant (Figure 5.15), the fraction of OAA arising from PEP and the fraction of OAA synthesized via the glyoxylate shunt can be calculated as 18% and 42%, respectively. The calculated values are in good agreement with the results of flux ratios shown in Table 5.12, despite the differences in the two methods employed.

Note that the abolishment of Pck has no significant impact on the specific activity of Ppc (Table 5.12), while flux analysis reveals that the intracellular flux through Ppc is significantly reduced in the *pck* mutant (Figure 5.15). These observations indicate that the flux reduction of Ppc in the *pck* mutant results from activity-level regulation, i.e., the flux is regulated by varying concentrations of substrates, activators, or inhibitors. The specific activity of Icl cannot be detected in the wild-type *E. coli*, while the *pck* mutant exhibits an Icl activity of 167 nmol mg protein $^{-1}$min^{-1} and a more than 6-fold decrease in the ICDH activity when compared to the wild-type strain (Table 5.13). From the table, no significant changes in the specific activities of the investigated enzymes are found, except the increased activity of anaplerotic Ppc at higher growth rates. In addition, the presence of AcCoA in the assay mixture increases the Ppc activity by 10–20-fold, which indicates that the activity of *E. coli* Ppc is low without any activator and AcCoA is the most powerful activator (Izui et al., 1981). The results in Table 5.14 show that knockout of Pck has no significant impact on the pool sizes of most intracellular metabolites. The only exceptions are a slight reduction in AcCoA pool size, a lower OAA concentration, and increased concentrations of isocitrate, MAL, and ADP in the *pck* mutant when compared to the wild type. However, the increase in the growth rates results in higher intracellular concentrations of ATP and ADP, almost unchanged malate pool size, and decreased concentrations of other metabolites. Isocitrate concentrations are low in *E. coli* W3110 and only an upper bound of its pool size (0.03 mM) is given.

Tables 5.12 and Figure 5.15 show a high *in vivo* activity of gluconeogenic Pck in the wild-type *E. coli* grown on glucose. This indicates the operation of an ATP-dissipating futile cycling via Ppc and ATP-consuming Pck. This futile cycle activity is reduced in faster growing *E. coli* cells. Assuming 1.3 ATP molecules form per atom of oxygen, i.e. P/O ratio to be 1.3 (Varma and Palsson, 1995), it can be calculated from the carbon fluxes presented in Figure 5.15 that the ATP dissipation for the futile cycling via Pck is 8.2%, 5.7%, and 3.5% of total ATP produced during the growth of wild-type

Table 5.13 Specific enzymatic activities in chemostat cultures of *E. coli* W3110 and *pck* mutant JWK3366

Enzyme activity (nmol min^{-1}mg protein^{-1})	*E. coli* strain at growth rate			
	W3110 (0.10 h^{-1})	W3110 (0.32 h^{-1})	W3110 (0.55 h^{-1})	Δpck (0.10 h^{-1})
Pck	*E. coli* strain (growth rate)	36 ± 6	33 ± 6	<1.2
Ppc	3.5 ± 0.6	19 ± 3	23 ± 4	2.9 ± 0.6
Ppc(AcCoA added)[a]	67 ± 12	270 ± 30	350 ± 40	56 ± 10
ICDH	630 ± 90	760 ± 120	720 ± 110	98 ± 14
Icl	0	0	0	170 ± 12

[a]Acetyl-coenzyme A (1 m*M*) was added to the assay mixtures

Table 5.14 Intracellular metabolite concentrations of E. coli W3110 and pck mutant JWK3366 in the continuous cultures

Metabolite conc. (mM)	E. coli strain at growth rate			
	W3110 (0.10 h^{-1})	W3110 (0.32 h^{-1})	W3110 (0.55 h^{-1})	Δpck (0.10 h^{-1})
FDP	0.92 ± 0.11	0.78 ± 0.14	0.46 ± 0.04	1.41 ± 0.21
3PG	1.67 ± 0.21	0.68 ± 0.06	0.42 ± 0.04	1.79 ± 0.16
PEP	0.88 ± 0.14	0.17 ± 0.04	0.06 ± 0.01	1.28 ± 0.18
PYR	1.64 ± 0.32	0.48 ± 0.07	0.28 ± 0.04	1.42 ± 0.28
AcCoA	1.42 ± 0.35	1.02 ± 0.21	0.68 ± 0.14	0.80 ± 0.12
ICT	<0.03	<0.03	<0.03	0.05 ± 0.01
AKG	2.54 ± 0.24	1.02 ± 0.09	0.30 ± 0.03	2.13 ± 0.27
MAL	0.07 ± 0.01	0.06 ± 0.01	0.07 ± 0.01	0.15 ± 0.02
OAA	1.07 ± 0.21	0.77 ± 0.14	0.49 ± 0.07	0.38 ± 0.06
Asp	3.95 ± 0.80	3.45 ± 0.67	2.28 ± 0.41	3.36 ± 0.65
ATP	0.94 ± 0.22	1.01 ± 0.21	1.20 ± 0.25	1.22 ± 0.31
ADP	0.32 ± 0.09	0.51 ± 0.14	0.63 ± 0.17	1.21 ± 0.35

E. coli at the dilution rates of 0.10, 0.32, and 0.55 h^{-1}, respectively. Deletion of *pck* leads to a significant decrease in CO_2 production and O_2 uptake (Table 5.11), which is consistent with the reduced energy demands by abolishment of futile cycling via Pck. A remarkable fraction of the available ATP is consumed for the futile cycling via Pck. To understand the significance of Pck activity, consider the *in vivo* regulation of the flux through Pck by integrating the measured results of *in vivo* fluxes and intracellular metabolite concentrations. The kinetics of the Pck in E. coli is assumed to follow rapid equilibrium mechanisms (Krebs and Bridger, 1980). Thus, we may express the rate equation for this enzyme by:

$$v_{Pck} = \frac{v_{max}[OAA][ATP]}{\left(\begin{array}{l} K_{i,ATP}K_{OAA}[ATP] + K_{ATP}[OAA] + [OAA][ATP] + \dfrac{K_{i,ATP}K_{OAA}[PEP]}{K_{i,PEP}} + \\ \dfrac{K_{i,ATP}K_{OAA}[ADP]}{K_{i,ADP}} + \dfrac{K_{i,ATP}K_{OAA}[PEP][ADP]}{K_{PEP}K_{i,ADP}} + \dfrac{K_{i,ATP}K_{OAA}[ATP][PEP]}{K_{i,PEP}K_{i,ATP}} \\ + \dfrac{K_{i,ATP}K_{OAA}[OAA][ADP]}{K_{i,ADP}K_{i,OAA}} \end{array}\right)}$$

(5.69)

where $K_{ATP} = 0.06\,mM, K_{OAA} = 0.67\,mM, K_{PEP} = 0.07\,mM, K_{i,ATP} = 0.04\,mM,$
$K_{i,PEP} = 0.06\,mM, K_{i,ADP} = 0.04\,mM, K_{i,ATP} = 0.04\,mM, K_{i,OAA} = 0.45\,mM$

Using this equation, the maximal rate, max of Pck is calculated from the measured metabolite concentrations (Table 5.14) and determines *in vivo* flux value. As shown in Figure 5.15, the carbon flux through Pck is 0.93, 1.70, and 1.46 mmol g^{-1} h^{-1} during the growth of the wild type at the dilution rates of 0.10, 0.32, and 0.55 h^{-1}, respectively. Thus, the maximum values calculated from Equation 5.69 are 13.6, 11.6, and 9.5 mmol g^{-1} h^{-1}, respectively. These maximum rates of Pck show no significant changes between different dilution rates, which is in agreement with the measured data of specific enzyme activities (Table 5.13). Furthermore, it may be seen that the *in vivo* Pck fluxes relative to their maximal rate values (/max) are less than 15%. These results indicate that the *in vivo* regulation of the flux through Pck according to the growth requirements is exerted at the activity level via the changes in metabolite concentrations.

Since the intracellular ATP and ADP concentrations in *E. coli* are much higher than their respective K and K_i values (Table 5.14), Equation 5.69 can be simplified as:

$$v_{Pck} = \frac{v_{max}[OAA]\frac{[ATP]}{[ADP]}}{\left(K_{i,ATP}K_{OAA}\frac{[ATP]}{[ADP]} + [OAA]\frac{[ATP]}{[ADP]} + \frac{K_{i,ATP}K_{OAA}}{K_{i,ADP}} + \frac{K_{i,ATP}K_{OAA}[PEP]}{K_{PEP}K_{i,ADP}} + \frac{K_{i,ATP}K_{OAA}}{K_{i,PEP}K_{i,ATP}}\frac{[ATP]}{[ADP]}[PEP] + \frac{K_{i,ATP}K_{OAA}[OAA]}{K_{i,ADP}K_{i,OAA}}\right)} \quad (5.70)$$

This equation can be used to assess the dependence of the Pck flux on the ATP/ADP concentration ratio and the sensitivities toward the PEP and OAA concentrations. It can be seen from Figure 5.16a that the flux is insensitive to alterations of the [ATP]/[ADP] ratio, unless the latter falls below 1.0, while the measured [ATP]/[ADP] ratio is greater than 1.9 in wild-type *E. coli* (Table 5.14). However, Figure 5.16b shows that the flux sensitivities toward the PEP and OAA pool sizes are high when the concentrations of PEP and OAA are less than 1 mM and 2 mM, respectively. Interestingly, the measured intracellular PEP and OAA pool sizes in wild-type *E. coli* lie in the most sensitive regulatory domain for Pck flux (Table 5.14). Especially for the changes in PEP concentration within the range of measurement data, the flux varies sharply. From the above analysis, it is clear that the *in vivo* regulation of the Pck flux in *E. coli* occurs mainly by modulation of enzyme activity and by the

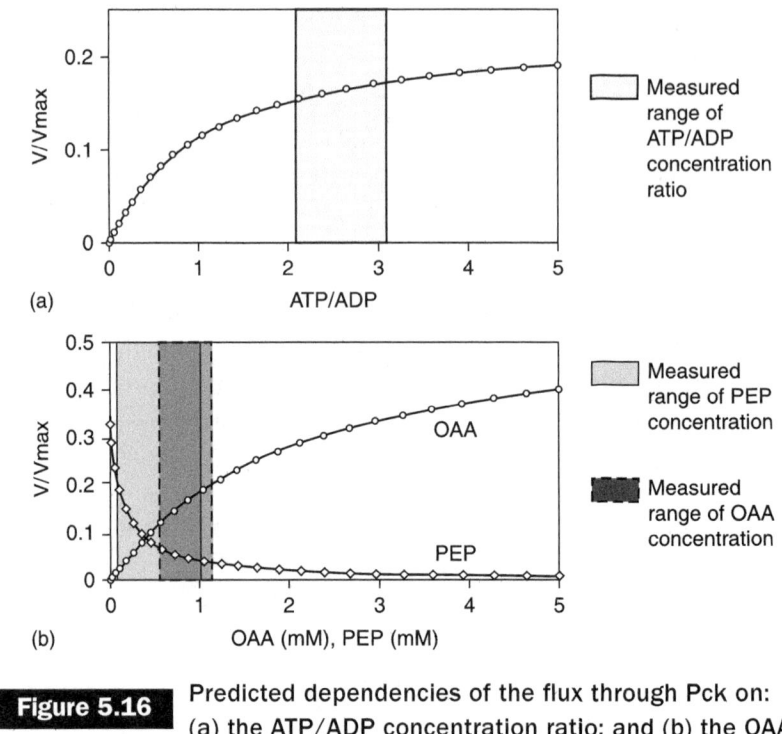

Figure 5.16 Predicted dependencies of the flux through Pck on: (a) the ATP/ADP concentration ratio; and (b) the OAA (○) and PEP (◊) concentrations

changes in PEP and OAA concentrations, rather than by the ATP/ADP concentration ratio. This regulatory mechanism implies that the reaction catalyzed by gluconeogenic Pck can respond flexibly to the availability of PEP and OAA. It is known that PEP is an important intermediate in *E. coli* metabolism, because it alone directly regulates not only the PTS system but also the enzymes Pfk and Pyk. Due to the activity of Pck, PEP can be generated gluconeogenically from the TCA cycle. It is thus suggested that the reaction catalyzed by Pck in *E. coli* serves to maintain the relative balance between the OAA and PEP pools and drain off excess carbon of the TCA cycle to supply PEP for cellular requirements. Moreover, the flux analysis of the *pck* mutant indicates that due to abolishment of Pck activity, the Ppc cannot solely fulfill the anaplerotic function and the glyoxylate shunt also participates in replenishing the TCA cycle (Figure 5.15). Consequently, an important inference is that the gluconeogenic Pck is not an enzyme that merely consumes ATP via the futile cycle without any biological purposes, but plays a key role in anaplerosis and metabolic regulations during growth on glucose.

Gluconeogenic PEP carboxylation is generally considered the sole anaplerotic reaction during growth of E. coli on glucose. However, the *in vivo* flux through Ppc decreases 7-fold in the *pck* mutant when compared to wild type (Figure 5.15). In contrast to the *in vivo* flux, no significant difference in the specific activity of Ppc is found for the mutant strain (Table 5.13), indicating that *in vivo* regulation of anaplerotic Ppc flux is exerted at the activity level via changes in metabolite concentrations. The Ppc activity in E. coli is low without any activator and AcCoA turns out to be a potent activator (Table 5.13). The pck mutant exhibits a slightly reduced AcCoA pool size (Table 5.14), which is consistent with the decrease in Ppc flux. The sensitivity index for the Ppc flux with respect to intracellular AcCoA concentration, defined as (Δppc/Δ[ACoA]) ([ACoA]/ppc), is 1.13 within the range of the measured concentrations of AcCoA, which is calculated based on the data of Izui et al. (1981). This indicates that the flux reduction of Ppc in the *pck* mutant may not be caused solely by a change in AcCoA concentration. Severe inhibition of Ppc activity can be caused by L-Asp and L-MAL. Particularly, L-Asp is a usual inhibitor of the Ppc from various sources (Kai et al., 1999).

However, according to the study of Izui et al. (1981), the Ppc flux is sensitive to the changes in aspartate and malate pool sizes, only if the concentrations of aspartate and malate fall below 1 mM. Since the intracellular aspartate concentration is found to be larger than 2 mM in all cases, and almost unchanged in both the wild type and the *pck* mutant (Table 5.14), aspartate can probably be ruled out as an essential inhibitor of Ppc under *in vivo* conditions. However, the measured malate concentrations are less than 0.15 mM in all cases (Table 5.14). Within this concentration range, the Ppc flux is sensitive to alterations of the malate concentration. Furthermore, the intracellular malate concentration increase in the pck deletion mutant, when compared to the wild-type strain (Table 5.14), is consistent with the decrease in the Ppc flux.

The results thus suggest that *in vivo* PEP carboxylation in *E. coli* may be regulated by the intracellular acetyl-CoA and malate rather than by aspartate. The glyoxylate bypass is activated due to abolishment of Pck activity in the *pck* deletion mutant (Table 5.12 and Figure 5.15). The regulation of the glyoxylate shunt in E. coli has been demonstrated to be the result of reversible phosphorylation of the isocitrate dehydrogenase (ICDH) (Walsh and Koshland, 1985). During growth on acetate, a large fraction of ICDH is inactivated to its phosphorylated form. Consistently, a drastic reduction in the ICDH activity of the mutant is found by the enzyme activity analyses (Table 5.13). The reversible phosphorylation/inactivation of ICDH is catalyzed by a bifunctional enzyme, ICDH-kinase/phosphatase. As has been pointed out by Holms (1986), the ICDH-kinase/phosphatase is regulated by a number of effectors, including OAA, which inhibits the ICDH-kinase and stimulates the phosphatase.

The current results show that the intracellular OAA concentration decreases upon *pck* deletion (Table 5.14), thereby suggesting that a shortage of intracellular OAA causes the phosphorylation/inactivation of ICDH and flux through the glyoxylate shunt to replenish OAA in the *pck* mutant. Another intermediate, 3PG, which acts as an important regulatory effector responsible for the phosphorylation of ICDH during growth on acetate (Cronan and LaPorte, 1996), is unlikely to play the same role in the *pck* deletion mutant, because the pool size of 3PG is approximately identical in both the wild-type strain and the mutant (Table 5.14). In *E. coli*, the genes encoding for Icl (*aceA*), MS (*aceB*), and ICDH-kinase/phosphatase (*aceK*) form the acetate operon (*aceBAK*). The expression of this operon is induced during growth on acetate or fatty acids, but repressed in the presence of glucose. The present operon is negatively controlled by IclR, and release of this repressor is responsible for induction of operon expression during growth on acetate (Cronan and LaPorte, 1996).

The intracellular metabolite pool sizes increase in intracellular isocitrate concentration in the *pck* mutant when compared to the wild-type strain (Table 5.14). This may be explained by the phosphorylation/inactivation of ICDH upon *pck* deletion, which has a very high affinity for isocitrate (i.e. $K_m = 8 \mu M$). With the increase in isocitrate concentration, the flux through Icl increases dramatically, because the intracellular isocitrate concentration is below the K_m of Icl (604 μM). The ICDH is less sensitive to the availability of isocitrate, as ICDH operates largely in the zero-order region (i.e. [ICT] > Km). Due to this branch point effect, the flux through the glyoxylate bypass is remarkably sensitive to the phosphorylation state of ICDH, and the inactivation of ICDH by phosphorylation and the simultaneous induction of Icl activity may thus be sufficient to direct 23% of isocitrate to the glyoxylate shunt in the *pck* mutant. Metabolic flux responses to Pck knockout in *C. glutamicum* have also been determined (Petersen et al., 2001). The reduced *in vivo* flux through anaplerotic Ppc is also found in the Pck-deficient *C. glutamicum*. However, deletion of *pck* in *C. glutamicum* does not lead to the activation of the glyoxylate shunt. This flux difference between *E. coli* and *C. glutamicum* is probably due to the different anaplerotic metabolism. *C. glutamicum* possesses two anaplerotic enzymes carboxylating C_3 metabolites, Ppc and pyruvate carboxylase (Pyc). During the growth of *C. glutamicum* on glucose, pyruvate carboxylase Pyc contributes a major fraction to C_3 carboxylation (Petersen et al., 2000). Hence, although abolishment of Pck activity in *C. glutamicum* causes a remarkably reduced Ppc flux, the remaining anaplerotic function can be fulfilled by Pyc.

However, in *E. coli*, there is no evidence for the presence of Pyc, and the glyoxylate bypass is activated to participate in anaplerosis upon *pck* deletion. Significant changes in the intracellular anaplerotic fluxes are observed in response to the Pck knockout in *E. coli* (Figure 5.15). However, the pool sizes of the intermediates (i.e. 3PG, PEP, PYR, AcCoA, αKG, Asp), which are precursors for biosynthesis, differ only slightly in the *pck* mutant when compared to the wild-type strain (Table 5.14). It may thus be hypothesized that *E. coli* changes the anaplerotic fluxes for homeostasis of these intermediate concentrations, since the availability of the precursor metabolites has an important impact on the cell growth. In addition, due to the capability of the *pck* mutant to operate the TCA cycle and the glyoxylate shunt simultaneously, it is likely that this mutant can grow on high concentrations of glucose at high yields without producing acetate. A similar metabolic flux pattern has been found in a low acetate producer, *E. coli* BL21, by Noronha et al. (2000), who conclude that the low acetate production is attributed to the activation of the glyoxylate shunt. Acetate accumulation is a common problem of aerobic high cell density *E. coli* cultures for recombinant protein production. Furthermore, the *pck* deletion mutant is found to exhibit a significantly increased biomass yield on glucose when compared to the wild type (Table 5.10). The effect of the *ppc* gene knockout on the metabolism will be explained in Chapter 6.

5.6.6 Metabolic flux analysis of pgi and zwf gene knockout E. coli

For quantification of carbon fluxes in the central metabolism of *E. coli*, a bioreaction network is shown in Figure 5.17. The reactions catalyzed by Ppc (v_{19}), Pck (v_{20}), and Mez (v_{21}) are also included. The following enzyme reactions are considered to be reversible: Pgi (v_2), Tkt (v_{10} and v_{12}), Tal (v_{11}), the sequence of glycolytic reactions leading from T3P (DHAP + GAP) to PEP (v_4 and v_5), SDH (v_{16}), Fum (v_{17}), and MDH (v_{18}).

The reversible transfer of reducing equivalents between NAD(H) and NADP(H) is included in the bioreaction network. The malic enzyme in *E. coli* may be NAD^+- or $NADP^+$-dependent. While the analysis cannot distinguish between fluxes through the two isoenzymes, the only influence of the enzyme is its influence on the flux of reducing equivalents via transhydrogenase. The flux distribution is calculated by assuming that the NAD$^+$- and NADP$^+$-dependent malic enzymes are equally active, and the influence of this assumption on the transhydrogenase flux is investigated.

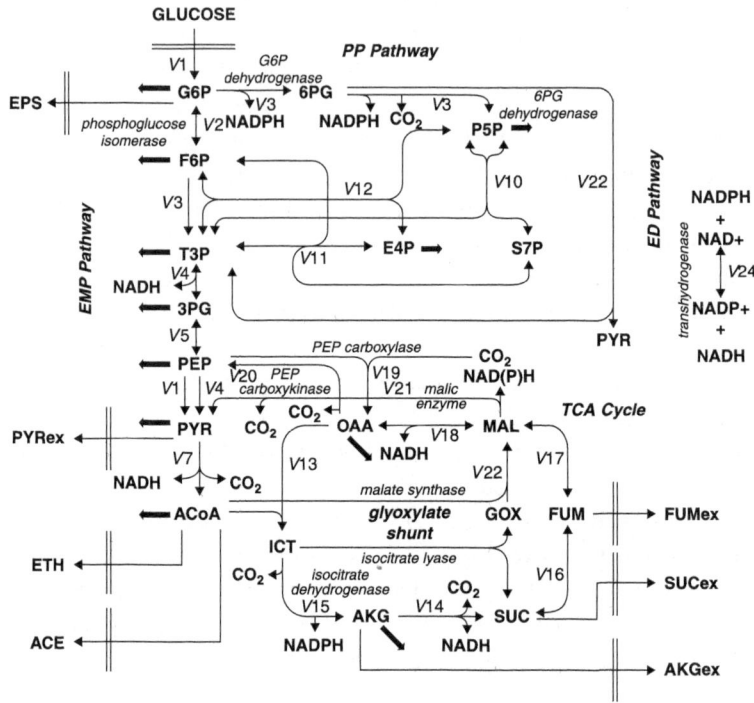

Figure 5.17 Bioreaction network of *E. coli* central carbon metabolism. The arrows indicate the physiological directions of reactions. Fluxes to biomass building blocks are indicated by bold arrows

To investigate quantitatively the effects of *pgi* or *zwf* gene deletion on *E. coli* physiology, *E. coli* wild-type strain W3110, strain BW25113, the *pgi* mutant (JW3985), and the *zwf* mutant (JW1841) are cultivated in the aerobic chemostats at a dilution rate of 0.1 h^{-1} under glucose- or ammonia-limited conditions (Hua et al., 2003). The growth parameters are summarized in Table 5.15. The parent strain BW25113 exhibits almost the same growth parameters as W3110. Under glucose-limited conditions,

Table 5.15 Growth parameters of glucose (C)- and ammonia (N)-limited chemostat cultures of *E. coli* wild-type strain W3110, the *pgi* mutant, and the *zwf* mutant[a]

Chemostat conditions	Strain	$Y_{X/S}$ (g g^{-1})	Glucose concn (mM)	Ammonium concn (mM)	$q_{glucose}$ (mmol g^{-1} h^{-1})	q_{O2} (mmol g^{-1} h^{-1})	q_{CO2} (mmol g^{-1} h^{-1})	$q_{acetate}$ (mmol g^{-1} h^{-1})	q_{EPS} (mmol g^{-1} h^{-1})	C balance (%)
C limited	W3110	0.40 ± 0.03	<0.1	38.0 ± 3.4	1.4 ± 0.1	4.0 ± 0.7	4.2 ± 0.4	0	0	99 ± 7
	pgi mutant	0.43 ± 0.03	<0.1	37.1 ± 3.3	1.3 ± 0.1	3.9 ± 0.7	3.7 ± 0.4	0	0	97 ± 7
	zwf mutant	0.42 ± 0.03	<0.1	37.4 ± 3.3	1.4 ± 0.1	3.7 ± 0.6	4.2 ± 0.4	0	0	103 ± 8
N limited	W3110	0.23 ± 0.02	8.3 ± 0.6	<0.1	2.9 ± 0.3	7.1 ± 1.4	7.2 ± 0.9	1.60 ± 0.14	0.04 ± 0	93 ± 9
	pgi mutant	0.39 ± 0.03	15.4 ± 1.3	<0.1	1.6 ± 0.2	3.8 ± 0.7	4.0 ± 0.4	0.31 ± 0	0	97 ± 7
	zwf mutant	0.20 ± 0.02	7.1 ± 0.5	<0.1	2.8 ± 0.3	3.0 ± 0.6	3.1 ± 0.4	2.71 ± 0.23	0.16 ± 0	95 ± 7

[a]Chemostat cultures were operated at a dilution rate of 0.1 h^{-1}
$Y_{X/S}$, biomass yield on glucose; q, specific substrate uptake rate or product formation rate
EPS, extracellular polysaccharide
The levels of formate and D-lactate were below the detection levels in all experiments.

all the strains convert glucose completely to biomass and CO_2 without any by-product formation, and the biomass yields on glucose are similar.

When ammonia- and glucose-limited cultures are compared, a significant increase in the specific glucose uptake rate and a drastically reduced biomass yield are observed for both wild-type *E. coli* and the *zwf* mutant, while only a slight reduction in biomass yield may be observed for the *pgi* mutant. The specific O_2 uptake rate and the CO_2 evolution rate of the wild type are higher under ammonia-limited conditions than under glucose-limited conditions. This indicates that increased respiration is the response to ammonia limitation. Formation of by-products by so-called overflow metabolisms is observed for the ammonia-limited cultures of all the strains. The primary by-product is acetate, but various amounts of pyruvate, fumarate, ethanol, α-ketoglutarate, succinate, and extracellular polysaccharide may also be found (Hua et al., 2003). The specific rates of production of acetate are the lowest in the *pgi* mutant and highest in the *zwf* mutant.

The relative fractions of the major biomass components of *E. coli*, such as protein, RNA, and glycogen, are shown in Table 5.16. As expected, the reserve carbohydrate glycogen content is markedly increased under ammonia-limited conditions. The remaining fraction of biomass is assigned to minor macromolecules, such as DNA, lipids, or peptidoglycan.

The cells from [U-^{13}C]glucose labeling experiments were harvested and subjected to hydrolysis, and the relative intensities of ^{13}C-^{13}C scalar coupling multiplet components of amino acids, (glycerol) and glucose in hydrolysates were analyzed by 2D [^{13}C, ^{1}H]-COSY (Figure 5.18). The data can be first interpreted by using flux ratio analysis, which yields

Table 5.16 Protein, RNA, and glycogen contents of glucose (C)- and ammonia (N)-limited chemostat cultures of *E. coli* wild-type strain W3110, the *pgi* mutant, and the *zwf* mutant

Chemostat conditions	Strain	% of total		
		Protein	RNA	Glycogen
C limited	W3110	70 ± 7	7 ± 1	1.4 ± 0.1
	pgi mutant	69 ± 8	8 ± 1	2.1 ± 0.2
	zwf mutant	67 ± 7	8 ± 1	1.8 ± 0.2
N limited	W3110	58 ± 6	9 ± 1	11 ± 1
	pgi mutant	56 ± 7	7 ± 1	15 ± 1
	zwf mutant	63 ± 8	10 ± 1	4.8 ± 0.5

¹³C-metabolic flux analysis

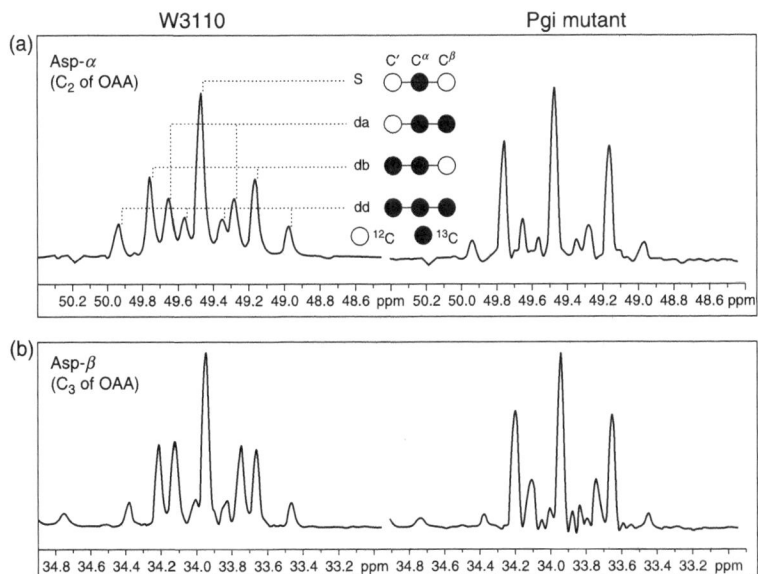

Figure 5.18 ¹³C–¹³C scalar coupling multiplets observed for aspartate from glucose limited chemostat cultures of *E. coli* W3110 (left) and the *pgi* mutant (right). (a) Asp-α; (b) Asp-β. As indicated in panel A, the multiplets consist of a singlet (s), a doublet with a small coupling constant (Da), a doublet split by a larger coupling constant (db), and a doublet of doublets (dd). Aspartate corresponds directly to its metabolic precursor, OAA

information on the origins of key metabolites in the central metabolism (Table 5.17). The results of flux ratio analysis allow us to identify the network of active reactions and to determine the ratios of some carbon fluxes.

The flux ratio analysis of W3110 shows the activities of two enzymes that are generally considered to be inactive in *E. coli* grown on glucose, Pck and Mez. These enzymes catalyze the gluconeogenic conversion of OAA to PEP and the conversion of MAL to PYR, respectively. The anaplerotic Ppc replenishes the TCA cycle in wild-type *E. coli*, and more than one-half of the OAA molecules are found to originate from PEP (Table 5.17).

The glyoxylate shunt consisting of Icl and MS is inactive in W3110. This result can be obtained based on the analysis of the intact carbon fragments for Asp-α, Asp-β, Phe-α, Glu-β, and Glu-γ (*f* values are shown

Table 5.17 Origins of metabolic intermediates in glucose (C)- and ammonia (N)-limited chemostat cultures of *E. coli* wild-type strain W3110, the *pgi* mutant, and the *zwf* mutant, as determined by flux ratio analysis

Chemostat conditions	Strain	% of total pool										
		G6P from F6P	P5P from G6P (lower boundary)	P5P via nonoxidative PP pathway	PEP from OAA	Pyruvate via ED pathway	Pyruvate from malate (lower boundary)	Pyruvate from malate (upper boundary)	Acetyl coenzyme A from pyruvate	AKG from OAA plus acetyl coenzyme A	OAA from PEP	OAA exchanged to fumarate	OAA via glyoxylate shunt
C limited	W3110	>58	9 ± 2	ND	28 ± 3	ND	2 ± 2	8 ± 4	>98	>97	55 ± 2	76 ± 10	—[b]
	pgi mutant	0 ± 2	47 ± 9	ND	24 ± 3	6 ± 4	0 ± 2	0 ± 2	>98	>99	29 ± 2	86 ± 14	57 ± 16
	zwf mutant	>35	0 ± 2	>94	23 ± 3	ND	4 ± 2	15 ± 4	>98	>98	43 ± 2	81 ± 12	—[b]
N limited	W3110	>56	4 ± 2	ND	10 ± 3	ND	4 ± 2	14 ± 4	>98	>97	53 ± 2	81 ± 10	—[b]
	pgi mutant	0 ± 2	37 ± 8	ND	8 ± 2	11 ± 7	0 ± 2	0 ± 2	>99	>98	18 ± 2	50 ± 15	20 ± 12
	zwf mutant	>38	0 ± 2	>95	0 ± 2	ND	4 ± 2	27 ± 8	>98	>98	85 ± 2	60 ± 6	—[b]

ND, not determined.

in Table 5.18), which are derived from C_2 and C_3 of OAA, C_2 of PEP, and C_3 and C_4 of α-KG, respectively. The f values for these carbon positions satisfy the relationship as mentioned before within experimental error, demonstrating that OAA is synthesized exclusively from the TCA cycle and the anaplerotic carboxylation of PEP. The abundances of intact carbon fragments originating from a single glucose source molecule (f values) are calculated from the observed relative ^{13}C multiplet intensities (I values). Equation 5.71 can be used for a terminal carbon atom, and Equation 5.72 can be used for the central carbon in a C_3 fragment:

$$\begin{bmatrix} f^{(1)} \\ f^{(2)} \end{bmatrix} = K_{term}^{-1} \cdot \begin{bmatrix} I_s \\ I_d \end{bmatrix} = \begin{bmatrix} 0.904 & 0.104 \\ 0.096 & 0.896 \end{bmatrix} \cdot \begin{bmatrix} I_s \\ I_d \end{bmatrix} \quad (5.71)$$

$$\begin{bmatrix} f^{(1)} \\ f^{(2a)} \\ f^{(2a)} \\ f^{(3)} \end{bmatrix} = K_{central}^{-1} \cdot \begin{bmatrix} I_s \\ I_{da} \\ I_{db} \\ I_{dd} \end{bmatrix} = \begin{bmatrix} 0.819 & 0.094 & 0.094 & 0.102 \\ 0.086 & 0.811 & 0.010 & 0.001 \\ 0.086 & 0.010 & 0.811 & 0.001 \\ 0.009 & 0.085 & 0.085 & 0.896 \end{bmatrix} \cdot \begin{bmatrix} I_s \\ I_{da} \\ I_{db} \\ I_{dd} \end{bmatrix} \quad (5.72)$$

Table 5.18 Relative abundances of intact carbon fragments at the carbon positions used for identification of the glyoxylate shunt activity in E. coli wild-type strain W3110 and the pgi mutant

	Relative abundance							
	W3110[b]				pgi mutant			
Carbon atom	$f^{(1)}$	$f^{(2a)}$	$f^{(2b)}$	$f^{(3)}$	$f^{(1)}$	$f^{(2a)}$	$f^{(2b)}$	$f^{(3)}$
Asp-α	0.24	0.23	0.32	0.21	0.28	0.14	0.50	0.08
Asp-β	0.25	0.33	0.29	0.13	0.28	0.20	0.46	0.06
Phe-α	0.11	0.18	0.09	0.62	0.11	0.31	0.12	0.46
Glu-βc	0.54	0.46		0.00	0.75	0.25		0.00
Glu-γ	0.22	0.00	0.78	0.00	0.20	0.00	0.80	0.00
Leu-α	0.22	0.00	0.78	0.00	0.19	0.00	0.81	0.00

[c]Glu-β has scalar coupling constants identical to those of the adjacent carbons.

When ammonia- and glucose-limited cultures are compared, ammonia limitation is found to induce the following metabolic responses in wild-type E. coli: (i) a decrease in the amount of PEP molecules originating from OAA, indicating decreased activity of the Pck; and (ii) a decrease in the amount of P5P molecules derived from G6P, which suggests low activity of the oxidative PP pathway, assuming that the exchanges via transketolase and transaldolase did not differ significantly (Table 5.17). The only exception is the fraction of OAA molecules originating from PEP, which is 47% in the glucose-limited strain BW25113. This difference is small compared to the changes arising from knockout mutations.

5.6.7 Identification of network structure in the pgi mutant

The flux ratio analysis of the Pgi mutant reveals the absence of G6P molecules derived from F6P through the Pgi reaction (Table 5.17). This information can be obtained from direct interpretation of the $^{13}C-^{13}C$ scalar coupling fine structure of glucose C_4 in the $[^{13}C, ^1H]$-COSY spectra. As Figure 5.19 shows, the scalar coupling multiplets of glucose C_4, which can be used to assess the labeling pattern of G6P, shows that the doublet is not present in the *pgi* mutant, demonstrating the lack of C_3-C_4 and C_4-C_5 carbon bond cleavage in G6P. However, in wild-type *E. coli*, a significant fraction of the C_3-C_4 carbon bonds in F6P are cleaved due to the action of the PP pathway, and this carbon bond cleavage is introduced into the G6P pool by the Pgi reaction. Hence, a significant contribution of the doublet

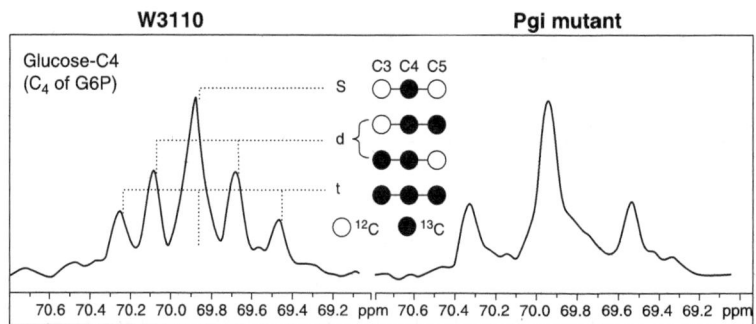

Figure 5.19 $^{13}C-^{13}C$ scalar coupling multiplets observed for C-4 of glucose from ammonia-limited chemostat cultures of *E. coli* W3110 (left) and the *pgi* mutant (right). As glucose C-4 exhibits scalar coupling constants identical to those of the adjacent carbons, the multiplets consist of a singlet (s), a doublet (d), and a triplet (t). The labeling data for glucose represent the labeling patterns of G6P

to the multiplets of glucose C_4 is observed for W3110 (Figure 5.19), reflecting the cleavage of C_3-C_4 bonds in G6P introduced by the active Pgi reaction. This result indicates that in the *pgi* mutant, the Pgi is inactive *in vivo*, and the labeles substrate, glucose, is the sole source of G6P.

A decrease in the fraction of OAA molecules originating from PEP can be observed for the *pgi* mutant compared to the fraction for the wild-type strain (Table 5.17). This can be qualitatively assessed from direct inspection of the ^{13}C-^{13}C scalar coupling fine structure of Asp-α (Figure 5.18). The abundance of the doublet with a small coupling constant and the doublet of doublets in the multiplets of Asp-α is related to the OAA molecules (the direct precursors of Asp) that possesses intact C_2-C_3 connections. Since the intact C_2-C_3 fragments in OAA can be introduced only by the anaplerotic reaction of PEP carboxylation, the abundance of the doublet with a small coupling constant and the doublet of doublets are reflected the *in vivo* activity of Ppc. The abundances of the doublet with a small coupling constant and the doublet of doublets are significantly lower for the *pgi* mutant than for the wild-type strain (Figure 5.18), showing that a smaller fraction of OAA molecules is synthesized from the anaplerotic reaction of PEP carboxylation in this mutant.

The glyoxylate shunt is found to be active in the *pgi* mutant by flux ratio analysis (Table 5.17). Visual inspection of the scalar coupling fine structures of Asp-α and Asp-β reveals a high abundance of the doublet split by a larger coupling constant for the *pgi* mutant (Figure 5.18). The analysis of the intact carbon fragments from Asp-α, Asp-β, Phe-α, Glu-β, and Glu-γ (*f* values are shown in Table 5.18) shows that the $f^{(2b)}$(Asp-α) and $f^{(2b)}$(Asp-β) values are significantly higher than the values calculated for $f^{(2b)}$(Phe-α), $f^{(2)}$(Glu-β), and $f^{(2b)}$(Glu-γ). This indicates that the intact C_1–C_2 and C_3–C_4 fragments in the OAA pool cannot be derived entirely from the TCA cycle and PEP carboxylation and that excess intact C_1–C_2 and C_3–C_4 connections are introduced via the glyoxylate shunt. Therefore, the flux ratio analysis provides evidence of the *in vivo* activity of the glyoxylate shunt, which is normally required for growth on carbon sources, such as acetate or fatty acids, and is generally considered to be repressed in *E. coli* grown on glucose. Consistently, *in vitro* enzyme activity analysis also confirms that no Icl can be detected in the wild type, while the *pgi* mutant exhibits an Icl activity of 197 nmol mg of protein^{-1} min^{-1}. Based on the *f* values of Asp-α, Asp-β, Phe-α, Glu-β, Glu-γ, and Leu-α (Table 5.18), more than one-half of the OAA molecules are found to be formed via the glyoxylate shunt in the glucose-limited culture of the *pgi* mutant (Table 5.17).

The flux ratios in the *pgi* mutant also show that the ED pathway makes a minor contribution to glucose catabolism (Table 5.17). Three carbon positions carry the information necessary to identify the activity of the ED pathway (Table 5.19): Phe-α is derived from C_2 of PEP, and Ala-α and Val-α are derived from C_2 of pyruvate. The f values of Ala-α and Val-α fulfill the following relationship: $f^{(2b)}$(Val-α) = $f^{(2b)}$(Ala-α) + $f^{(3)}$(Ala-α). As described above, the G6P molecules in the *pgi* mutant originate solely from the labeled substrate, glucose, so that $f^{(3)}$(G6P C_2) = 1. The flux through the malic enzyme is negligible in the *pgi* mutant (Table 5.17). Thus, the ED pathway accounts for about 6% and 11% of the pyruvate molecules synthesized in the glucose- and ammonia-limited *pgi* mutant, respectively (Table 5.17).

5.6.8 Identification of network structure in the zwf mutant

The flux ratio analysis of the *zwf* mutant shows that two biosynthetic precursors, P5P and E4P, are synthesized from GAP and F6P via the non-oxidative PP pathway (Table 5.17). The labeling patterns of P5P and E4P are assessed by using the labeling data for histidine and the aromatic ring of tyrosine, respectively. Generally, both the oxidative and non-oxidative PP pathways may contribute to the synthesis of these two precursor metabolites. Analysis of the intact carbon fragments for His-α, His-β, Tyr-δ^x, glycerol C_2, and Tyr-β (f values are shown in Table 5.20) indicates that the non-oxidative branch of the PP pathway is used only for synthesis of the precursors and that the oxidative PP pathway is inactive.

When the *zwf* mutant is compared to W3110 in glucose-limited cultures, similar flux ratio results are obtained (Table 5.17). The glyoxylate shunt is inactive in the *zwf* mutant. These results suggest that inactivation of G6PDH

Table 5.19 Relative abundances of intact carbon fragments at the carbon positions used for identification of the ED pathway activity in glucose (C)- and ammonia (N)-limited cultures of the *pgi* mutant

Carbon atom	Relative abundance							
	C-limited culture				N-limited culture			
	$f^{(1)}$	$f^{(2a)}$	$f^{(2b)}$	$f^{(3)}$	$f^{(1)}$	$f^{(2a)}$	$f^{(2b)}$	$f^{(3)}$
Phe-α	0.11	0.31	0.12	0.46	0.08	0.28	0.04	0.60
Ala-α	0.10	0.29	0.12	0.49	0.08	0.24	0.04	0.64
Val-α	0.39	0.00	0.61	0.00	0.30	0.00	0.70	0.00

^{13}C-metabolic flux analysis

Table 5.20 Relative abundances of intact carbon fragments at the carbon positions used for identification of the origin of P5P and E4P pools in glucose (C)- and ammonia (N)-limited cultures of the *zwf* mutant

Carbon atom	Relative abundance							
	C-limited culture				N-limited culture			
	$f^{(1)}$	$f^{(2a)}$	$f^{(2b)}$	$f^{(3)}$	$f^{(1)}$	$f^{(2a)}$	$f^{(2b)}$	$f^{(3)}$
His-α	0.10	−0.01	0.20	0.71	0.09	−0.01	0.14	0.78
His-β	0.26	0.69	0.00	0.05	0.16	0.78	0.00	0.06
Tyr-$\delta x^{b,c}$	0.14	0.86		0.00	0.06	0.94		0.00
Glycerol C-2^c	0.12	0.19		0.69	0.08	0.13		0.79
Tyr-β	0.22	0.78	0.00	0.00	0.08	0.92	0.00	0.00

^{b}The two carbons of Tyr, δ^1 and δ^2, give rise to only one ^{13}C fine structure
^{c}Tyr-δ^x and glycerol C-2 have scalar coupling constants identical to those of the adjacent carbons

does not have significant influence on the central carbon metabolism. However, under ammonia-limited conditions, the *zwf* mutant exhibits a flux ratio pattern different from that of the wild-type strain (Table 5.17):

(i) A significant increase in the fraction of OAA molecules originating from PEP is observed, demonstrating that there is an increased contribution from anaplerotic PEP carboxylation and a corresponding decrease in TCA cycle activity. As described above, this information is obtained from inspection of the scalar coupling multiplets of Asp-α. The levels of the doublet with a small coupling constant and the doublet of doublets in the multiplets of Asp-α are significantly higher in the ammonia-limited culture of the *zwf* mutant than in the wild type (Figure 5.20). This result suggests that in the ammonia-limited *zwf* mutant, the TCA cycle operates predominantly for the generation of biosynthetic precursor metabolites and to a lesser extent for ATP generation via oxidative phosphorylation.

(ii) The PEP molecules arising from OAA are absent, indicating that there is negligible activity of the Pck *in vivo*.

(iii) There is an increase in the amount of pyruvate molecules derived from malate, indicating the increased activity of the malic enzyme.

(iv) The fraction of OAA molecules that are reversibly interconverted to fumarate is reduced (Table 5.17).

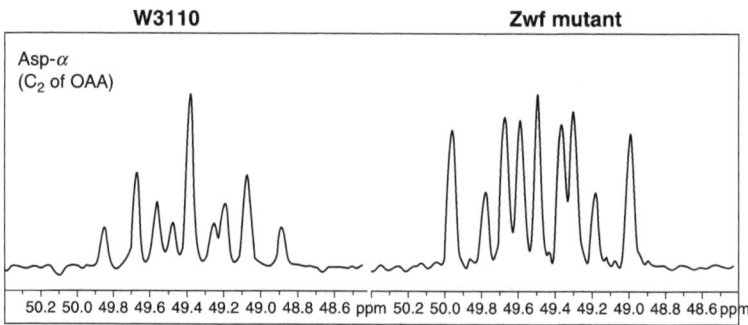

Figure 5.20 ^{13}C-^{13}C scalar coupling multiplets observed for Asp-α from ammonia-limited chemostat cultures of *E. coli* W3110 (left) and the *zwf* mutant (right)

5.6.9 ^{13}C-Metabolic flux analysis of pgi and zwf gene knockout mutant E. coli

The intracellular flux distributions in the wild type depend strongly on the limiting nutrient (Figures.5.21a and b). Changing the limiting nutrient from glucose to ammonia results in a reduced flux through the oxidative branch of the PP pathway, so that the contribution of the EMP pathway to glucose catabolism is increased (Sauer et al., 1999; Emmerling et al., 2002). Moreover, the gluconeogenic flux from OAA to PEP via the Pck is significantly lower, whereas the Mez flux from MAL to PYR is higher under ammonia-limited conditions than under glucose-limited conditions.

In the *pgi* mutant, the ED pathway accounts for 5% and 13% of the glucose catabolic flux under glucose- and ammonia-limited conditions, respectively (Figures 5.21c and d). Hence, the ED pathway is used to a limited extent, and the PP pathway is the primary route for glucose catabolism. A similar observation is made with other *pgi*-deficient *E. coli* strains (Fraenkel and Levisohn, 1967; Canonaco et al., 2001; Fischer and Sauer, 2003). The transhydrogenase flux converting NADPH to NADH is found to be significantly higher in the *pgi* mutant than in the wild type (Figure 5.21c and d), suggesting that transhydrogenase plays an important role in redox metabolism in this mutant.

The flux distributions in the glucose- and ammonia-limited *zwf* mutant are shown in Figures 5.21e and f. Under both limiting conditions, the *zwf* knockout results in exclusive glucose catabolism through the EMP pathway, synthesis of P5P and E4P via the non-oxidative PP pathway,

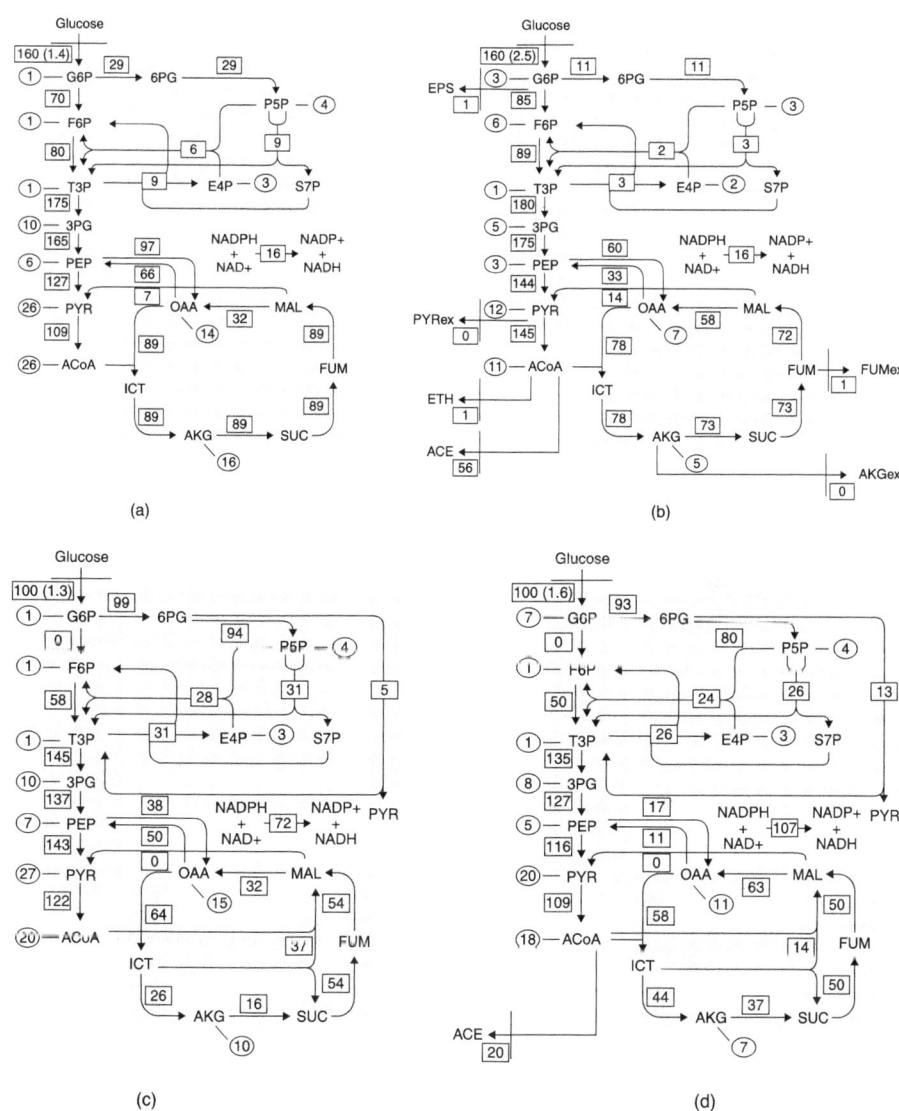

Figure 5.21 Metabolic flux distribution in chemostat cultures of *E. coli* W3110 under glucose-limited conditions (a) and ammonia-limited conditions (b), the *pgi* mutant under glucose-limited conditions (c) and ammonia-limited conditions (d), and the *zwf* mutant under glucose-limited conditions

(*Continued*)

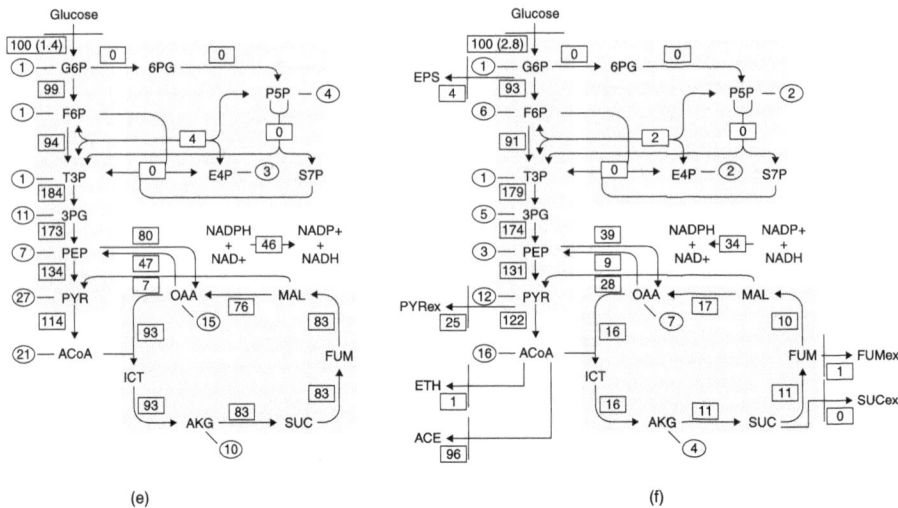

Figure 5.21 *(continued)* (e) and ammonia-limited conditions (f). The chemostats were operated at a dilution rate of 0.1 h^{-1}. The numbers in rectangles are the net fluxes. The flux values are expressed relative to the specific glucose uptake rate, which is indicated in parentheses (in millimoles per gram [dry weight] per hour). The arrows indicate the directions of the fluxes. The numbers in ellipses are the fluxes for withdrawal of precursor metabolites for biomass formation

and the transhydrogenase flux converting NADH to NADPH. The fluxes through other parts of metabolism are similar in the glucose-limited cultures of the *zwf* mutant and the wild type (Figure 5.21e). However, under ammonia-limited conditions, the inactivation of *zwf* drastically alters the flux distribution (Figure 5.21f). Most strikingly, the TCA cycle flux is reduced to an extremely low level, and a reverse flux through the MDH converts OAA to MAL. Moreover, the ammonia-limited *zwf* mutant exhibits surprisingly high fluxes of secretion of acetate (and pyruvate). In addition, the flux through the Pck is negligible, while the Mez flux is increased in the ammonia-limited *zwf* mutant.

The flux solutions with the lowest χ^2 values are shown in Figure 5.21, under glucose- and ammonia-limited conditions, where χ^2 values were 62 and 136, respectively, for W3110, 76 and 185, respectively, for the *pgi* mutant, and 132 and 89, respectively, for the *zwf* mutant. Since the 95% confidence level of the χ^2 value in this type of experiment is around 120 (Dauner et al., 2001), these χ^2 values are good for analysis of a biological system. Therefore, the flux distributions determined can provide a reliable description of the behavior of the *E. coli* metabolic network. The flux solutions are also subject to statistical error analysis based on Monte Carlo simulations (Schmidt et al., 1999a; Yang et al., 2002a). For most fluxes obtained, 90% confidence intervals are less than 8% of the estimated flux, but the 90% confidence intervals for the oxidative PP and ED pathway fluxes are less than 25%.

The flux results provide a holistic perspective on intracellular metabolism and thus provide a unique insight into the generation and consumption of the anabolic reducing power, NADPH (Figure 5.22). In wild-type *E. coli*, the ICDH reaction is the major producer of NADPH,

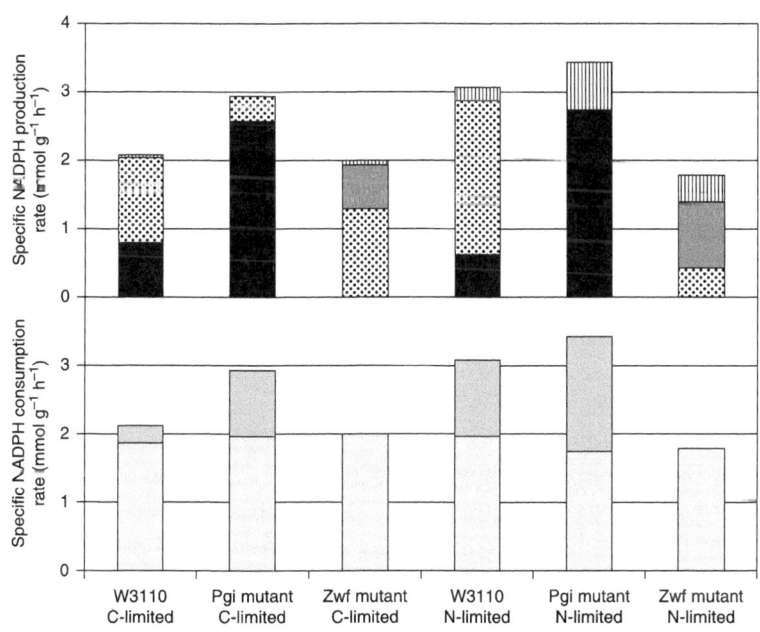

Figure 5.22 Specific rates of NADPH production and consumption in glucose (C)- and ammonia (N)-limited chemostat cultures of *E. coli* W3110, the *pgi* mutant, and the *zwf* mutant. NADPH production was contributed to by the oxidative PP pathway (solid bars), isocitrate dehydrogenase (cross-hatched bars), transhydrogenase (hatched bars), and malic enzyme (stippled bars). NADPH was consumed via biomass formation (stippled bars) and the transhydrogenase reaction (hatched bars)

accounting for more than 60% of NADPH production. In the *pgi* mutant, the PP pathway, which is used as the primary glucose catabolic pathway, generates a large amount of NADPH. Over-production of NADPH is deleterious, as a limited capacity for re-oxidation of NADPH is one reason for the low growth rate of *pgi*-deficient *E. coli* (Canonaco et al., 2001). The *pgi* mutant bypasses the ICDH reaction by redirecting carbon flow through the glyoxylate shunt, so that the amount of NADPH produced by the ICDH reaction is markedly decreased (Figure 5.22). This reaction contributes to less than 20% of the NADPH production in the *pgi* mutant. Therefore, activation of the glyoxylate shunt can significantly reduce the production of excess NADPH in the *pgi* mutant that is limited in re-oxidizing NADPH. NADPH and $NADP^+$ have been identified as effectors that regulate the reversible phosphorylation or inactivation of ICDH in *E. coli* (Holms, 1986).

5.7 ^{13}CMFA for cyanobacteria based on GC-MS and NMR

Consider how metabolic flux analysis can be made based on the isotopomer data obtained by both GC-MS and NMR. Let us consider the ^{13}C MFA for the application to cyanobacterium *Synechocystis* sp. PCC6803 (Yang et al., 2002b,c). Cyanobacteria constitute a large group of prokaryotes that perform oxygenic photosynthesis similar to plants and are regarded both as the source of the Earth's oxidizing atmosphere and as the ancestor of chloroplasts (Gray, 1993). Cyanobacteria are widely distributed in various ecological habitats, and their potential use in biotechnology has been investigated (Hansel and Lindblad, 1998; Deng and Coleman, 1999). *Synechocystis* sp. PCC6803 is commonly used as a model system in studies of green plant CO_2 metabolism and energy metabolism. In early works, specific aspects of the cyanobacterial metabolism have been described qualitatively by label distribution studies of radio labeling experiments (Pelroy and Bassham, 1972, 1976). Here we consider the intracellular metabolic fluxes in *Synechocystis* under different cultivation conditions to gain a better understanding of its metabolism, where 2D NMR spectroscopy and GC-MS are used to analyze the labeling patterns of the amino acids in biomass hydrolysates of *Synechocystis*. By integrating these labeling measurement data with metabolite balancing, the intracellular flux distributions in *Synechocystis* grown mixotrophically and heterotrophically can be quantified.

^{13}C-metabolic flux analysis

Network reactions in the central metabolism of *Synechocystis* can be constructed from the biochemistry in the literature (Smith, 1983; Tabita, 1994). The central carbon metabolic network in *Synechocystis*, with the stoichiometric reactions and the corresponding carbon atom transitions, are shown in Table 5.21. The metabolic reactions considered include those of the glycolysis, pentose phosphate pathway, Calvin cycle, TCA

Table 5.21 Stoichiometric reactions for cyanobacteria

Glycolysis[a]

hxk:	GLC → G6P		#ABCDEF → #ABCDEF	
hxi:	G6P ⇔ F6P		#ABCDEF ⇔ #ABCDEF	
pfk:	F6P ⇔ 2GAP		#ABCDEF ⇔ #CBA + #DEF	
eno:	GAP ⇔ PEP		#ABC ⇔ #ABC	
pyk:	PEP ⇔ PYR		#ABC ⇔ #ABC	
pdh:	PYR → ACA + CO_2		#ABC → #BC + #A	

Pentose phosphate pathway

gdh:	G6P → RU5P + CO_2	#ABCDEF → #BCDEF + #A
ppi:	RU5P ⇔ R5P	#ABCDE ⇔ #ABCDE
ppe:	RU5P ⇔ X5P	#ABCDE ⇔ #ABCDE
tk1:	X5P + R5P ⇔ S7P + GAP	#ABCDE + #abcde ⇔ #ABabcde + #CDE
tal:	S7P + GAP ⇔ F6P + E4P	#abcdefg + #ABC ⇔ #abcABC + #defg
tk2:	X5P + E4P ⇔ F6P + GAP	#ABCDE + #abcd ⇔ #ABabcd + #CDE

Calvin cycle

rbc:	CO_2 + RUDP → 2 GAP	#A + #BCDEF → #ACB + #DEF
fbp:	2GAP ⇔ F6P	#ABC + #abc ⇔ #CBAabc
tk2:	F6P + GAP ⇔ X5P #pl E4P	#abcdef + #ABC ⇔ #abABC + #cdef
sbp:	E4P + GAP → S7P	#abcd + #ABC → #CBAabcd
tk1:	S7P + GAP ⇔ R5P + X5P	#abcdefg + #ABC ⇔ #cdefg + #abABC
ppi:	R5P ⇔ RU5P	#ABCDE ⇔ #ABCDE
ppe:	X5P ⇔ RU5P	#ABCDE ⇔ #ABCDE
prk:	RU5P → RUDP	#ABCDE → #ABCDE

TCA cycle

cis:	OAA + ACA → ICI	#abcd + #AB → #dcbaBA
iod:	ICI → AKG + CO_2	#abcdcf → #abccf + #d
icl:	ICI + ACA → FUM + OAA	#abcdef + #AB → #dcef + #abBA

(Continued)

Table 5.21 Stoichiometric reactions for cyanobacteria (*Continued*)

TCA cycle (*Continued*)

    mdh: FUM ⇔ OAA   #abcd ⇔ #abcd #αabcdβχδ ⇔#dcba

    ppc: PEP + CO_2 → OAA   #abc + #A → #abcA

    me: OAA → CO_2 + PYR   #ABCD → #D + #ABC

C1 metabolism

    Serine hydroxymethyltransferase: Ser ⇔ Gly +C1 #ABC ⇔ #AB + #C

    glycine cleavage: Gly ⇔ CO_2 + C1 #AB ⇔ #A + #B

Biomass synthesis

    0.460 G6P + 0.362 F6P + 0.399 R5P + 0.406 E4P + 0.238 GAP + 1.53 PEP + 2.64 PYR + 4.64 ACA + 1.23 OAA +1.04 AKG − BIOMASS H+ 1.8346 CO_2

    0.891 G6P + 0.337 F6P + 0.382 R5P + 0.376 E4P + 0.208 GAP + 1.42 PEP + 2.44 PYR + 3.96 ACA + 1.14 OAA +0.886 AKG − BIOMASS M + 1.8346 CO_2

[a]The reactions pfk and pyk were considered to be bidirectional, and the backward ones represented the reactions catalyzed by fructose-1,6-bisphosphatase and phosphoenolpyruvate synthase, respectively.

cycle, and the C_1 metabolism. Cyanobacteria have an incomplete TCA cycle lacking KGDH (Pearce *et al.*, 1969), and this is also consistent with genetic experiments of Vazquez-Bermudex *et al.* (2000). Thus this reaction can be excluded in the network. The enzymes, Icl and MS, which form the glyoxylate cycle, have been known to be present in cyanobacteria (Pearce and Carr, 1967). Since the entire *Synechocystis* genome contains the genes for malic enzyme and Pps, these two enzymes are considered to be responsible for the initial gluconeogenetic steps rather than Pck, which is absent in *Synechocystis*.

Synechocystis cells grown on labeled glucose were harvested from the mid-exponential growth phase of the heterotrophic and mixotrophic cultures, subjected to hydrolysis, and the labeling patterns of amino acids in biomass hydrolysates were analyzed using NMR and GC-MS.

The degree of coupling between adjacent carbon atoms (Table 5.22) (Yang et al., 2002) can be determined from the NMR spectra. Table 5.23 shows the measurements of the relative abundance of the mass isotopomers, as determined by GC-MS (Yang et al., 2002a). The mass spectrum of a compound can contain fragment ions that are caused by partial fragmentation of the analyzed compound. As shown in Table 5.23, the mass spectrum of ECF-derived glycine includes two ion clusters, which contain C_1–C_2 and C_2 of glycine, respectively. The mass isotopomer distribution of the fragment ion containing C_2 of glycine provides additional information for the labeling pattern analysis. The GC-MS data in Table 5.23 can be corrected for the occurrence of natural isotope distribution in amino acid derivatives. Moreover, since the biomass originating from the inoculum is unlabeled, the NMR and MS measurement data can be corrected for the unlabeled biomass. The effect of the unlabeled biomass can be also taken into account by correcting the isotopomer distribution vector for the input substrate in the flux calculation.

Before using the isotopomer measurement data of NMR and GC-MS for flux estimations, it is important to know the actual number of independent constraints on the isotopomer distribution that can be obtained from the NMR data, GC-MS data, and the combination of these two methods. Since the measured values are relative contributions, n measurements for 1 carbon atom or for 1 amino acid give only n − 1 constraints on the isotopomer distribution. Then by the detailed analysis of the linear equation system (Noronha et al., 2000), the number of independent constraints from the combination of both methods can be deduced. As illustrated in Table 5.24, NMR spectroscopy and GC-MS yields a large amount of independent constraints on isotopomer distribution of amino acid. The isotopomer fractions of glycine can be completely resolved. Moreover, the information available from NMR spectroscopy measurements is shown to be complementary to the information available from GC-MS analysis.

The reaction networks of central carbon metabolism in *Synechocystis* grown heterotrophically and mixotrophically are shown in Figure 5.23. The PP pathway operates for glucose catabolism in the heterotrophic network of reactions, while the Calvin cycle is active in the mixotrophic culture. The CO_2 fixation by the Calvin cycle under the mixotrophic conditions is reported by Pelroy and Bassham (1972) and also supported by the data of CO_2 evolution (Table 5.25). Moreover, the operation of the Calvin cycle in the mixotrophic culture can also be identified through the observation of the labeling pattern of histidine that is derived from R5P. As can be seen from Table 5.22, the ^{13}C multiplet patterns of β-His in both cultures are clearly different. Using the probabilistic analysis (Szyperski, 1995), the fraction of α- and β-carbons of histidine originating from the same source molecule decrease significantly under mixotrophic conditions, indicating a marked decrease in the C_3–C_4 connectivities of R5P. This kind of labeling suggests that R5P is synthesized through the reactions of the Calvin cycle in the mixotrophic culture, because the CO_2 fixation in the Calvin cycle results in the cleavage of C_3-C_4 bonds in R5P (Figure 5.23).

Although it is usually considered that the growth of cyanobacteria under heterotrophic conditions is slow, heterotrophic growth of

Table 5.22 Relative intensities of ^{13}C multiplet components of amino acids

| Carbon position | Heterotrophic conditions ||||||||| Mixotrophic conditions |||||||||
| | Measured |||| Estimated |||| | Measured |||| Estimated ||||
	S	D1	D2	DD	S	D1	D2	DD	S	D1	D2	DD	S	D1	D2	DD
α-Ala	0.19	0.06	0.18	0.57	0.23	0.10	0.13	0.54	0.39	0.31	0.14	0.16	0.34	0.32	0.10	0.24
β-Ala	0.33	0.67	—	—	0.36	0.64	—	—	0.46	0.54	—	—	0.44	0.56	—	—
α-Asp	0.26	0.15	0.29	0.30	0.31	0.09	0.25	0.35	0.37	0.30	0.24	0.09	0.42	0.22	0.21	0.16
βAsp	0.29	0.39	0.28	0.04	0.25	0.35	0.31	0.09	0.40	0.28	0.21	0.11	0.36	0.31	0.26	0.07
α-Glu	0.19	0.42	0.36	0.03	0.25	0.35	0.31	0.09	0.39	0.26	0.22	0.13	0.36	0.31	0.26	0.07
β-Glu	0.54	0.39	—	0.07	0.50	0.45	—	0.05	0.51	0.42	—	0.07	0.55	0.41	—	0.04
γGlu	0.26	0.61	0.02	0.11	0.32	0.57	0.04	0.07	0.43	0.52	0.02	0.03	0.39	0.50	0.05	0.06
αGly	0.37	0.63	—	—	0.38	0.62	—	—	0.70	0.30	—	—	0.68	0.32	—	—
βHis	0.24	0.42	0.05	0.29	0.20	0.52	0.02	0.25	0.51	0.19	0.01	0.29	0.41	0.18	0.05	0.36
δ²-His	0.63	0.37	—	—	0.69	0.31	—	—	0.53	0.47	—	—	0.50	0.50	—	—
αIle	0.27	0.02	0.61	0.10	0.35	0.04	0.54	0.07	0.63	0.04	0.29	0.04	0.57	0.07	0.32	0.04
γ²-Ile	0.35	0.65	—	—	0.36	0.64	—	—	0.44	0.56	—	—	0.44	0.56	—	—
δ-Ile	0.58	0.42	—	—	0.60	0.40	—	—	0.69	0.31	—	—	0.67	0.33	—	—
α-Leu	0.23	0.07	0.56	0.14	0.32	0.04	0.57	0.07	0.31	0.04	0.56	0.09	0.39	0.05	0.50	0.06
β-Leu	0.82	0.15	—	0.03	0.79	0.20	—	0.01	0.85	0.14	—	0.01	0.79	0.20	—	0.01
δ¹-Leu	0.38	0.62	—	—	0.36	0.64	—	—	0.40	0.60	—	—	0.44	0.56	—	—

Published by Woodhead Publishing Limited, 2013

	S	D1	D2	DD	S	D1	D2	DD	S	D1	D2	DD	S	D1	D2	DD
δ²-Leu	0.91	0.09	—	—	0.89	0.11	—	—	0.86	0.14	—	—	0.89	0.11	—	—
β-Lys	0.22	0.65	—	0.13	0.29	0.63	—	0.08	0.34	0.65	—	0.01	0.38	0.56	—	0.06
δ-Lys	0.24	0.70	—	0.06	0.29	0.63	—	0.08	0.35	0.55	—	0.10	0.38	0.56	—	0.06
εLys	0.43	0.57	—	—	0.46	0.54	—	—	0.51	0.49	—	—	0.53	0.47	—	—
α-Phe	0.11	0.07	0.02	0.80	0.16	0.11	0.01	0.71	0.27	0.47	0.07	0.19	0.28	0.40	0.03	0.29
β-Phe	0.14	0.62	0.07	0.17	0.16	0.73	0.02	0.09	0.19	0.74	0.03	0.04	0.28	0.61	0.03	0.08
α-Pro	0.21	0.42	0.27	0.10	0.25	0.35	0.31	0.09	0.41	0.27	0.23	0.09	0.36	0.31	0.26	0.07
β-Pro	0.56	0.33	—	0.09	0.50	0.45	—	0.05	0.53	0.46	—	0.01	0.55	0.40	—	0.04
γ-Pro	0.27	0.69	—	0.04	0.32	0.61	—	0.07	0.38	0.50	—	0.12	0.39	0.55	—	0.06
β-Ser	0.58	0.42	—	—	0.58	0.42	—	—	0.56	0.44	—	—	0.56	0.44	—	—
δˣ-Tyr	0.19	0.78	—	0.03	0.14	0.76	—	0.09	0.19	0.80	—	0.11	0.18	0.73	—	0.09
α-Val	0.22	0.02	0.63	0.13	0.30	0.04	0.59	0.07	0.52	0.08	0.36	0.04	0.59	0.07	0.30	0.04
γ¹-Val	0.36	0.64	—	—	0.36	0.64	—	—	0.42	0.58	—	—	0.44	0.56	—	—
γ²-Val	0.87	0.13	—	—	0.39	0.11	—	—	0.85	0.15	—	—	0.89	0.11	—	—

The relative contributions of singlet (S), doublet (D1, D2), and doublet of doublets or triplets (DD) to the overall multiplet pattern were determined from the 2D ^{1}H-^{13}C NMR spectra of the amino acids in biomass hydrolysates of *Synechocystis*.

Table 5.23 Mass isotopomer distribution of ECF-derived amino acids

Heterotrophic conditions

Amino acid	Ion cluster (C atoms in fragments)	Origin	m	m+1	m+2	m+3	m+4	m+5	m+6	m+7	m+8
Ala	116	Exp.	.842	0.093	0.065						
	(2, 3)	Est.	0.849	0.081	0.070						
Asp	188	Exp.	0.762	0.164	0.064	0.010					
	(2–4)	Est.	0.764	0.150	0.077	0.009					
Gly	102	Exp.	0.886	0.114							
	(2)	Est.	0.883	0.117							
	175	Exp.	0.861	0.065	0.074						
	(1, 2)	Est.	0.859	0.069	0.072						
Ile	158	Exp.	0.644	0.201	0.128	0.020	0.007	0.001			
	(2–6)	Est.	0.649	0.189	0.131	0.025	0.006	0.001			
Leu	158	Exp.	0.639	0.185	0.131	0.027	0.008	0.000			
	(2–6)	Est.	0.641	0.202	0.127	0.024	0.006	0.001			
Lys	156	Exp.	0.645	0.201	0.125	0.021	0.006	0.002			
	(2–6)	Est.	0.649	0.189	0.131	0.025	0.006	0.001			
Phe	192	Exp.	0.586	0.156	0.140	0.064	0.034	0.014	0.006	0.000	0.000

Mixotrophic conditions

Amino acid	Ion cluster (C atoms n fragments)	Origin	m	m+1	m+2	m+3	m+4	m+5	m+6	m+7	m+8
Ala	116	Exp.	0.845	0.091	0.064						
	(2, 3)	Est.	0.841	0.097	0.061						
Asp	188	Exp.	0.759	0.164	0.075	0.002					
	(2–4)	Est.	0.752	0.173	0.068	0.007					
Gly	102	Exp.	0.892	0.108							
	(2)	Est.	0.888	0.112							
		Exp.	0.844	0.106	0.050						
	(1, 2)	Est.	0.840	0.108	0.052						
Ile	158	Exp.	0.633	0.208	0.124	0.022	0.008	0.000			
	(2–6)	Est.	0.633	0.219	0.120	0.024	0.005	0.000			
Leu	158	Exp.	0.640	0.209	0.124	0.016	0.010	0.001			
	(2–6)	Est.	0.630	0.224	0.119	0.023	0.005	0.000			
Lys	156	Exp.	0.636	0.232	0.084	0.035	0.012	0.001			
	(2–6)	Est.	0.633	0.219	0.120	0.024	0.005	0.000			
Phe	192	Exp.	0.615	0.146	0.111	0.025	0.082	0.016	0.005	0.000	0.000

Table 5.24 Independent constraints on the isotopomer distribution of amino acids available from labeling measurements

Amino acid	Precursor	Constraints on isotopomer distribution		
		2D-NMR	GC-MS	Combined information
Ala	PYR	4	2	5
Asp	OAA	6	3	9
Glu	AKG	8		8
Gly	3PG	1	3	3
His	R5P	5		5
Ile	OAA, PYR	5	5	10
Leu	PYR, ACA	7	5	12
Lys	OAA, PYR	5	5	10
Phe	E4P, PEP	6	8	14
Pro	AKG	7	4	11
Ser	3PG	1		1
Thr	OAA		3	3
Tyr	E4P, PEP	2		2
Val	PYR	5	4	9

Synechocystis with a generation time as short as 19 h has been reported previously (Astier *et al.*, 1984). CO_2 in the inlet gas during cultivation was removed in the experiment. In the mixotrophic culture, the overall degree of ^{13}C labeling, P_1, will be decreased by 6% if cyanobacterial cells fix 10% of the atmospheric CO_2. P_1 can be determined from the NMR spectra and the mass isotopomer distribution of glycine. Since P_1 is found to be 11% in both cultures, after the errors arising from the occurrence of natural isotopes and unlabeled biomass are corrected, it is unlikely that atmospheric CO_2 plays any significant role. Moreover, it can be seen from Table 5.25 that the substrate carbon is almost completely recovered in biomass (~43% carbon content) and in CO_2.

As shown in Table 5.25, the biomass yields on glucose are 0.50 g DW/g glucose in the heterotrophic culture and 0.87 g DW/g glucose in the mixotrophic culture, respectively. Fluxes to biomass formation can be calculated from the measurements of the biomass composition of the green algal cells cultivated under the same conditions (Yang et al., 2000c). The measurement data of CO_2 evolution rate is not used for flux estimation,

^{13}C-metabolic flux analysis

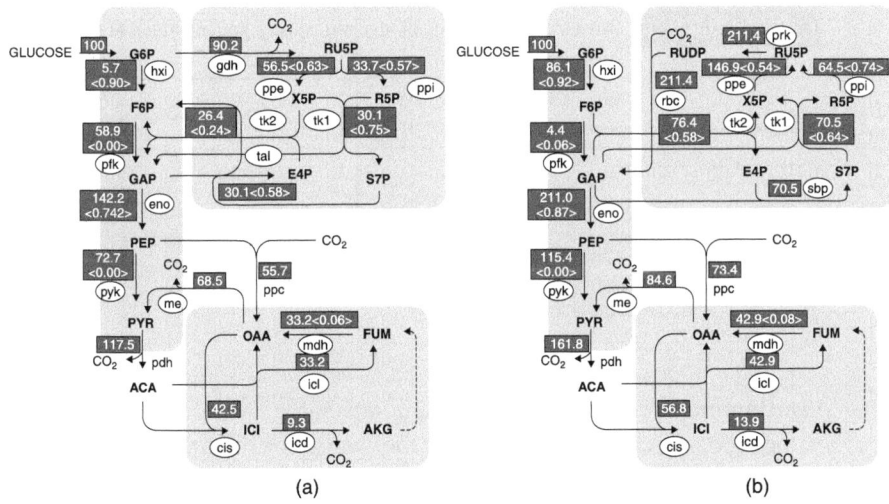

Figure 5.23 *In vivo* flux distributions in the central metabolism of *Synechocystis* sp. PCC6803 cultivated (a) heterotrophically and (b) mixotrophically. The flux names are given in white circles. Biomass and CO_2 effluxes are not shown for simplicity. Numbers in dark boxes represent the estimated net fluxes normalized to the glucose uptake rates, which were 0.85 mmol/g DW/h in the heterotrophic culture and 0.38 mmol/g DW/h in the mixotrophic culture. Arrows indicate the direction of the estimated fluxes, and the exchange coefficients in reversible reactions are given in angle brackets. The flux distributions were obtained from the best fit to the measured fluxes and the constraints derived from the NMR and MS measurements

since the inclusion of the CO_2 balance in the metabolic model does not decrease the degrees of freedom for net fluxes and the measuring errors for these data are larger than other measured extracellular fluxes. Since the substrate uptake and the biomass effluxes are directly measured, the metabolite balances now leave two degrees of freedom for net fluxes. As for free net fluxes, we choose the flux through the oxidative PP pathway (gnd^{net}) and the gluconeogenic flux of malic enzyme (me^{net}) for the heterotrophic culture, and the flux of CO_2 fixation in the Calvin cycle (rbc^{net}), and the gluconeogenic flux of malic enzyme (me^{net}) for the mixotrophic culture. The set of free fluxes can be completed by all the exchange coefficients of reaction steps that are assumed to be bidirectional based on thermodynamic considerations. The net fluxes and exchange coefficients of the best fit flux

Table 5.25 Growth parameters of exponentially growing *Synechocystis*

Culture	μ (h^{-1})	$Y_{X/S}$ (g/g)	q_{glc} (mmol/g/h)	q_{CO2} (mmol/g/h)
Heterotrophic culture	0.076	0.50	0.85	1.99
Mixotrophic culture	0.059	0.87	0.38	0.00

Data are determined at an OD$_{730}$ of 0.6. μ, specific cell growth rate
$Y_{X/S}$, biomass yield on glucose
q_{glc}, specific glucose consumption rate
q_{CO2}, specific CO_2 production rate

distributions are given in Figure 5.23 and Table 5.26. The minimum χ^2 values (i.e. the criterion for judging the quality of the fit) are 204 for the heterotrophic culture and 317 for the mixotrophic culture, respectively. As can be seen from Tables 5.22 and 5.23, the labeling measurements are in good agreement with the corresponding estimated values.

As shown in Figure 5.23a, 90.2% of the G6P enters into the PP pathway (gdh^{net}) and 5.7% is catabolized through glycolysis (hxi^{net}) in the heterotrophic flux distribution. The flux through the Ppc (ppc^{net}) is estimated as 55.7% of the glucose uptake, which accounts for 62.6% of the OAA synthesized. The reaction catalyzed by Mez is identified by labeling experiments, and the backward flux from the TCA cycle to glycolysis is 68.5% (me^{net}). The estimated values for the exchange coefficients are shown in angle brackets in Figure 5.23. High exchange rates in the Pgi (hxi$^{exch[0,1]}$), Tkt (tk1$^{exch[0,1]}$), and GAP to PEP conversions (eno$^{exch[0,1]}$) are found, while the Fbp (pfk$^{exch[0,1]}$) and Pps (pyk$^{exch[0,1]}$) are inactive during heterotrophic growth on glucose.

From Figure 5.23b, the flux of CO_2 assimilation through the Calvin cycle (rbc^{net}) is 211.4% of the glucose input flux in the mixotrophic flux distribution. The reaction mediated by Fbp is present in *Synechocystis* grown mixotrophically (pfk$^{exch[0,1]}$). The CO_2 fixation through the Ppc (ppc^{net}) is 73.4%, which represents about 25% of the assimilated CO_2. The reaction catalyzed by the Mez is still active (me^{net}). Therefore, the reactions directing the backward flux from the TCA cycle to glycolysis, which has been reported in mammalian and bacteria cells (Marx et al., 1996; Katz et al., 1993), are also present in photosynthetic organism. Figure 5.24 shows the identification of the operation of Calvin cycle by observing the labeling pattern of histidine.

In order to check the reliability of the flux estimates, a statistical analysis must be made (Schmidt et al., 1999b). A total of 500 datasets were generated by addition of normally distributed measurement noise to

^{13}C-metabolic flux analysis

Table 5.26 Estimated values and 90% confidence regions for estimated free fluxes

	Heterotrophic culture			Mixotrophic culture		
Flux	Estimated value	90% confidence region	Flux	Estimated value	90% confidence region	
gnd^{net}	90.2	[87.6,	92.7]	rbc^{net}	211.4	[208.8, 214.0]
me^{net}	68.5	[65.7,	71.3]	me^{net}	84.6	[81.0, 88.2]
$hxi^{exch[0,1]}$	0.90	[0.86,	0.93]	$hxi^{exch[0,1]}$	0.92	[0.00, 0.95]
$pfk^{exch[0,1]}$	0.00	[0.00,	0.01]	$pfk^{exch[0,1]}$	0.06	[0.06, 0.07]
$eno^{exch[0,1]}$	0.74	[0.03,	0.95]	$eno^{exch[0,1]}$	0.87	[0.00, 0.95]
$pyk^{exch[0,1]}$	0.00	[0.00,	0.02]	$pyk^{exch[0,1]}$	0.00	[0.00, 0.02]
$ppi^{exch[0,1]}$	0.57	[0.49,	0.65]	$ppi^{exch[0,1]}$	0.74	[0.72, 0.76]
$ppe^{exch[0,1]}$	0.63	[0.61,	0.66]	$ppe^{exch[0,1]}$	0.54	[0.44, 0.63]
$tk1^{exch[0,1]}$	0.75	[0.72,	0.78]	$tk1^{exch[0,1]}$	0.64	[0.00, 0.95]
$tal^{exch[0,1]}$	0.58	[0.57,	0.59]	$tk2^{exch[0,1]}$	0.58	[0.00, 0.95]
$tk2^{exch[0,1]}$	0.24	[0.19,	0.30]	$mdh^{exch[0,1]}$	0.08	[0.00, 0.11]
$mdh^{exch[0,1]}$	0.06	[0.04,	0.07]			

The exchange coefficients are limited to values below 0.95, because the presence of large exchange rates significantly affect the speed of convergence

the estimated isotopomer labeling data given in Tables 5.22 and 5.23, and then 500 flux distributions were estimated from these datasets using the same parameter fitting approach. Thus the confidence regions for the individual flux estimates are calculated from the probability distribution of the 500 flux distributions. The 90% confidence intervals for the estimated net fluxes and exchange coefficients are given in Table 5.26. The variance estimates obtained for all non-free net fluxes are omitted in Table 5.26 for the sake of brevity, since all linearly dependent fluxes have confidence regions in the same order of magnitude as involved free net fluxes. As becomes clear from Table 5.26, all free net fluxes are well determined from the measured data with small confidence regions. However, relatively large confidence intervals are found for the exchange coefficients. The 90% confidence intervals for the exchange coefficients of eno reaction in the heterotrophic flux distribution and hxi, eno, tk1, and tk2 reactions in the mixotrophic flux distribution span almost the whole range of parameter values, indicating these parameters are not identifiable by the available measurement data.

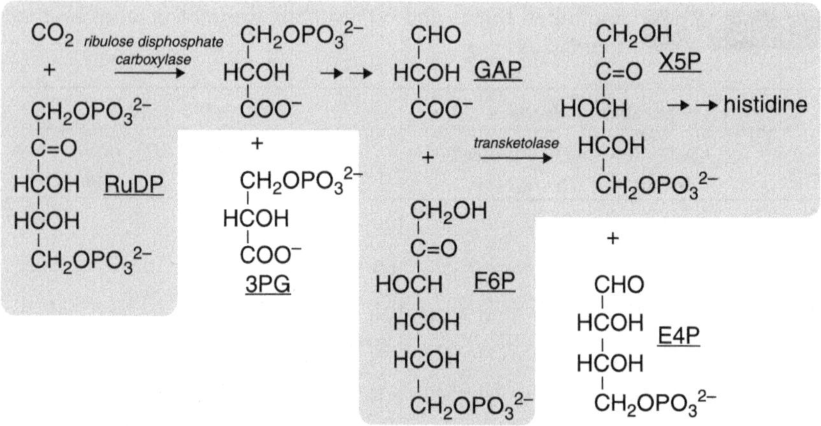

Figure 5.24 Synthesis of the precursor for histidine (R5P or X5P) via the reactions of Calvin cycle. Thick lines indicate carbon–carbon connections arising from a single-source molecule, and thin lines connect fragments arising from different source molecules. For simplicity, we assume the intact fragments of RuDP or F6P originating from a single-source molecule. The CO_2 fixation catalyzed by RuDP carboxylase results in the cleavage of C_3–C_4 connections of the precursor for histidine, hence the operation of Calvin cycle can be identified by the observation of the ^{13}C multiplet pattern of histidine

The heterotrophic metabolism is likely to reflect the dark metabolism of cyanobacteria in the natural environment. Although glycogen is the major endogenous reserve for the naturally growing cyanobacterial cells and is degraded to provide the energy in the absence of light, it can be replaced by labeled glucose because the degradation of glycogen begins with the conversion to G6P. In addition, the concentration of glucose used (5 mM) can sustain normal growth requirements.

The steady-state isotope labeling experiment provides a powerful approach for measuring the fluxes of heterotrophic metabolism, but it is not suited for analysis of photoautotrophic metabolism. This is due to the fact that photosynthesis relies on CO_2 as the only carbon substrate and steady state positional labeling with a one-carbon substrate is impossible, that is, the carbon enrichment in all intermediates is equal to that of CO_2, independent of the fluxes. The dynamic positional labeling experiment may be made for autotrophic flux analysis (Roessner et al., 2000; Young et al., 2000; Young et al., 2011).

5.7.1 Pentose phosphate pathway and Calvin cycle

In the heterotrophic flux distribution of *Synechocystis*, more than 90% of glucose is channeled through the PP pathway. This result is in agreement with Pelroy et al. (1972), who found that the yield of $^{14}CO_2$ from [1-^{14}C] glucose is much higher than that from [6-^{14}C]glucose supplied to cyanobacterial cells. Hence, the PP pathway is the major pathway of glucose catabolism in *Synechocystis* grown heterotrophically. The high flux through the oxidative PP pathway yields a large amount of NADPH, as well as biosynthetic precursors (i.e. R5P and E4P). Since the generated precursor metabolite exceeds the biosynthetic requirement, the excess pentoses are catabolized via the non-oxidative PP pathway and then join to glycolysis.

In the mixotrophic culture, CO_2 is fixed through the Calvin cycle. As described above, the operation of the Calvin cycle can be identified by the observation of the labeling pattern of histidine. The Calvin cycle is closely related to the PP pathway, except many of the reactions proceed in the reverse direction. Indeed, the addition of two unique enzymatic steps (rbc and prk in Figure 5.23b), along with some other idiosyncrasies, allows the operation of the Calvin cycle. Therefore, cyanobacteria have a flexible central metabolism in the switch from the Calvin cycle to the PP pathway during the diurnal light-dark cycles. From the estimated mixotrophic flux distribution, the *in vivo* activity of gluconeogenic Fbp is identified. Hence, an ATP-dissipating futile cycle via ATP-consuming Pfk and Fbp is found to be present in the mixotrophic culture of *Synechocystis*.

5.7.2 C_1 metabolism

C_1 metabolism can be investigated based on the labeling data of serine and glycine. The cleavage of serine into glycine and a C_1 unit constitutes the main route to fulfill the cellular requirement of C_1 units. Since cyanobacteria have negligible photorespiration and produce little or no glycolate during photosynthesis, it is unlikely that serine is synthesized, as in higher plants, from glycine by the glycolate pathway. It is reported that serine is synthesized directly from 3PG through a phosphorylated route in cyanobacteria (Colman and Norman, 1997). Assuming that the isotopomer distribution of 3PG is identical to that of PEP, the ^{13}C multiplet pattern of β-Ser shows that a significantly higher than expected fraction of C_2-C_3 bonds in serine have been cleaved. This is due to the interconversion of serine to glycine and a C_1 unit. It can be seen that the fraction of the serine molecules, which have been reversibly converted to glycine and a C_1 unit, is 56% in the heterotrophic culture and 42% in the mixotrophic culture.

From Tables 5.22 and 5.23, it is observed that there is a lower contribution than expected of the doublet signal to the α-Gly multiplet pattern and m+2 signal to the mass spectrum of glycine, if glycine is derived exclusively from serine. A candidate reaction is the glycine cleavage pathway, which balances the cell's glycine and C_1 requirements and converts glycine into CO_2 and C_1 units. It is found that 17% and 3.7% of glycine have been cleaved at least once for the heterotrophic and mixotrophic cultures, respectively.

5.7.3 Ppc and Mez

In both cultures, the relative flux through Ppc is high. The reaction catalyzed by the Ppc contributes to about 25% of the assimilated CO_2 in the mixotrophic culture, indicating that Ppc is important for the metabolism of CO_2 in cyanobacterial cells (Owittrim and Colman, 1988), where cyanobacterial cells fix significant amounts of carbon as C_4 acids (primarily as aspartate and malate) under light conditions.

The metabolic flux distribution of *Synechocystis* grown in both cultures also reveals a surprisingly high activity of the Mez. This is why the cleavage of the C_2-C_3 carbon bond in pyruvate is increased when comparing with the C_2-C_3 connectivity in PEP (see the ^{13}C multiplet patterns of β-Ala and β-Phe in Table 5.22). The flux through malic enzyme is also identified in *E. coli* (Sauer et al., 1999), *B. subtilis* (Sauer et al., 1997), and possibly *C. glutamicum* (Marx et al., 1996). However, the activity levels of this gluconeogenic flux in these organisms are much less than that in *Synechocystis*.

The estimated high flux through the Ppc pathway challenges the classical hypothesis that the principal function of the PEP carboxylating reaction is to replenish the loss of TCA cycle metabolites for anabolic processes. Actually, there is a substantial outflow from the TCA cycle to glycolytic pathway carried by Mez. Considering that the Mez in cyanobacteria is NADP-linked (Pearce et al., 1969), it is more likely that the Ppc and malic enzyme serve as a peculiar metabolic device to fix a large amount of CO_2 as C_4 acids and then release CO_2 and produce NADPH by the decarboxylation of malate. This is similar to the carbon metabolism in C_4 plants (Figure 5.23), for which the CO_2 and NADPH generated by the malic enzyme are utilized by the Calvin cycle. In fact, although the major pathway of CO_2 fixation in cyanobacteria in the Calvin cycle is similar to that in C_3 plants, cyanobacteria have many of the physiological characteristics of C_4 plants, for example, low CO_2 compensation points and limited photorespiration. The operation of a C_4 pathway in cyanobacteria, as indicated here, is consistent with the fact that in cyanobacteria, C_3 and C_4 plants have a common ancestor in evolution.

5.7.4 Energy metabolism

From the estimated *in vivo* flux distributions, the cellular energetics in cyanobacterial cells can be investigated. The amount of energy and reducing equivalents used and generated during metabolism can be calculated more reliably based on the estimated fluxes. The calculated result for NADPH synthesis and requirement is shown in Table 5.27. Apparently, cyanobacterial metabolism has a high requirement of NADPH for biosynthesis, especially for the reductive assimilation of

Table 5.27 Estimated production and consumption of NADPH

Heterotrophic conditions		Mixotrophic conditions	
NADPH production		NADPH production	
PP pathway (gnd^{net})	1.804	Photosynthesis[a]	8.808
Malic enzyme (me^{net})	0.685	Malic enzyme (me^{net})	0.846
Isocitrate dehydrogenase (icd^{net})[b]	0.093	Isocitrate dehydrogenase (icd^{net})[b]	0.139
NADPH consumption		NADPH consumption	
Biosynthesis	1.101	Biosynthesis	2.029
Nitrate reduction[c]	1.349	Nitrate reduction[c]	2.514
		Calvin cycle (rbc^{net})	4.228

Data were calculated from the flux distribution shown in Figure 5.25, and values are expressed in mol NADPH/mol glucose consumed. In parentheses are given the fluxes used for calculation.
[a]NADPH produced by the photosynthetic electron transport was estimated based on the reported quantum yield (0.06 mol CO_2 fixed per quanta absorbed) and the widely accepted two-step model of photosynthesis (4 mol quanta of light is required to generate 1 mol NADPH).
[b]Isocitrate dehydrogenase was considered to be NADP-dependent (Pearce et al., 1969).
[c]NADPH consumption for nitrate reductive assimilation was calculated on the basis of ammonia requirement for biomass synthesis

nitrate, which consumes three molecules of NADPH per molecule of nitrate. In the mixotrophic culture, copious amounts of reducing power are required in the Calvin cycle to reduce CO_2 to organic carbon molecules. Hence, there has to be a supply of NADPH in large amounts to fulfill the biosynthetic demands. The PP pathway in the heterotrophic culture and the photosynthetic electron transport in the mixotrophic culture account for a major fraction of NADPH production. A significant amount of NADPH is also provided by Mez. According to the calculation, the NADPH synthesis is in excess relative to its requirement. That is, excess NADPH must be reoxidized in other ways. In fact, cyanobacteria have a unique capacity to utilize NADPH as an electron donor of the respiratory electron transport chain. Therefore, the excess NADPH might be reoxidized in cyanobacterial respiration to provide the energy for cellular requirements.

Similarly, from the estimated flux distributions, the amount of ATP generated and utilized for both cultures can be obtained (Table 5.28). Not surprisingly, the Calvin cycle is the main ATP sink in the mixotrophic culture, since reducing the most oxidized source of carbon on Earth does

Table 5.28 Estimated production and consumption of ATP

Heterotrophic conditions		Mixotrophic conditions	
ATP production		ATP production	
Direct ATP	0.560	Direct ATP	2.220
NADH oxidative Phosphorylation[a]	6.695	NADH oxidative Phosphorylation[a]	8.942
NADPH oxidative Phosphorylation[a]	0.331	NADPH oxidative Phosphorylation[a]	2.560
		Photo-phosphorylation[b]	11.451
ATP consumption		ATP consumption	
Biosynthesis	1.858	Biosynthesis	3.176
		Calvin cycle	6.342

Data were calculated from the flux distribution shown in Figure 5.25, and values are expressed in mol ATP/mol glucose consumed.
[a]ATP formed from the oxidative phosphorylation was estimated by assuming the P/O value was 2.5 for NADH and NADPH.
[b]ATP produced from the photo-phosphorylation was calculated by assuming the P/2e⁻ ratio (the number of ATP molecules formed per pair of electrons moving through the photosynthetic electron transport chain) was 1.3

not come cheaply to photosynthetic cells. In cyanobacterial respiration, both NADH and NADPH serve as electron donors. The utilization of NADPH for ATP generation is consistent with the report that the activity of respiration in cyanobacterial cells is inhibited in the presence of the photosystem II inhibitor DCMU, since the inhibitor prevents the formation of NADPH by photosynthetic electron transport (Schmetterer, 1994). In the mixotrophic culture, although the photo-phosphorylation generates much energy, a significant fraction of ATP is also supplied by the oxidative phosphorylation, indicating that the activity of the respiratory chain in *Synechocystis* is not inhibited by light. This result is supported by the finding that cyanobacteria contain another complete respiratory chain besides the one sharing common components with the photosynthetic electron transport chain (Schmetterer, 1994). Similar results have also been obtained for microalgal cells (Yang et al., 2000).

5.8 Appendix 5.A Λ_{AP} and b_{PEP} in Equation 5.41

The following equation shows the relationship between the isotopomers of PEP and AcCoA. For convenience, Λ_{PEP} is obtained by substituting (A-1) into Λ_{AP}, as given below (Matsuoka and Shimizu, 2010):

$$\begin{bmatrix} A_0 \\ A_1 \\ A_2 \\ A_{12} \end{bmatrix} = \begin{bmatrix} 1 & 1 & 0 & 0 & 0 & 0 & 0 & 0 \\ 0 & 0 & 1 & 0 & 1 & 0 & 0 & 0 \\ 0 & 0 & 0 & 1 & 0 & 1 & 0 & 0 \\ 0 & 0 & 0 & 0 & 0 & 0 & 1 & 1 \end{bmatrix} \begin{bmatrix} P_0 \\ P_1 \\ P_2 \\ P_3 \\ P_{12} \\ P_{13} \\ P_{23} \\ P_{123} \end{bmatrix} \quad (A-1)$$

$$b_{PEP}^{T} = -y[P_0C_0 \ P_1C_0 \ P_2C_0 \ P_3C_0 \ P_0C_1 \ P_{12}C_0 \ P_{13}C_0 \ P_1C_1 \\ P_{23}C_0 \ P_2C_1 \ P_3C_1 \ P_{123}C_0 \ P_{12}C_1 \ P_{13}C_1 \ P_{23}C_1 \ P_{123}C_1]$$

$$I_{OAA}^{T} = [O_0 \ O_1 \ O_2 \ O_3 \ O_4 \ O_{12} \ O_{13} \ O_{14} \ O_{23} \ O_{24} \ O_{34} \\ O_{123} \ O_{124} \ O_{134} \ O_{234} \ O_{1234}]$$

Λ_{AP} is given next, where $f \equiv 1 + 2z + y$.

$$\Lambda_{AP} =$$

0	0	0	0	0	$2^{-1}A_0+2^{-1}zA_0$	0	0	0	A_0	0	zA_0	A_0+zA_0-f
0	0	0	0	0	0	0	0	$2^{-1}A_0+2^{-1}zA_0$	$2^{-1}A_1+2^{-1}zA_0$	zA_1	zA_1	$2^{-1}A_1+2^{-1}zA_1-f$
0	$2^{-1}zA_0$	0	0	$2^{-1}zA_0$	0	$2^{-1}A_0+2^{-1}zA_0$	$2^{-1}zA_0$	0	$2^{-1}A_2$	$2^{-1}A_2+2^{-1}zA_2$	$2^{-1}A_0+2^{-1}zA_2-f$	$2^{-1}A_2+zA_2$
0	$2^{-1}zA_0$	0	zA_0	0	zA_0	$2^{-1}A_0+2^{-1}zA_0$	zA_0	$2^{-1}zA_2$	$2^{-1}A_2$	$2^{-1}A_2+2^{-1}zA_2$	$2^{-1}A_2+2^{-1}zA_2$	$2^{-1}A_2$
0	0	$2^{-1}A_0+2^{-1}zA_0$	0	$2^{-1}A_0+2^{-1}zA_0$	0	0	$2^{-1}A_0+2^{-1}zA_0$	$2^{-1}A_1+zA_0+2^{-1}zA_0$	$2^{-1}A_1+2^{-1}zA_1$	$2^{-1}A_1+2^{-1}zA_1$	$2^{-1}A_0+2^{-1}zA_0$	$2^{-1}A_1+2^{-1}zA_1$
0	$2^{-1}A_0$	zA_0	$2^{-1}A_0$	0	zA_0	$2^{-1}A_0+2^{-1}zA_0$	0	$2^{-1}A_{12}$	$2^{-1}A_{12}$	$2^{-1}A_1+2^{-1}zA_{12}$	$2^{-1}zA_{12}+2^{-1}zA_{12}$	$2^{-1}A_{12}+2^{-1}zA_{12}+zA_{12}$
0	$2^{-1}zA_1$	$2^{-1}A_1$	zA_1	$2^{-1}A_1$	$2^{-1}A_0+2^{-1}zA_0$	$2^{-1}zA_{12}$	$2^{-1}A_2$	$zA_1+2^{-1}zA_1$	$2^{-1}zA_2$	0	$2^{-1}A_2+zA_1$	$2^{-1}zA_2$
0	$2^{-1}A_1$	zA_1	0	0	$2^{-1}A_0$	$2^{-1}A_2$	zA_1	0	$2^{-1}A_1+zA_1-f$	A_1+zA_1	0	zA_1
0	A_2	zA_2	0	zA_2	A_1+zA_1	A_1	A_2+zA_2	0	0	0	A_2+zA_2	0
$2^{-1}A_0+2^{-1}zA_0$	0	0	0	0	A_2	0	0	zA_2+zA_2-f	0	zA_2	0	0
0	zA_1	$2^{-1}zA_1$	$2^{-1}A_2$	zA_1	0	A_1+zA_1	A_2	zA_2+zA_2-f	0	0	0	0
0	0	0	0	0	0	0	0	0	0	0	0	0

$$\Lambda_{AP} = \begin{pmatrix}
0 & 2^{-1}zA_1 & 2^{-1}A_2+2^{-1}zA_2 & 2^{-1}A_1+zA_2 & 0 & 2^{-1}A_2 & 2^{-1}A_1+2^{-1}zA_2+zA_1-f & 2^{-1}zA_1 & 2^{-1}A_1+2^{-1}zA_2 & zA_2 & 2^{-1}A_2+2^{-1}zA_2 & 2^{-1}A_1 & 2^{-1}A_1+2^{-1}zA_1 & 0 & 2^{-1}zA_2 \\
2^{-1}A_0+2^{-1}zA_0 & 2^{-1}A_0 & zA_0 & 2^{-1}zA_0 & 2^{-1}A_0+2^{-1}zA_0 & zA_0+2^{-1}zA_{12}-f & 0 & 2^{-1}A_2 & 2^{-1}A_{12}+2^{-1}zA_{12} & 2^{-1}zA_{12} & 0 & 2^{-1}zA_0 & 0 & 2^{-1}zA_{12} & 2^{-1}A_1-2^{-1}zA_{12} \\
2^{-1}A_2+2^{-1}zA_2 & 2^{-1}A_2+2^{-1}zA_2 & 0 & 2^{-1}A_{12}+2^{-1}zA_2 & 2^{-1}A_0+2^{-1}zA_{12}+zA_2-f & 0 & 2^{-1}A_{12}-zA_{12}+2^{-1}zA_{12} & 2^{-1}A_{12}+2^{-1}zA_{12} & 0 & zA_{12}+2^{-1}zA_{12} & 2^{-1}A_2+2^{-1}zA_2 & 2^{-1}A_{12}+2^{-1}zA_2 & 2^{-1}A_{12}+2^{-1}zA_{12} & 0 & 0 \\
2^{-1}A_1+2^{-1}zA_1 & 2^{-1}A_1 & 2^{-1}A_{12}+2^{-1}zA_{12} & 2^{-1}A_1+zA_2-f & 2^{-1}A_1+zA_1 & 2^{-1}A_{12} & 2^{-1}zA_{12} & 2^{-1}A_{12} & zA_{12} & 0 & 2^{-1}A_{12}+2^{-1}zA_{12} & 2^{-1}zA_1 & 0 & 2^{-1}zA_{12} & 0 \\
2^{-1}A_1+2^{-1}zA_1 & zA_1 & 2^{-1}A_{12}+2^{-1}zA_{12} & zA_1 & 2^{-1}A_1+zA_1 & 2^{-1}A_{12}+zA_1 & 0 & 2^{-1}A_{12}+2^{-1}zA_{12} & 2^{-1}A_{12}+2^{-1}zA_{12} & 2^{-1}zA_{12} & 0 & 2^{-1}A_1+zA_1 & 2^{-1}A_{12}+2^{-1}zA_{12} & 0 & 0 \\
2^{-1}A_2+2^{-1}zA_2 & 2^{-1}A_1+zA_{12}-f & zA_2 & 2^{-1}zA_1 & 2^{-1}A_2+2^{-1}zA_2 & zA_2 & 2^{-1}A_{12}+2^{-1}zA_{12} & A_{12} & 0 & 0 & 0 & zA_{12} & 2^{-1}A_{12}+2^{-1}zA_{12} & 0 & 0 \\
A_{12}+zA_{12}-f & A_{12}+zA_{12} & zA_{12} & zA_{12} & A_{12}+zA_{12} & zA_{12} & 0 & A_{12} & 0 & 0 & 0 & zA_{12} & 0 & 0 & 0
\end{pmatrix}$$

5.9 References

Antoniewicz, M.R., Kelleher, J.K., and Stephanopoulos, G. (2007) 'Elementary metabolite units (EMU): a novel framework for modeling isotopic distributions', *Metabolic Engineering*, 9: 68–86.

Arauzo-Bravo, M.J. and Shimizu, K. (2003) 'An improved method for statistical analysis of metabolic flux analysis using isotopomer mapping matrices with analytical expressions', *Journal of Biotechnology*, 105: 117–33.

Astier, C., Elmorjani, K., Meyer, I., Joset, F., and Herdman, M. (1984) 'Photosynthetic mutants of the cyanobacteria *Synechocystis* sp. Strains PCC6714 and PCC6803: Sodium p-hydroxymercuribenzoate as a selective agent', *Journal of Bacteriology*, 158: 659–64.

Canonaco, F., Hess, T.A., Heri, S., Wang, T., Szyperski, T., and Sauer U. (2001) 'Metabolic flux response to phosphoglucose isomerase knockout *Escherichia coli* and impact of overexpression of the soluble transhydrogenase UdhA', *FEMS Microbiology Letters*, 204:247–252.

Cronan, J.E. Jr. and LaPorte, D. (1996) 'Tricarboxylic acid cycle and glyoxylate bypass', in: Escherichia coli *and* Salmonella: *Cellular and Molecular Biology*, edited by F.C. Neidhardt et al. Washington, DC: American Society of Microbiology Press. pp. 206–16.

Christensen, B. and Nielsen, J. (1999) 'Isoto[pomer analysis using GC–MS', *Metabolic Engineering*, 1: 282–90.

Colman, B. and Norman, E.G. (1997) 'Serine synthesis in cyanobacteria by a non-photorespiratory pathway', *Plant Physiology*, 100: 133–6.

Dauner, M. and Sauer, U. (2001) 'Stoichiometric growth model for riboflavin-producing *Bacillus subtilis*', *Biotechnology and Bioengeering*, 76: 132–43.

Dauner, M., Bailey, J.E., and Sauer, U. (2001) 'Metabolic flux analysis with a comprehensive isotopomer model in *Bacillus subtilis*', *Biotechnology and Bioengeering*, 76: 144–56.

Davis, L. (1991) *The Handbook of Genetic Algorithms*. New York: Van Nostrand Reinhold.

Deng, M.D. and Coleman, J.R. (1999) 'Ethanol synthesis by genetic engineering in cyanobacteria', *Applied Environmental Microbiology*, 65: 523–8.

Emmerling, M., Dauner, M., Ponti, A., Fiaux, J., Hochuli, M. et al. (2002) 'Metabolic flux responses to pyruvate kinase knockout in *Escherichia coli*', *Journal of Bacteriology*, 184: 152–64.

Fischer, E. and Sauer, U. (2003) 'Metabolic flux profiling of *Escherichia coli* mutants in central carbon metabolism using GC-MS', *European Journal of Biochemistry*, 270: 880–91.

Forbes, N.S., Clark, D.S., and Blanch, H.W. (2001) 'Using isotopomer path tracing to quantify metabolic fluxes in pathway models containing reversible reactions', *Biotechnology and Bioengeering*, 74: 196–211.

Fraenkel, D.G. and Levisohn, S.R. (1967) 'Glucose and gluconate metabolism in an *Escherichia coli* mutant lacking phosphoglucose isomerase', *Journal of Bacteriology*, 93: 1571–8.

Gray, M.W. (1993) 'Origin and evolution of organelle genomes', *Current Opinion in Genetic Development*, 3: 884–890.

Hansel, A. and Lindblad, P. (1998) 'Towards optimization of cyanobacteria as biotechnologically relevant producers of molecular hydrogen, a clean and renewable energy source, *Applied Microbiology and Biotechnology*, 50: 153–60.

Holms, W.H. (1986) 'The central metabolic pathways of *Escherichia coli*: Relationship between flux and control at a branch point, efficiency of conversion to biomass, and excretion of acetate', *Current Topics in Cellular Regulation*, 28: 69–105.

Hua, Q. Yang, C., Baba, Mori, H., and Shimizu, K. (2003) 'Responses of the central carbon metabolism in *Escherichia coli* to phosphoglucose isomerase and glucose-6-phosphate dehydrogenase knockouts', *Journal of Bacteriology*, 185: 7053–67.

Izui, K., Taguchi, M., Morikawa, M., and Katsuki, H. (1981) 'Regulation of *Escherichia coli* phosphoenolpyruvate carboxylase by multiple effectors *in vivo*. II. Kinetic studies with a reaction system containing physiological concentrations of ligands', *Journal of Biochemistry*, 90: 1321–31.

Jeffrey, F.M., Rajagopal, A., Malloy, C.R., and Sherry, A.D. (1991) '^{13}C-NMR: a simple yet comprehensive method for analysis of intermediary metabolism', *Trends in Biochemical Science*, 16: 5–10.

Kai, Y., Matsumura, H., Inoue, T., Terada, K., Nagara, Y. et al. (1999) 'Three-dimensional structure of phosphoenolpyruvate carboxylase: A proposed mechanism for allosteric inhibition', *Proceedings of the National Academy of Science USA*, 96: 823–8.

Katz, J., Wals, P. and Lee, W.N.P. (1993) 'Isotopomer studies of gluconeogenesis and the Krebs cycle with ^{13}C-labeled lactate', *Journal of Biological Chemistry*, 268: 25509–21.

Kelleher, J.K. (2001) 'Flux estimation using isotopic tracers: Common ground for metabolic physiology and metabolic engineering', *Metabolic Engineering*, 3: 100–10.

Klapa, M.I., Park, S.M., Sinskey, A.J., and Stephanopoulos, G. (1999) 'Metabolite and isotopomer balancing in the analysis of metabolic cycles: I Theory', *Biotechnology and Bioengineering*, 62: 375–91.

Krebs, A. and Bridger, W.A. (1980) 'The kinetic properties of phosphoenolpyruvate carboxykinase of *Escherichia coli*', *Canadian Journal of Biochemistry*, 58: 309–18.

Lee, W.N.P., Byerley, L.O., Bergner, E.A., and Edmond, J. (1991) 'Mass isotopomer analysis: Theoretical and practical considerations', *Biological Mass Spectrometry*, 20: 451–8.

Lee, W.N.P., Bergner, E.A., and Guo, Z.K. (1992) 'Mass isotopomer pattern and precursor-product relationship', *Biological Mass Spectrometry*, 21: 114–22.

Marx, A., de Graaf, A.A., Wiechert, W., Eggeling, L., and Sahm, H. (1996) 'Determination of the fluxes in the central metabolism of *Corynebacterium glutamicum* by nuclear magnetic resonance spectroscopy combined with metabolite balancing', *Biotechnology and Bioengeering*, 49: 111–29.

Matsuoka, Y. and Shimizu, K. (2010) 'The relationships between the metabolic fluxes and ^{13}C-labeled isotopomer distribution for the flux analysis of the main metabolic pathways', *Biochemical Engineering Journal*, 49: 326–36.

Mollney, M., Wiechert, W., Kownatzki, D., and de Graaf, A.A. (1999) 'Bidirectional reaction steps in metabolic networks: IV. Optimal design of isotopomer labeling experiments', *Biotechnology and Bioengineering*, 66: 86–103.

Moré, J.J. (1977) 'The Levenberg-Marquardt algorithm: implementation and theory', in: *Numerical Analysis*, edited by G.A. Watson. New York: Springer.

Noronha, S.B., Yeh, H.J.C., Spande, T.F., and Shiloach, J. (2000) 'Investigation of the TCA cycle and the glyoxylate shunt in *Escherichia coli* BL21 and JM109 using ^{13}C-NMR/MS', *Biotechnology and Bioengeering*, 68: 316–27.

Owittrim, G.W. and Colman, B. (1988) 'Phosphoenolpyruvate carboxylase mediated carbon flow in a cyanobacterium', *Biochemistry and Cell Biology*, 66, 93–9.

Park, S.M., Klapa, M.I., Sinskey A.J., and Stephanopoulos, G. (1999) 'Metabolite and isotopomer balancing in the analysis of metabolic cycles: II Applications', *Biotechnology and Bioengineering*, 62: 392–401.

Pearce, J. and Carr, N.G. (1967) 'The metabolism of acetate by the blue-green algae, *Anabaena variabilis* and *Anacystis nidulans*', *Journal of General Microbiology*, 49: 301–13.

Pearce, J., Leach, C.K., and Carr, N.G. (1969) 'The incomplete tricarboxylic acid cycle in the blue-green alga *Anabaena variabilis*', *Journal of General Microbiology*, 55: 371–8.

Pelroy, R.A. and Bassham, J.A. (1972) 'Photosynthetic and dark carbon metabolism in unicellular blue-green algae', *Archives of Microbiology*, 86: 25–38.

Pelroy, R.A., Rippka, R., and Stanier, R.Y. (1972) 'Metabolism of glucose by unicellular blue-green algae', *Archives of Microbiology*, 87: 303–22.

Pelroy, R.A. and Bassham, J.A. (1976) 'Kinetics of light-dark CO_2 fixation and glucose assimilation by *Aphanocapsa*', *Journal of Bacteriology*, 128: 633–43.

Petersen, S., de Graaf, A.A., Eggeling, L., Mollney, M., Wiechert, W., and Sahm, H. (2000) '*In vivo* quantification of parallel and bidirectional fluxes in the anaplerosis of *Corynebacterium glutamicum*', *Journal of Biological Chemistry*, 275: 35932–841.

Petersen, S., Mack, C., de Graaf, A.A., Riedel, C., Eikmanns, B.J., and Sahm, H. (2001) 'Metabolic consequences of altered phosphoenolpyruvate carboxykinase activity in *Corynebacterium glutamicum* reveal anaplerotic regulation mechanisms *in vivo*', *Metabolic Engineering*, 3: 344–61.

Roessner, U., Wagner, C., Kopka, J., Trethewey, R.N., and Willmitzer, L. (2000) 'Simultaneous analysis of metabolites in potato tuber by gas chromatography–mass spectrometry', *Plant Journal*, 23: 131–42.

Sauer, U., Hatzimanikatis, V., Bailey, J.E., Hochuli, M., Szyperski, T., and Wutrich, K. (1997) 'Metabolic fluxes in riboflavin-producing *Bacillus subtilis*', *Nature Biotechnology*, 15: 448–52.

Sauer, U., Lasko, D.R., Fiaux, J., Hochuli, M., Glaser, R. et al. (1999) 'Metabolic flux ratio analysis of genetic and environmental modulations of *Escherichia coli* central carbon metabolism', *Journal of Bacteriology*, 181: 6679–88.

Sauer, U. (2006) 'Metabolic networks in motion: ^{13}C-based flux analysis', *Molecular Systems Biology*, doi: 10.1038/msb4100109.

Sauer, U., Hatzimanikatis, V., Hohmann, H.P., Mannebegr, M., and Bailey, J.E. (1996) 'Physiology and metabolic fluxes of wild-type and Riboflavin-producing, *Bacillus subtilis*', *Applied Environmental Microbiology*, 62: 3687–95.

Schmetterer, G. (1994) 'Cyanobacterial respiration', in: *The Molecular Biology of Cyanobacteria*, edited by D.A. Bryant. The Netherlands: Kluwer Academic. pp. 409–35.

Schmidt, K., Carlsen, M., Nielsen, J., and Villadsen, J. (1997) 'Modeling isotopomer distributions in biochemical networks using isotopomer mapping matrices', *Biotechnology and Bioengineering*, 55: 831–40.

Schmidt, K., Nielsen, J., and Villadsen, J. (1999a) 'Quantitative analysis of metabolic fluxes in *Escherichia coli* using two-dimensional NMR spectroscopy and complete isotopomer models,' *Journal of Biotechnology*, 71: 175–90.

Schmidt, K., Norregaard, L.C., Pedersen, B., Meissner, A., Duus, J.O. et al. (1999b) 'Quantification of intracellular fluxes from fractional enrichment and ^{13}C-^{13}C coupling constraints on the isotopomer distribution in labeled biomass components', *Metabolic Engineering*, 1: 166–79.

Sherry, A.D., Jeffrey, F.M.H., and Malloy, C.R. (2004) 'Analytical solutions for ^{13}C isotopomer analysis of complex metabolic conditions: substrate oxidation, multiple pyruvate cycles, and gluconeogenesis,' *Metabolic Engineering*, 6: 12–24.

Smith, A.J. (1983) 'Modes of cyanobacterial carbon metabolism,' *Annals of Microbiology*, 134B: 93–113.

Szyperski, T. (1995) 'Biosynthetically directed fractional ^{13}C-labeling of proteinogenic amino acids: An efficient analytical tool to investigate intermediary metabolism', *European Journal of Biochemistry*, 232: 433–48.

Tabita, F.R. (1994) 'The biochemistry and molecular regulation of carbon dioxide metabolism in cyanobacteria', in: *The Molecular Biology of Cyanobacteria*, edited by D.A. Bryant. The Netherlands: Kluwer Academic. pp. 437–67.

Toya, Y., Ishii, N., Hirasawa, T., Naba, M., Hirai, K. et al. (2007) 'Direct measurement of isotopomer of intracellular metabolites using capillary electrophoresis time-of-flight mass spectrometry for efficient metabolic flux analysis', *Journal of Chromatography*, 1159: 134–41.

Toya, Y., Ishii, N., Nakahigashi, K., Hirasawa, T., Soga, T. et al. (2010) '^{13}C-Metabolic Flux Analysis for batch culture of *Escherichia coli* and its *pyk* and *pgi* gene knockout mutants based on mass isotopomer distribution of intracellular metabolites', *Biotechology Progress*, DOI 10.1002/(ISSN) 1520-6033

van Winden, W.A., Heijnen, J.J., and Verheijen, P.J.T (2002) 'Cumulative bondomers: new concept in flux analysis from 2D [^{13}C, ^{1}H]-COSY NMR data', *Biotechnology and Bioengineering*, 80: 731–45.

Varma, A. and Palsson, B.O. (1995) 'Parametric sensitivity of stoichiometric flux balance models applied to wild-type *Escherichia coli* metabolism', *Biotechnology and Bioengineering*, 45: 69–79.

Vazquez-Bermudex, M.F., Herrero, A., and Flores, E. (2000) 'Uptake of 2-oxoglutarate in *Synechococcus* strain transformed with the *Escherichia coli kgtP* gene', *Journal of Bacteriology*, 182: 211–15.

Walsh, K. and Koshland, D.E. Jr. (1984) 'Determination of flux through the branch point of two metabolic cycles', *Journal of Biological Chemistry*, 259: 9646–54.

Walsh, K. and Koshland, D.E. Jr. (1985) 'Branch point control by the phosphorylation state of isocitrate dehydrogenase', *Journal of Biological Chemistry*, 260: 8430–7.

Wiechert, W. and de Graaf, A.A. (1997) 'Bidirectional reaction steps in metabolic networks: I. Modeling and simulation of carbon isotope labeling experiments', *Biotechnology and Bioengineering*, 55: 101–17.

Wiechert, W., Siefke, C., de Graaf, A.A., and Marx, A. (1997) 'Bidirectional reaction steps in metabolic networks: II. Flux estimation and statistical analysis', *Biotechnology and Bioengineering*, 55, 118–35.

Wiechert, W., Mollney, M., Isermann, N., Wurzel, M., and de Graaf, A.A. (1999) 'Bidirectional reaction steps in metabolic networks: III Explicit solution and analysis of isotopomer labeling systems', *Biotechnology and Bioengineering*, 66: 69–85.

Wittmann, C. and Heinzle, E. (1999) 'Mass spectrometry for metabolic flux analysis', *Biotechnology and Bioengineering*, 62: 739–50.

Wittmann, C. (2007) 'Fluxome analysis using GC-MS', *Microbial Cell Factories*, 6: 6.

Yang, C., Hua, Q., and Shimizu, K. (2000) 'Energetic and carbon metabolism during growth of microalgal cells under photoautotrophic, mixotrophic and cyclic light-autotrophic/dark-heterotrophic conditions', *Biochemical Engineering Journal*, 6: 87–102.

Yang, C., Hua, Q., and Shimizu, K. (2002a) 'Quantitative analysis of intracellular metabolic fluxes using GC-MS and two-dimensional NMR spectroscopy', *Journal of Bioscience and Bioengineering*, 93: 78–87.

Yang, C., Hua, Q., and Shimizu, K. (2002b) 'Integration of the information from gene expression and metabolic fluxes for the analysis of the regulatory mechanisms in *Synechocystis*', *Applied Microbiology and Biotechnology*, 58: 813–22.

Yang, C., Hua, Q., and Shimizu, K. (2002c) 'Metabolic flux analysis in Synechocystis using isotope distribution from ^{13}C-labeled glucose', *Metabolic Engineering*, 4, 202–16.

Yang, C., Hua, Q., Baba, T., Mori, H., and Shimizu, K. (2003) 'Analysis of *E. coli* anaplerotic metabolism and its regulation mechanism from the metabolic responses to alter dilution rates and *pck* knockout', *Biotechnology and Bioengineering*, 84: 129–44.

Young, J.D., Shastri, A.A., Stephanopoulos, G., and Morgan, J.A. (2011) 'Mapping photoautotrophic metabolism with isotopically nonstationary 13C flux analysis', *Metabolic Engineering*, 13: 656–65.

Zhao, J. and Shimizu, K. (2003) 'Metabolic flux analysis of *Escherichia coli* K12 grown on ^{13}C-labeled acetate and glucose using GC-MS and powerful flux calculation method', *Journal of Biotechnology*, 101: 101–17.

Zupke, C. and Stephanopoulos, G. (1994) 'Modeling of isotope distributions and intracellular fluxes in metabolic networks using atom mapping matrices', *Biotechnology Progress*, 10: 489–98.

6

Effect of a specific-gene knockout on metabolism

Abstract: The metabolic regulation of the specific gene knockout *Escherichia coli* is explained in terms of ^{13}C-metabolic fluxes, enzyme activities, intracellular metabolite concentrations, and gene expressions. It is shown that the glyoxylate pathway is activated in *pckA* and *ppc* mutants, where the cell yield can be improved as compared to the wild type. In the case of the *gnd* mutant *E. coli*, the ED pathway is activated where Mez is activated for both *zwf* and *gnd* mutants to back up the production of NADPH. The regulation mechanism of the *pykF* mutant *E. coli* is analyzed, where it is shown to be robust against such genes knockout by rerouting the pathways. Moreover, the metabolic regulations of *lpdA*, *sucA*, and C mutants as well as *icdA* and *ldhA* mutants are also explained, based on ^{13}C metabolic flux analysis and other information.

Key words: ^{13}C-metabolic flux analysis; *pckA* mutant; *ppc* mutane; metabolic regulation; *gnd* mutant; *pykF* mutant; *lpdA* mutant; *sucA*, C mutant; *icdA* mutant; *ldhA* mutant.

6.1 Introduction

In order to understand the role of metabolic pathway genes in metabolism, it is useful to study the effects of a specific gene knockout (or a change in culture environment) on metabolism, based on the metabolic flux distribution obtained by ^{13}C labeling experiments as well as gene expressions, protein expressions (enzyme activities), and intracellular metabolite concentrations (Shimizu, 2004). Here, the change in metabolism due to a single-gene mutation is discussed, focusing on

E. coli metabolism. The different levels of information, such as ^{13}C-metabolic flux analysis, gene expressions by DNA microarray with qRT-PCR, protein expressions by 2D DIGE with shotgun proteomics, metabolite concentrations by CE-TOFMS with LC-MS/MS, have been reported for 24 single-gene knockout *E. coli* mutants, such as *galM, glk, pgm, pgi, pfkA, pfkB, fbp, fbaB, gapC, gpmA, gpmB, pykA, pykF, ppsA, zwf, pgl, gnd, rpe, rpiA, rpiB, tktA, tktB, talA,* and *talB* mutants for a continuous culture at a dilution rate of 0.2 h^{-1}, as well as the wild-type *E. coli* BW25113 for a continuous culture at different dilution rates (Ishii et al., 2007). From such data, it is found that while some single-gene knockout mutations in the central metabolism preclude growth on glucose, the majority of such variations seem to be potentially compensated by either using alternative enzymes or by the re-routing of carbon fluxes through alternative pathways, resulting in the robustness of the cell's phenotype such as cell growth (Ishii et al., 2007). This chapter considers how metabolism changes in response to specific pathway gene mutations.

Basically, the regulation of metabolic processes ultimately depends upon the control of protein synthesis and enzyme activity (Martin, 1987). There are three general mechanisms by which the activity of enzymes can be regulated, such as control by reversible binding of effectors, by covalent modification, and by alteration of enzyme concentration. In the first case, the enzyme is activated or inhibited by binding of a signal molecule, which may or may not be the substrate or product of the enzyme reaction, to the specific regulatory site, producing a conformational change. Substrate effect and allosteric control may be the example. In covalent modification, the structure of the enzyme can be altered by the action of other enzymes. For instance, regulation by phosphorylation, i.e. phosphate, is incorporated into the enzyme by a protein kinase using ATP, and is removed by a protein phosphatase. The third mechanism that regulates enzyme activity is the alteration of the concentration of the enzyme protein in the cell. The concentration of a protein in the cell is governed by the rate of the synthesis and the rate of degradation. The rate of synthesis of a particular protein may be controlled at several different levels, such as the rate of transcription of the gene, etc. Other possible sites of control are the processing of the transcript to give mRNA, the transfer of mRNA out of the nucleus, the rate of degradation of mRNA in the cytoplasm, or the rate of translation of mRNA to make the protein on the ribosome. There is strong evidence that the rate of transcription is under rigorous control, which is important in determining the enzyme profile of a particular cell type (Martin, 1987).

6.2 Effect of *ppc* and *pck* gene knockout on metabolism

As stated in Chapter 5, if the *pckA* gene, which codes for the gluconeogenic PEP carboxikinase, is knocked out, the formation of OAA is reduced, which activates the glyoxylate pathway. The regulation mechanism is also explained in Chapter 5. The reversible phosphorylation/inactivation of ICDH is catalyzed by the bifunctional enzyme ICDH-kinase/phosphatase, where it is regulated by a number of effectors including OAA, which inhibits the ICDH-kinase and stimulates phosphatase. Thus the decrease in OAA concentrations due to the *pckA* gene knockout gene causes the phosphorylation/inactivation of ICDH, which results in an increase in isocitrate concentration, and the flux through Icl also significantly increases (Yang et al., 2003). Moreover, the *pckA* gene knockout causes accumulation of PEP, which in turn inhibits Pfk activity, and thus reduces the glucose consumption rate. This mutant produces less acetate and CO_2, resulting in higher cell yield with less growth rate than the wild type (Yang et al., 2003). It is shown that *in vivo* regulation of the Pck flux occurs mainly by modulation of enzyme activity and by the changes in PEP and OAA concentrations rather than by the ATP/ADP ratio. This indicates that the reaction catalyzed by Pck can respond flexibly to the availability of PEP and OAA, which may form the metabolic cycle at low glucose concentrations (Fischer and Sauer, 2003). It is known that PEP is an important intermediate in metabolism, since it alone directly regulates the phosphotransferase (PTS) system but also affects Pfk and Pyk activities. Since PEP can be formed gluconeogenetically through Pck from the TCA cycle, Pck serves to maintain the relative balance between OAA and PEP pools and drain off excess carbon of the TCA cycle to supply PEP for cellular requirements (Yang et al., 2003).

In relation to the *pckA* gene knockout, consider the effect of the *ppc* gene knockout on metabolism. Among the central metabolic pathway enzymes of *E. coli*, PEP carboxylase (Ppc) plays an anaplerotic role in replenishing OAA consumed in biosynthetic reactions and keeping the TCA cycle intermediates from starvation. Chang et al. (1999) reported increased D-lactate production by a *ppc* mutant of *E. coli* under anaerobic conditions. Farmer and Liao (1997) over-expressed Ppc and/or induced a glyoxylate shunt by the *fadR* knockout gene, to elevate TCA cycle activity for the reduction of acetate excretion under aerobic conditions, since acetate excretion is a major obstacle in recombinant protein production.

Here, we consider the effects of *ppc* mutation on metabolism based on ^{13}C-labeling experiments, together with enzyme activity and intracellular metabolite concentrations, where *E. coli* BW25113 and its *ppc* mutant (JWK3928) are considered (Peng et al., 2004).

Growth parameters at the exponential growth phase in the aerobic batch culture are shown in Table 6.1. The specific cell growth rate and specific glucose consumption rate of the *ppc* mutant are lower than those of the wild type. The CO_2 evolution rate (CER) is also reduced in the *ppc* mutant. Note that little acetate is excreted in the *ppc* mutant during cultivation, whereas the wild-type *E. coli* produces acetate significantly. As a consequence, an improvement of biomass yield on glucose can be observed in the *ppc* mutant.

Some key enzyme activites of central metabolism, for the cells grown at the exponential growth phase, are shown in Table 6.2. The wild-type *E. coli* shows high Ppc activity, indicating that this anaplerotic pathway plays an important role in metabolism. The activity of Pck, which catalyzes the reverse reaction to Ppc, is considerably lower in the *ppc* mutant. The activities of glycolytic enzymes, such as Hxk, Pgi, Pfk, Fba, and GAPDH and the pentose phosphate (PP) pathway enzymes, such as G6PDH, 6PGDH, and Tal, are all significantly decreased in the *ppc* mutant (Table 6.2). These data correspond to the slower growth and lower glucose utilization rates of the *ppc* mutant (Table 6.1). However, Pyk is an exception, where it is increased 2.4-fold (Table 6.2). The up-regulation of Pyk is expected to channel more carbon flux from PEP to pyruvate due to the blockage of Ppc in the *ppc* mutant. Accordingly,

Table 6.1 Cell growth parameters of the wild-type *E. coli* and its *ppc* mutant grown on glucose under aerobic conditions

Parameter	Wild type	*ppc* mutant
μ_{max} (h^{-1})	0.41	0.34
q_{glc} (mmol g^{-1} h^{-1})[a]	5.61	3.16
q_{CO2} (mmol g^{-1} h^{-1})[b]	3.38	2.47
Q_{ace} (mmol g^{-1})[c]	0.22	0.00
$Y_{biomass/glc}$ (g g^{-1})[d]	0.48	0.55

[a]: specific glucose consumption rate
[b]: specific carbon dioxide evolution rate
[c]: specific acetate production rate
[d]: yield of biomass on glucose.

Effect of specific-gene knockout

Table 6.2 Specific enzyme activities of the wild-type *E. coli* and its *ppc* mutant grown on glucose under aerobic conditions

Enzyme	Wild type	ppc mutant
Hexokinase (Hxk)	0.0752 ± 0.0001	0.0466 ± 0.0001
Glucose phosphate isomerase (Pgi)	2.078 ± 0.004	0.436 ± 0.002
Phosphofructose kinase (Pfk)	0.564 ± 0.003	0.170 ± 0.002
Fructose bisphosphate aldolase (Fba)	0.675 ± 0.003	0.0607 ± 0.0005
Glyceraldehyde-3-phosphate dehydrogenase (GAPDH)	0.692 ± 0.004	0.0899 ± 0.0005
Pyruvate kinase (Pyk)	0.220 ± 0.005	0.534 ± 0.004
Phosphoenolpyruvate carboxylase (Ppc)	0.0448 ± 0.0005	—[a]
Phosphoenolpyruvate carboxykinase (Pck)	0.0427 ± 0.0004	0.0021 ± 0.0004
NADP+-specific malic enzyme (Mae)	0.067 ± 0.002	0.082 ± 0.002
NAD+-specific malic enzyme (Sfc)	0.088 ± 0.005	—[a]
Glucose-6-phosphate dehydrogenase (G6PDH)	0.245 ± 0.002	0.170 ± 0.002
6-Phosphogluconate dehydrogenase (6PGDH)	0.426 ± 0.003	0.00085 ± 0.00004
Citrate synthase (CS)	0.319 ± 0.002	0.369 ± 0.003
Aconitase (Acn)	0.115 ± 0.002	0.179 ± 0.002
Isocitrate lyase (Icl)	0.0361 ± 0.0003	0.113 ± 0.004
Isocitrate dehydrogenase (ICDH)	2.230 ± 0.002	0.971 ± 0.004
Malate dehydrogenase (MDH)	0.121 ± 0.007	0.193 ± 0.006

The unit of the enzyme activity is $\mu mol\ min^{-1}\ mg\ (protein)^{-1}$
Overall ED pathway activity is represented as $mg\ (pyruvate)\ min^{-1}\ mg\ (protein)^{-1}$
[a] Not detected.

CS, the first enzyme of the TCA cycle, is up-regulated 1.8-fold in the *ppc* mutant (Table 6.2). This up-regulation is expected to increase the carbon flux from pyruvate into the TCA cycle for the replenishment of OAA in response to Ppc deficiency. The activities of Acn and MDH, but not ICDH, in the TCA cycle, increase in a co-ordinated manner. It is reported that CS and Acn, but not ICDH, are regulated co-ordinately, which might be due to the fact that citrate is an activator of Acn (Nakano et al., 1998). Note that Icl, encoded by *aceA* involved in the glyoxylate shunt, is significantly induced by about 3.1-fold in the *ppc* mutant (Table 6.2). This regulation pattern clearly demonstrates that the *ppc* mutant utilizes an alternative anaplerotic pathway, the glyoxylate shunt, to replenish OAA in response to the blockage through Ppc.

The concentrations of some key intracellular metabolites are shown in Table 6.3, where the glycolytic intermediates, such as G6P, F6P, F1, 6P, and PEP and the PP pathway intermediate, such as 6PG, accumulate in the *ppc* mutant due to inhibition of enzyme activity in their related pathways. In contrast, the intracellular concentration of AcCoA decreases in the *ppc* mutant, implying higher activity of the TCA cycle (or glyoxylate pathway) relative to glycolysis. OAA is not detected in the *ppc* mutant due to its low abundance, but it is detected in the wild type at low concentrations, implying that intracellular OAA is limited in the *ppc* mutant.

Figures 6.1a and b show metabolic flux distributions in the wild-type *E. coli* and its *ppc* mutant. Three notable differences between the two strains can be observed. First, 50.7% of carbon flux channeled through Ppc, and the backflow through Pck, is also high, accounting for 23.6% of the total glucose consumed in the wild-type strain. The flux from OAA to PEP catalyzed by Pck is significantly decreased to only 0.05% in the *ppc* mutant. These flux results are consistent with those of enzyme activity

Table 6.3 Intracellular metabolite concentrations in the wild-type *E. coli* and its *ppc* mutant grown on glucose under aerobic conditions

Intracellular metabolite (mM)	Wild type	ppc mutant
G6P	0.68 ± 0.09	1.213 ± 0.13
F6P	0.29 ± 0.06	0.48 ± 0.07
F1,6P	1.01 ± 0.1	3.1 ± 0.21
PEP	0.07 ± 0.006	0.2 ± 0.06
6PG	0.38 ± 0.07	0.56 ± 0.09
AcCoA	0.13 ± 0.04	0.077 ± 0.005
OAA	0.03 ± 0.001	—[a]

Cell volume: 2.55 μl mg $(DCW)^{-1}$
[a] Not detected.

Effect of specific-gene knockout

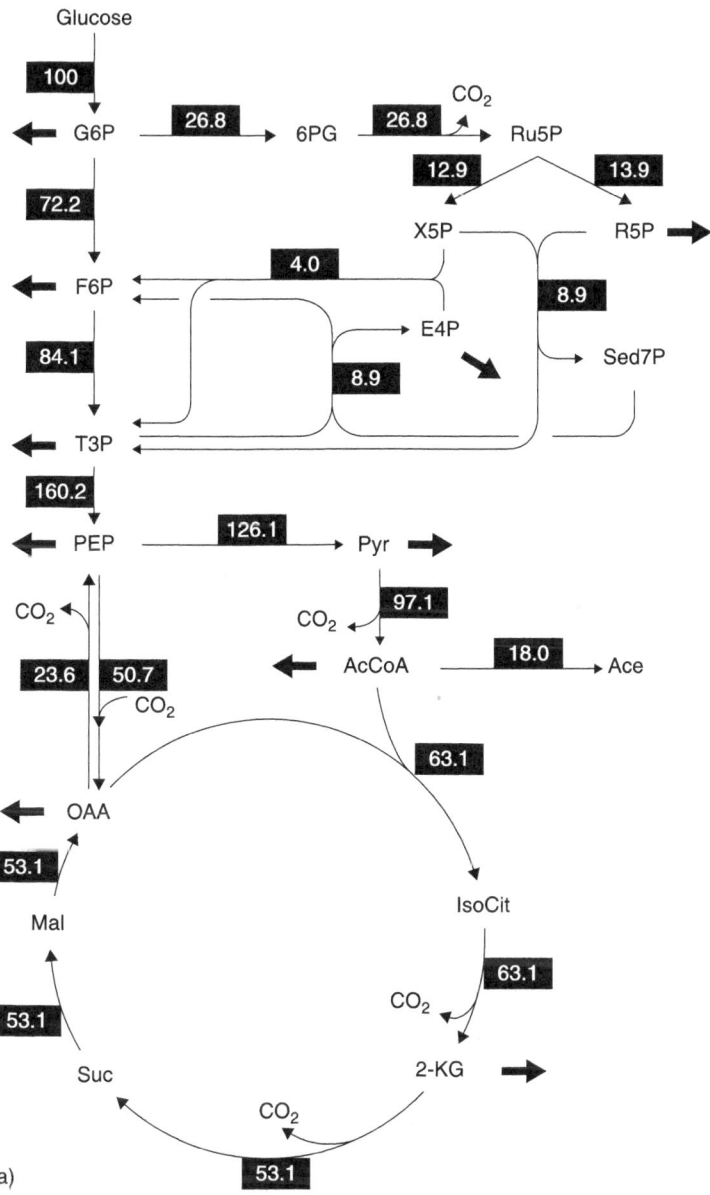

Figure 6.1 Metabolic flux distribution in the chemostat culture of the wild type *E. coli* (a) and ppc mutant (b).

Bacterial cellular metabolic systems

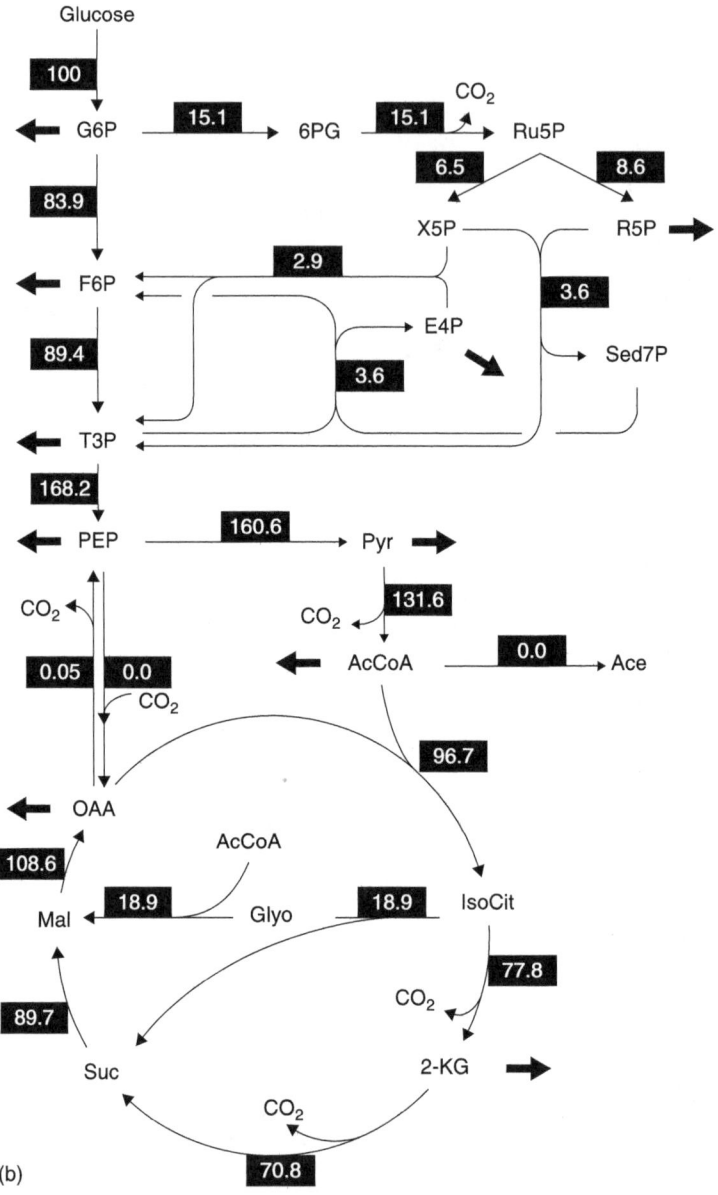

(b)

Figure 6.1 (Continued) Numbers in black boxes represent the estimated net fluxes at a dilution rate of 0.2 h^{-1}. Flux values are given in parentheses in mmol gDCW^{-1} h^{-1} relative to the specific glucose uptake rate. Arrowheads indicate the primary direction of the determined fluxes. Arrows without destination indicate the withdrawal of the precursors for biosynthesis

Effect of specific-gene knockout

(Table 6.2), which is also indicated by the visual observation of the $^{13}C-^{13}C$ scalar coupling multiplet patterns of Asp2 and Phe2 (Table 6.4). The da and dd components in the multiplets of Asp2 relate to the intact C_2-C_3 connectivity of the OAA molecule (the precursor of Asp). Since the intact C_2-C_3 fragments of OAA can be introduced only by the anaplerotic reaction of Ppc, the absence of da and dd components of Asp2 in the *ppc* mutant reflect the lack of *in vivo* activity of Ppc. In addition, the lower abundance of the db component in Phe2 in the *ppc* mutant implies less intact C_1-C_2 fragments of the PEP molecule due to the low activity of Pck. Second, the glyoxylate shunt channels 18.9% of the carbon flux in the *ppc* mutant. Visual inspection of the $^{13}C-^{13}C$ multiplets of Asp2 reveal a higher abundance of the db component in the *ppc* mutant (Table 6.4), which reflects the contribution of the glyoxylate shunt in excessive intact C_1-C_2 and C_3-C_4 connections (Maaheimo et al., 2001; Yang et al., 2003). Up-regulation of Icl activity in the *ppc* mutant (Table 6.2) is consistent with this flux analysis result. The remaining carbon flux through ICDH via the TCA cycle is still larger than that of the wild type, implying that more NADPH required for biosynthesis can be generated via the TCA cycle in the *ppc* mutant. Third, 26.8% of the carbon flux is channeled through the oxidative PP pathway in the wild type, in contrast to only 15.1% in the *ppc* mutant.

Table 6.4 The NMR spectra of cellular amino acids in the wild-type *E. coli* and its *ppc* mutant

Atom	Wild type				*ppc* mutant			
Carbon	s	da	db	dd	s	da	db	dd
Ala2	0.10	0.12	0.06	0.71	0.08	0.09	0.05	0.78
Ala3	0.42	0.58	–	–	0.25	0.75	–	–
Asp2	0.48	0.14	0.29	0.08	0.43	0.00	0.57	0.00
Glu4	0.38	0.01	0.54	0.07	0.28	0.01	0.65	0.06
Gly2	0.37	0.63	–	–	0.23	0.77	–	–
Ile2	0.50	0.01	0.37	0.11	0.39	0.07	0.49	0.04
Ile6	0.44	0.56	–	–	0.29	0.71	–	–
Phe2	0.16	0.16	0.10	0.58	0.13	0.10	0.02	0.75
Thr3	0.47	0.49	–	0.04	0.37	0.57	–	0.06
Thr4	0.55	0.45		–	0.59	0.41	–	–
Val2	0.35	0.05	0.51	0.06	0.20	0.05	0.67	0.05

Published by Woodhead Publishing Limited, 2013

PEP is a critical metabolite in *E. coli*, because of its role not only in the PTS system as a phosphoryl donor, but also in the regulation of many enzymes as an effecter. It accumulates in the *ppc* mutant and allosterically inhibits some of the glycolytic enzymes, such as Pgi and Pfk (Fraenkel, 1999). Inhibition of these enzymes leads to higher intracellular concentrations of their intermediates, such as F6P and F1,6BP (FDP), which in turn affect some other enzymes. For instance, G6PDH is allosterically inhibited by FDP and PRPP and induced by glucose, while 6PGDH is inhibited by FDP, PRPP, GAP, Ru5P, E4P, and NADPH and induced by gluconate (Sugimoto and Shiio, 1987). Apparently, the higher concentrations of intracellular F1,6BP (FDP) in the *ppc* mutant partially cause down-regulation of both enzymes. In addition, the higher flux through the TCA cycle produces more NADPH in the *ppc* mutant, which may also be considered the reason for the down-regulation of 6PGDH. The transcript of the glucose transport gene, *ptsG*, is also associated with accumulation of the glycolytic intermediates, such as G6P and F6P, which degrade the mRNA of *ptsG* by activating the RNaseP enzyme (Morita et al., 2003). Both down-regulation of the glycolytic and PP pathway enzymes results in slower growth rates and lower glucose uptake rates in the *ppc* mutant (Table 6.1). The remarkably reduced Pck activity in the *ppc* mutant may be caused by the higher intracellular concentration of PEP, since Pck is allosterically inhibited by nucleotides ATP and PEP (Krebs and Bridger, 1980). Note that the activation of the glyoxylate shunt contributes to the reduction in CO_2 production in the *ppc* mutant.

6.3 Effect of *zwf* and *gnd* genes knockout on metabolism

The effect of the *zwf* gene knockout on the fluxes is explained in Chapter 5, when compared with *pgi* gene knockout. Although the *zwf* gene knockout shows little influence on central metabolism under glucose-limited continuous culture by flux re-routing via the non-oxidative PP pathway (Zhao et al., 2004a), this mutant shows significant overflow metabolism and extremely low TCA cycle fluxes under ammonia limited conditions (Hua et al., 2003). The effect of the *gnd* gene knockout on metabolism is different from the *zwf* gene knockout mutation. The *gnd* gene knockout activates the ED pathway and causes a decrease in the flux through G6PDH, which reduces NADPH production through the oxidative PP pathway. This decreased NADPH production is backed up by activating Mez using MAL together with transhydrogenase Pnt. The reduced OAA due to the utilization of MAL for Mez, Ppc is up-regulated while Pck is down-regulated. The shortage of R5P, due to the decreased flux through the oxidative PP pathway, is backed up by reversing the non-oxidative PP pathway as compared to the wild type. This back-up system has little effect on the phenotype such as the cell growth rate, while glucose consumption rates are increased (Zhao et al., 2004b). The detailed metabolic regulation analysis is explained in the following.

The batch cultivation results of *zwf* and *gnd* mutants for glucose and pyruvate as carbon sources are given in Table 6.5. As can be seen in Table 6.6, the specific substrate uptake rate and specific CO_2 evolution

Table 6.5 Exponential growth rates of *E. coli* wild-type (*WT*) and mutant cultures on glucose/pyruvate media

Growth rate	Glucose culture			Pyruvate culture		
	WT	gnd	zwf	WT	gnd	zwf
(μ)	0.62	0.60	0.56	0.38	0.39	0.36

Table 6.6 Metabolic parameters of *E. coli* continuous cultures at $D = 0.2\ h^{-1}$

Parameter	Glucose culture			Pyruvate culture		
	Wild	gnd	zwf	Wild	gnd	zwf
q_S	3.20±0.10	3.75±0.15	3.82±0.07	7.05±0.08	6.68±0.13	6.28±0.10
q_{CO2}	8.17±0.30	9.90±0.50	11.00±0.61	10.82±0.58	9.69±0.51	8.35±0.40
q_{ace}	0.58±0.07	1.31±0.10	1.11±0.07	0.33±0.03	0.37±0.06	0.56±0.05
$Y_{X/S}$	0.35±0.01	0.30±0.02	0.29±0.01	0.26±0.02	0.27±0.03	0.29±0.01
$Y_{CO2/S}$	2.55±0.09	2.64±0.03	2.88±0.12	1.53±0.05	1.45±0.04	1.33±0.07
$Y_{ace/S}$	0.18±0.03	0.35±0.04	0.29±0.02	0.05±0.01	0.06±0.01	0.09±0.02

q_S : Carbon source consumption rate (mmol $g^{-1}\ h^{-1}$)
q_{CO2} : CO_2 evolution rate (mmol $g^{-1}\ h^{-1}$)
q_{ace} : acetate formation rate (mmol $g^{-1}\ h^{-1}$)
$Y_{X/S}$: biomass yield on substrate (g g^{-1})
$Y_{CO2/S}$: CO_2 yield on substrate (mmol $mmol^{-1}$)
$Y_{ace/S}$: acetate yield on substrate (mmol $mmol^{-1}$)

rate (CER) obtained for either mutant grown on glucose are somewhat higher than those obtained for the parent strain. This is contrary to what is observed for mutants grown on pyruvate, where these two rates are markedly decreased. The results also imply the differences in acetate excretion between the parent strain and the mutants. Gene deletion leads to increased acetate production; and this effect is more significant when glucose is used as the sole carbon source.

The enzyme activity in Table 6.7, shows a remarkable increase in Pgi activity for the glucose-grown mutant, as compared to the parent strain. Although a significant increase in the specific activity of Pgi is observed for the glucose-grown *gnd* mutant, the activity of TCA-related enzymes (e.g. ICDH) is only a little higher than that of the parent strain. However, in the case of the *zwf* mutant, a remarkable increase in the activity of ICDH is observed.

The specific activity of Pgi decreases 3.7-fold in the parent strain grown on pyruvate, as compared to that grown on glucose. Moreover, Pgi activity is significantly lower in both mutants than in the parent strain. It is worth noting that a high level of ED pathway enzymes is detected for the *gnd* mutant (398 ± 29 units of activity). Likewise, an active pathway through Mez is observed for both mutants.

Let us now consider metabolic flux analysis. Here, only those that are located at key branch points of the metabolic pathways are shown as the absolute net fluxes with lower and upper bounds (Table 6.8). Others are shown in Figures 6.2 and 6.3 as the best-fit flux distributions relative to the specific uptake rates of substrates (Zhao et al., 2004b).

Table 6.7 Activities of enzymes located at key branch points and involved in NADPH formation. Activities are given in nmol min^{-1} mg^{-1} protein

Enzymes	Glucose culture			Pyruvate culture		
	Wild	gnd	zwf	Wild	gnd	zwf
6PGDH	381±34	ND	126±10	137±18	ND	114±10
Pgi	1277±89	1675±98	1905±86	343±45	65±9	55±7
ICDH	1205±96	1229±61	1631±98	1390±70	1414±81	1371±69
G6PDH	354±31	248±12	ND	145±10	83±6	ND
Mez	ND	70±6	15±2	<5	19±3	66±5

ND: not determined

Table 6.8 Absolute metabolic fluxes at several key branch points in the central metabolic pathways, when glucose or pyruvate were used as sole carbon source. Fluxes are expressed as mmol g^{-1} dry cell weight h^{-1} with 95% confidence limits obtained from statistic analysis. Negative values indicate the reversed pathway direction

Pathway	Wild		gnd		zwf	
	Optimal estimate	95% confidence limit	Optimal estimate	95% confidence limit	Optimal estimate	95% confidence limit
Glucose culture						
G6P → F6P	2.52	2.32, 2.72	3.33	3.16, 3.56	3.78	3.67, 3.91
G6P → 6PG	0.64	0.57, 0.69	0.33	0.23, 0.41	0.00	–
X5P+E4P → F6P+T3P	0.09	0.08, 0.10	−0.13	−0.12, −0.14	−0.12	−0.11, −0.13
AcA+OAA → CIT	2.34	2.11, 2.55	2.89	2.73, 3.01	3.32	3.03, 3.63
MAL → OAA	2.10	2.01, 2.25	2.31	2.18, 2.46	2.97	2.72, 3.16
Pyruvate culture						
G6P → F6P	0.26	0.24, 0.28	0.04	0.03, 0.06	0.04	0.02, 0.07
G6P → 6PG	0.22	0.19, 0.27	0.00	–	0.00	–
X5P+E4P → F6P+T3P	−0.05	−0.03, −0.07	−0.12	−0.10, −0.13	−0.12	−0.11, −0.15
AcA+CAA → CIT	3.44	3.15, 3.68	3.13	2.94, 3.37	2.63	2.39, 2.88
MAL → OAA	3.12	2.98, 3.32	2.75	2.54, 2.93	1.87	1.63, 2.05

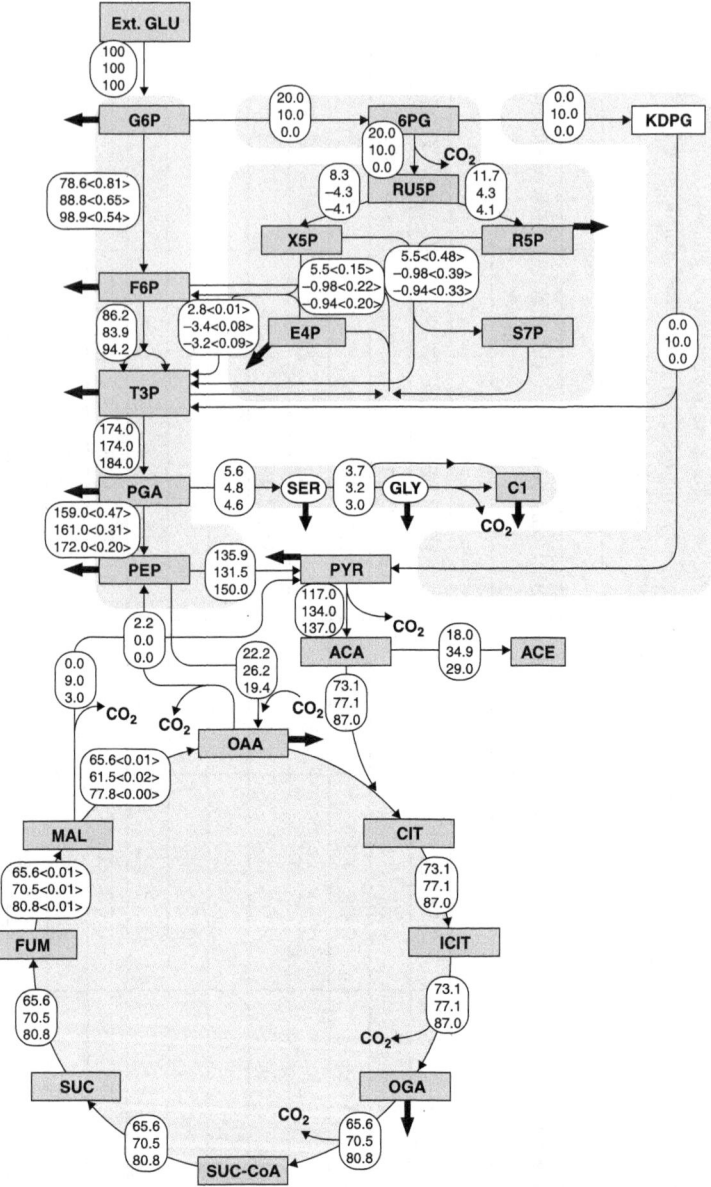

Figure 6.2 Metabolic flux distributions in chemostat culture of glucose-grown *E. coli* parent strain (*upper values*), *gnd* (*middle values*), and *zwf* (*lower values*) mutants at $D = 0.2\ h^{-1}$. Fluxes are given relative to the specific glucose consumption rate and are expressed as the net fluxes. The exchange coefficients are *shown in brackets* for the reactions that were considered reversible. *Negative values* indicate the reversed pathway direction

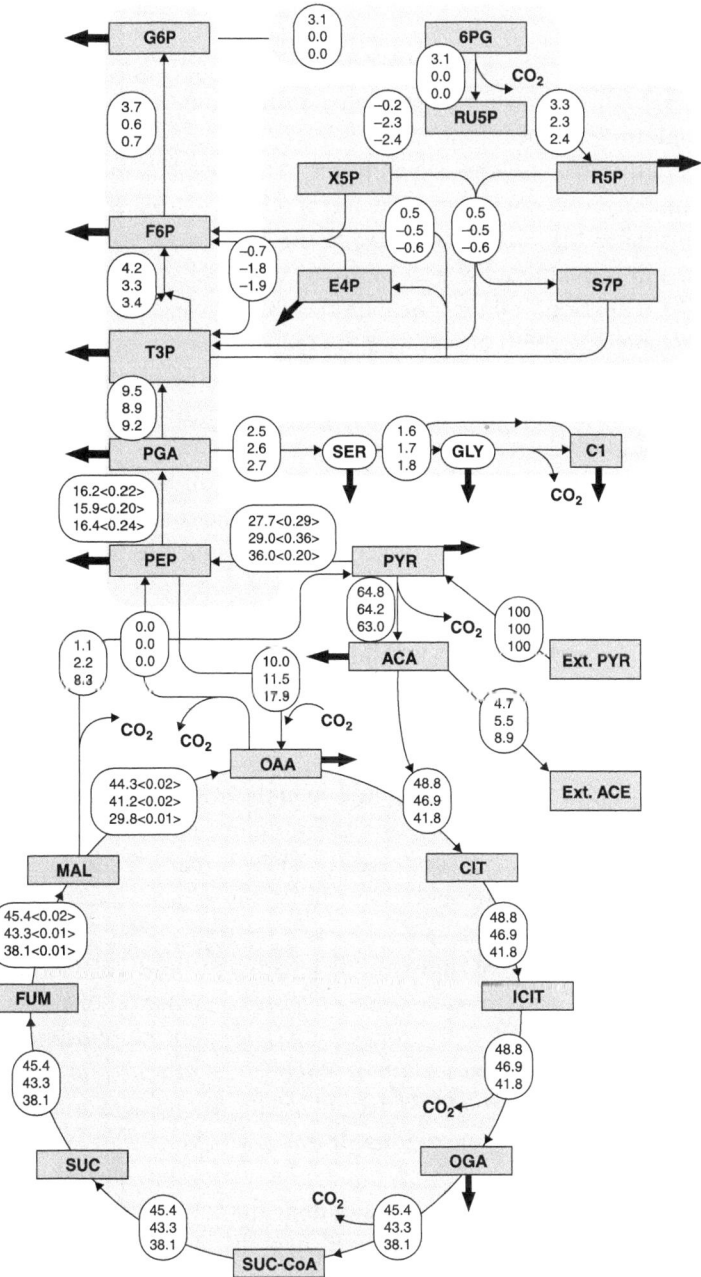

Figure 6.3 Metabolic flux distribution in chemostat culture of pyruvate-grown *E. coli* parent strain (*upper values*), *gnd* (*middle values*), and *zwf* (*lower values*) mutants at $D = 0.2\ h^{-1}$. Fluxes are given relative to the specific pyruvate consumption rate and are expressed as the net fluxes

While the flux through G6PDH is decreased in the *gnd* mutant grown on glucose, this pathway is totally blocked in mutants grown on pyruvate, although the enzyme, G6PDH, remain expressed. Since the flux through the oxidative PP pathway is significantly decreased or totally blocked in the mutants grown on glucose, the metabolic network has to give a higher flux through EMP or ED pathways to the TCA cycle. Opposite relations are seen in the mutants grown on pyruvate, where the deletion of genes causes a reduction in fluxes through the TCA cycle.

Corresponding with observations in the enzymatic study, the metabolic flux redistribution indicates an activation of the ED pathway in the *gnd* mutant grown on glucose and an increase in the flux through Mez in both glucose- and pyruvate-grown mutants. Further inspection of NADPH-generating reactions reveals a negative correlation between the Mez and ICDH pathways in the mutants, i.e. a higher flux of the Mez pathway is always accompanied by a lower flux through the ICDH pathway. Moreover, the conversion of carbon skeletons from the TCA cycle through Mez enables the cells to respond to TCA carbon depletion by regulating the carbon flux through Ppc. This can be shown in the positive correlation between the Mez and Ppc pathways.

Growth characteristics indicate that neither *zwf* nor *gnd* is essential (Fraenkel et al., 1968). However, further inspection of the metabolic parameters indicates a unique alteration in the utilization of the metabolic pathways for optimal growth of the mutants under different culture conditions. This is further supported by the enzymatic study, in which the enzyme activities are strongly dependent on genetic background and environmental conditions.

The carbon can be fed into the ED pathway through two routes in *E. coli* grown on glucose. One route consists of oxidation of D-glucose to D-gluconate by glucose dehydrogenase (GDH) and the phosphorylation of D-gluconate to 6PG by gluconate kinase. The other route is such that glucose is utilized via G6PDH to 6PG. The mechanism for activating GDH in *E. coli* remains intriguing. It is shown (Matsushita et al., 1997) that *E. coli* may not be able to produce gluconate from glucose via GDH, since it cannot synthesize pyrroloquinoline quinone (PQQ), a cofactor of glucose dehydrogenase. Although it may be possible for *E. coli* to scavenge its surroundings for PQQ, the enzyme analysis excludes this possibility when cells are grown in minimal medium.

A negative effect of the *gnd* deletion has been observed for *Saccharomyces cerevisiae* grown on glucose, since intermediates such as 6PG are reported to be toxic in high concentrations (Holger et al., 1996). It is clear that the ED pathway, which is not possessed by *S. cerevisiae*, serves as the route to relieve the toxic level of 6PG in the *E. coli* mutant grown on glucose. Because of the activation of this potential bypass reaction, the flux via G6PDH is reduced but not blocked. The activity of the oxidative PP pathway is therefore partly maintained.

Metabolic flux analysis suggests significant differences in carbon flux distribution over the TCA cycle between glucose- and pyruvate-grown mutants. The deletion of genes causes an enhanced activity of the TCA cycle in the former but a decrease in the latter. Using the flux redistribution and growth parameters presented in Table 6.6, it is reasonable to postulate that although the PP pathway flux is considerably lower during gluconeogenesis, expression of the *zwf* gene may co-ordinate with other genes to control the substrate uptake. The mutant directs a lower carbon flux through the TCA cycle and, therefore, produces CO_2 at relatively low rates, compared to those of the parent strain. This cellular response is necessary because the network has to conserve more carbon for biosynthesis through the reduction in CO_2 production, so that the mutant can grow at the same rate as the parent strain when the carbon source uptake rate decreases.

In *E. coli*, the oxidative PP pathway plays a major role in the generation of NADPH. ICDH is also important in producing NADPH (Choi et al., 2003). Although the shortage of NADPH due to disruption of the oxidative PP pathway can be partially compensated for by an increased flux through ICDH, it appears that the flux via ICDH alone cannot enable the apparent shortage of NADPH to be met under certain circumstances. For this, malate is deviated out of the TCA cycle through Mez to function as the route in which an adequate supply of NADPH is generated to meet the biosynthesis requirements. This postulation can explain the negative correlation between the fluxes through Mez and ICDH, as occurring in the mutants. When glucose is used as a sole carbon source, a higher flux through ICDH is observed for the *zwf* mutant. This can generate more NADPH than in the *gnd* mutant, thus rendering the pathway via Mez less active. However, in the case of using pyruvate as a carbon source, *zwf* gene deletion leads to a lower flux through ICDH, as compared to the case of *gnd* gene deletion. Therefore, the flux via Mez significantly increases to complement the shortage of NADPH. Activation of the Mez pathway changes the flux distribution around the OAA node. The drain of carbon skeletons from the TCA cycle through Mez enables the cells to replenish the OAA pool by regulating the carbon flux through Ppc. To increase the synthesis of OAA from PEP, Ppc is therefore up-regulated in accordance with the activity of the Mez pathway.

In the mutants grown on glucose, all directions of the fluxes through non-oxidative PP pathways are reversed, indicating that the mutant tries to compensate for the lack of E4P and R5P through the glycolytic metabolites GAP and F6P. In mutants grown on pyruvate, an increased flux through Tkt (F6P + T3P → X5P + E4P) is found, suggesting a similar function for the non-oxidative PP pathway as in the glucose-grown mutants. In this way, the non-oxidative branch can function as an important metabolic route without the participation of the oxidative PP pathway.

One of the goals of a gene-knockout study is the engineering of metabolic pathways for enhanced production of industrial chemicals. However, genetically altering metabolic pathways often cause undesirable changes, such as reduced growth, decreased glycolytic flux, and the creation of futile cycles, which may limit its utility (Holger et al., 1996). The results indicate that *E. coli* grown on glucose with genetic disruption in the oxidative PP pathway can serve as a good candidate for industrial production, since the mutations have little effect on cell growth, and the increased activities of EMP and TCA cycle may benefit the production of useful chemicals that are synthesized from precursors involved in these two branches. The genetic alteration of *zwf* is preferred, since mutation in the other gene, *gnd*, activates the ED pathway from which only one molecule of ATP is produced. In view of energy metabolism, the ED pathway is less efficient than EMP, with a net gain of two molecules of ATP for each molecule of glucose.

The problem in utilizing the *zwf* mutant in industrial production is that gene deletion causes an enhanced production of acetate. Acetate accumulation should be avoided in industrial-scale production, since a high acetate concentration in the culture medium has been found to decrease the yield of recombinant protein. This type of overflow metabolism usually occurs when high fluxes from the pyruvate pool exceed the capacity for respiratory metabolism and the balance is therefore excreted as acetate. This can be manifested in the *zwf* mutant, where the flux entering the pyruvate pool is extremely high due to the high activity of EMP. Note that the reason for acetate excretion in the *gnd* mutant is somewhat different from the *zwf* mutant when glucose is used as a sole carbon source. The flux profile of the former demonstrates that the increase in the flux through the early step of EMP is actually counteracted later by enhanced fluxes entering into the non-oxidative PP pathway through GAP and F6P. The increase in the flux through PDHc is mainly due to the supply from the ED pathway and Mez pathway. Thereafter, the increased flux is partly channeled toward acetate excretion as a result of overflow metabolism.

6.4 Effect of *pyk* gene knockout on metabolism

Let us consider how the cell regulates the *pyk* gene knockout (Siddiquee et al., 2004a,b; Toya et al., 2010). The growth rate of the *pykF* mutant is a little lower compared to that of the wild type. During exponential growth, acetic acid is also produced but its formation rate is lower compared to that of the wild type, and the CO_2 evolution rate (CER) is lower for the *pykF* mutant, where the decrease in CO_2 evolution rate in the *pykF* mutant may be partly due to an increase in CO_2 fixation through the Ppc reaction. Another possibility may be decreased glucose consumption rate.

The GC-MS and NMR data for the *E. coli pykF* mutant are shown in Tables 6.9 and 6.10. Figure 6.4 shows the metabolic flux distributions of the wild type (upper figure) and for the *pykF* mutant (lower figure). The flux distribution in the *pykF* mutant reveals a high flux from PEP to OAA. The flux through the anaplerotic reaction catalyzed by Ppc in the *pykF* mutant is also high (44%) in the *pykF* mutant as compared to 17% for the wild-type strain. It should be noted that the flux value in brackets for Ppc is also high, which indicates that the flux through Pck is also high. The flux from PEP to PYR for the *pykF* mutant is significantly lower compared to the wild type and is consistent with enzyme activity measurements. It can also be seen that the flux through Mez is high, about 21% for the *pykF* mutant, while it is very low for wild type. Moreover, the glycolytic flux from G6P to F6P is 20% for the *pykF* mutant, while it is 65% in the wild-type strain. However, the flux through the oxidative PP pathway is 79% for the *pykF* mutant, while it is 34% in the wild type. The flux from AcCoA to acetate reduces to 0.82% in the *pykF* mutant, as compared to 20% in the wild type.

The enzyme activity for glycolytic pathway enzymes, such as Pgi, and Tpi, decreases by about 2.65-, 2.45-, and 2.08-fold at high D (0.5 h^{-1}) compared to those at low D (0.1 h^{-1}). Figure 6.5 shows that the enzyme activity of Pfk decreases slightly as D increases, which may be due to the increase in the intracellular concentration of PEP (Figure 6.6). The decreased activity of Pfk at high D increases the concentrations of G6P and F6P, which in turn increase the enzyme activity of G6PDH and 6 PGDH, by about 3- and 9-fold, respectively. The enzyme activity of Pyk at both D values is very low. The enzyme activity of Mez decreases by about 3.5-fold at high D. In accordance with this phenomenon, Pck activity decreases at high D. The enzyme activity of the TCA cycle

Table 6.9 Fragment mass distribution of t-butyldimethylsilyl (TBDMS)-derived amino acids from the *pykF* mutant. Considering standard deviation in the range 0.002 to 0.006 for mass distribution, only mass signals of m, $m+1$, and $m+2$ were compared

Amino acids	Fragments	m	$m+1$	$m+2$
Ala	$[M-57]^+$	0.699^a	0.181	0.099
		0.693^b	0.180	0.097
Asx	$[M-57]^+$	0.650^a	0.250	0.078
		0.666^b	0.043	0.031
Gly	$[M-57]^+$	0.744^a	0.160	0.096
		0.735^b	0.156	0.093
Glx	$[M-57]^+$	0.540^a	0.170	0.162
		0.539^b	0.167	0.153
Ile	$[M-57]^+$	0.685^a	0.079	0.085
		0.681^b	0.075	0.080
Ile	$[M-159]^+$	0.641^a	0.079	0.071
		0.639^b	0.077	0.084
Leu	$[M-57]^+$	0.677^a	0.129	0.084
		0.671^b	0.126	0.081
Leu	$[M-57]^+$	0.614^a	0.112	0.112
		0.610^b	0.102	0.106
Phe	$[M-57]^+$	0.629^a	0.116	0.110
		0.627^b	0.113	0.097
Phe	$[M-159]^+$	0.568^a	0.112	0.077
		0.565^b	0.110	0.075
Met	$[M-57]^+$	0.608^a	0.041	0.029
		0.606^b	0.039	0.028
Thr	$[M-57]^+$	0.587^a	0.242	0.133
		0.586^b	0.238	0.127
Tyr	$[M-57]^+$	0.617^a	0.115	0.099
		0.616^b	0.111	0.095
Tyr	$[M-159]^+$	0.544^a	0.108	0.076
		0.543^b	0.106	0.072
Val	$[M-57]^+$	0.640^a	0.081	0.067
		0.642^b	0.078	0.063

[a] Experimentally determined
[b] Calculated

Table 6.10 Measured and simulated values of the NMR spectra of cellular amino acids. s – Singlet; d_1 – doublet with larger scalar coupling; d_2 – doublet with smaller scalar coupling; dd – doublet of doublets

Atom	Measured				Simulated			
	s	d_1	d_2	dd	s	d_1	d_2	dd
Arg γ	0.39	0.61	–	–	0.35	0.63	–	–
Asx α	0.39	0.61	–	–	0.38	0.52	–	–
Gly α	0.34	0.66	–	–	0.32	0.60		
Ile α	0.19	0.11	0.61	0.09	0.17	0.01	0.62	0.02
Ile δ	0.35	0.65	–	–	0.30	0.56	–	–
Leu α	0.17	0.08	0.66	0.09	0.12	0.01	0.63	0.02
Leu β	0.43	0.57	–	0.00	0.42	0.56	–	0.00
Lys ε	0.32	0.68	–	–	0.34	0.63	–	–
Pro α	0.11	0.15	0.19	0.55	0.08	0.12	0.14	0.55

enzyme CS is lower at high D, while that of MDH changes little at both D values. The enzyme activity of the ED pathway is higher at high D. This corresponds to the increase in carbon flow in the oxidative PP pathway. The enzyme activities of Ack and ICDH change little.

Acetate production is substantially lower for the *pykF* mutant compared to the wild type (Ponce et al., 1998; Zhu et al., 2001). The enzyme activities of glycolytic enzymes, such as Pgi, Pfk, Tpi, and Ppc, are higher during 2–4 h of fermentation. Oh et al. (2002) found that Mez and Pck enzymes correlated when acetate was utilized. They explain that the mutation of *mez* and *pck*A genes prevents growth on acetate. As the glucose is consumed and acetate accumulates, cells switch smoothly to co-metabolism, utilizing both glucose and acetate. This switch involves induction of the TCA cycle enzymes and glyoxylate bypass enzymes required to provide energy and to replenish intermediates used for amino acid biosynthesis (Clark and Cronan, 1996). Icl also becomes more active under acetate conditions. TCA cycle genes are known to be under the control of cAMP and Crp, which mediate catabolic repression (Cronan and LaPorte, 1999).

In the case of the *pykF* mutant, the enzyme activities of glycolytic pathway enzymes decreases compared to those of the wild type, while the enzyme activity of oxidative PP pathway enzymes increases for the

Bacterial cellular metabolic systems

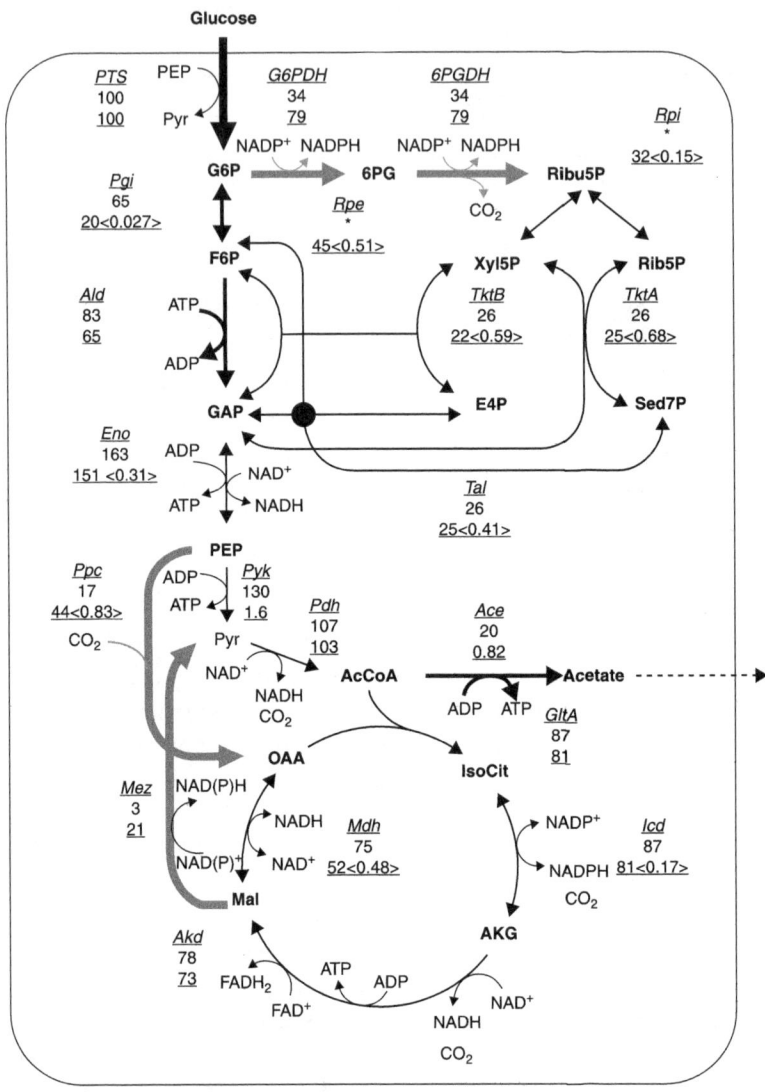

Figure 6.4 Metabolic flux distribution of wild type (upper value) and $pykF^-$ mutant (lower value) at dilution rate (D) of 0.1 h^{-1}. Values in < > indicate exchange coefficient

Effect of specific-gene knockout

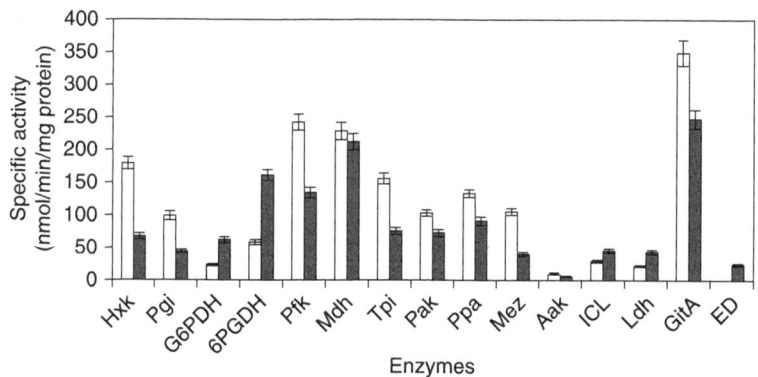

Figure 6.5 Enzyme activities for the *pyk F* mutant at two dilution rates (*D* values): ☐ = 0.1 h^{-1}, ■ = 0.5 h^{-1}

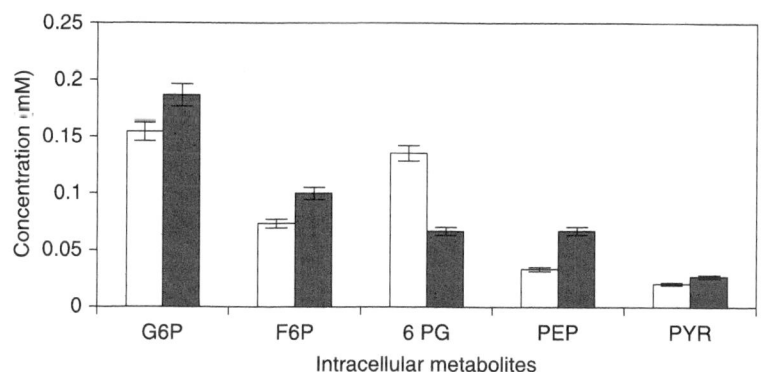

Figure 6.6 Comparison of concentrations of intracellular metabolites in the *pykF* mutant at two *D* values: ☐ = 0.1 h^{-1}, ■ = 0.5 h^{-1}

former. Ponce et al. (1998) also reported increased activity of the oxidative PP pathway enzymes in the *pykF* mutant. The most significant differences between the wild type and the *pyk F* mutant are observed in Ppc and Mez activities. In the presence of glucose, these two enzyme activities are significantly increased in the *pykF* mutant. This can be explained by blockage of PEP to the PYR pathway activating both Ppc and Mez to supply PYR by forming an alternative pathway. It can be seen that blockage of PEP to PYR will increase the PEP pool in the mutant, which might inhibit Pfk enzyme activity in the glycolytic pathway. The accumulation of PEP and the activation of the PP pathway (or increased E4P concentration) may activate the aromatic amino acids synthetic pathways (Kedar et al., 2007). The decreased activity of Pfk in turn increases the G6P and F6P concentrations. Increased activity of the oxidative PP pathway is observed in the *pykF* mutant. This might be due to the increased G6P concentration.

Although futile cycling could be induced by simultaneous overexpression of *Ppc* and *Pck* in *E. coli* (Chao and Laio, 1994), it is generally maintained at a low level (Chambost and Fraenkel, 1980; Dalal and Fraenkel, 1983; Chao and Laio, 1994). The enzyme activity results for Ppc and Pck support the futile cycle results obtained by flux calculation.

6.5 Effect of *lpdA* gene knockout on metabolism

The lipoamide dehydrogenase (LPD) encoded by the *lpdA* gene is a component of PDHc, KGDH, and glicine cleavage multi-enzyme (GCV) system. This *lpdA* gene knockout produces more pyruvate and L-glutamate under aerobic conditions than the wild-type strain. It is shown that AcCoA is considered to be formed by the combined rerouting pathways through PoxB, Acs, Ack, and Pta in the *lpdA* mutant. Metabolic flux analysis of the *lpdA* knockout gene mutant indicates that the ED pathway and glyoxylate pathway are activated, while the glycolysis and the oxidative PP pathway, as well as the TCA cycle, are down-regulated (Li et al., 2006a). If only KGDH was blocked by the *sucA* gene knockout, metabolism turns out to be a little different from the *lpdA* gene knockout mutant (Li et al., 2006b). The regulation of the anaplerotic pathway, such as the glyoxylate shunt, plays a role in these mutants. Let us now consider the detailed metabolic regulation analysis.

Figure 6.7 shows the aerobic batch cultivation results using an LB medium containing glucose, where the results indicate that the cell growth of the mutant is slow compared to its parent strain, and it indicates that a fair amount of pyruvate is produced during exponential growth phase of *lpdA* mutant, while little pyruvate is produced in the parent strain. Moreover, 1 g/l of D-lactate and 0.5 g/l of succinate accumulate in the culture of the mutant by the time of complete glucose utilization. The concentration of acetate is almost undetectable in the cultivation of the mutant when glucose is utilized. The accumulated pyruvate begins to decrease rapidly at 6 h after glucose exhaustion. Then the cell starts to

Effect of specific-gene knockout

grow again using pyruvate as another carbon source, and produces acetate, which is subsequently utilized after the pyruvate is consumed. Although cell growth is depressed in the mutant, the final cell concentrations are higher than that in the parent strain (Figure 6.7).

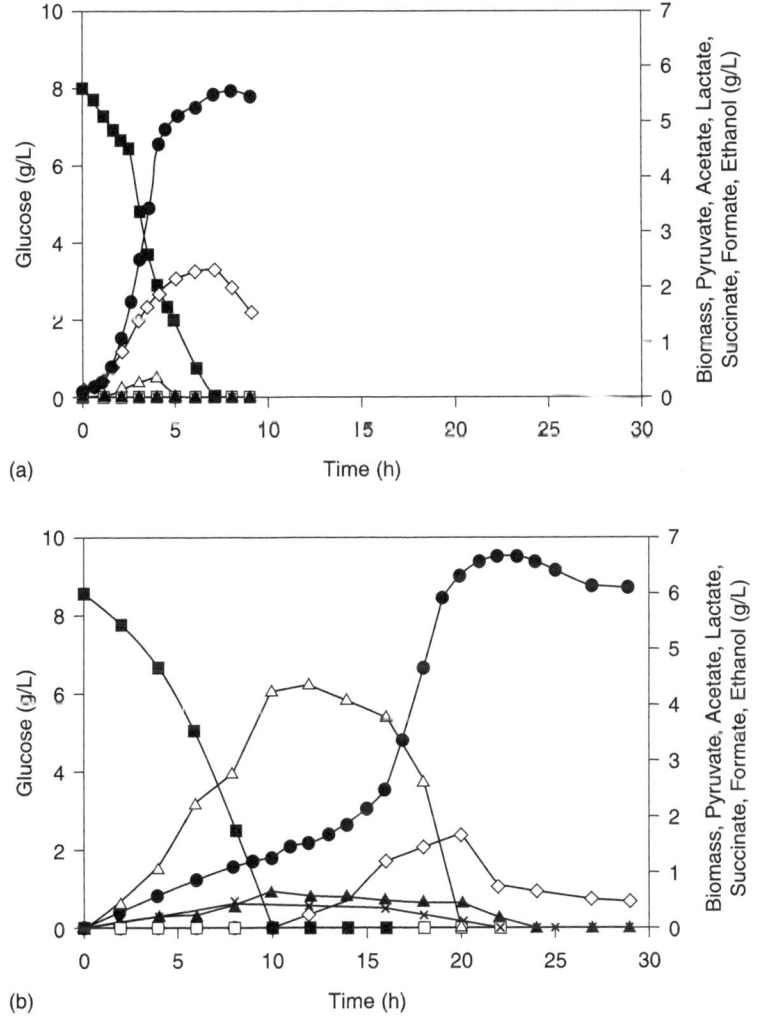

Figure 6.7 Aerobic batch cultivation of (a) *E. coli* BW25113 and (b) *E. coli lpdA* mutant using glucose as carbon source in LB medium: (■) glucose; (●) biomass; (△) pyruvate; (◊) acetate; (▲) D-lactate; (×) succinate; (○) formate; (□) ethanol

Another two cultivations were conducted to investigate the PoxB function in the *lpdA* mutant. Sodium pyrophosphate (10 mM), one of the inhibitors of PoxB, was added to the medium 2 h after the glucose was depleted in the batch culture. It is found that addition of sodium pyrophosphate inhibits PYR assimilation in the *lpdA* mutant and the cell growth almost ceases, while the pattern of the assimilation of acetate is little affected in the parent strain with the addition of this inhibitor. These phenomena indicate that PoxB is essential for the *lpdA* mutant, while it is not so for the wild type.

Figure 6.8 shows that the activity of KGDH in the *lpdA* mutant is negligible compared to the parent strain. Since the *lpdA* gene encodes for the subunit of both KGDH and PDHc, and this subunit plays the similar role in both enzymes, the enzyme activity of PDHc is considered to be small. Figure 6.8 shows that the enzyme activity of some of the glycolytic enzymes, such as Pyk, are up-regulated in the *lpdA* mutant. As for TCA cycle enzymes, the CS activity is down-regulated in the *lpdA* mutant, while the activity of MDH and Fum is up-regulated in this mutant as compared to those of the parent strain. The activity of the oxidative PP pathway enzyme, such as G6PDH, is up-regulated in the mutant and also the activity of the ED pathway enzymes and the glyoxylate shunt enzyme

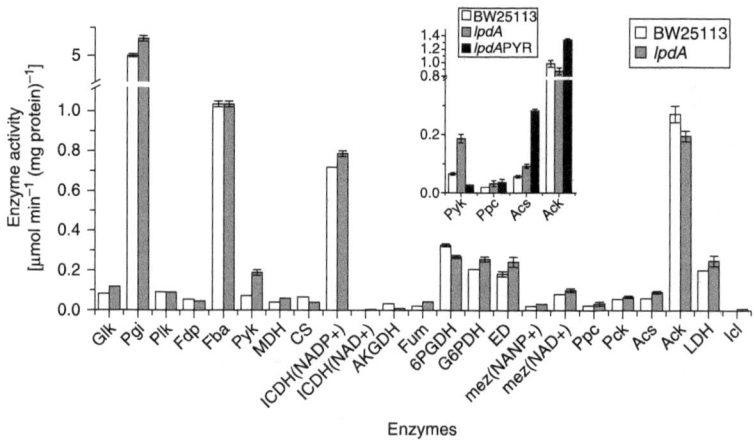

Figure 6.8 Enzyme activities of *E. coli* BW25113 and *E. coli* lpdA mutant under aerobic conditions in LB medium. The upper smaller graph shows some key enzyme activities of *lpdA* mutant when produced pyruvate was reconsumed

Icl is higher in the *lpdA* mutant. It can be seen that the activities of the anaplerotic pathway enzyme Ppc and $NADP^+$ (NAD^+)-specific Mez are all activated in the *lpdA* mutant. The activity of the fermentative pathway enzymes, such as Acs and LDH, increase in the *lpdA* mutant, while Ack activity decreases in the *lpdA* mutant.

Some of the intracellular metabolite concentrations are compared between the parent and the mutant (Figure 6.9). PYR concentrations are about 10-fold higher in the *lpdA* mutant as compared to that in the parent strain, which may be due to the blockage of PDHc. In addition, accumulation of L-Glu is 20 times higher than in the parent strain. The concentrations of αKG also increase in the mutant. The concentrations of G6P and F6P are slightly higher in the *lpdA* mutant as compared to those of the parent strain, whereas AcCoA is almost undetectable in the mutant.

Table 6.11 shows the growth parameters of the *lpdA* mutant and the parent strain in continuous cultures. It is found that the biomass yield of the *lpdA* mutant is higher than that of the parent strain, while glucose uptake rates are lower for the mutant and the specific CO_2 evolution rate is much reduced in the mutant.

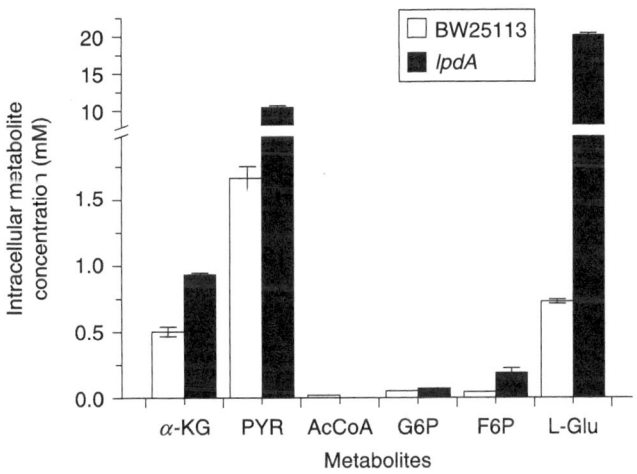

Figure 6.9 Comparison of the intracellular metabolite concentrations of *E. coli* BW25113 and *E. coli lpdA* mutant in the batch cultivation. (The intracellular concentrations were calculated assuming the cellular volume to be 2.15 ml (g dry cell weight)/l.)

Table 6.11 Growth characteristics of *E. coli* BW25113 at the dilution rate of 0.2 h^{-1} and its *lpdA* mutant at the dilution rate of 0.22 h^{-1} in continuous culture

Variables	BW25113	*lpdA* mutant
Biomass yield (g g^{-1})	0.40	0.49
Glucose uptake rate (mmol g^{-1} h^{-1})	3.04	2.48
Specific CO_2 evolution rate (mmol g^{-1} h^{-1})	8.17	4.08

Figure 6.10 shows the comparison of the metabolic flux distributions between the *lpdA* mutant (lower values underlined) and the parent strain (Li et al., 2006a). All the flux values are normalized with respect to glucose uptake rates. Figure 6.10 shows that with the blockage of the flux through the reactions of PDHc and KGDH, the mutant exhibits a slight reduction in the flux values of glycolysis and PP pathways, except for a slight increase in the flux through G6PDH. However, the ED pathway is activated in the mutant strain with a flux value of 13.0%. The blockage of the PDHc reaction has also induced the PEP synthase (Pps), which catalyzes the reversed reaction from PYR to PEP, where the exchange coefficient is relatively high at 0.75. Since the *lpdA* mutant cannot utilize the pathway through PDHc to produce AcCoA, the alternative pathway through PoxB is activated. In the TCA cycle, all the net flux values are reduced when compared to the parent strain. With the blockage of the pathway between α-KG and SUC, the flux through ICDH is limited, and this flux value is determined by the biomass formation through α-KG. The glyoxylate shunt is also activated in order to replenish OAA for biomass synthesis. For the reactions through Fum and MDH, both forward and backward fluxes are high in the case of the mutant. The fluxes through anaplerotic reactions increase drastically in the case of the mutant strain, i.e. the flux through Pck increases from 2.2% to 37.3%, the flux through Ppc increases from 22.2% to 64.9%, and the flux through Mez increases to 47.8%.

Since PoxB is known to produce acetate *in vivo* (Lemaster and Cronan, 1982; Grabau and Cronan, 1984), the *poxB* knockout mutant is also cultivated using minimal medium. The *poxB* mutant shows almost the same fermentative properties as those of the parent strain, except that the glucose-uptake rate is slightly lower for the mutant. The enzyme activity of the *poxB* knockout mutant indicates that the activities of glycolytic enzymes, such as Pfk, Fdp, and Fba, decrease in the *poxB* mutant, and the TCA cycle enzymes are also slightly down-regulated in the mutant

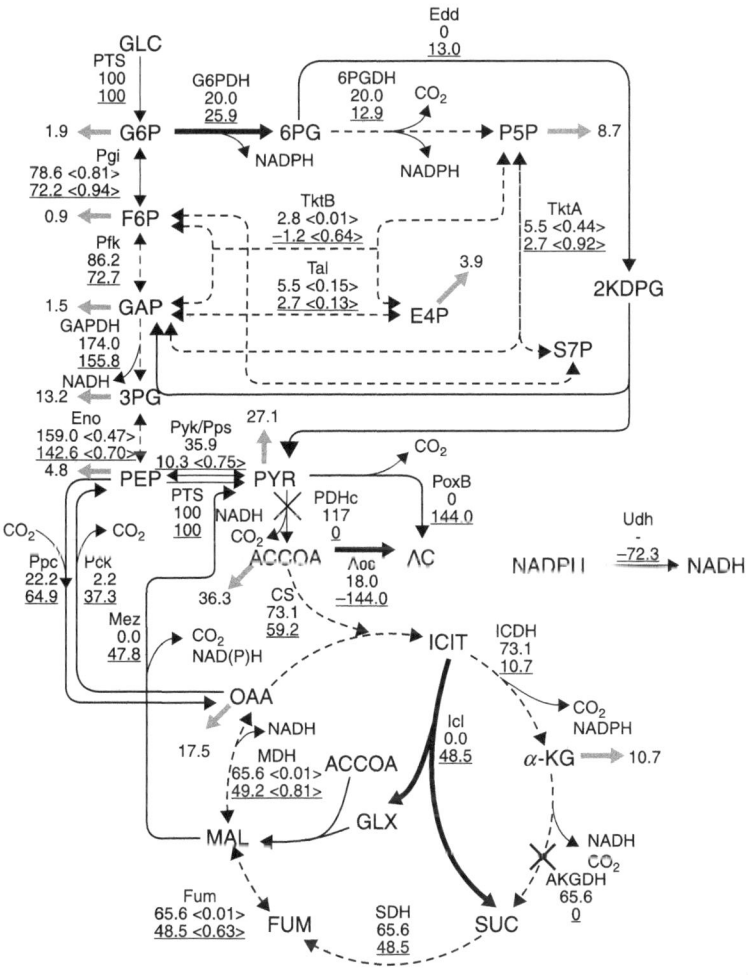

Figure 6.10 Metabolic flux distributions of wild type (upper values) at dilution rate of 0.2 h^{-1} and *lpdA* mutant (lower values, underlined) at dilution rate of 0.22 h^{-1}. Bold solid lines indicate increased fluxes while dash lines indicate reduced flux. Values in ⟨·⟩ indicate exchange coefficient. Gray arrows indicate biomass formation, but only the biomass formation flux for *lpdA* mutant is shown

(Figure 6.11). Interestingly, Ppc, NAD(P)+-specific Mez, Acs, LDH, and the oxidative PP pathway enzyme in the *poxB* mutant, show a similar pattern to that in the *lpdA* mutant.

The knockout of *lpdA* gene causes deactivation of the PDHc, which is the main enzyme catalyzing the reaction from PYR to produce AcCoA. The accumulation of PYR in the *lpdA* mutant may be due to a defect in PDHc activity. To maintain growth, the cell can generate AcCoA from PYR using another less efficient route by first converting PYR to acetate using PoxB, and then convert acetate to AcCoA via Acs and Pta–Ack. The activation of PoxB can be detected in the PDHc-deficient mutant by *in vitro* PoxB enzyme assay (Abdel-Hamid et al., 2001).

Figure 6.7 shows that the *lpdA* mutant has diauxic growth. It has been shown that both *poxB* gene transcription and PoxB activity depend on the *rpoS* gene product, σ^{38} (Chang et al., 1994). Furthermore, bacteria increase RpoS protein levels and RpoS-dependent protective functions during the stationary phase, when the glucose concentration drops below about 10^{-7} M (Notley and Ferenci, 1996). PoxB in the *lpdA* mutant might be activated due to the increased level of σ^{38} factor after the glucose is depleted. Also, during pyruvate assimilation, the Ppc activity increases and Pyk activity is much decreased in the *lpdA* mutant (Figure 6.8). It is assumed that the PYR assimilation in the *lpdA* mutant might be done via PoxB and Ppc. It has been reported that the double mutants lacking the

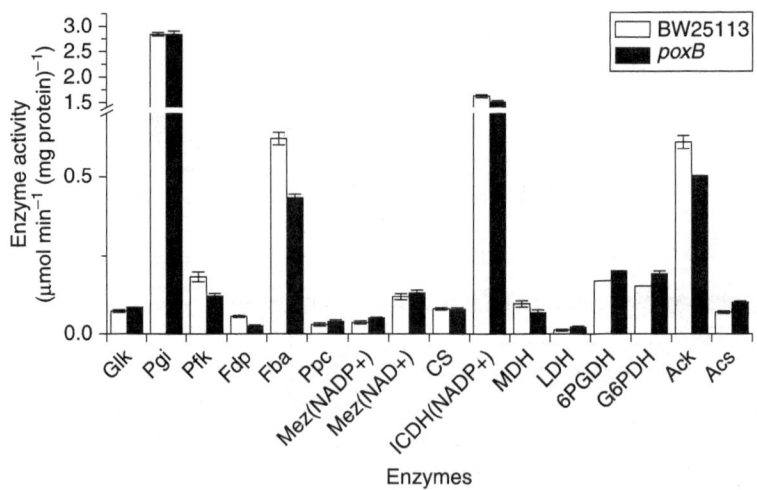

Figure 6.11 Comparison of enzyme activities of *E. coli* BW25113 and *E. coli poxB* mutant under aerobic conditions using synthetic medium

PDHc and Pps, or the PDHc and Ppc, could not metabolize PYR accumulated during growth on glucose (Dietrich and Henning, 1970). Thus, the direct utilization of PYR accumulated by this mutant is made by the combining reactions catalyzed by PoxB and Pps–Ppc.

As mentioned above, D-lactate is also detected in the culture medium. Normally, D-lactate is produced under anaerobiosis. The LDH in *E. coli* is an allosteric enzyme with PYR acting as an activator at the site distinct from the catalytic site, and the activity of this enzyme appears to be controlled by the cellular concentrations of PYR (Tarmy and Kaplan, 1968a,b). However, it has been reported that the addition of PYR results in a 2- to 4-fold increase in the expression of *ldhA–lacZ* fusions (Jiang et al., 2001). Thus, the increase in the PYR concentrations enhances the production of lactate via gene expression as well as enzyme activity. The inactivation of KGDH in the *lpdA* mutant causes the TCA cycle to be interrupted, which may activate the glyoxylate shunt to supply the precursor for cell growth.

E. coli converts acetate to AcCoA by two distinct pathways, one catalyzed by Acs, mainly functioning under low acetate concentrations, and the other catalyzed by Pta and Ack, which functions under higher acetate concentrations (Kumari et al., 1995). In the *lpdA* mutant, the activity of Acs is higher when glucose is depleted. Furthermore, its activity significantly increases when the pyruvate is assimilated through PoxB. Ack activity is higher only in the second period in Figure 6.8, where pyruvate is used as a carbon source. To supply the AcCoA needed for cell growth, both enzymes for acetate utilization are activated in the *lpdA* mutant. Acs functions both under glucose utilization and pyruvate assimilation, while Pta–Ack functions mainly for growth under pyruvate assimilation.

The low concentration of AcCoA causes down-regulation of CS in the TCA cycle. Another important phenomenon is that the deficiency of KGDH causes the accumulation of α-KG, which is a precursor of Glu. The synthesis of Glu needs NADPH, which may be supplied by the overproduction of NADPH in the PP pathway due to the up-regulated activity of the oxidative PP pathway enzymes.

6.6 Effect of *sucA* and *sucC* gene knockout on metabolism

In *E. coli*, *sucA* and *sucC* genes are located in the same operon *sucABCD*. Park et al. (1997) studied how the *sucABCD* genes are expressed in response to the changes of oxygen and carbon source availability in *E. coli*, and found that *sucABCD* gene expressions are repressed by *arcA* and *fnr* gene products under anaerobic conditions. The *sucA* gene encodes E1 subunit for the KGDH, while *sucC* gene encodes α subunit for the succinyl-CoA synthetase (SCS). KGDH catalyzes the oxidative decarboxylation of α-ketoglutarate (2-oxoglutarate) (αKG) to generate SucCoA and CO_2 along with the production of NADH (Park et al., 1997). SCS catalyzes the substrate-level phosphorylation step of the TCA cycle to convert succinyl-CoA into succinate, accompanying the production of ATP (Wolodko and Bridger, 1987). Both enzymes participate in the generation of the cellular biosynthetic intermediates and the energy generation in the TCA cycle. The TCA cycle is critical for the oxidation of AcCoA and for the production of reducing equivalents that are used by respiratory chains to produce ATP. To understand the role of the TCA cycle in the yeast cell, McCammon et al. (2003) studied some gene expressions in response to the disruption of the TCA cycle, caused by the defects in each of 15 genes encoding subunits of the 8 TCA cycle enzymes, by using a DNA microarray.

Here, consider the effect of *sucA* or *sucC* genes knockout on metabolism in *E. coli*, based on batch and continuous cultivations with gene expressions, enzyme activity, intracellular metabolite concentrations, and metabolic flux distributions based on ^{13}C-labeling experiments. To determine how *sucA* or *sucC* gene knockout affect cell growth characteristics, *sucA* and *sucC* genes knockout mutants are, respectively, grown in a minimal medium with glucose as a carbon source under aerobic conditions (Figure 6.12). The maximum specific growth rates of the wild type strain, *sucA* mutant, and *sucC* mutant, are 0.83, 0.66 and 0.84 h^{-1}, respectively. Acetate production is significantly higher in the *sucC* mutant, as compared to that in the parent strain (Figure 6.12). The produced acetate is not assimilated in the *sucC* mutant. Unlike the parent strain, a little accumulation of L-glutamate is detected in the *sucA* mutant.

Table 6.12 shows the fermentation characteristics of the continuous culture at the dilution rate of 0.2 h^{-1}, where glucose concentrations in the reactor are less than the detectable level for all the cultivations of the parent stain and the *sucA* mutant, while the glucose concentrations in the cultivation of *sucC* mutant are detectable but significantly low (0.03 mM). As shown in Table 6.12, the biomass yields of both mutants are lower than that of the parent strain. The absence of KGDH or SCS activity significantly increases the specific oxygen consumption rate, and

Effect of specific-gene knockout

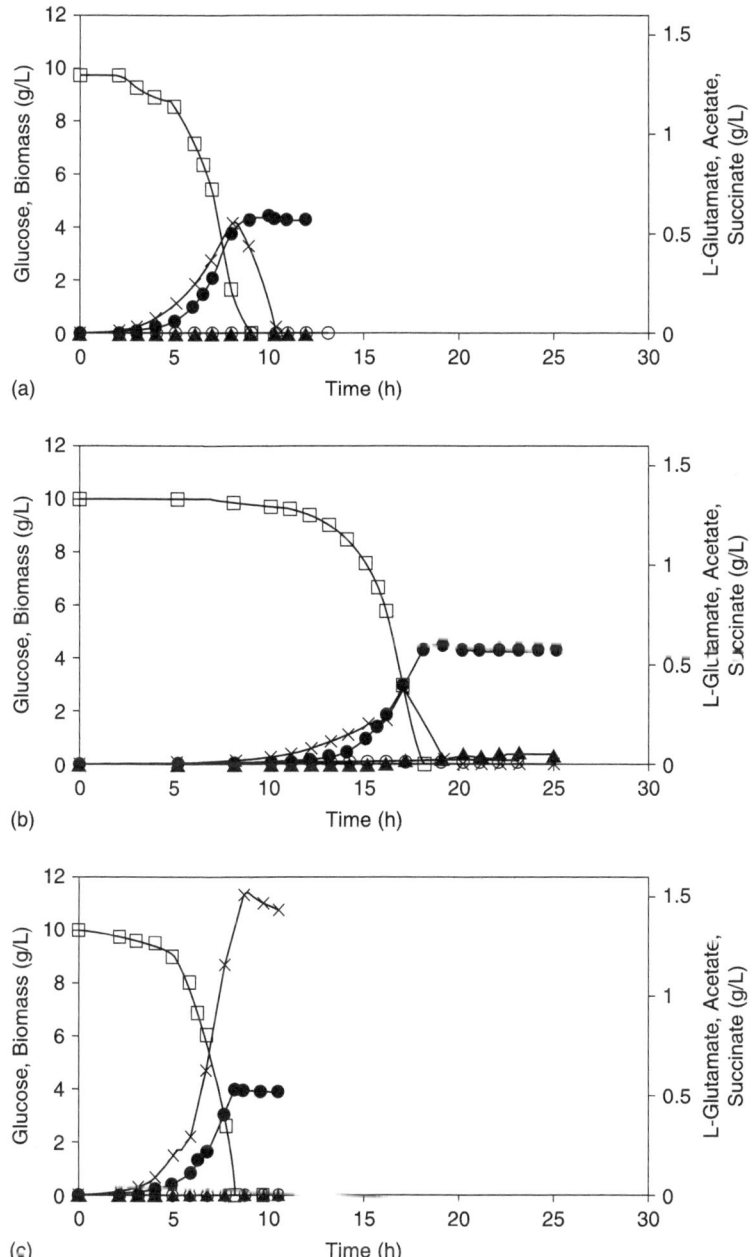

Figure 6.12 Aerobic batch cultivation of: (a) *E. coli* BW25113; (b) *E. coli sucA* mutant; and (c) *E. coli sucC* mutant using glucose as carbon source. Symbols: (□) glucose; (•) biomass; (×) acetate; (▲) l-glutamate; (◊) succinate

Table 6.12 Growth characteristics of parent strain *E. coli* BW25113 and its *sucA*, *sucC* mutants in the continuous culture at the dilution rate of 0.2 h^{-1}

Variables	BW25113	sucA mutant	sucC mutant
Biomass yield (g g^{-1})	0.45	0.42	0.40
Glucose uptake rate (mmol g^{-1} h^{-1})	2.48	2.65	2.77
Specific CO2 evolution rate (mmol g^{-1} h^{-1})	6.20	6.80	5.45
Specific O2 consumption rate (mmol g^{-1} h^{-1})	9.11	16.29	17.42
Acetate production rate (mmol g^{-1} h^{-1})	–	–	0.35

the specific CO_2 evolution rate (CER) is increased in the *sucA* mutant, while it is reduced in the *sucC* mutant.

Figure 6.13 shows comparison of the enzyme activity at a dilution rate of 0.2 h^{-1}. The enzyme activity of glycolytic enzymes, such as Pgi, Tpi, GAPDH, and Pyk, increase in *sucA* and *sucC* mutants, while the activity of Pfk, the rate-limiting enzyme in glycolysis, is down-regulated in these two mutants. The glycolytic enzyme, such as Pgk, and the PP pathway enzymes, such as G6PDH and 6PGDH, are up-regulated in the *sucA* mutant. The glyoxylate shunt enzymes, such as Icl and MS, are significantly up-regulated for the *sucA* mutant, while this is not the case for the *sucC* mutant. As for the TCA cycle enzymes, the CS, Fum, and MDH are up-regulated in both mutants, while the activity of ICDH is down-regulated in the *sucA* mutant but up-regulated in the *sucC* mutant. It can be seen that fermentative pathway enzyme Ack activity increases, and LDH activity decreases in both mutants.

Figure 6.14 compares intracellular metabolite concentrations between the parent and the mutants in continuous cultivations. The intracellular concentration of G6P is considerably higher in *sucA* and *sucC* mutants, and the concentration of FDP is also higher in *sucA* and *sucC* mutants, as compared to that in the parent strain. PEP concentrations tend to increase, while PYR concentrations decrease in both mutants. The concentration of AcCoA is higher in the mutants, especially in the *sucC* mutant. ATP concentrations are significantly higher in both mutants.

As shown in Figure 6.15, the transcript levels of *arcA* and *fnr* genes are down-regulated in the mutants, and the *rpoS* gene is up-regulated in the mutants. The expressions of *iclR* and *fadR* are down-regulated in the *sucA* mutant but not in the *sucC* mutant, and also the transcript level

Effect of specific-gene knockout

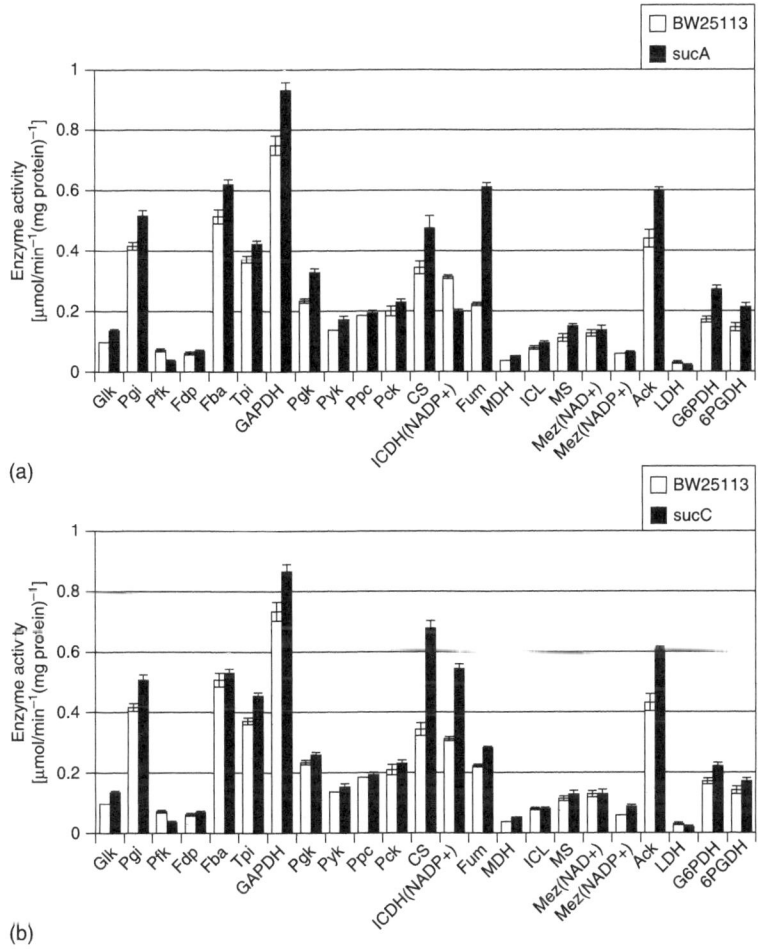

Figure 6.13 Comparison of enzyme activities at a dilution rate of 0.2 h^{-1} in a continuous culture: (a) *E. coli* BW25113 and *E. coli sucA* mutant; and (b) *E. coli* BW25113 and *E. coli sucC* mutant. (Pgi, ICDH (NADP+) × 5; Pgk × 2; Tpi × 20)

of the *aceA* gene shows the same trend as the enzyme activity in the mutants.

Figure 6.16 compares metabolic flux distributions of the parent strain and its *sucA* mutant (Li et al., 2005). The flux analysis results indicate that the *sucA* mutant exhibits reduction in glycolysis flux. In the TCA cycle, all the net flux values of the *sucA* mutant are significantly reduced when compared to the parent strain. With the blockage of

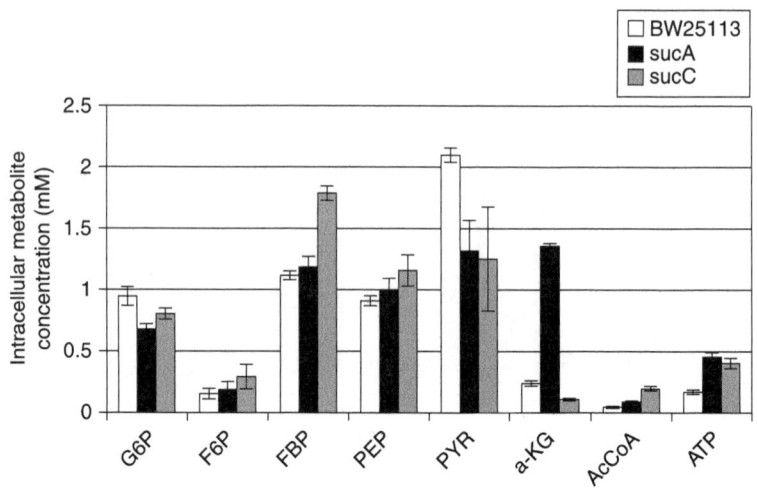

Figure 6.14 Intracellular metabolite concentrations of *E. coli* BW25113, *sucA* and *sucC* gene knockout mutants at 0.2 h^{-1} specific growth rate in chemostat cultures

the pathway between α-KG and SUC in the *sucA* mutant, the flux through ICDH is limited. The glyoxylate shunt is activated in order to replenish OAA for biomass synthesis in the *sucA* mutant. Figure 6.16 also shows that the PP pathway fluxes are significantly up-regulated, which is consistent with the enzyme activity results, and the anaplerotic pathways, such as Ppc-Pck and Mez, are also up-regulated in the *sucA* mutant.

In batch cultivations, acetate production patterns show significantly different trends in *sucA* and *sucC* mutants. Acetate production in *E. coli* is generally attributed to the 'overflow metabolism' (El-Mansi and Holms, 1989), which is believed to be the consequence of an imbalance between glucose uptake and demand for energy and biosynthesis (Han et al., 1992; Kleman and Strohl, 1994; Aristidou et al., 1995; Farmer and Liao, 1997; van de Walle and Shiloach, 1998). Acetate formation in *E. coli* can be reduced either by decreasing the specific glucose uptake rate or by enhancing the maximum capacity of the oxidative metabolism (Han et al., 1992). The maximum glucose uptake rate for *sucA* mutant is lower than that of the parent strain, which causes a decrease in the acetate production of the *sucA* mutant in batch cultivation. Although TCA cycle activity is reduced in the *sucA* mutant, acetate production does not increase due to the activation of the

Effect of specific-gene knockout

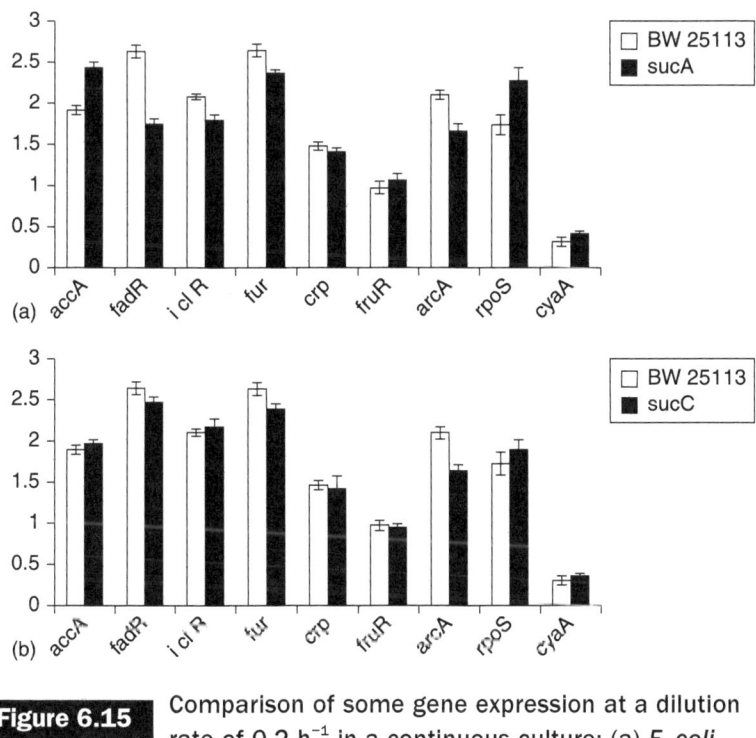

Figure 6.15 Comparison of some gene expression at a dilution rate of 0.2 h^{-1} in a continuous culture: (a) *E. coli* BW25113 and *E. coli sucA* mutant; and (b) *E. coli* BW25113 and *E. coli sucC* mutant

glyoxylate shunt in the *sucA* mutant. However, the *sucC* mutant shows high acetate production rates, caused by low TCA cycle activity without glyoxylate shunt activity. Acetate production reduces cell yield in the *sucC* mutant. The *sucC* mutant cannot utilize acetate, which is consistent with other investigation (Mat-Jan et al., 1989). Mat-Jans et al. (1989) shows that the deficiency in SCS prevents aerobic growth on acetate and αKG, but it can grow on TCA cycle intermediates, such as succinate, malate, and fumarate. It may be considered that the SCS mutant makes SucCoA from αKG via KGDH, and makes succinate from fumarate by reversing the succinate dehydrogenase reaction, as under anaerobic conditions. This may explain the decreased CO_2 evolution rate (CER) in the *sucC* mutant. As such, all the TCA cycle intermediates can be fully generated in the *sucC* mutant, which shows a similar growth rate to the parent strain. The activity of Pfk, a rate-limiting enzyme in glycolysis, is down-regulated in both mutants, which are responsible for the decreased

Bacterial cellular metabolic systems

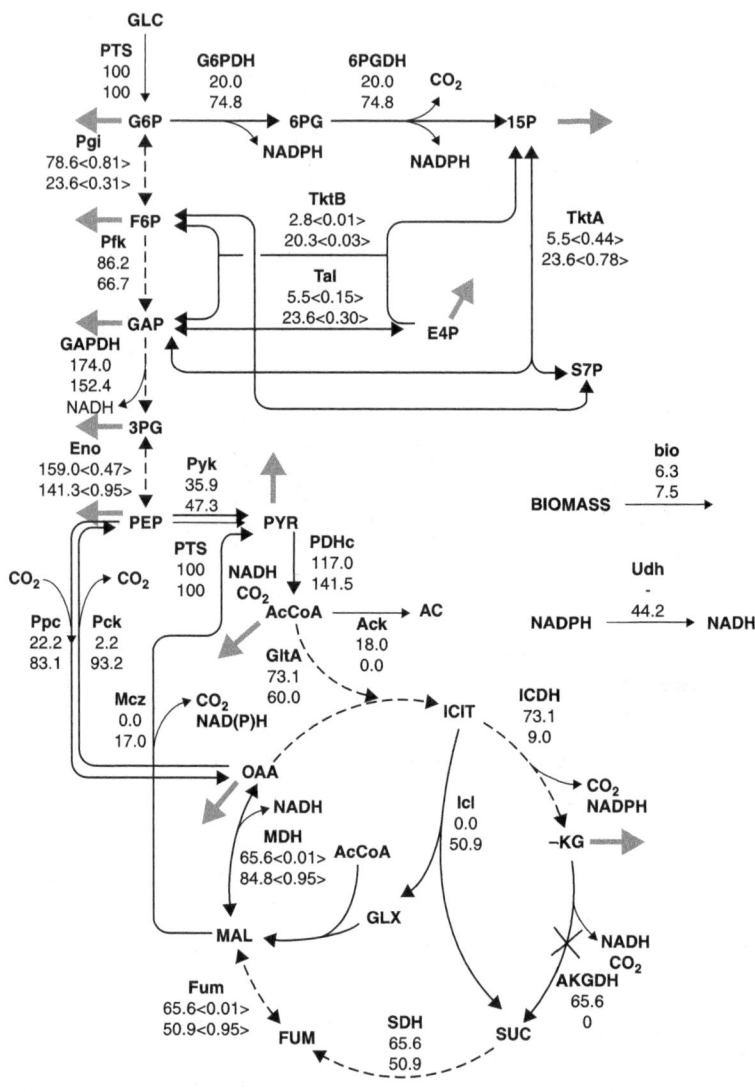

Figure 6.16 Metabolic flux distributions of wild-type (upper values), *sucA* mutant (lower values) at dilution rate of 0.2 h^{-1}. Solid lines indicate increased fluxes, while dash lines indicate reduced flux. Values in '⟨·⟩' indicate exchange coefficient. Gray arrows indicate biomass formation

glycolytic flux (Chassagnole et al., 2002). Since it has been demonstrated that Pfk is allosterically inhibited by PEP and ATP, the increase in PEP and ATP concentrations in the mutants cause decreased Pfk activity, though not significantly (Niedhardt, 1996; Chassagnole et al., 2002; Siddiquee et al., 2004a,b).

The accumulation of AcCoA leads to up-regulation of CS in both mutants. Figure 6.14 shows that AcCoA is more accumulated in the *sucC* mutant, and CS is more activated (Figure 6.13). Down-regulation of the global regulatory genes, such as *arcA* and *fnr* (Figure 6.15), also implies the activation of TCA cycle enzymes (Park et al., 1994, 1995a; Perrenoud and Sauer, 2005), such as CS, Fum, and MDH. It is known that the $NADP^+$-specific ICDH is responsible for partitioning isocitrate between the TCA cycle and the glyoxylate shunt, by reversible phosphorylation catalyzed by a bifunctional kinase/phosphatase (Hoyt and Reeves, 1988; Laporte et al., 1985; Nimmo et al., 1987). Figures 6.13 and 6.15 show that the increase in enzyme activity of Icl and MS and the increase in gene expression levels of *aceA* indicate activation of the glyoxylate shunt in the *sucA* mutant but not in the *sucC* mutant. The expression of this *aceBAK* operon is negatively controlled by an adjacent regulatory gene, IclR (Brice and Kornberg, 1968; Maloy and Nunn, 1981), as well as FadR (Maloy et al., 1981). Figure 6.15 shows a slight decrease in the expression of *iclR* and *fadR* in the *sucA* mutant, but not in the *sucC* mutant. It is reported that the expression of *fumAC-lacZ* is up-regulated about 3-fold when acetate is used as a carbon source instead of glucose (Park et al., 1995b). Figure 6.13 shows the activity of Fum as almost 2-fold higher in the *sucA* mutant as compared to that of the parent strain, which also indicates that the *sucA* mutant utilizes acetate through the glyoxylate shunt. Similar to *fumAC* genes, it is reported that *mdh-lacZ* expression is enhanced more than 2-fold when acetate is used as a carbon source instead of glucose (Park et al., 1995b). Figure 6.13 shows that MDH activity is higher in the *sucA* mutant as compared to the parent strain, while the activities of Fum and MDH change very little in the *sucC* mutant.

In both mutants, the reduction of TCA cycle activity also affects the redox balance in the cell, where the reductive NAD^+ and $NADP^+$ are significantly increased. The glyoxylate shunt combining with Pck operates as a cycle, which catalyzes complete oxidation of carbohydrates to CO_2 (Fischer and Sauer, 2003). Figure 6.16 indicates the activation of this cycle in the *sucA* mutant, and the enzyme results also confirm this. This causes CER to increase a little in the *sucA* mutant in the continuous culture. The oxidative PP pathway is enhanced in both mutants, especially in the *sucA* mutant. The activation of this pathway over-supplies NADPH to the cell, and the NADPH can be interconverted between NADH by the transhydrogenation reaction. Figure 6.16 implies that the *sucA* mutant converts NADPH to NADH by this reaction. Lower pyruvate concentrations in *sucA* and *sucC* mutants cause decreased LDH activity in both mutants (Tamy et al., 1968; Jiang et al., 2001; Li et al., 2006b).

6.7 Effect of *icdA* gene knockout on metabolism

Although there has been a long history for the selection of *E. coli* mutants lacking activity of TCA cycle enzymes, most research has focused on analysis of their growth phenotype (Miles and Guest, 1989; Guest and Russell, 1992). Their research reveals that the mutants lacking any of the TCA cycle enzymes have a common phenotype: all fail to grow with acetate as a sole carbon source. Some TCA cycle mutants (*sucAB*, *mdh*, *sdhCDAB*, and *fumA*) grow normally on glucose minimal medium under anaerobiosis, which is attributed either to the lack of function of the enzyme in the branched mode or to the other enzymes that supply the missing biosynthetic intermediates. The enzymes CS (*gltA*), Acn (*acnA,B*), and ICDH (*icdA*) are responsible for glutamate synthesis and thus should be essential for both aerobic and anaerobic growth on minimal media. Considering several central roles of the TCA cycle, it is expected that the blockage in the TCA cycle pathway may cause flux re-routing, which absolutely exerts a different type of metabolic regulation and triggers the compensatory mechanism.

Both *icdA* mutant *E. coli* (JW 1122), and its parent strain *E. coli* BW 25113, were cultivated using LB medium with different carbon sources such as glucose and acetate, in order to elucidate the physiological responses to *icd* gene deletion (Kabir and Shimizu, 2004). Batch cultivation results of both strains grown on glucose under aerobic conditions are shown in Figure. 6.17. Batch cultivation under microaerobic conditions was also conducted with glucose as a sole carbon source, to observe the effects of the *icd* knockout gene on metabolism at a limited DO level (<1 ppm) (Figure 6.18). Figure 6.19 shows the batch cultivation results for the use of acetate as a carbon source under aerobic conditions. The growth parameters are summarized in Table 6.13, based on three different culture conditions. As can be seen in Figure 6.17, the wild-type *E. coli* utilizes glucose for cell growth with accumulation of acetate as the

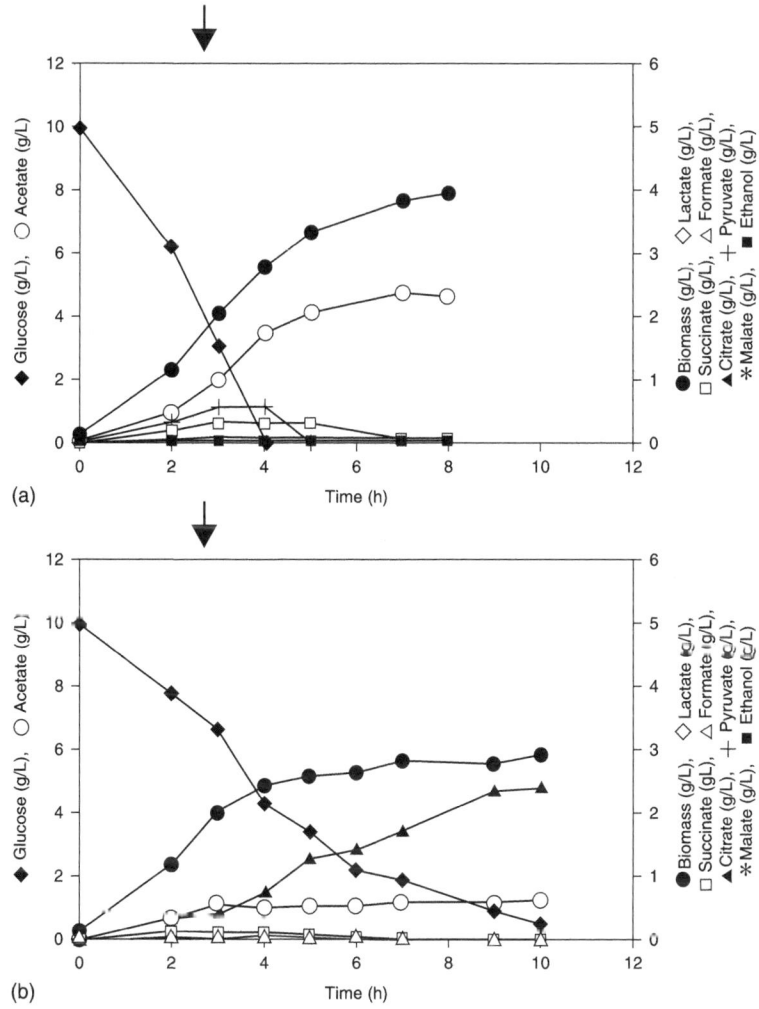

Figure 6.17 Batch cultivation results of: parent *E. coli*; and (b) *icd* mutant, grown on glucose under aerobic conditions. The arrow indicates the sampling time for proteome analysis and for measurement of enzyme activity

major by-product (Figure 6.17a), whereas the *icd* mutant consumes glucose for cell growth with slight accumulation of acetate (Figure 6.17b). Moreover, citrate accumulation (~2.40 g/l) is observed in the *icd* mutant. Other organic acids, such as lactate, succinate, formate, pyruvate, malate, and ethanol, are almost negligible in both the *icd* mutant and the parent

Bacterial cellular metabolic systems

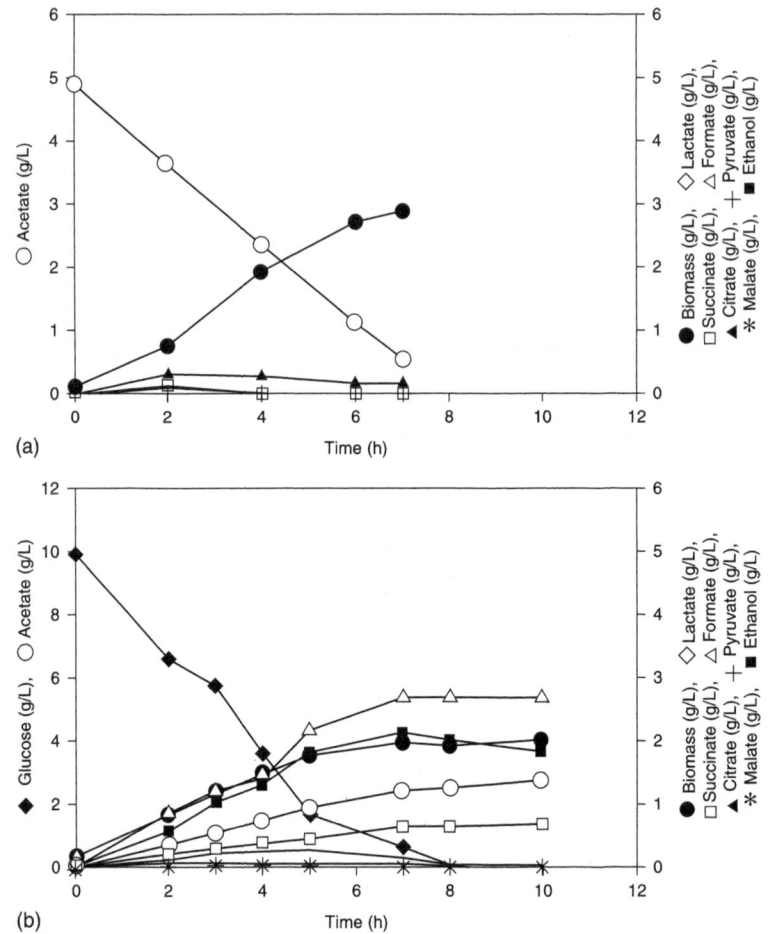

Figure 6.18 Batch cultivation results of: (a) parent *E. coli*; and (b) *icd mutant*, grown on glucose under microaerobic conditions

E. coli. Furthermore, the *icd* mutant shows higher specific CO_2 evolution rates (0.16 mM/gDCW/h), which is 60% higher compared to its parent *E. coli* (0.10 mM/gDCW/h) (Table 6.13). However, approximately 27% increase in the biomass yield on glucose is observed in the *icd* mutant compared to that of the parent *E. coli*. When cells are grown on glucose under microaerobic conditions, little difference in fermentation characteristics (Table 6.13) is observed where formate is the main by-product in both the *icd* mutant and the wild type *E. coli* followed by

Effect of specific-gene knockout

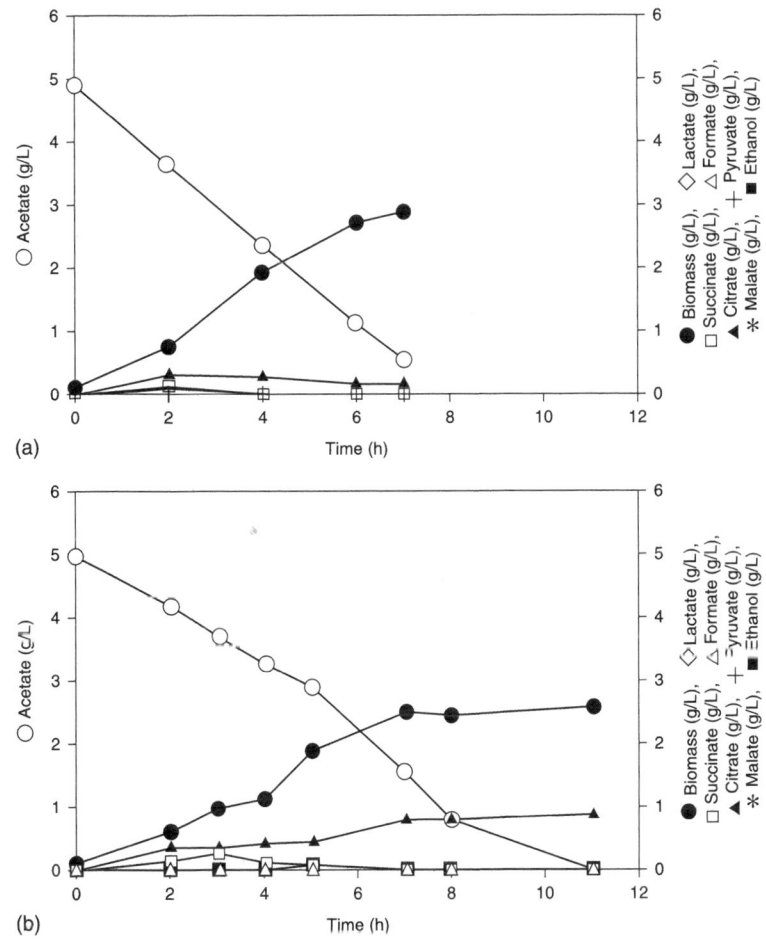

Figure 6.19 Batch cultivation results of: parent *E. coli* (a) and *icd* mutant (b) grown on acetate under aerobic conditions

ethanol, acetate, and succinate (Figures 6.18a,b). No citrate accumulation is observed in the *icd* mutant *E. coli* under microaerobic conditions. Similarly, when acetate is used as a sole carbon source, similar fermentation characteristics are observed between the two strains, except for a slight citrate accumulation in the *icd* mutant (Figures 6.19a,b).

Table 6.14 represents a summary of the mass spectrometry data, including predicted molecular masses, pI values, and protein identification. As shown in Table 6.15, the enzyme activity levels vary significantly in the *icd* mutant compared to the parent. To see how each enzyme activity

Table 6.13 Specific rate of parent and *icd* mutant grown on different carbon sources under different culture conditions. Standard deviations were calculated from four independent measurements (duplicate measurements from duplicate experiments). q_S – specific substrate uptake rate; μ – specific growth rate; $Y_{X/S}$ – biomass yield; q_{Ace} – specific acetate production rate; q_{O_2} – specific oxygen uptake rate; qCO_2 – specific carbon dioxide evolution rate

Experiment		qS [g/gDCW/h]	μ [h-1]	$Y_{X/S}$ [gDCW/g substrate]	q_{Ace} [g/DCWg/h]	q_{O_2} [mM/gDCW/h]	q_{CO_2} [mM/gDCW/h]
Glucose (aerobic)	Parent	1.53(±0.05)	0.44(±0.03)	0.26(±0.04)	0.78(±0.05)	0.14(±0.03)	0.10(±0.03)
	Mutant	0.88(±0.03)	0.34(±0.01)	0.33(±0.02)	0.15(±0.02)	0.17(±0.04)	0.16(±0.01)
Glucose (microaerobic)	Parent	1.58(±0.11)	0.42(±0.06)	0.17(±0.01)	0.56(±0.11)	0.09(±0.02)	0.05(±0.02)
	Mutant	1.32(±0.04)	0.29(±0.05)	0.21(±0.05)	0.59(±0.13)	0.04(±0.01)	0.03(±0.01)
Acetate (aerobic)	Parent	0.33(±0.02)	0.32(±0.07)	0.79(±0.03)	—	0.37(±0.05)	0.29(±0.06)
	Mutant	0.36(±0.05)	0.32(±0.08)	0.79(±0.07)	—	0.29(±0.02)	0.22(±0.01)

is affected, the ratio of enzyme activity is calculated for the *icd* mutant over the corresponding value in the parent strain and the ratio values in the central metabolic map are annotated. Figure 6.20 shows such a metabolic map of *E. coli* grown on glucose under aerobic conditions, where the solid lines indicate up-regulation and the dotted lines represent down-regulation. The thickness of the lines is proportional to the magnitude of the ratio, that is, the thicker the line, the more regulated.

The Pgi and Fba are up-regulated significantly in the *icd* mutant compared to the parent, except Pfk. The two enzymes involved in the oxidative PP pathway, such as G6PDH and 6PGDH, are also affected significantly in the *icd* mutant grown on glucose under aerobic conditions. G6PDH and 6PGDH are increased by 1.77- and 1.42-fold in the *icd* mutant compared to that in the parent, respectively (Table 6.15

Table 6.14 Summary of MALDI-TOF mass spectrometry data for protein spots showing altered expression levels on 2D gels for parent *E. coli* (WT) and *icd* mutant

Spot number	Theoretical pI/Mw (kDa)	Experimental pI/Mw (kDa)	Gene name with protein identity	Abundance (icd-/WT) ratio
41	5.26/51.8	5.19/54.7	glnA (Glutamine synthetase)	0.161
35	6.05/56.7	6.23/55.9	trpD (Anthranilate synthase component II)	0.082
46	6.35/58.2	6.34/54.0	cydA (Cytochrome BD-I oxidase subunit I)	0.128
100	5.37/41.4	5.40/42.4	sucC (Succinyl-CoA synthetase beta chain)	0.152
103	6.01/41.2	6.03/41.9	metB (Cystathionine gamma-synthase)	0.089
115	6.58/35.4	6.72/38.7	gapA (Glyceraldehyde 3-phosphate dehydrogenase A)	0.059
123	6.90/36.2	6.89/37.9	nuoH (NADH dehydrogenase I, chain H)	0.032
170	5.99/32.2	6.77/32.2	cyoA (Cytochrome O subunit 2)	0.142

(*Continued*)

Table 6.14 Summary of MALDI-TOF mass spectrometry data for protein spots showing altered expression levels on 2D gels for parent *E. coli* (WT) and *icd* mutant (*Continued*)

Spot number	Theoretical pI/Mw (kDa)	Experimental pI/Mw (kDa)	Gene name with protein identity	Abundance (*icd*-/WT) ratio
381	6.31/29.6	6.16/29.8	sucD (Succinyl-CoA synthetase alpha chain)	0.071
202	–	6.70/27.1	not detected	0.034
210	6.31/26.8	6.80/26.2	sdhB (Succinate dehydrogenase iron-sulfur protein)	0.046
316	–	8.01/25.4	not detected	0.15
426	–	6.79/24.4	not detected	19.118
465	5.40/18.6	5.41/19.6	nuoE (NADH dehydrogenase I, chain E)	0.139
510	6.58/16.1	6.40/15.4	rpiB (Ribose 5-phosphate isomerase B)	0.159
644	6.56/12.0	6.93/11.8	cyoD (Cytochrome O ubiquinol oxidase protein cyoD)	6.305
761	–	6.83/10.6	not detected	0.156
616	5.46/14.9	5.42/14.8	atpC (ATP synthase epsilon chain)	6.997
498	4.72/19.2	4.59/16.1	aroL (Shikimate kinase II)	0.057
606	5.03/17.6	4.97/15.6	bcp (Bacterioferritin comigratory protein)	0.071
719	4.71/17.6	4.62/14.4	greA (Transcription elongation factor greA)	0.033

and Figure 6.20). The up-regulation of these two enzymes in the *icd* mutant is contrary to the slower growth rates, since both enzymes are known to be growth rate dependent (Wolf et al. 1979; Pease and Wolf, 1994).

Both enzymes G6PDH and 6PGDH in the PP pathway are known to be growth rate dependent. The reason for up-regulation of these

Table 6.15 Specific enzyme activities in cell extracts of parent and icd mutant grown on glucose under aerobic conditions. The unit of enzyme activity is nM/min/mg protein. Overall ED pathway enzyme was represented as mg pyruvate/min/mg protein. The ratio of enzyme activity was calculated for icd mutant over the corresponding value in the parent strain. Therefore, a value of 1.0 indicates no change. The mean value from four independent measurements (duplicate measurements from duplicate experiments) is presented with standard deviation

Enzyme with their corresponding gene name	Parent	icdA mutant	Ratio
Glucokinase (glk)	257.02 ± 25.13	175.25 ± 23.33	0.66
Phosphoglucose isomerase (pgi)	14.99 ± 2.84	31.83 ± 4.51	2.25
Phosphofructose kinase (pfk)	105.18 ± 13.09	112.11 ± 15.62	1.05
Fructose diphosphate aldolase (fba)	45.66 ± 17.05	52.14 ± 11.45	1.42
Glyceraldehyde-3-phosphate dehydrogenase (gapA)	64.26 ± 13.46	46.5 ± 10.73	0.70
3-phosphoglycerate kinase (pgk)	85.49 ± 12.32	82.18 ± 13.68	0.94
Pyruvate kinase (pyk)	161.36 ± 21.78	152.46 ± 23.17	0.93
Pyruvate dehydrogenase complex (aceE, aceF & lpdA)	14.69 ± 3.55	14.72 ± 3.82	0.98
Phosphoenol pyruvate carboxylase (ppc)	64.26 ± 5.24	50.07 ± 4.25	0.78
Phosphoenol pyruvate carboxykinase (pckA)	28.56 ± 11.37	42.92 ± 8.19	2.02
NADP⁺-specific malic enzyme (mae1)	1.42 ± 1.31	11.8 ± 9.42	21.64
NAD⁺-specific malic enzyme (mae2)	1.22 ± 1.15	1.25 ± 1.19	0.86
Glucose-6-phosphate dehydrogenase (zwf)	71.4 ± 7.81	125.18 ± 12.67	1.77
6-phosphogluconate dehydrogenase (gnd)	23.56 ± 9.94	33.98 ± 7.58	1.42
ED pathway enzyme (edd & eda)	3.64 ± 3.41	3.61 ± 3.43	0.78

(Continued)

Table 6.15 Specific enzyme activities in cell extracts of parent and *icd* mutant grown on glucose under aerobic conditions. The unit of enzyme activity is nM/min/mg protein. Overall ED pathway enzyme was represented as mg pyruvate/min/mg protein. The ratio of enzyme activity was calculated for *icd* mutant over the corresponding value in the parent strain. Therefore, a value of 1.0 indicates no change. The mean value from four independent measurements (duplicate measurements from duplicate experiments) is presented with standard deviation (Continued)

Enzyme with their corresponding gene name	Parent	*icdA* mutant	Ratio
Acetate kinase (*ack*)	113.16 ± 19.46	61.51 ± 6.37	0.59
Transhydrogenase (*udhA, pntA & pntB*)	26.59 ± 6.35	27.13 ± 5.18	1.08
Citrate synthase (*gltA*)	10.12 ± 3.27	38.28 ± 9.48	4.20
Isocitrate dehydrogenase (*icd*)	154.22 ± 26.55	1.43 ± 1.35	0.00
α-Ketoglutarate dehydrogenase complex (*sucA, sucB & lpdA*)	135.45 ± 24.11	33.19 ± 16.44	0.15
Fumarase (*fumA*)	105.73 ± 22.06	52.11 ± 17.51	0.41
Malate dehydrogenase (*mdh*)	35.56 ± 8.77	77.21 ± 14.74	2.33
Isocitrate lyase (*aceA*)	8.01 ± 2.93	149.65 ± 22.69	24.99
Malate synthase (*aceB*)	2.11 ± 1.73	48.90 ± 38.57	27.18

enzymes in the *icd* mutant may be due to the demand for NADPH, since intracellular NADPH concentrations are much lower in the mutant than in the parent strain (Table 6.16). Moreover, the extent of the regulation of these two enzymes may be accounted for by their

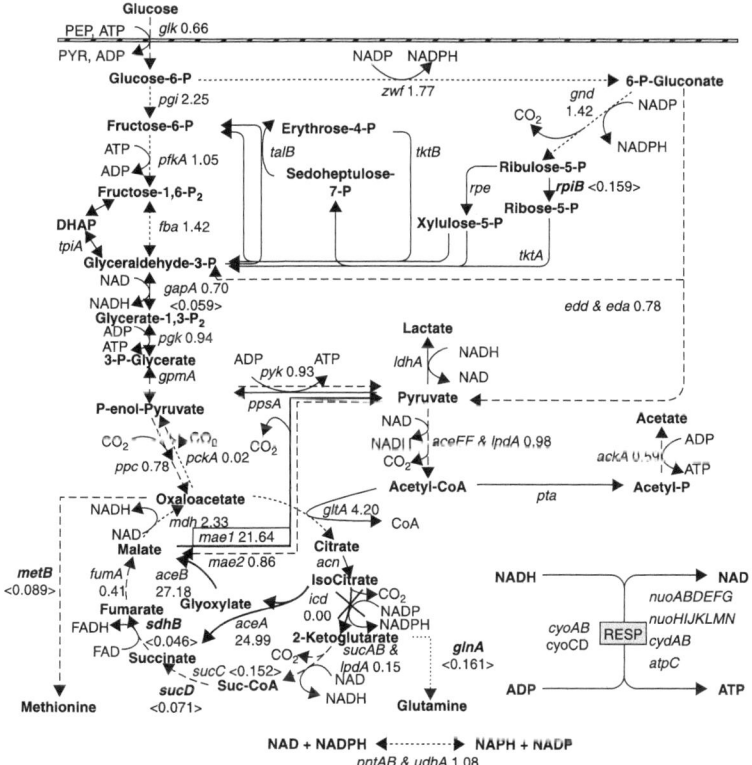

Figure 6.20 Relative protein and enzyme levels of *icd* mutant JW1122 from exponential growth phase in comparison to the parent *E. coli* BW25113 grown on glucose under aerobic conditions. The enzymes are annotated by their gene names. The numbers represent the up-regulation ratios in *icdA* mutant (JW1122) as compared to its parent. The genes with numbers in parentheses indicate the protein abundance ratios deduced from proteome analysis. The solid lines/genes indicate up-regulation, whereas the broken lines/genes for down-regulation.

individual allosteric control. Sugimoto and Shiio (1987) reported that G6PDH was allosterically inhibited by FDP and PRPP, and induced by glucose. However, 6PGDH is inhibited by FDP, PRPP, GAP, Ru5P, E4P, and NADPH, and induced by gluconate. Apparently, the shortage of FDP is facilitated and functions as a major activator to 6PGDH, due to significant up-regulation of Fba activity in the *icd* mutant (Table 6.13 and Figure 6.20). However, the protein encoded by *rpiB* is down-regulated in the *icd* mutant (Table 6.14). *E. coli* hosts two discernible Rpi activities (David and Wiesmeyer, 1970; Esseenberg and Cooper, 1975): a constitutive activity (*rpiA*) that accounts for at least 99% of the total Rpi and an inducible one (*rpiB*). The expression of *rpiB* is very low when cells are grown on complex media, even in the presence of D-ribose. The mode of its regulation is not known, but the involvement of a negative regulator (product of *rpiR*) is more likely (Skinner and Cooper 1971). The significant down-regulation of *rpiB* in the *icd* mutant may be due to inactivation of ICDH, which can efficiently induce *rpiR*, which in turn negatively regulates expression of the *rpiB* protein.

Among the TCA cycle enzymes, CS and MDH are up-regulated significantly in the *icd* mutant *E. coli* compared to its parent strain. It has been revealed that the enzymes at the branch point of the TCA cycle and glyoxylate shunt are regulated in the opposite way (LaPorte and

Table 6.16 Measurement of intracellular metabolites for parent *E. coli* and *icd* mutant grown on glucose under aerobic conditions. Standard deviations were calculated from four independent measurements (duplicate measurements from duplicate cultures). DCW denotes dry cell weight

	Concentration [μM/gDCW]			Ratio	
	Parent	icdA⁻ mutant		Parent	icdA⁻ mutant
ATP	97.01 ± 13.55	43.05 ± 18.35			
ADP	231.02 ± 11.29	169.07 ± 15.24	ATP/ADP	0.380	0.161
NADH	72.11 ± 9.27	64.03 ± 7.41			
NAD$^+$	181.15 ± 15.49	28.06 ± 6.76	NADH/NAD$^+$	0.379	2.658
NADPH	83.04 ± 8.32	49.05 ± 3.83			
NADP$^+$	232.06 ± 14.67	196.02 ± 16.52	NADPH/NADP$^+$	0.344	0.252

Koshland, 1983; Nimmo and Nimmo, 1984). It is shown that the knockout of the *icd* gene may induce glyoxylate shunt significantly, as is evidenced by the significant up-regulation of Icl and MS activity in the mutant compared to that in the parent strain (Table 6.15 and Figure 6.20).

Anaplerotic enzyme Ppc is down-regulated in the *icd* mutant compared to the parent strain (Table 6.15 and Figure. 6.20). However, Pck is up-regulated in the mutant. The reason may be due to the shortage of PEP in the *icd* mutant. Two other anaplerotic enzymes, such as the NAD^+ and $NADP^+$-specific malic enzymes, are regulated differently in the *icd* mutant. The activity of Mez ($NADP^+$-specific) is up-regulated significantly, whereas Sfc (NAD^+-specific) is down-regulated in the *icd* mutant compared to that in the parent (Table 6.15 and Figure 6.20). Acetate kinase (Ack) activity is down-regulated in the *icd* mutant, a result which is consistent with lower acetate production rates (Table 6.13).

Table 6.15 and Figure 6.20 imply that re-oxidation of the reducing power are more or less balanced in the mutant *E. coli*. The intracellular concentrations of ATP, ADP, NADH, NAD^+, NADPH, and $NADP^+$ are shown in Table 6.16. The intracellular concentrations of these metabolites are lower in the *icd* mutant than those in the parent. However, the $NADH/NAD^+$ ratio is higher in the *icd* mutant, whereas the ATP/ADP and $NADPH/NADP^+$ ratios are much lower in the mutant than in the parent strain.

Basically, ICDH is involved in the conversion of isocitrate to αKG, with concomitant generation of NAD(P)H and CO_2. Therefore, it is expected that the *icd* mutant can produce less CO_2. However, it is shown that specific CO_2 evolution rates are higher in the *icd* mutant than in the parent strain (Table 6.13). Enzyme activity measurements show that the *icd* gene knockout results in the activation of the glyoxylate shunt that ultimately increases the $NADP^+$-dependent malic enzyme, PckA, and Gnd (6PGDH) activity significantly in the mutant compared to the parent (Table 6.15 and Figure 6.20). This may be why higher CO_2 evolution rates are observed in the *icd* mutant than in the parent strain. The significant up-regulation of CS (GltA) in the *icd* mutant is consistent with other results (Park et al., 1994), where GltA activity levels are inversely proportional to growth rates. The NAD^+-specific malic enzyme is repressed by glucose and inhibition can be released by malate, whereas glucose and AcCoA repress the $NADP^+$-specific malic enzyme. Previous studies have shown that the NAD^+-specific malic enzyme takes part in catabolism of malate, controlling C_4-dicarboxylic acids as well as amino acids in the cell, while the NADP$^+$-specific malic enzyme supplies AcCoA from MAL via PYR, which is utilized for biosynthesis and maintains the action of the TCA cycle (Murai et al., 1971). By this mechanism, the down-regulation of the NAD$^+$-specific malic enzyme may explain why the intracellular malate concentrations decrease due to a significant increase of MDH activity in the *icd* mutant (Table 6.15), which diverts more carbon flux from MAL to OAA as compared to the parent strain. The significant up-regulation of the NADP$^+$-specific malic enzyme may be due to the reduced intracellular AcCoA pool, which can relieve the inhibition to this enzyme. This speculation might be reflected by the down-regulation of GAPDH, Pgk, Pyk, PDHc, and Ack activity, which are directly and/or indirectly linked to AcCoA synthesis. The AcCoA pool is expected to decrease, since the entry to the TCA cycle seems to increase in the influx from glycolysis, which is inferred by increased CS (GltA) activity in the *icd* mutant compared to the parent strain (Table 6.15 and Figure 6.20). Moreover, down-regulation of GapA (GAPDH) activity in the *icd* mutant strain may be the effect of lower intracellular NAD$^+$ concentrations, which have been reported to restrict GAPDH activity in *E. coli* (Seta et al., 1997).

The protein encoded gene *atpC* (ATP synthase) is up-regulated significantly in the *icd* mutant compared to its parent strain (Table 6.14 and Figure 6.20). The reason for up-regulation may be due to lower ATP concentrations and lower ATP/ADP ratios in the *icd* mutant strain (Table 6.16), since the expression of *atpC* does not appear to be subject to substrate or growth rate control (Pedersen et al., 1978; Nielsen, 1984), or to respond to anaerobic or aerobic shifts (Smith and Neidhardt, 1983). A lower ATP/ADP ratio is responsible for a lower growth rate in the *icd* mutant than in the parent *E. coli*, since Brian et al. (2002) have shown that growth rates of the cells are modulated by the intracellular ATP/ADP ratio, i.e. the ATP/ADP ratio is proportional to cell growth rate. This result indicates that the *icd* mutant gains enough energy for growth that ultimately reduces specific growth rates in the *icd* mutant strain. Moreover, significant down-regulation of the proteins encoded by genes *nuoE*, *nuoH*, *cydA*, and *cyoA* is consistent with lower ATP, ADP, NADH, and NAD$^+$ concentrations, respectively (Table 6.16), in the *icd* mutant than in the parent strain, indicating that the *icd* gene knockout significantly affects the respiratory system and electron transport chain. Therefore, compensation of cellular respiratory components can be mediated by the protein encoded gene *cyoD*, which is up-regulated significantly in the *icd* mutant compared to the parent strain (Table 6.14 and Figure 6.20).

Down-regulation of the protein encoded genes *glnA*, *sucC*, *sucD*, *sdhB*, and *metB* in the *icd* mutant compared to the parent strain (Table 6.14 and Figure 6.20) may be due to the effects of *icd* gene deletion. Moreover, significant down-regulation of the protein encoded genes *trpD*, *aroL*, *bcp*, and *greA* in the *icd* mutant are also observed in Table 6.14. Down-regulation of the *trpD* gene encoding protein in the *icd* mutant might be the effect of the Trp repressor. The *E. coli* Trp repressor is one of the smallest proteins that shows sequence-specific DNA binding and is also allosterically controlled (Somerville, 1992), and this protein has been studied extensively by physical as well as genetic methods, yet its mode of DNA recognition remains controversial. It is of particular interest, as it binds to five diverse sequences in the *E. coli* genome. Four of these sequences are in promoters, in the *trpEDCBA* operon (Rose et al., 1973; Bennett et al., 1976), the *trpR* gene (Gunsalus and Yanofsky, 1980; Kelley and Yanofsky, 1982), the *aroH* gene (Zurawski et al., 1981; Grove and Gunsalus 1987), and the *mtr* gene (Heatwole and Somerville, 1991; Sarsero et al., 1991), respectively. The fifth site is within the leader region of the *aroL* gene (Heatwole and Somerville, 1992; Lawley and Pittard, 1994). These five operons control the biosynthesis and uptake of the amino acid tryptophan. The repressor binds to the operators only in the presence of tryptophan, repressing gene expression, hence controlling the intracellular level of its co-repressor. Consequently, significant down-regulation of the *aroL* gene encoding protein is reasonable in the *icd* mutant compared to the parent, since higher tryptophan synthesis is expected in the *icd* mutant strain as a compensatory function due to blockage of the *icd* pathway in the TCA cycle.

Furthermore, bacterioferritin co-migratory protein encoded gene *bcp* is down-regulated significantly in the *icd* mutant compared to the parent strain (Table 6.14). This result indicates that *E. coli* lacking the *icd* gene is less efficient to protect against toxicity of reactive oxygen species (ROS), since *E. coli* bacterioferritin co-migratory protein (BCP) has been categorized as a new member of the thiol-specific antioxidant protein (TSA)/alkyl hydroperoxide peroxidase C (*ahpC*) family, because BCP exhibits thioredoxin-dependent hydroperoxide peroxidase activity (Jeong et al., 2000). This might be another reason for lower growth rates of the *icd* mutant than in the parent strain grown on glucose (Table 6.13). However, the function of BCP has not yet been fully clarified, despite the wide distribution of BCP in most pathogenic bacteria, including *Haemophilus influenzae*, *Helicobacter pylori*, and *Mycobacterium tuberculosis*. Down-regulation of the protein encoded gene *greA* (transcription elongation factor) in the *icd* mutant indicates that inactivation of ICDH greatly affects the regulation of transcription elongation and cleavage, since *greA* stimulates intrinsic nucleolytic activity of RNA polymerase (RNAP) and also has a role on RNA hydrolysis, which includes transcription proofreading, suppression of transcriptional pausing and arrest, and facilitation of RNAP transition from transcription initiation to transcription elongation (Fish and Kane, 2002).

6.8 Effect of *pfl* gene knockout on metabolism

In the case of anaerobic or microaerobic cultivation, NADH regeneration and ATP formation control metabolic fluxes at the branch points, such as PEP, PYR, and AcCoA (Zhu and Shimizu, 2004). As a result, the *pflA,B* gene knockout causes over-production of lactate (Zhu and Shimizu, 2004, 2005). Figure 6.21 shows the fermentation results of *E. coli* BW25113 and *E. coli pfl*B mutant grown on glucose. It indicates that the fermentation patterns of these two strains are similar under aerobic conditions, where the *E. coli pfl*B mutant produces less acetate. However, under microaerobic conditions (Figure 6.22), the results are different, and the *pfl*B mutant strain shows an almost homolactic acid fermentation as compared to that of *E. coli* BW25113. The concentrations of acetate and formate are about 3 g l^{-1} in the parent strain. However, no formate is detected in the culture of the *pfl*B mutant, and the acetate concentration is kept low at about 0.2–0.3 g l^{-1}. A similar fermentation pattern is found in the *pfl*A mutant (JW0885). Two other mutant strains, *pfl*C and *pfl*D, are also cultivated under microaerobic conditions, where the fermentation patterns of these two mutants are similar to that of the parent strain.

Specific glucose uptake rates and metabolite formation rates in these cultures are compared in Figure 6.23. It indicates that the glucose uptake rate is higher in the *pfl*B and *pfl*A mutants than in the parent strain, under either aerobic or microaerobic conditions. The lactate formation rate of the mutant strains of *pfl*A and *pfl*B is about 70-fold higher than in the parent strain. This results in the high yields of lactate production of more than about 70% from glucose (Figure 6.24B).

The *pfl*A mutant is also cultivated using several other carbon sources, i.e. gluconate, pyruvate, fructose, and glycerol, under microaerobic conditions. The results are shown in Tables 6.17 and 6.18. When using glucose as a carbon source, the *pfl*A mutant shows extensive homolactic

Effect of specific-gene knockout

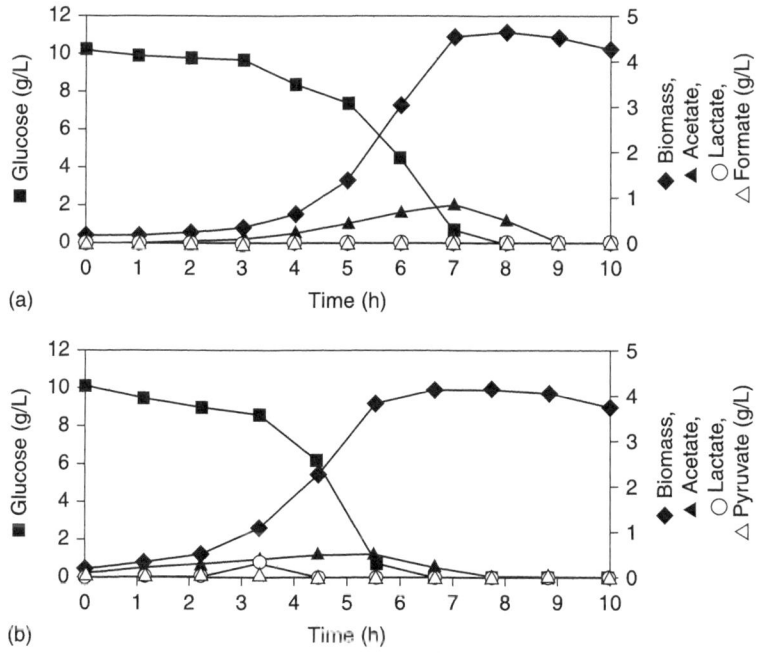

Figure 6.21 The cultivation results of: (a) *E. coli* BW25113; and (b) the *pfl*B⁻ mutant (**B**) grown on glucose under the aerobic conditions. **A** *E. coli* BW25113, **B** *E. coli* *pfl*B⁻ mutant. ■ Glucose, ◆ biomass, ○ lactate, ▲ acetate, △ formate, ◇ pyruvate

acid fermentation in the microaerobic culture. However, a large amount of acetate is produced together with lactate, when gluconate or pyruvate is used as a carbon source. In particular, acetate instead of lactate is the major product when pyruvate is used as a carbon source. The fermentation patterns, when using fructose or glycerol as a carbon source, are similar to those when using glucose. However, fructose and glycerol uptake rates are significantly lower than glucose uptake rates (Table 6.17), and the lactate yields vary from 60% to more than 70% (Table 6.18). Succinate concentration is always very low. For the *pfl*A and *pfl*B mutants, the specific growth rate is lower under microaerobic conditions, especially when using glycerol as a carbon source.

Pfl activities in BW25113, *pfl*A, *pfl*B, *pfl*C, and *pfl*D mutants, using glucose as a carbon source under microaerobic conditions, are shown in Figure 6.25. It can be seen that Pfl is inactivated in *pfl*A and *pfl*B mutants.

Bacterial cellular metabolic systems

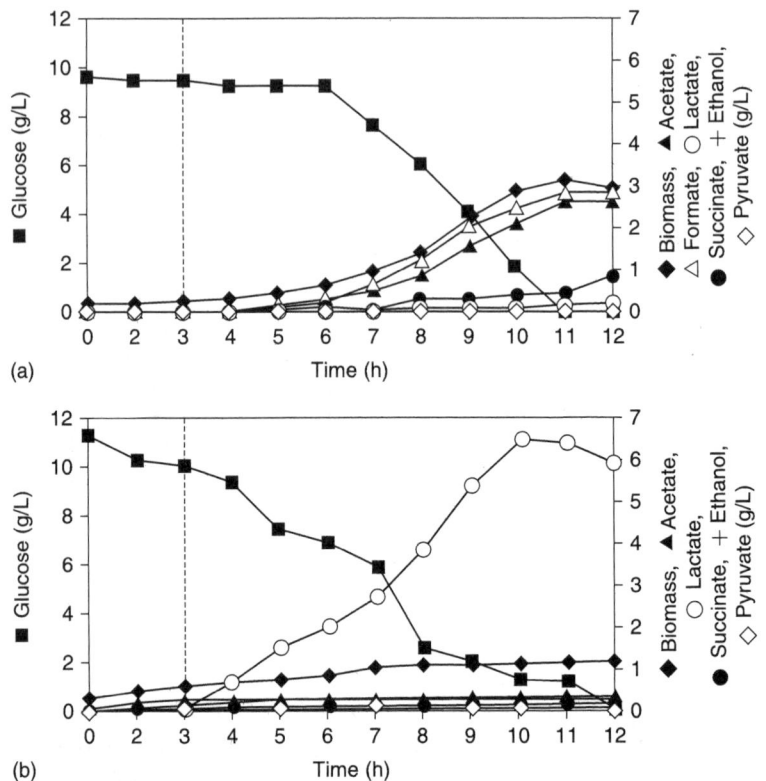

Figure 6.22 The cultivation results of: (a) *E. coli* BW25113; and (b) the *pfl*B⁻ mutant, grown on glucose under the microaerobic conditions. The dotted line indicates the time when the culture was shifted from the aerobic conditions to the microaerobic conditions. **A** *E. coli* BW25113, **B** *pfl*B⁻ mutant. ■ Glucose, ◆ biomass, ● succinate, ○ lactate, ▲ acetate, △ formate, ◇ pyruvate, + ethanol

However, Pfl activity in *pfl*C and *pfl*D mutants is comparable to those in the parent strain. The enzyme activities of the strains *E. coli* BW25113, and *pfl*A and *pfl*B mutants, grown on glucose under microaerobic conditions, are shown in Figure 6.26. The enzyme activity indicates that glycolytic enzymes, such as GAPDH and Pyk, are up-regulated in the *pfl*A and *pfl*B mutants as compared to the parent strain. By comparing the higher glucose uptake rates in the mutant strains, it can be seen that

Effect of specific-gene knockout

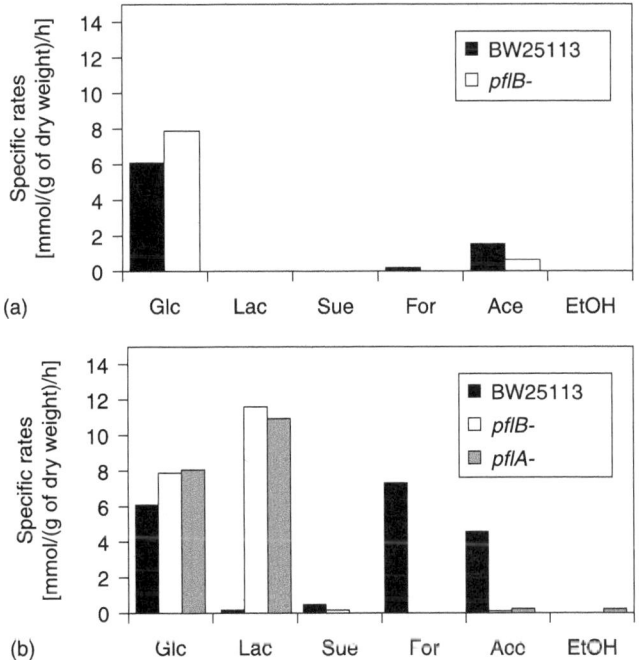

Figure 6.23 The specific glucose uptake rates and the specific product formation rates for: (a) the aerobic conditions; and (b) the microaerobic conditions. *Glc* Glucose uptake rate; *Lac* lactate production rate; *Suc* succinate production rate; *For* formate production rate; *Ace* acetate production rate; *Etoh* ethanol production rate

the glycolytic flux increases in the mutants. Higher glycolytic flux may be used to fulfill energy requirements. To maintain the glycolytic flux, PYR has to be assimilated, and the NADH produced by GAPDH must be reoxidized to NAD^+. LDH activity is 3-fold higher in the *pfl*A mutant, which is consistent with the higher lactate production. Another metabolic enzyme, ADH, is also slightly up-regulated for the mutants as compared to the parent strain. It is interesting to see that the activities of Ppc and Ack increase more than 100-fold in the mutants, suggesting that these two enzymes play important roles in the microaerobic growth of these two mutants. Ppc is the first enzyme for succinate production. Based on the activity of this enzyme, significant succinate production is expected in *pfl*A and *pfl*B mutants. However,

Bacterial cellular metabolic systems

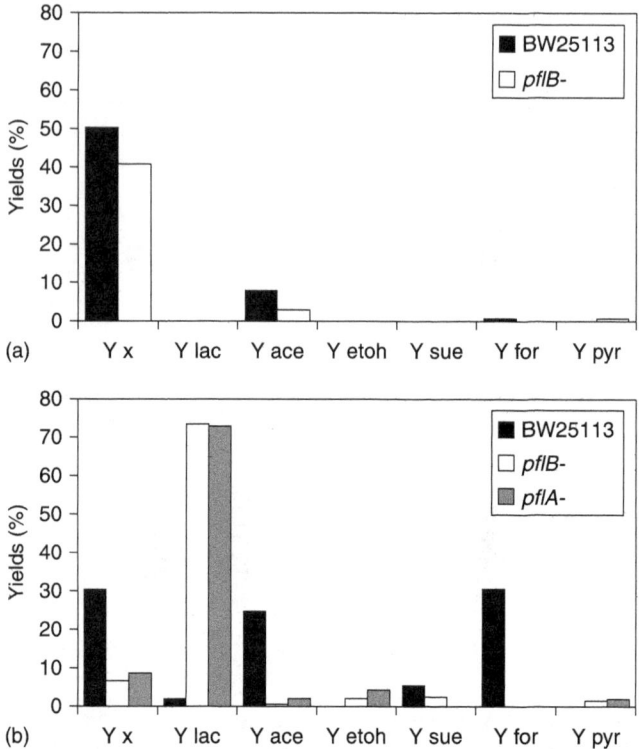

Figure 6.24 Biomass and metabolite yields on glucose for: (a) the aerobic conditions; and (b) the microaerobic conditions. Y_x Biomass yield on glucose; Y_{lac} lactate yield; Y_{ace} acetate yield; Y_{etoh} ethanol yield; Y_{suc} succinate yield; Y_{for} formate yield; Y_{pyr} pyruvate yield

Table 6.17 The specific carbon source uptake rates and product formation rates for the *E. coli pfl*A mutant using different carbon sources under microaerobic conditions. q is the specific rate calculated as the average rate during the exponential growth phase

Carbon source	q *(mmol (g dry weight)$^{-1}$ h^{-1}) for:					
	Substrate	Lactate	Succinate	Formate	Acetate	Ethanol
Glucose	8.15	11.01	0.0	0.0	0.27	0.2
Gluconate	6.57	8.22	0.21	0.0	5.63	0.32
Pyruvate	6.87	1.78	0.25	0.0	4.58	0.11
Fructose	3.21	4.42	0.0	0.0	0.23	0.0
Glycerol	1.33	0.83	0.0	0.0	0.05	0.0

Effect of specific-gene knockout

Table 6.18 The yields (Y) of cell mass (x) and metabolites for different carbon sources for the *E. coli pfl*A mutant grown under microaerobic and the anaerobic conditions. AN – The *pfl*A mutant grown on glucose under anaerobic conditions filled with CO_2. Y values were calculated based on data in the exponential growth phase

Carbon source	Yields on carbon source (g g^{-1})%						
	$Y_{x/s}$	$Y_{lactate/s}$	$Y_{acetate/s}$	$Y_{ethanol/s}$	$Y_{succinate/s}$	$Y_{formate/s}$	$Y_{pyruvate/s}$
Glucose	8.1	72.5	1.1	3.8	0.0	0.0	1.1
Gluconate	8.1	57.4	26.2	1.1	1.9	0.0	1.0
Pyruvate	5.9	26.5	45.4	0.2	1.3	0	–
Fructose	9.7	69.0	3.7	0.0	0.0	0.0	1.2
Glycerol	0.0	61.1	2.6	0.0	0.0	0.0	0.0
Glucose (AN)	6.7	48.2	0.6	–	3.6	0.0	–

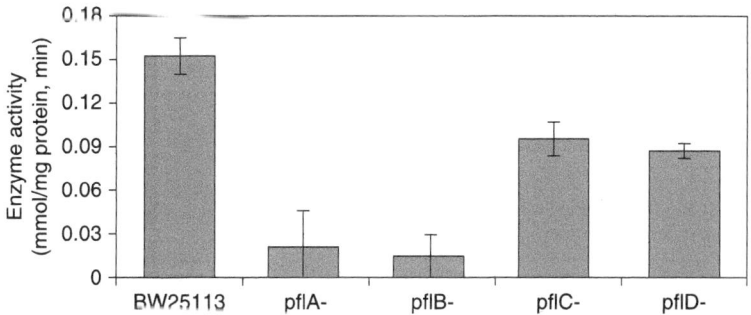

Figure 6.25 The pyruvate formate lyase activities of *E. coli pfl* mutants and the parent strain *E. coli* BW25113

succinate production is even lower in the mutant strains. One possible reason may be due to a CO_2 shortage in the microaerobic cultivation of the mutants, since CO_2 is required for the Ppc reaction to form OAA. By supplying CO_2, succinate production is slightly increased in the *pfl*A mutant (Table 6.18). However, it is still lower than that in the parent strain. Therefore, the shortage of PEP may be the most plausible reason for the lower succinate production.

Enzyme activities are also assayed for the *pfl*A mutant grown on different carbon sources (Table 6.19). Enzyme activities of GAPDH and LDH show almost the same behavior, which means that higher LDH

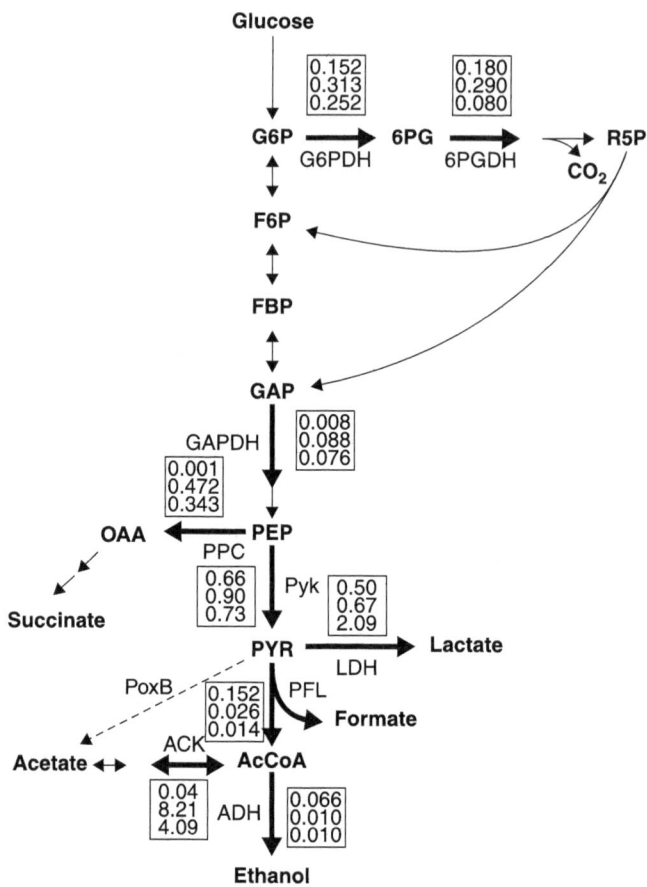

Figure 6.26 Enzyme activities for strains grown on glucose in the microaerobic conditions. The values in the *boxes* show enzyme activities. First row *E. coli* BW25113; second row *pfl*Bmutant; third row *pfl*A mutant. The *red lines* indicate the up-regulation of the enzymes of *pfl*A and *pfl*B mutants, compared to that of the parent strain

activity causes higher GAPDH activity when using different carbon sources, implying a coupling between these two enzymes. Another interesting phenomenon is that Ack activity varies dramatically, depending on the carbon source. Compared to using glucose as a carbon source, Ack activity is about 10-fold higher in cells using gluconate or PYR as a carbon source, while this enzyme activity is reduced 50- to 100-fold when using glycerol or fructose as a carbon source. Ppc activity is highest when using

Table 6.19 Enzyme activities for the E. coli pflA mutant grown on different carbon sources in the microaerobic conditions

Carbon source	Enzyme activity (mmol (mg protein)$^{-1}$ min^{-1})							
	G6PDH	6PGDH	GAPDH	PYK	LDH	PPC	ACK	ADH
Glucose	0.252±0.019	0.080±0.003	0.076±0.003	0.73±0.04	2.09±0.12	0.343±0.032	4.09±0.77	0.010±0.001
Glucorate	0.449±0.02	0.244±0.009	0.126±0.005	0.90±0.04	4.38±0.42	0.084±0.010	38.15±2.5	ND
Pyruvate	0.264±0.019	ND	ND	ND	0.42±0.04	ND	39.82±2.8	ND
Fructose	0.170±0.019	0.066±0.004	0.030±0.006	0.88±0.09	0.6±0.07	0.013±0.001	0.036±0.006	0.015±0.002
Glycerol	0.141±0.013	0.098±0.005	0.026±0.002	0.44±0.03	0.52±0.10	0.025±0.002	0.072±0.009	0.006±0.001

glucose as a carbon source, but it is significantly lower when using fructose or glycerol as a carbon source. ADH activity is low when using fructose, glycerol, gluconate, glucose, or PYR as the carbon source.

Table 6.20 shows the intracellular metabolite concentrations for the three strains grown on glucose in aerobic and microaerobic conditions. In aerobic conditions, intracellular FDP(FBP) and PYR are significantly accumulated in the *pflB* mutant, where the intracellular FBP concentrations increase to 35.89 μ mol (g dry weight)$^{-1}$, while PYR increase to 34.60 μ mol (g dry weight)$^{-1}$. Under microaerobic conditions, ATP/AMP ratios in the *pflA* and *pflB* mutants are significantly lower than in the parent strain. This suggests that the mutant strains require more ATP than the parent strain. Intracellular FBP and PYR concentrations are significantly increased and the NADH/NAD$^+$ ratios are 4- to 7-fold higher compared to those in the parent strain (Zhu and Shimizu, 2004).

Table 6.21 shows intracellular metabolite concentrations in the *pflA* mutant grown on different carbon sources under microaerobic conditions. It indicates that the intracellular PYR concentrations are related to FBP concentrations. When using PYR as a carbon source, the high intracellular PYR concentrations are found together with high FBP concentrations, which are 10-fold higher than when using glucose as a carbon source. The NADH/NAD$^+$ ratio is highest when using glycerol as a carbon source and this ratio is relatively low when using gluconate or PYR as a carbon source.

Production of pure D-lactate using the *E. coli pfl* mutant strain has been reported (Pascal et al., 1981; Zhou et al., 2003), and the main characteristics are shown in Figure 6.27. The *pfl* gene knockout blocks the main PYR assimilation pathway under microaerobic or anaerobic conditions. This mutant may cause the shortage of AcCoA. In *E. coli*, the Ack–Pta pathway is related to the AcCoA pool. This pathway has two directions. One direction is to excrete acetate and produce 1 mol ATP. The other direction is to utilize acetate by consuming 1 mol ATP to produce intracellular AcCoA (Brown et al., 1977; Kumari et al., 2000). Since AcCoA formation through Pfl is deficient in the two *pfl* mutants, Ack–Pta reactions may occur in the direction of forming AcCoA using acetate and ATP for biomass formation. This may be the reason why Ack activity is up-regulated in the *pflA* and *pflB* mutants under microaerobic conditions (Figure 6.26).

Since ATP can only be generated through glycolysis in oxygen-limited conditions, and it is known that the glycolytic flux is controlled by the demand for ATP (Koebmann et al., 2002), glycolysis has to show a stepping-up to fulfill the energy requirement in microaerobic or anaerobic conditions. In the *pfl* gene-knockout cells, energy generation is further reduced since the cell cannot use Ack to produce ATP through acetate

Table 6.20 Intracellular metabolite concentrations in cells grown on glucose

	Intracellular concentration [μmol (g dry weight)$^{-1}$ cell^{-1}]					
	Aerobic conditions			Microaerobic conditions		
Metabolite	Wild	pflB		Wild	pflB	pflA
G6P	0.591±0.001	0.753±0.050		0.049±0.009	0.133±0.020	1.093±0.006
FBP	1.347±0.010	35.89±0.12		4.59±0.021	41.32±0.11	16.41±0.11
PEP	1.35±0.30	0.67±0.11		0.32±0.12	0.84±0.14	0.12±0.04
PYR	0.53±0.01	34.60±0.20		8.21±0.01	67.97±0.20	25.49±0.01
AcCoA	0.144±0.006	0.012±0.001		0.066±0.002	0.070±0.001	0.050±0.001
ATP	0.375±0.015	0.442±0.045		0.087±0.013	0.044±0.014	0.052±0.012
ADP	1.253±0.014	0.922±0.011		0.380±0.006	0.342±0.006	0.195±0.010
AMP	0.392±0.010	0.515±0.012		0.214±0.006	0.193±0.006	0.197±0.004
NADH	0.026±0.001	0.028±0.005		0.013±0.002	0.029±0.003	0.056±0.001
NAD$^+$	0.018±0.002	0.312±0.002		0.143±0.001	0.050±0.001	0.060±0.002
ATP/AMP	0.957	0.858		0.407	0.228	0.264
NADH/NAD$^+$	1.44	0.09		0.13	0.58	0.93

Table 6.21 Intracellular metabolite concentrations in the E. coli pflA mutant grown on different carbon sources in microaerobic conditions

Metabolite	Intracellular concentration [μ mol (g dry weight)$^{-1}$ cell^{-1}]				
	Glucose	Gluconate	Pyruvate	Fructose	Glycerol
G6P	1.093±0.006	0.123±0.001	0.120±0.001	0.240±0.001	0.045±0.003
FBP	16.41±0.11	27.08±0.10	164.9±2.7	28.92±0.04	33.02±0.04
PEP	0.12±0.04	0.22±0.11	0.50±0.13	0.45±0.13	0.10±0.05
PYR	25.49±0.01	41.81±0.70	250.6±0.85	47.22±0.50	52.98±0.30
AcCoA	0.050±0.001	0.003±0.001	0.039±0.004	0.132±0.035	0.037±0.001
ATP	0.052±0.012	0.008±0.004	0.034±0.011	0.027±0.006	0.056±0.026
ADP	0.195±0.010	0.143±0.003	0.835±0.011	0.458±0.092	0.531±0.091
AMP	0.197±0.004	0.226±0.007	0.932±0.049	0.127±0.013	0.499±0.006
NADH	0.056±0.001	0.032±0.002	0.016±0.002	0.092±0.012	0.085±0.020
NAD$^+$	0.060±0.002	0.046±0.001	0.032±0.001	0.244±0.003	0.037±0.001
ATP/AMP	0.264	0.035	0.036	0.213	0.112
NADH/NAD$^+$	0.93	0.7	0.5	0.38	2.3

Effect of specific-gene knockout

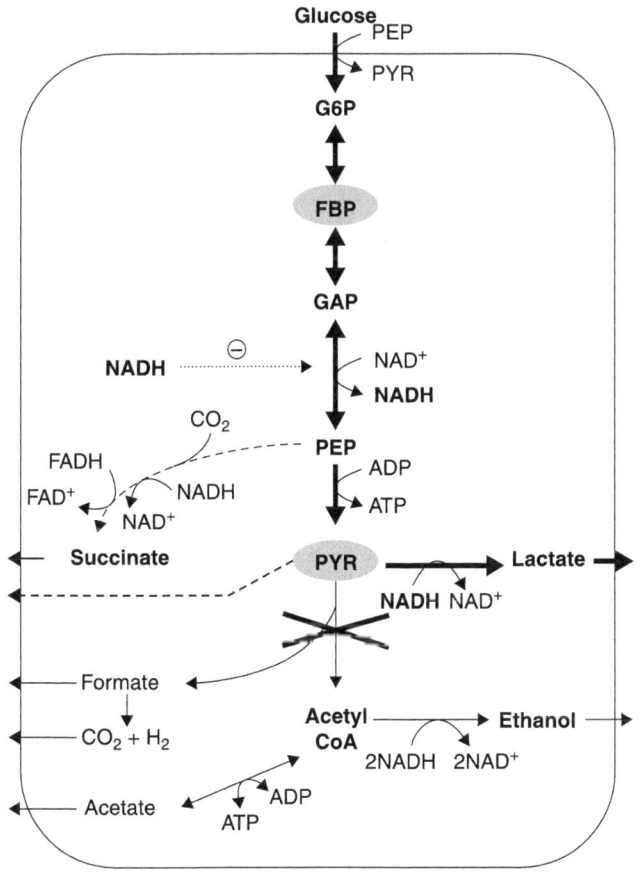

Figure 6.27 Metabolic regulation mechanisms of lactate production in *E. coli pfl*A and *pfl*B mutants

formation under microaerobic conditions. However, the up-regulated Pta–Ack pathway consumes more ATP to form AcCoA. This results in a considerably lower ATP/AMP ratio in the cell than in the parent strain (Table 6.20). The energy requirement will further promote glycolytic flux, which is reflected by the higher glucose uptake rate in the *pfl* mutant under microaerobic conditions (Figure 6.23b).

During glycolysis, the reaction through GAPDH produces NADH. This NADH must be regenerated to maintain the glycolytic flux, which is essential for ATP generation. In aerobic conditions, this regeneration can be achieved by the respiratory chain. However, the NAD⁺ can only be regenerated through fermentative pathways in the anaerobic conditions in *E. coli*. However, PYR produced from glycolysis has to be assimilated.

Several pathways can be used to maintain the stoichiometric balance and the intracellular redox balance; one of these is to produce ethanol through ADH. However, this pathway needs AcCoA, which is not present in sufficient amounts in the *pfl* mutant strains.

Another pathway is to produce succinate through Ppc and the reducing arm of the TCA cycle. Succinate production is attained by using 1 mol of NADH and 1 mol of $FADH_2$, which is favorable from a redox balance point of view. Note that the reaction through Ppc requires PEP as well as CO_2, which may limit the flux through this pathway for *pfl* mutants. With supplementation of CO_2 in the medium, succinate production is slightly enhanced but not as much as expected (Table 6.18), which indicates that the Ppc reaction is not mainly limited by CO_2. The reactions through Ppc, Pyk, and PTS, as well as cell synthesis, compete for PEP as the substrate. Note that the reaction through Pyk produces 1 mol ATP, and PTS is essential for energy generation. It can be seen that the enzyme activity of both Ppc and Pyk increases in the mutants as compared to the wild type (Figure 6.26). The low succinate production through Ppc indicates that ATP demand is favored over intracellular redox balance. This is consistent with the previous reports, indicating that instead of the amplification of Ppc, introduction of an exogenous PYR carboxylase or malic enzyme into *E. coli* can significantly improve succinate production under anaerobic conditions, since Pyc is not competitive with Pyk and PTS in such a case (Gokarn et al., 1998, 2000; Hong and Lee, 2001; Vemuri et al., 2002).

Another possibility is that PYR can be directly excreted, but this is unfavorable from the intracellular redox balance point of view. The accumulated PYR will allosterically activate LDH. Finally, the *E. coli pfl* mutant uses LDH exclusively to regenerate NAD^+. In conclusion, glycolysis is promoted by *pfl* mutation under microaerobic conditions due to the energy requirement for cell growth, and lactate production is enhanced from the points of stoichiometric balance and intracellular redox balance to maintain glycolysis.

It is shown that PYR and FBP are significantly accumulated in *pfl*A mutants, when using different carbon sources. PYR accumulation may be considered as the result of *pfl* knockout, since Pfl is the main PYR assimilation pathway under microaerobic conditions. As previously discussed, the *pfl* gene-knockout results in a significantly higher intracellular $NADH/NAD^+$ ratio. Since GAPDH is competitively inhibited by NADH (Garrigues et al., 1997), FBP cannot be adequately assimilated through glycolysis, thus the intracellular FBP concentrations are high.

The NADH/NAD$^+$ ratio also affects the metabolite production pattern (Riondet et al., 2000). Since most of the NADH is produced by the reaction through GAPDH, NADH production is variable when using different carbon sources. It is clear that the cell will produce less NADH when using gluconate or PYR as a carbon source. This is confirmed by the lower NADH/NAD$^+$ ratio when using gluconate or PYR as a carbon source than when using glucose (Table 6.21). The activity of GAPDH is significantly higher when using gluconate as a carbon source (Table 6.19), which indicates the partial release of NADH inhibition on this enzyme. Correspondingly, the lactate yields, when using gluconate or PYR, are lower, which again confirms that lactate production in *pfl* mutants is promoted by the intracellular redox balance.

The results using gluconate or pyruvate as a carbon source may also be distinguished from those using other carbon sources, because of the significantly lower ATP/AMP ratio in the *pfl*A mutant (Table 6.21), which is less than one-seventh of that when using glucose as a carbon source. One reason may be due to the difference in the metabolic networks used; when using gluconate as a carbon source, some of the carbon atoms will go directly to PYR through the Entner-Doudoroff pathway, which reduces the flux through glycolysis where ATP is generated. The other reason may be due to the activation of Ack. It should be noted that the activity of Ack is up-regulated about 10-fold when using gluconate or pyruvate as a carbon source. As previously discussed, the Ack–Pta pathway will occur in the direction of AcCoA synthesis. It may utilize ATP for this pathway to form AcCoA, and finally result in a lower ATP/AMP ratio. There is significant acetate production when using gluconate or PYR as a carbon source. One possibility is that some other back-up systems of Pfl are used to produce AcCoA using PYR. However, since there was no formate detected in the medium, and at the same time the CO_2 concentration in the effluent gas is not higher than the parent strain, the acetate may not be produced from the Ack–Pta pathway. The other possibility may be the activation of PoxB, which is shown to have some relationship with Ack (Abdel-Hamid et al., 2001). When PoxB is inactivated, Ack activity is also relatively low. However, the activity of Ack increases several fold when PoxB is activated and the increase is 13- to 20-fold in the PDH-E1P-null strains. Consistent with this research, the significant induction of Ack in the *pfl*A mutant may be related to the activation of PoxB, which produces acetate and the cell utilizes the Ack–Pta pathway to supplement the AcCoA pool. In comparison, Ack activity is much lower when using glycerol as a carbon source. Considering the significantly higher NADH/NAD$^+$ ratio when using glycerol, the result suggests that the redox balance is less important for metabolism.

6.9 Effect of *ldhA* gene knockout on metabolism

When *E. coli* converts sugars to pyruvate under anaerobic conditions, two major alternative pathways exist for the formation of terminal fermentation products from PYR. The first pathway involves the conversion of PYR to AcCoA and formate by the enzyme Pfl (Gottschalk, 1985). Subsequently, AcCoA is transformed to an approximately equal mixture of ethanol and acetate (Stokes, 1949). The other alternative pathway involves the direct conversion of PYR to lactate in a single step by means of LDH (Tarmy and Kaplan, 1968a,b). The conversion of L- or D-lactate to PYR is catalyzed by two separate isomer-specific flavoproteins, which are often called LDHs but better regarded as lactate oxidases. Basically, *E. coli* has two such enzymes to use when lactate is the sole carbon source and when oxygen or nitrate is available as terminal electron acceptors (Haugaard, 1959; Kline and Mahler, 1965). These flavoproteins are in fact membrane-bound components in the electron transport chain. The conversion of pyruvate to lactate under anaerobic conditions is catalyzed by the third enzyme, LDH, which is a soluble and NADH-linked enzyme, and produces D-lactate (Tarmy and Kaplan, 1968a,b). The fermentative LDH has been purified, is allosterically activated by its substrate PYR, and is induced approximately 10-fold in anaerobically grown cultures at acidic pH (Tarmy and Kaplan, 1968a,b).

Mutants deficient in the fermentative NAD-linked LDH have been studied (Mat-Jan et al., 1989). It is shown that *ldhA* mutants have little effect on cell growth, and convert sugars to a variety of other metabolites. The mutants are tested for aerobic and anaerobic growth on a variety of sugars, such as glucose, fructose, maltose, rhamnose, xylose, sorbitol, and gluconate in minimal medium and the results are almost identical. However, introduction of plasmids carrying *ldhA* genes into the *ldhA* mutant and parent *E. coli* have also been studied (Yang et al., 1999). Over-expression of LDH in the parent strain increases the lactate synthesis rate from 0.19 to 0.40 mM/g/h, when LDH activities increase from 1.3 to 15.3 units. Even an increase of more than 10 times in the LDH activity fails to divert a large fraction of the carbon flux to lactate; the majority of the flux is channeled

Effect of specific-gene knockout

through the AcCoA branch. Here, the effect of the *ldhA* gene knockout on metabolism of *E. coli* under anaerobic conditions is explained (Figure 6.28).

Both *E. coli* BW25113 and its *ldhA* mutant (JW1375) are cultivated anaerobically on glucose as a sole carbon source (Kabir et al., 2005). Batch cultivation results for both strains grown on glucose under

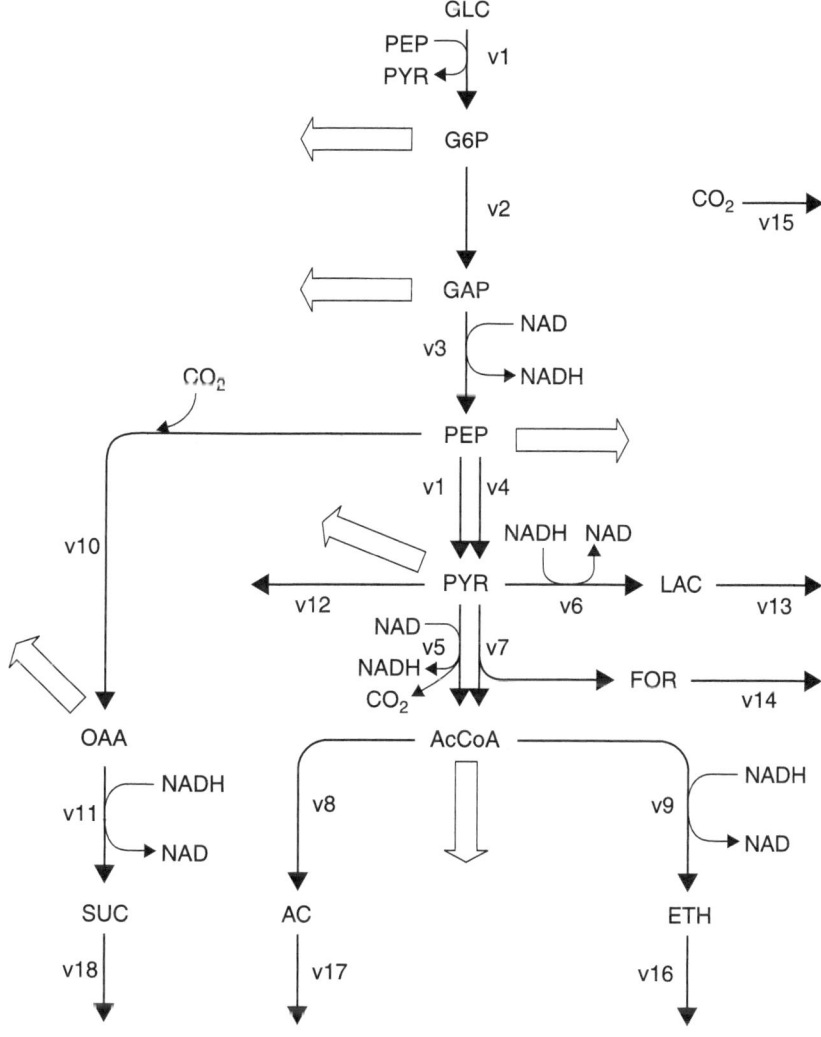

Figure 6.28 Fermentative pathways of *E. coli* grown on glucose. ⇨ indicates biomass formation flux. The fluxes through each pathway are designated v_1 through v_{18}

anaerobic conditions are shown in Figure 6.29. Growth parameters at the exponential growth phase are shown in Table 6.22. The growth rate of the *ldhA* mutant is slightly lower (13.33% decrease) at a lower glucose consumption rate compared to the parent strain. However, the biomass yield on glucose changed little in the *ldhA* mutant as compared to the parent strain. Table 6.22 shows that the extracellular formate, acetate, ethanol, pyruvate, and glutamate production rates increase by about 13.2, 17.7, 13.2, 55.6, and 156.0%, respectively, in the *ldhA* mutant as compared to the parent strain. However, extracellular production rates of citrate, malate, and succinate are reduced in the *ldhA* mutant as compared to the parent strain.

As shown in Table 6.23, enzyme activity levels vary significantly in the *ldhA* mutant, as compared to those in the parent strain. Although the glycolytic enzymes, such as Glk and Pgi, are down-regulated, Pfk, GAPDH, Pgk, Pyk, and PDHc are up-regulated significantly in the *ldhA* mutant as compared to the parent strain, although Fba is not. Although the conversion of PYR to AcCoA through Pfl is the major pathway under anaerobic conditions, the PDH enzyme is up-regulated (Spencer and Guest, 1985; Kaiser and Sawer, 1994).

The activity of G6PDH increases by 38.70%, whereas 6PGDH activity decreases by 28.75% in the *ldhA* mutant compared to the parent stain. The down-regulation of 6PGDH agrees with the slower growth rate of the *ldhA* mutant, since this enzyme is known to be growth rate dependent (Pease and Wolf, 1994).

Citrate synthase (CS) is up-regulated by 1.45-fold in the *ldhA* mutant compared to the parent strain (Table 6.23). This up-regulation causes an increase in the carbon flux from citrate to glutamate via α-KG, as is evidenced by the up-regulation of ICDH, higher glutamate production rate (Table 6.22), and high concentration of intracellular α-KG (Figure 6.30) in response to *ldhA* gene deletion. Moreover, the activities of MDH and Fum are also up-regulated in the *ldhA* mutant compared to those in the parent strain.

The anaplerotic enzyme Ppc, which is also known to contribute to catabolic succinate formation under anaerobic conditions, is down-regulated in the *ldhA* mutant compared to the parent strain (Table 6.24). However, Pck is up-regulated significantly in the mutant. This is mainly due to the shortage of PEP in the *ldhA* mutant, as evidenced by the measurement of intracellular PEP concentration (Figure 6.30). The activity of $NADP^+$-specific malic enzyme ($NADP^+$-Mez) remains unchanged, whereas the NAD^+-specific malic enzyme (NAD^+-Mez) is up-regulated significantly in the *ldhA* mutant compared to that in the

Effect of specific-gene knockout

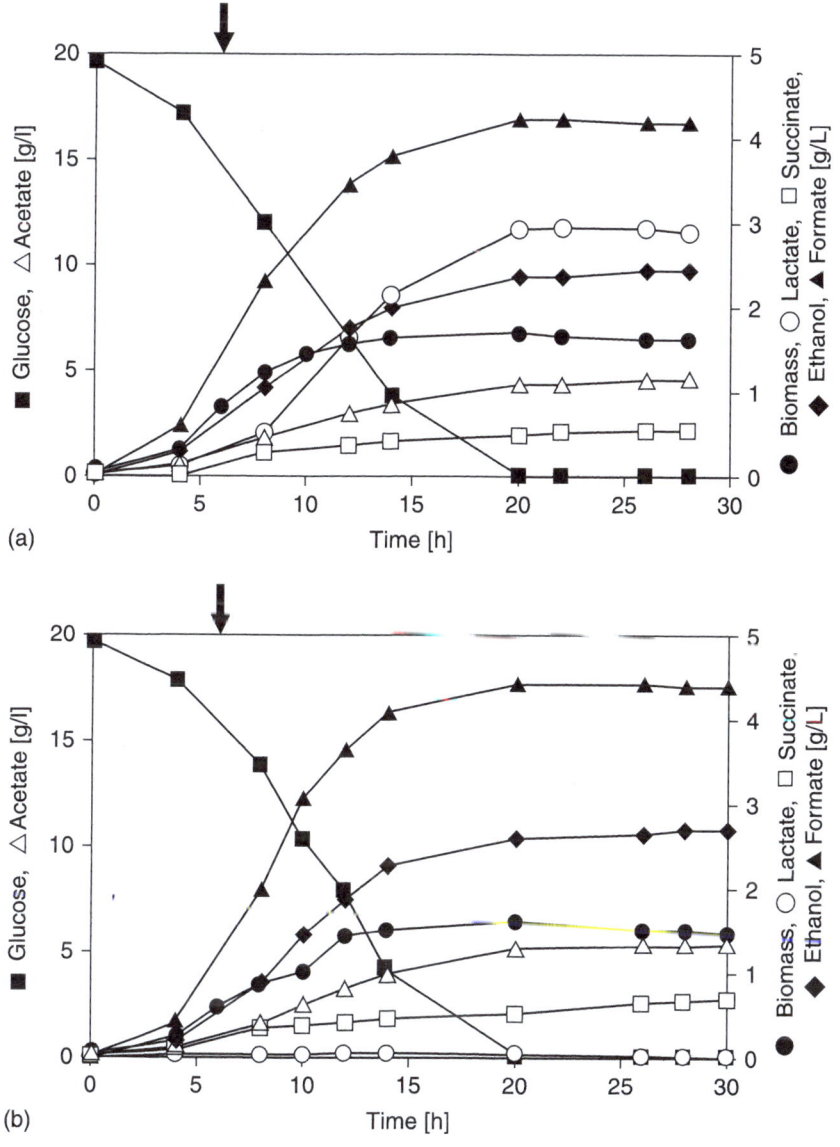

Figure 6.29 Batch cultivation results of: (a) parent *E. coli*; and (b) *ldhA* mutant *E. coli*; grown on glucose under anaerobic conditions. The arrow indicates the sampling time for RT-PCR analysis, measurement of enzyme activities, and intracellular metabolites

Table 6.22 Specific rates of parent and *ldhA* mutant *E. coli* grown on glucose under anaerobic conditions. Standard deviations were calculated from four independent measurements (duplicate measurements from duplicate experiments). μ – specific growth rate; q – specific uptake/production rate; $Y_{X/S}$ – biomass yield

Parameter	BW 25113 (Parent)	JW 1375 (*ldhA*⁻)	Ratio (*ldhA*⁻/Parent)
μ [h^{-1}]	0.16 ± 0.01	0.14 ± 0.01	0.87
$Y_{X/S}$ [g.g^{-1}]	0.12 ± 0.01	0.12 ± 0.01	1.00
q_{GLC} [mmol.gDCW^{-1}.h^{-1}]	7.49 ± 0.75	6.85 ± 0.68	0.92
q_{ACE} [mmol.gDCW^{-1}.h^{-1}]	4.70 ± 0.47	5.53 ± 0.55	1.18
q_{LAC} [mmol.gDCW^{-1}.h^{-1}]	1.84 ± 0.18	UD*	—
q_{FOR} [mmol.gDCW^{-1}.h^{-1}]	7.46 ± 0.74	8.45 ± 0.84	1.13
q_{ETH} [mmol.gDCW^{-1}.h^{-1}]	3.79 ± 0.38	4.29 ± 0.43	1.13
q_{PYR} [mmol.gDCW^{-1}.h^{-1}]	0.10 ± 0.01	0.16 ± 0.02	1.56
q_{SUC} [mmol.gDCW^{-1}.h^{-1}]	0.38 ± 0.04	0.31 ± 0.03	0.82
q_{CIT} [μmol.gDCW^{-1}.h^{-1}]	1.74 ± 0.20	1.48 ± 0.20	0.83
q_{GLU} [μmol.gDCW^{-1}.h^{-1}]	1.60 ± 0.20	4.08 ± 0.50	2.56
q_{MAL} [μmol.gDCW^{-1}.h^{-1}]	19.1 ± 2.01	16.1 ± 2.03	0.82
Carbon recovery [%]	88.37 ± 8.1	90.87 ± 14.3	

*UD: undetected

parent strain (Table 6.23), implying that the inactivation of the *ldhA* gene leads to the production of malate from pyruvate, while the parent strain produces MAL from PEP in two-step reactions catalyzed by Ppc and MDH. This result is consistent with the study of Stols and Donnelly (1997), where inactivation of the *ldhA* gene produces malic acid from pyruvate. Acetate kinase (Ack) activity is up-regulated in the *ldhA* mutant, compared to the parent strain. This result is consistent with the higher acetate production rates (Table 6.22). The activity of soluble transhydrogenase (THD) encoded by *udhA*, which catalyzes the conversion of NAD$^+$ and NADPH to NADH and NADP^{+}, is almost unchanged in the *ldhA* mutant compared to the parent strain (Table 6.23), implying that re-oxidation of reducing power is balanced in the mutant.

The relative expressions of *ptsG*, *pgi*, *fbaA*, *gapA*, *pgk*, *aceE*, and *aceF* involved in the glycolysis are significantly up-regulated in the

Table 6.23 Specific enzyme activities in cell extracts of parent and *ldhA* mutant *E. coli* grown on glucose under anaerobic conditions. The unit of enzyme activity is nmol/min/mg protein. Overall ED pathway enzyme was represented as mg pyruvate/min/mg protein. The ratio of enzyme activity was calculated for *ldhA* mutant over the corresponding value in the parent strain. Therefore, a value of 1.0 indicates no change. ND indicates not detected

Enzyme (abbreviated name)	Parent	$\Delta ldhA$	Ratio
Glucokinase (Glk)	187 ± 13	138 ± 12	0.7
Phosphoglucose isomerase (Pgi)	684 ± 48	537 ± 35	0.8
Glucose-6-phosphate dehydrogenase (G6PDH)	61 ± 9	84 ± 12	1.4
6-phosphogluconate dehydrogenase (6PGDH)	105 ± 7	75 ± 5	0.7
ED pathway enzyme (6PGD & 2KDPGA)	1 ± 0.1	1 ± 0.1	1.0
Phosphofructose kinase (Pfk)	158 ± 11	191 ± 13	1.2
Fructose diphosphate aldolase (Fba)	175 ± 12	178 ± 12	1.0
GAP dehydrogenase (GAPDH)	301 ± 21	358 ± 25	1.2
3-phosphoglycerate kinase (Pgk)	74 ± 5	84 ± 6	1.1
Phosphoenol pyruvate carboxylase (Ppc)	97 ± 7	40 ± 3	0.4
Phosphoenol pyruvate carboxykinase (Pck)	34 ± 2	62 ± 5	1.8
Pyruvate kinase (Pyk)	265 ± 19	330 ± 12	1.3
Pyruvate dehydrogenase complex (PDH)	13 ± 1	23 ± 2	1.8
Acetate kinase (Ack)	522 ± 36	636 ± 44	1.2
Lactate dehydrogenase (LDH)	151 ± 11	ND	—
Citrate synthase (CS)	49 ± 10	71 ± 15	1.4
Isocitrate dehydrogenase (ICDH)	43 ± 9	72 ± 15	1.7
Isocitrate lyase (Icl)	1 ± 0.1	1 ± 0.1	1.0
Malate synthase (MS)	1 ± 0.1	1 ± 0.1	1.0
α-Ketoglutarate dehydrogenase (α-KGDH)	ND	ND	—
Malate dehydrogenase (MDH)	37 ± 5	59 ± 8	1.6
$NADP^+$-specific malic enzyme ($NADP^+$-Mez)	1 ± 0.1	1 ± 0.1	1.0
NAD^+-specific malic enzyme (NAD^+-Mez)	2 ± 0.1	16 ± 1	7.9
Fumarase (Fum)	15 ± 2	20 ± 2	1.4
Transhydrogenase (THD)	23 ± 2	23 ± 2	1.0

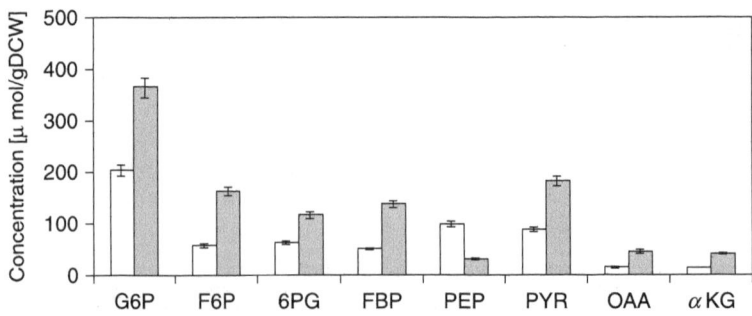

Figure 6.30 Intracellular metabolite concentrations in the central metabolic pathways at the exponential growth phase for both *ldhA* mutant and parent *E. coli* (see Materials and Methods for details). ☐, parent; ■, mutant

mutant compared to those in the parent strain (Figure 6.31a). The down-regulation of *pykA*, but not *pykF*, in the *ldhA* mutant, indicates that these two isozymes are differentially regulated. The *ppc* gene is down-regulated by almost 25%, whereas *pckA* is up-regulated by about 52% in the *ldhA* mutant compared to the parent strain, a result that is consistent with the measurement of enzyme activity (Table 6.23). Among the PP pathway genes, the relative expression level of *zwf* is slightly up-regulated, whereas all other genes, except *gnd*, are almost unchanged in the *ldhA* mutant compared to the parent strain (Figure 6.31b). Along the fermentative pathway, the expression levels of *ack*, *adhE*, and *pflA* are also up-regulated significantly in the *ldhA* mutant.

Of the TCA cycle, *gltA*, *acnB*, *icdA*, and *mdh* transcripts are up-regulated significantly in the *ldhA* mutant compared to those in the parent strain (Figure 6.31c). This up-regulation is consistent with the fact that most of the TCA cycle genes, such as *gltA* (Park et al., 1994), *mdh* (Park et al., 1995b), and *icdA* (Saier and Ramseier, 1996), are known to be regulated by several global regulators, such as Fnr, ArcA, and Cra (or FruR), whose encoded genes are also up-regulated in the *ldhA* mutant (Figure 6.31d). The expression levels of the genes involved in protein secretion and disulfide bond formation, such as *secA*, *lepB*, and *dsbA*, are up-regulated slightly in the *ldhA* gene knockout *E. coli* (Figure 6.31c). This result suggests that *ldhA* gene deletion may modulate the activity of protein secretion and disulfide bond formation. The transcript levels of *rpoE* and *rpoH* (encoding σ^E and σ^{32}, respectively) are increased by 1.2-fold, whereas *rpoD* (encoding σ^{70}) is almost unregulated in the mutant (Figure 6.31d). The significant up-regulation of heat shock genes, such as *dnaJ, grpE, groS*, and *fkpA*, in the mutant correspond to the fact that the heat shock gene expression is largely proportional to the amount of σ^{32}, though the level of σ^{32} activity is known to be regulated at multiple stages, including translational efficiency and protein stability (Straus et al., 1987).

Intracellular metabolite concentrations of G6P, F6P, 6PG, FBP, PYR, OAA, and α-KG are increased in the *ldhA* mutant, as compared to those of the parent strain. In contrast, the intracellular concentration of PEP decreases in the *ldhA* mutant, implying that the intracellular PEP is limiting in the mutant. The reason is mainly due to the significant up-regulation of Pyk and PDH activities (Table 6.23) that channel more carbon flux toward the formate, acetate, and ethanol pathways, from the PYR and AcCoA pool, as evidenced by their increased specific rate (Table 6.22), as well as increased NAD^+-Mez activity to replenish C_4 intermediates in the TCA cycle, such as OAA.

Deletion of the LDH pathway blocks carbon flow from pyruvate and results in a higher rate of pyruvate excretion. Metabolic flux analysis shows that most of the carbon flux in *ldhA* mutant *E. coli* is forced through formate, acetate, and ethanol production pathways, resulting in a concomitant increase in these fluxes, though little increase in fluxes through the glycolytic pathway is observed (Figure 6.32). This result is consistent with the study of Yang et al. (1999), where they observed that an *E. coli* mutant strain deficient in *ackA, pta*, and *ldhA* major carbon flux was forced through ethanol and formate production pathways.

When PEP is converted to MAL via OAA by Ppc and MDH, the free energy is wasted. In contrast, one ATP is produced when PEP is directly converted to pyruvate by Pyk (Figure 6.32). Therefore, a large fraction of PEP is channeled through the Pyk pathway and this might be why the *ldhA* mutant induces the sfcA pathway to supply the C_4 intermediates in the TCA cycle for conserving the free energy of PEP, as evidenced by the significant up-regulation of the NAD^+-Mez enzyme (Table 6.23). However, the up-regulation of this pathway cannot increase the specific production rate of malic acid and succinic acid (Table 6.22). The reason is mainly due to lower PEP concentrations in the *ldhA* mutant strain (Figure 6.30). Since PEP is an essential intermediate to supply the precursor metabolites for biomass production as well as amino acid biosynthesis, the demand for intracellular PEP in the *ldhA* mutant is maintained through redirecting the flux from OAA.

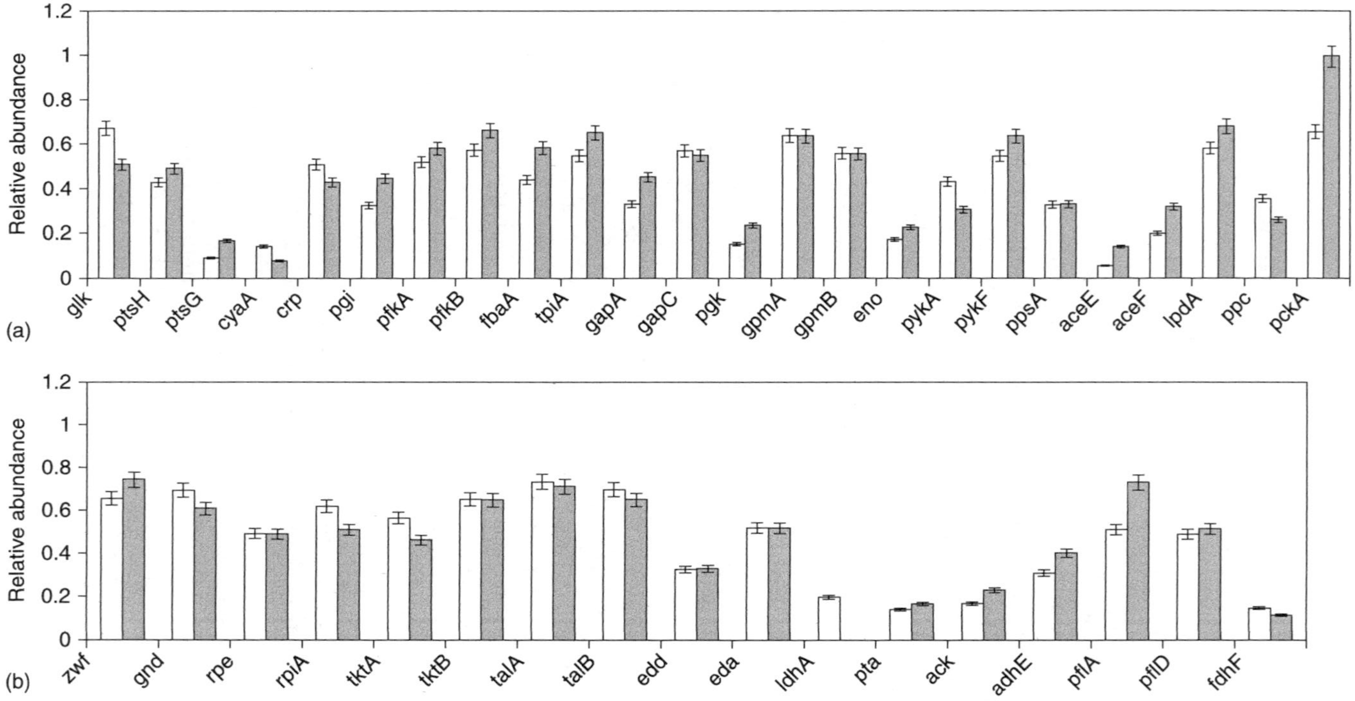

Figure 6.31 Comparison of gene expressions for parent (□) and *ldhA* mutant *E. coli* (■):(a) Transport and glycolytic pathway genes; (b) PP pathway and fermentative genes;

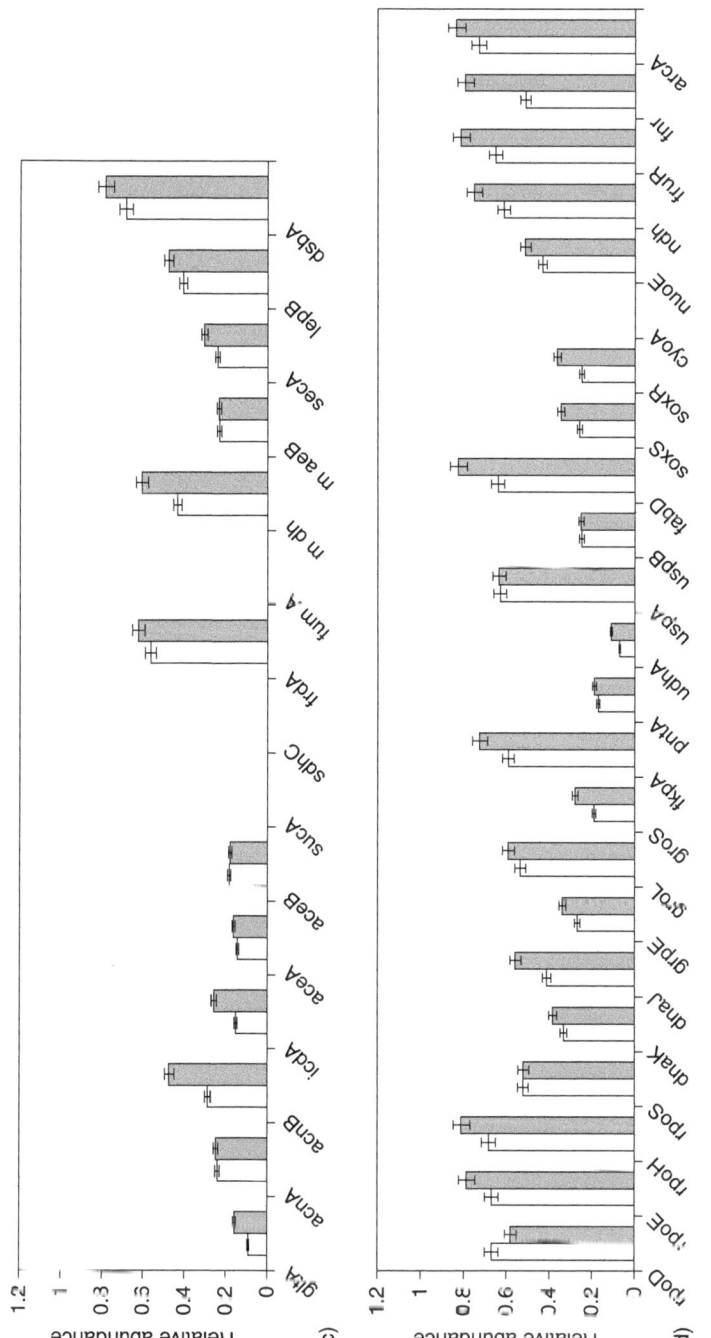

Figure 6.31 *(Continued)* (c) Genes involved in TCA cycle and protein synthesis; and (d) Genes involved in sigma factor, heat shock, NADPH re-oxidation, respiration, global, and other regulations (d). No RT-PCR amplifications were obtained for *sucA*, *sdhC*, *fumA*, and *cyoA*

Bacterial cellular metabolic systems

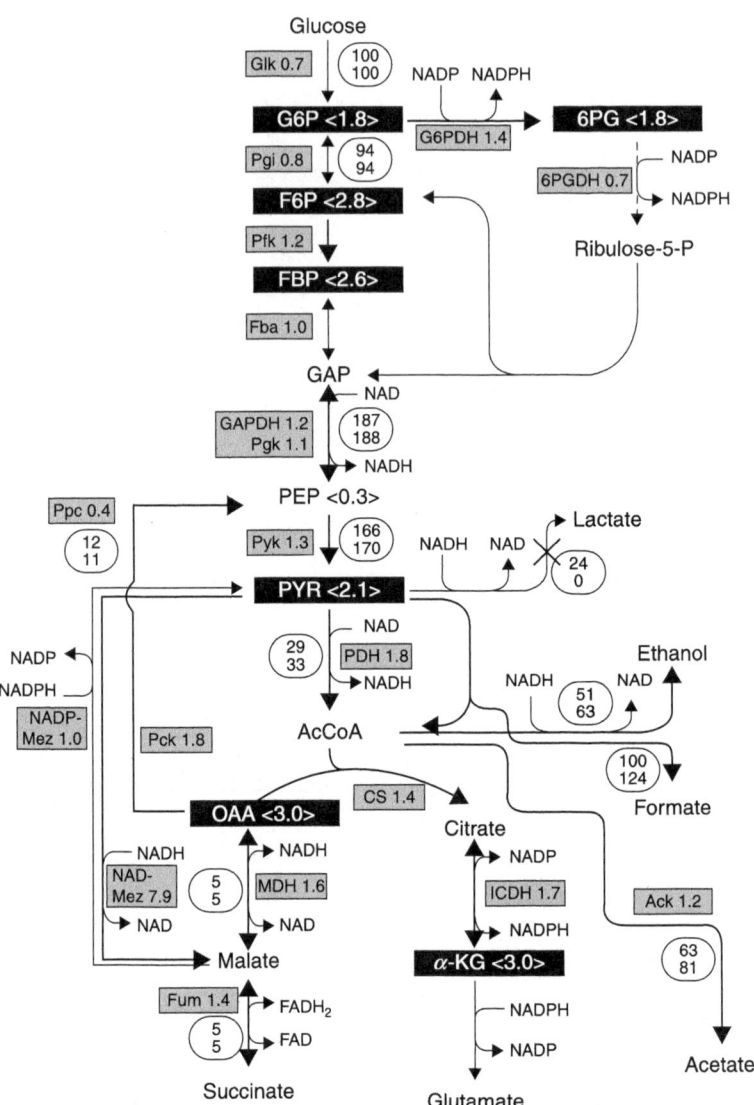

Figure 6.32 Anaerobic metabolism of glucose in *E. coli*. The value beside the intracellular metabolite and the enzyme in gray box is the relative value for *ldhA* mutant compared to the parent strain. The oval box represents the flux data; upper value, parent strain and lower value, *ldhA* mutant. The cross represents knockout of *ldhA* gene. Thick line, up-regulation; dotted line, down-regulation

Effect of specific-gene knockout

The product of the global regulatory gene *cra* (*fruR*) is known to control the transcriptional expressions of numerous genes concerned with carbon and energy metabolism. The genes *ptsHI*, *pfkA*, *adhE*, and *pykF* are regulated negatively, whereas the genes *ppsA*, *pckA*, *icdA*, and *cydAB* are regulated positively by the product of *cra* (*fruR*) gene, as explained in Chapter 3. Gene expression analysis shows a slight up-regulation of *ptsH*, *pfkAB*, *pykF* (Figure 6.31a), and *adhE* (Figure 6.31b) genes in the *ldhA* mutant, compared to those in the parent strain. The reason is due to the higher concentration of FDP in the *ldhA* mutant (Figure 6.30), since the effect of the Cra protein on the transcription is counteracted by a high concentration of FDP, which promotes catabolite repression of Cra-activated operons and catabolite activation of Cra-repressed operons (Saier and Ramseier, 1996). However, although the higher concentrations of FDP are observed in the mutant, the genes *icdA* (Figure 6.31c) and *pckA* (Figure 6.31a) are found to be up-regulated, implying that *cra* (*fruR*) does not act alone to exert its multifarious effects but interacts with other pleiotropic regulators to create a network of transcriptional effects that serve to co-ordinate the effects of various bacterial sensing devices.

Although the activity of Pgi is down-regulated, the gene *pgi* is up-regulated by 1.4-fold in the *ldhA* mutant compared to the parent strain. This is mainly due to the higher concentrations of 6PG in the *ldhA* mutant, since 6PG is known to inhibit Pgi activity (Schreyer and Bock, 1980); an artifact that the activity of enzymes are regulated by reversible binding of effectors, covalent modification, and alteration of enzyme concentration, whereas transcript levels are influenced by both transcriptional activity and mRNA stability (Martin, 1987). Although a slight up-regulation of the *zwf* gene is observed, the enzyme activity of G6PDH increases significantly in the *ldhA* mutant. The reason is due to the demand for NADPH for glutamate synthesis, of which specific production rates are increased by 2.5-fold (150%) in the *ldhA* mutant compared to the parent strain (Table 6.22). A slight up-regulation of the *zwf* gene may be due to the effect of *soxS* and *soxR*, which was found to be up-regulated in the *ldhA* mutant (Figure 6.31d), since the *soxS* gene product has been found to act as a transcriptional activator of at least 10 genes, including *zwf* (Li and Demple, 1994). Based on the study of gene expressions, enzyme activities, and intracellular metabolite concentrations, together with metabolic flux analysis, the overall regulation mechanisms for *E. coli* grown on glucose minimal medium under anaerobic conditions in response to *ldhA* gene deletion are shown in Figure 6.32.

Bacterial cellular metabolic systems

The soluble NADH-linked LDH is responsible for the formation of lactate from pyruvate (Tarmy and Kaplan, 1968a,b). This reaction consumes one NADH per three carbons, thus balancing out the NADH produced through GAPDH in glycolysis. Therefore, deletion of the *ldhA* gene may result in a reducing power imbalance in *E. coli*. However, the overall NADH balance calculated, based on the metabolic flux distribution, indicates that NADH production and consumption are almost the same in both the *ldhA* mutant and the parent strain (Figure 6.33). The reason is mainly due to the increased carbon flux through the ethanol pathway in the *ldhA* mutant from pyruvate node via AcCoA, since 4 moles of NADH are needed to metabolize 2 moles of AcCoA for the production of 2 moles of ethanol.

Table 6.24 indicates that enzyme activity is correlated approximately with gene expressions, whilst flux values do not necessarily correlate with enzyme activities. Basically, the flux is determined not only by the enzyme activity but also by the concentration of substrates, effectors, and inhibitors. The lower flux ratio through Pyk, as compared to the enzyme activity and mRNA transcript ratio, is mainly due to the decreased concentration of PEP, which is the substrate of pyruvate kinase reaction.

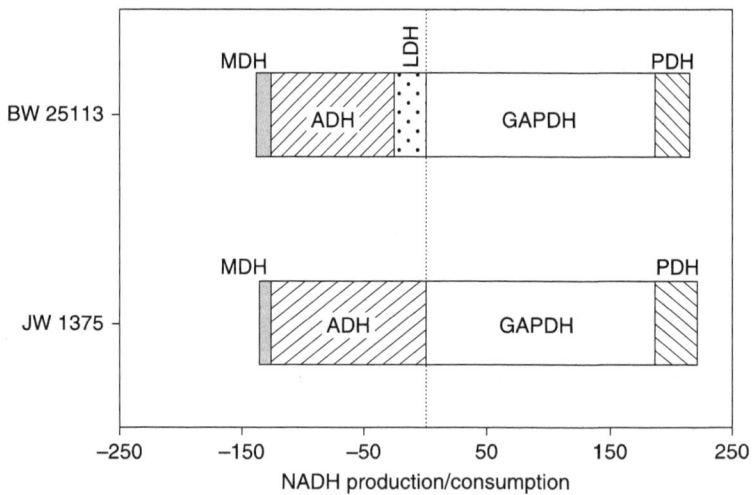

Figure 6.33 NADH balances in both *ldhA* mutant and parent *E. coli* calculated based on the metabolic fluxes. +: NADH production; −: NADH consumption

Table 6.24 Comparison between the ratios of gene expressions, enzyme activities, and metabolic fluxes in *E. coli* grown on glucose under anaerobic conditions

Gene	*ldhA* mutant versus parent E. coli		
	Flux Ratio	Enzyme Ratio	Transcript Ratio
glk	1.00	0.72	0.76
pgi	1.01	0.79	1.41
gapA	1.01	1.19	1.36
pykF	1.05	1.29 (pykAF)	1.16
aceEF & lpdA	1.16	1.84 (PDH)	1.86[a]
pfl	1.24	–	1.25[b]
ack	1.28	1.22	1.35
adhE	1.23	–	1.29
ppc	0.89	0.41	0.74
mdh	0.89	1.60	1.39

[a] average value of *aceE, aceF* and *lpdA* gene expressions which encode pyruvate dehydrogenase complex (PDH)
[b] average value of *pflA* and *pflD* gene expressions

6.10 References

Abdel-Hamid, A.M., Attwood, M.M., and Guest, J.R. (2001) 'Pyruvate oxidase contributes to the aerobic growth efficiency of *Escherichia coli*', *Microbiology*, 147: 1483–98.

Aristidou, A.A., San, K.Y., and Bennett, G..N. (1995) 'Metabolic engineering of *Escherichia coli* to enhance recombinant protein production through acetate reduction,' *Biotechnology Progress*, 11: 475–8.

Baba, T., Ara, T., Hasegawa, M., Takai, Y., Okumura, Y. et al. (2006) 'Construction of *Escherichia coli* K-12 in frame, single-gene knockout mutants: the Keio collection,' *Molecular Systtems Biology*, msb 4100050-E1.

Bennett, G.N., Schweingruber, M.E., Brown, K.D., Squires, C., and Yanofsky, C. (1976) 'Nucleotide sequence of region preceding *trp* mRNA initiation site and its role in promoter and operator function', *Proceedings of the National Academy of Sciences of USA*, 73: 2351–5.

Brian, J.K., Westerhoff, H.V., Snoep, J.L., Nilsson, D., and Jensen, P.R. (2002) 'The glycolytic flux in *Escherichia coli* is controlled by the demand for ATP', *Journal of Bacteriology*, 184: 3909–16.

Brice, C.B. and Kornberg, H.L. (1968) 'Genetic control of isocitrate lyase activity in *Escherichia coli*', *Journal of Bacteriology*, 96: 2185–6.

Brown, T.D.K., Jones-Mortimer, M.C., and Kornberg, H.L. (1977) 'The enzymic interconversion of acetate and acetyl-coenzyme A in *Escherichia coli*', *Journal of General Microbiology*, 102: 327–36.

Chambost, J.P. and Fraenkel, D.G. (1980) 'The use of 6-labeled glucose to assess futile cycling in *Escherichia coli*', *Journal of Biological Chemistry*, 255: 2867–9.

Chang, Y.Y., Wang, A.Y., and Cronan, J.E. Jr. (1994) 'Expression of *Escherichia coli* pyruvate oxidase (PoxB) depends on the sigma factor encoded by the *rpoS* (*katF*) gene', *Molecular Microbiology*, 11: 1019–28.

Chang, D., Jung, H., Rhee, J., and Pan, J. (1999) 'Homofermentative production of D-lactate or L-lactate in metabolically engineered *Escherichia coli* RR1', *Applied Environmental Microbiology*, 65: 1384–9.

Chao, Y.P. and Laio, J.C. (1994) 'Metabolic responses to substrate futile cycling in *Escherichia coli*', *Journal of Biological Chemistry*, 269: 5122–6.

Chassagnole, C., Noisommit-Rizzi, N., Schmid, J.W., Mauch, K., and Reuss, M. (2000) 'Dynamic modeling of the central carbon metabolism of *Escherichia coli*', *Biotechnology and Bioengineering*, 79: 53–73.

Choi, I.Y., Sup, K.I., Kim, H.J., and Park, J.W. (2003) 'Thermosensitive phenotype of *Escherichia coli* mutant lacking NADP(+)-dependent isocitrate dehydrogenase', *Redox Report*, 8: 51–6.

Clark, D.P. and Cronan, J.E. (1996) 'Two-carbon compounds and fatty acids as carbon sources', in: Escherichia coli *and* Salmonella typhimurium: *Cellular and Molecular Biology*, 2nd edition, edited by F.C. Neidhardt et al. Washington, DC: American Society of Microbiology Press. pp. 343–57.

Cronan, J.E. Jr. and LaPorte, D. (1999) 'Tricarboxylic acid cycle and glyoxalate bypass', in: Escherichia coli *and* Salmonella typhimurium: *Molecular Biology*, 2nd edition, edited by F.C. Neidhardt et al. Washington, DC: American Society of Microbiology Press. pp. 343–57.

Dalal, F. and Fraenkel, D.G. (1983) 'Assessment of a futile cycle involving reconversion of fructose-6-phosphate to fructose-1,6-biphosphate during gluconeogenic growth of *Escherichia coli*', *Journal of Bacteriology*, 153: 390–4.

David, J. and Wiesmeyer, H. (1970) 'Regulation of ribose metabolism in E. coli II: Evidence for two ribose-5-phosphate isomerase activities', *Biochimica et Biophysica Acta*, 208: 56–67.

Dessein, A., Schwartz, M., and Ullmann, A. (1978) 'Catabolite repression in *Escherichia coli* mutants lacking cyclic AMP', *Molecular and General Genetics*, 162: 83–7.

Dietrich, J. and Henning, U. (1970) 'Regulation of pyruvate dehydrogenase complex synthesis in *Escherichia coli* K12', *European Journal of Biochemistry*, 14: 258–69.

El-Mansi, E.M. and Holms, W.H. (1989) 'Control of carbon flux to acetate excretion during overflow in *Escherichia coli*', *Biotechnology and Bioengineering*, 35: 732–8.

Esseenberg, M.K. and Cooper, R.A. (1975) 'Two ribose-5-phosphate isomerase from *Escherichia coli* K-12: Partial characterization of the enzymes and consideration of their possible physiological roles,' *European Journal of Biochemistry*, 55: 323–32.

Farmer, W.R. and Liao, J.C. (1997) 'Reduction of aerobic production by *Escherichia coli*', *Applied Environmental Microbiology*, 63: 3205–10.

Fischer, E. and Sauer, U. (2003) 'A novel metabolic cycle catalyzes glucose oxidation and anaplerosis in hungry *Escherichia coli*,' *Journal of Biological Chemistry*, 278: 46446–51.

Fish, R.N. and Kane, C.M. (2002) 'Promoting elongation with transcript cleavage stimulatory factors', *Biochimica et Biophysica Acta*, 1577: 287–307.

Fraenkel, D.G. (1968) 'Selection of mutants lacking glucose-6-phosphate dehydrogenase or gluconate-6-phosphate dehydrogenase', *Journal of Bacteriology*, 95: 1267–71

Fraenkel, D.G. (1999) 'Glycolysis', in: Escherichia coli *and* Salmonella: *Cellular and Molecular Biology*, edited by F.C. Neidhardt et al. Washington, DC: American Society of Microbiology Press, Mira Digital Publishing.

Garrigues, C., Loubiere, P., Nic, D., Lindley, N.D., and Cocaign-Bousquet, M. (1997) 'Control of the shift from homolactic acid to mixed-acid fermentation in *Lactococcus lactis*: Predominant role of the NADH/NAD$^+$ ratio', *Journal of Bacteriology*, 179: 5282–7.

Gokarn, R.R., Altman, E., and Eiteman, M.A. (1998) 'Metabolic analysis of *Escherichia coli* in the presence of and absence of pyruvate carboxylase', *Biotechnology Letters*, 20: 795–8.

Gokarn, R.R., Eiteman, M.A., and Altman, E. (2000) 'Metabolic analysis of *Escherichia coli* in the presence and absence of carboxylating enzymes phosphoenolpyruvate carboxylase and pyruvate carboxylase', *Applied Environmental Microbiology*, 66: 1844–50.

Gottschalk, G. (1985) *Bacterial Metabolism*, 2nd edition. New York: Springer Verlag.

Grabau, C. and Cronan, J.E. Jr. (1984) 'Molecular cloning of the gene (poxB) encoding the pyruvate oxidase of *Escherichia coli*, a lipid-activated enzyme', *Journal of Bacteriology*, 160: 1088–92.

Grove, C.L. and Gunsalus, R.P. (1987) 'Regulation of the *aroH* operon of *Escherichia coli* by the Tryptophan repressor', *Journal of Bacteriology*, 169: 2158–64.

Guest, J.R. and Russell, G.C. (1992) 'Complexes and complexities of the citric acid cycle in *Escherichia coli*', *Current Topics in Cellular Regulation*, 33: 231–47.

Gunsalus, R.P. and Yanofsky, C. (1980) 'Nucleotide sequence and expression of *Escherichia coli trpR*, the structural gene for the *trp* aporepressor', *Proceedings of the NationalAcademy of.Sciences of the USA*, 77: 7117–21.

Han, K., Lim, H.C., and Hong, J. (1992) 'Acetic acid formation in *Escherichia coli* fermentation', *Biotechnology and Bioengineering*, 39: 663–71.

Haugaard, N. (1959) 'D- and L-lactic acid oxidases of *Escherichia coli*', *Biochimica et Biophysica Acta*, 31: 66–77.

Heatwole, V.M. and Somerville, R.L. (1991) 'The tryptophan-specific permease gene, *mtr*, is differentially regulated by the tryptophan and tyrosine repressors in *E. coli* K12', *Journal of Bacteriology*, 174: 3601–4.

Heatwole, V.M. and Somerville, R.L. (1992) 'Synergism between the *trp* repressor and *tyr* repressor in repression of the *aroL* promoter of *Escherichia coli* K12', *Journal of Bacteriology*, 174: 331–5.

Holger, J., Bernhard, K., Peter, K., and Karl-Dieter, E. (1996) 'Mutants that show increased sensitivity to hydrogen peroxide reveal an important role for the pentose phosphate pathway in protection of yeast against oxidative stress', *Molecular and General Genetics*, 252: 456–64.

Hong, S.H. and Lee, S.Y. (2001) 'Metabolic flux analysis for succinic acid production by recombinant *Escherichia coli* with amplified malic enzyme activity', *Biotechnology and Bioengineering*, 74: 89–95.

Hoyt, J.C. and Reeves, H.C. (1988) '*In vivo* phosphorylation of isocitrate lyase from *Escherichia coli* D5H3G7', *Biochemical and Biophysical Research Communications*, 153: 875–80.

Hua, Q. Yang, C., Baba, T., Mori, H., and Shimizu, K. (2003) 'Responses of the central carbon metabolism in *Escherichia coli* to phosphoglucose isomerase and glucose-6-phosphate dehydrogenase knockouts', *Journal of Bacteriology*, 185: 7053–67.

Ishii, N., Nakahigashi, K., Baba, T., Robert, M., Soga, T. et al. (2007) 'Multiple high throughput analyses monitor the response of *E. coli* to perturbations', *Science*, 316: 593–7.

Iuchi, S. and Lin, E.C.C. (1992) 'Purification and phosphorylation of the Arc regulatory components of *Escherichia coli*', *Journal of Bacteriology*, 174: 5617–23.

Jeong, W. Cha, M.K., and Kim, I.H. (2000) 'Thioredoxin-dependent hydroperoxide peroxidase activity of bacterioferritin co-migratory protein (BCP) as a new member of the thiol-specific antioxidant protein (TSA)/alkyl hydroperoxide peroxidase C (*ahpC*) family', *Journal of Biological Chemistry*, 275: 2924–30.

Jiang, G.R., Nikolova, S., and Clark, D.P. (2001) 'Regulation of *ldhA* gene, encoding the fermentative lactate dehydrogenase of *Escherichia coli*', *Microbiology*, 147: 2437–46.

Kabir, M. and Shimizu, K. (2004) 'Metabolic regulation analysis of *icd*-gene knockout *Escherichia coli* based on 2D electrophoresis with MALDI-TOF mass spectrometry and enzyme activity measurements', *Appl. Microbiol. Biotechnol.*, 65: 89–96.

Kabir, M., Ho, P.Y., and Shimizu, K. (2005) 'Effect of *ldhA* gene deletion on the metabolism of *E. coli* based on gene expression, enzyme activities, intracellular metabolite concentrations, and metabolic flux distribution', *Biochemistry Engineering Journal*, 26: 1–11.

Kaiser, M. and Sawer, G. (1994) 'Pyruvate formate-lyase is not essential for nitrate respiration by *Escherichia coli*', *FEMS Microbiology Letters*, 117: 163–8.

Kedar, P., Colah, R., and Shimizu, K. (2007) 'Proteomic investigation on the *pyk-F* gene knockout *Escherichia coli* for aromatic amino acid production', *Enzyme and Microbial Technology*, 41: 455–65.

Kelley, R.L. and Yanofsky, C. (1982) 'Trp aporepressor production is controlled by autogenous regulation and inefficient translations', *Proceedings of the National Academy of Sciences of USA*, 79: 3120–4.

Kleman, G.L. and Strohl, W.R. (1994) 'Acetate metabolism by *Escherichia coli* in high-cell-density fermentation', *Applied Environmental Microbiology*, 60: 3952–8.

Kline, E.S. and Mahler, E.R. (1965) 'The lactic acid dehydrogenases of *Escherichia coli*', *Annual New York Academic Science*, 119: 905–17.

Koebmann, B.J., Westerhoff, H.V., Snoep, J.L., Nilsson, D., and Jensen, P.R. (2002) 'The glycolytic flux in *Escherichia coli* is controlled by the demand for ATP', *Journal of Bacteriology*, 184: 3909–16.

Krebs, A. and Bridger, W.A. (1980) 'The kinetic properties of phosphoenolpyruvate carboxykinase of *Escherichia coli*', *Canadian Journal of Biochemistry*, 58: 309–18.

Kumari, S., Tishel, R., Eisenbach, M., and Wolfe, A.J. (1995) 'Cloning, characterization, and functional expression of *acs*, the gene which encodes acetyl coenzyme A synthetase in *Escherichia coli*', *Journal of Bacteriology*, 177: 2878–86.

Kumari, S., Beatty, C.M., Browning, D.F., Busby, S.J.W., Simel, E.J. et al. (2000) 'Regulation of acetyl coenzyme A synthetase in *Escherichia coli*', *Journal of Bacteriology*, 182: 4173–9.

LaPorte, D.C. and Koshland, D.E. Jr. (1983) 'Phosphorylation of isocitrate dehydrogenase as a demonstration of enhanced sensitivity in covalent regulation', *Nature*, 305: 286–90

LaPorte, D.C., Thorsness, P.E., and Koshland, D.E. Jr. (1985) 'Compensatory phosphorylation of isocitrate dehydrogenase. A mechanism for adaptation to the intracellular environment', *Journal of Biological Chemistry*, 260: 10563–8.

Lawley, B. and Pittard, A.J. (1994) 'Regulation of *aroL* expression by TyrR protein and Trp repressor in *E. coli* K12', *Journal of Bacteriology*, 176: 6921–30.

Lemaster, D.M. and Cronan, J.E. Jr. (1982) 'Biosynthetic production of ^{13}C-labeled amino acids with site-specific enrichment', *Journal of Biological Chemistry*, 257: 1224–30.

Li, M., Yao, S., and Shimizu, K. (2006a) 'Effect of *lpdA* gene knockout on the metabolism in *Escherichia coli* based on enzyme activities, intracellular metabolite concentrations and metabolic flux analysis by ^{13}C-labeling experiments', *Journal of Biotechnology*, 122: 254–66.

Li, M., Ho, P.Y., Yao, S., and Shimizu, K. (2006b) 'Effect of *sucA* or *sucC* gene knockout on the metabolism in *Escherichia coli* based on gene expressions, enzyme activities, intracellular metabolite concentrations and metabolic fluxes by ^{13}C-labeling experiments', *Biochemical Engineering Journal*, 30: 286–96.

Li, Z. and Demple, B. (1994) '*SoxS* an activator of superoxide stress genes in *Escherichia coli*: purification and interaction with DNA', *Journal of Biological Chemistry*, 269: 18371–7.

Maaheimo, H., Fiaux, J., Cakar, Z.P., Bailey, J.E., Sauer, U., and Szyperski, T. (2001) 'Central carbon metabolism of *Saccharomyces cerevisiae* explored by biosynthetic fractional of common amino acids', *European Journal of Biochemistry*, 268: 2464–79.

Maloy, S.R. and Nunn, W.D. (1981) 'Role of gene fadR in *Escherichia coli* acetate metabolism', *Journal of Bacteriology*, 148: 83–90.

Martin, B.R. (1987) 'The regulation of enzyme activity', in: *Metabolic Regulation: A Molecular Approach*, edited by B.R. Martin. Oxford: John Wiley & Co. pp 12–27.

Mat-Jan, F., Alam, K.Y., and Clark, D.P. (1989) 'Mutants of *Escherichia coli* deficient in the fermentative lactate dehydrogenase', *Journal of Bacteriology*, 171: 342–8.

Mat-Jan, F., Williams, C.R., and Clark, D.P. (1989) 'Anaerobic growth defects resulting from gene fusions affecting succinyl-CoA synthetase in *Escherichia coli* K12', *Molecular and General Genetics*, 215: 276–80.

Matsushita, K., Arents, J.C., Bader, R., Yamada, M., Adachi, O., and Postma, P.W. (1997) '*Escherichia coli* is unable to produce pyrroloquinoline quinine (PQQ)', *Microbiology*, 143: 3149–56.

McCammon, M.T., Epstein, C.B., Przybyla-Zawislak, B., McAlister-Henn, L., and Butow, R.A., (2003) 'Global transcription analysis of Krebs tricarboxylic acid cycle mutants reveals an alternating pattern of gene expression and effects on hypoxic and oxidative genes', *Molecular Biology of the Cell*, 14: 958–72.

Miles, J.S. and Guest, J.R. (1989) 'Molecular genetic aspects if the citric acid cycle of *Escherichia coli*', *Biochemical Society Symposium*, 54: 45–65.

Morita, T., Waleed, E., Yuya, T., Toshifumi, I., and Hiroji, A. (2003) 'Accumulation of glucose 6-phosphate or fructose 6-phosphate is responsible for destabilization of glucose transporter mRNA in *Escherichia coli*', *Journal of Biological Chemistry*, 278: 15246–608.

Murai, T.M., Tokushige, M., Nagai, J., and Katsuki, H. (1971) 'Physiological functions of NAD+- and NADP+-linked malic enzymes in *E. coli*', *Biochemical and Biophysical Research Communications*, 43: 875–81.

Nakano, M.M., Zuber, P., and Sonenshein, A. (1998) 'Anaerobic regulation of *Bacillus subtilis* krebs cycle genes', *Journal of Bacteriology*, 180: 3304–11.

Neidhardt, F.C. (1996) Escherichia coli *and* Salmonella: *Cellular and Molecular Biology*, 2nd edition. Washington, DC: American Society of Microbiology Press. pp. 189–98.

Nielsen, J. (1984) 'Structure and expression of the ATP synthase operon of *Escherichia coli*', Ph.D. thesis, Technical University of Denmark, Lyngby, Denmark.

Nimmo, G.A. and Nimmo, H.G. (1984) 'The regulatory properties of isocitrate dehydrogenase kinase and isocitrate dehydrogenase phosphatase from *Escherichia coli* ML308 and the roles of these activites in the control of isocitrate dehydrogenase', *European Journal of Biochemistry*, 141: 409–14.

Nimmo, H.G., Borthwick, A.C., el-Mansi, E.M., Holms, W.H., MacKintosh, C., and Nimmo, G.A. (1987) 'Regulation of the enzymes at the branchpoint between the citric acid cycle and the glyoxylate bypass in *Escherichia coli*', *Biochemical Society Symposium*, 54: 93–101.

Notley, L. and Ferenci, T. (1996) 'Induction of RpoS-dependent functions in glucose-limited continuous culture: What level of nutrient limitation induces the stationary phase of *Escherichia coli*', *Journal of Bacteriology*, 178: 1465–8.

Oh, M.K., Rohlon, L., Kao, K., and Liao, J.C. (2002) 'Global expression profiling of acetate-grown *E. coli*', *Journal of Biological Chemistry*, 277: 13175–83.

Park, S.J., McCabe, J., Turna, J., and Gunsalus, R.P. (1994) 'Regulation of citrate synthase (*gltA*) gene of *Escherichia coli* in response to anaerobiosis and carbon supply: Role of the *arcA* gene product', *Journal of Bacteriology*, 176: 5086–92.

Park, S.J. and Gunsalus, R.P. (1995) 'Oxygen, iron, carbon, and superoxide control of the fumarase fumA and fumC genes of *Escherichia coli*: Role of the arcA, fnr, and soxR gene products', *Journal of Bacteriology*, 177: 6255–62.

Park, S.J., Cotter, P.A., and Gunsalus, R.P. (1995b) 'Regulation of malate dehydrogenase (mdh) gene expression in *Escherichia coli* in response to oxygen, carbon, and heme availability', *Journal of Bacteriology*, 177: 6652–6.

Park, S.J., Chao, G., and Gunsalus, R.P. (1997) 'Aerobic regulation of the sucABCD genes of *Escherichia coli*, which encode α-ketoglutarate dehydrogenase and succinyl coenzyme A synthetase: Roles of arcA, fnr, and the upstream sdhCDAB promoter', *Journal of Bacteriology*, 179: 4138–42.

Pascal, M.C., Chippaux, M., Abou-Jaoudé, A., Blaschkowski, H.P., and Knappe, J. (1981) 'Mutants of *Escherichia coli* K12 with defects in anaerobic pyruvate metabolism', *Journal of General Microbiology*, 124: 35–42.

Pease, A.J. and Wolf, J.R.E. (1994) 'Determination of the growth rate-regulated steps in expression of the *Escherichia coli* K-12 *gnd* gene', *Journal of Bacteriology*, 176: 115–22.

Pedersen, S., Block, P.L., Reeh, S., and Neihardt, F.C. (1978) 'Patterns of protein synthesis in *E. coli*: A catalog of the amount of 140 individual proteins in different growth rates', *Cell*, 14: 179–86.

Peng, L., Arauzo, M., and Shimizu, K. (2004) 'Metabolic flux analysis for a *ppc* mutant *Escherichia coli* based on 13C-labeling experiments together with enzyme activity assays and intracellular metabolite measurements', *FEMS Microbiology Letters*, 235: 17–23.

Perrenoud, A. and Sauer, U. (2005) 'Impact of global transcriptional regulation by arcA, arcB, Cra, Crp, Cya, Fnr, and Mlc on glucose catabolism in *Escherichia coli*', *Journal of Bacteriology*, 187: 3171–9.

Ponce, E., Martinez, A., Bolivar, F., and Valle, F. (1998) 'Stimulation of glucose catabolism through the pentose pathway by the absence of the two pyruvate kinase isoenzymes in *Escherichia coli*', *Biotechnology and Bioengineering*, 58: 292–5.

Ramseier, T.M., Chie, S.Y., and Saier, M.H. Jr. (1996) 'Co-operative interaction between Cra and Fnr in the regulation of the *cydAB* operon of *Escherichia coli*', *Current Microbiology*, 33: 270–4.

Riondet, C., Cachon, R., Waché, Y., Alcaraz, G., and Diviès, C. (2000) 'Extracellular oxido-reduction potential modifies carbon and electron flow in *Escherichia coli*', *Journal of Bacteriology*, 182: 620–6.

Rose, J.K., Squires, C.L., Yanofsky, C., Yang, H., and Zubay, G. (1973) 'Regulation of *in vitro* transcription of the tryptophan operon by purified RNA polymerase in the presence of partially purified repressor and tryptophan', *Nature New Biology Archive*, 245: 133–7.

Saier, M.H. Jr., and Ramseier, T.M. (1996) 'The catabolite repressor/activator (Cra) protein of enteric bacteria', *Journal of Bacteriology*, 178: 3411–7.

Sarsero, J.P., Wookey, P.J., and Pittard, A.J. (1991) 'Regulation of the expression of *Escherichia coli* K12 *mtr* gene by TyrR protein and *trp* repressor', *Journal of Bacteriology*, 173: 4133–43.

Schreyer, R. and Bock, A. (1980) 'Phosphoglucose isomerase from *Escherichia coli* K10: Purification, properties and formation under aerobic and anaerobic conditions', *Archives of Microbiology*, 127: 289–98.

Seta, F.D., Boschi-Muller, S., Vignais, M.L., and Branlant, G. (1997) 'Characterization of *Escherichia coli* strains with *gapA* and *gapB* genes deleted', *Journal of Bacteriology*, 179: 5218–21.

Shimizu, K. (2004) 'Metabolic flux analysis based on ^{13}C labeling experiments and integration of the information with gene and protein expression patterns', *Advances in Biochemical Engineering/Biotechnology*, 91: 1–49.

Siddiquee, K.A.Z., Arauzo-Bravo, M., and Shimizu, K. (2004a) 'Metabolic flux analysis of *pykF* gene knockout *Escherichia coli* based on 13C-labeled experiment together with measurements of enzyme activities and intracellular metabolite concentrations', *Applied Microbiology and Biotechnology*, 63: 407–17.

Siddiquee, K.A.Z., Arauzo-Bravo, M., and Shimizu, K. (2004b) 'Effect of pyruvate kinase (*pykF* gene) knockout mutation on the control of gene expression and metabolic fluxes in *Escherichia coli*', *FEMS Microbiology Letters*, 235: 25–33.

Skinner, A.J. and Cooper, R.A. (1971) 'The regulation of ribose-5-phosphate isomerization in *Escherichia coli* K12', *FEBS Letters*, 12: 293–6.

Smith, M.W. and Neidhardt, F.C. (1983) 'Proteins induced by anaerobiosis in *Escherichia coli*', *Journal of Bacteriology*, 154: 336–43.

Somerville, R. (1992) 'The *trp* repressor, a ligand activated regulatory protein', *Progress in Nucleic Acids Research & Molecular Biology*, 42: 1–38.

Spencer, M.E. and Guest, J.R. (1985) 'Transcription analysis of the *sucAB*, *aceEF*, and *lpd* genes of *Escherichia coli*', *Molecular and General Genetics*, 200: 145–54.

Stokes, J.L. (1949) 'Fermentation of glucose by suspensions of *Escherichia coli*', *Journal of Bacteriology*, 57: 147–58.

Stols, L. and Donnelly, M.I. (1997) 'Production of succinic acid through overexpression of NAD$^+$-dependent malic enzyme in an *Escherichia coli* mutant', *Applied Environmental Microbiology*, 63: 2695–701.

Straus, D.B., Walter, W.A., and Gross, C.A. (1987) 'The heat shock response of *Escherichia coli* is regulated by changes in the concentration of σ^{32}', *Nature*, 329: 348–51.

Sugimoto, S. and Shiio, I. (1987) 'Regulation of 6-phosphogluconate dehydrogenase in *Brevibacterium flavum*', *Agriculture and Biological Chemistry*, 51: 1257–63.

Tarmy, E.M. and Kaplan, N.O. (1968a) 'Kinetics of *Escherichia coli* B D-lactate dehydrogenase and evidence for pyruvate-controlled change in conformation', *Journal of Biological Chemistry*, 243: 2587–96.

Tarmy, E.M. and Kaplan, N.O. (1968b) 'Chemical characterization of D-lactate dehydrogenase from *Escherichia coli* B', *Journal of Biological Chemistry*, 243: 2579–86.

Toya, Y., Ishii, N., Nakahigashi, K., Hirasawa, T., Soga, T. et al. (2010) '^{13}C-Metabolic flux analysis for batch culture of *Escherichia coli* and its *pyk* and *pgi* gene knockout mutants based on mass isotopomer distribution of intracellular metabolites', *Biotechnology Progress*, 26: 975–92.

Van de Walle, M. and Shiloach, J. (1998) 'Proposed mechanism of acetate accumulation in two recombinant *Escherichia coli* strains during high density fermentation', *Biotechnology and Bioengineering*, 57: 71–8.

Vemuri, G.N., Eiteman, M.A., and Altman, E. (2002) 'Effects of growth mode and pyruvate carboxylase on succinic acid production by metabolically engineered strains of *Escherichia coli*', *Applied Environmental Microbiology*, 68: 1715–27.

Wolf, J.R.E., Prather, D.M., and Shea, F.M. (1979) 'Growth-rate dependent alteration of 6-phosphogluconate dehydrogenase and glucose-6-phosphate dehydrogenase levels in *Escherichia coli* K-12', *Journal of Bacteriology*, 139: 1093–6.

Wolodko, W.T. and Bridger, W.A. (1987) 'Studies of the process of renaturation and assembly of *Escherichia coli* succinyl-CoA synthetase from its α and βa subunits', *Biochemistry and Cell Biology*, 65: 452–7.

Yang, C., Hua, Q., Baba, T., Mori, H., and Shimizu, K. (2003) 'Analysis of *E. coli* anaplerotic metabolism and its regulation mechanism from the metabolic responses to altered dilution rates and phosphoenolpyruvate carboxykinase knockout', *Biotechnology and Bioengineering*, 84: 129–44.

Yang, Y.T., San, K.Y., and Bennett, G.N. (1999) 'Redistribution of metabolic fluxes in *Escherichia coli* with fermentative lactate dehydrogenase over-expression and deletion', *Metabolic Engineering*, 1: 141–52.

Zhao, J., Baba, T., Mori, H., and Shimizu, K. (2004a) 'Effect of *zwf* gene knockout on the metabolism of *Escherichia coli* grown on glucose or acetate', *Metabolic Engineering*, 6: 164–74.

Zhao, J., Baba, T., Mori, H., and Shimizu, K. (2004b) 'Analysis of metabolic and physiological responses to *gnd* knockout in *E. coli* by using C-13 tracer experiment and enzyme activity measurement', *FEMS Microbiology Letters*, 220: 295–301.

Zhou, S., Causey, T.B., Hasona, A., Shanmugam, K.T., and Ingram, L.O. (2003) 'Production of optically pure D-lactic acid in mineral salts medium by metabolically engineered *Escherichia coli* W3110', *Applied Environmental Microbiology*, 69: 399–407.

Zhu, J. and Shimizu, K. (2004) 'The effect of *pfl* genes knockout on the metabolism for optically pure d-lactate production by *Escherichia coli*', *Applied Microbiology and Biotechnology*, 64: 367–75.

Zhu, J. and Shimizu, K. (2005) 'Effect of a single-gene knockout on the metabolic regulation in *E. coli* for d-lactate production under microaerobic conditions', *Metabolic Engineering*, 7: 104–15.

Zhu, T., Phalakornkule, C., Koepsel, R.R., Domach, M.M., and Ataai, M.M. (2001) 'Cell growth and by-product formation in pyruvate kinase mutant of *E. coli*', *Biotechnology Progress*, 17: 624–8.

Zurawski, G., Gunsalus, R.P., Brown, K.D., and Yanofsky, C. (1981) 'Structure and regulation of *aroH*, the structural gene for the tryptophan-repressible 3-deoxy-D-arabino-heptulosonic acid-7-phosphate synthetase of *Escherichia coli*', *Journal of Molecular Biology*, 145: 47–73.

Appendices

Appendix A: Global regulators and their regulated genes

ArcA/B:
+: *pfl, cydAB*
−: *aceBAK, aceEF, acnA, fumAC, gltA, icdA, lpdA, mdh, ptsG, sdhCDAB, cyo*

pdhR:
+: *aceEF, ndh, yfiD*
−: *cyoABCD*

Fur:
+: *hmp*
−: *cyoABCD, ompF*

Cra:
+: *aceA, acnA, fbp, icdA, pckA, ppsA, cydB*
−: *acnB, eda, edd, eno, gapA, pfkA, ptsHI, pykF*

Crp/Cya:
+: *mlc, aceEF, acnAB, crr, fumA, gltA, mdh, pckA, ptsG, ptsHI, sdhABCD, sucABCD, tpiA, ompF, rpoS*
−: *lpdA, aceBAK, acnA, cyoA, gdhA, glnAL, gltA, mdh, sdhCDAB, sodA, sucABCD, ugpA,*

Fnr:
+: *frd, pfl, ackA, ndh, nuoA, pstSCAB-phoU, yfiD*
−: *acnA, fumAC, icdA, lpdA, ptsG, sdhCDAB, talA, cyoABCD, cydAB*

PhoB:
+: *phoBR, phoA-psiE, asr, pstSCAB-phoU*
−: *phoH, phnCHN, ugpA, argP*

RpoS:
+: *gadA, gadB, osmB, sodC, talA, tktB, acs, poxB, acnA, fumC*

SoxR/S:
+: *sodA, zwf, rpoD, rpoS, fumC, tolC, micF, marA*
– :*rob*

Mlc
+: *ldhA*
– : *crr, ptsG, ptsH1, manXYZ, malT*

GadE
+: *gadE, gadXW*
–: *cyoABCDE, gltB*

Appendix B: Metabolic pathway reactions and the corresponding enzymes and genes

	enzyme	gene	reactions
PTS			Glc+PEP→PYR+G6P
	EI	*ptsI*	
	HPr	*ptsH*	
	EIIA	*crr*	
	EIICB	*ptsG*	
Glk	Glk	*glk*	
Glycolysis			
	Pgi	*pgi*	G6P⇔F6p
	Pfk	*pfkA,B*	F6P+ATP⇒F16BP+ADP
	Fbp	*fbp*	F16BP+Pi⇒F6P
	Fba	*fba*	F16BP⇔GAP+DHAP
	Tpi	*tpi*	GAP⇔DHAP
	GAPDH	*gapA,C*	GAP+Pi+NAD+⇒13BPG+NADH
	Pgk	*pgk*	13BPG+ADP⇔3PG+ATP
	Pgm	*pgm*	3PG⇔2PG
	Eno	*eno*	2PG⇔PEP
	Pyk	*pykF,A*	PEP+ADP⇒PYR+ATP
	PDH	*AceE,F,lpdA*	PYR+CoA+NAD+⇒AcCoA+CO_2+NADH
	Ppc	*ppc*	PEP+CO_2⇒OAA+Pi
	Pck	*pckA*	OAA+ATP⇒PEP+CO_2+ADP
	Pps	*ppsA*	PYR+ATP⇒PEP+AMP+Pi

PP pathway			
	G6PDH	zwf	G6P+NADP+$\Rightarrow$6PGL+NADPH
	6Pgl	pgl	6PGL$\Rightarrow$6PG
	6PGDH	gnd	6PG+NADP+$\Rightarrow$RU5P+NADPH+CO_2
	Rpi	rpiA,B	Ru5P$\Leftrightarrow$R5P
	Rpe	rpe	Ru5P$\Leftrightarrow$X5P
	Tkt1	tktA	R5P+X5P$\Leftrightarrow$GAP+S7P
	Tal	talA,B	GAP+S7P$\Leftrightarrow$E4P+F6P
	Tkt2	tktB	X5P+E4P$\Leftrightarrow$F6P+GAP
ED pathway			
		edd	6PG$\Rightarrow$KDPG
	KDPG	eda	KDPG$\Rightarrow$GAP+PYR
Fermentative pathway			
	LDH	ldhA	PYR+NADH$\Leftrightarrow$NAD+Lac
	ADH	adhE	AcAld+NADH$\Leftrightarrow$NAD+ETOH
	AcAldDH	adhE	Ace+NADH$\Leftrightarrow$NAD+AcAld
	Pfl	pflABCD,foc	PYR+CoA$\Rightarrow$AcCoA+formic acid
	Fhl	fhl	HCOOH$\Rightarrow$$CO_2$
	Pta	pta	AcCoA+Pi$\Leftrightarrow$AcP+CoA
	Ack	ackA	AcP+ADP$\Leftrightarrow$ATP+Ace
TCA cycle			
	CS	gltA	AcCoA+OAA$\Rightarrow$CoQ+CIT
	Acn	acnA,B	CIT$\Rightarrow$ICIT
	ICDH	icdA	ICIT+NAD$\Rightarrow$$CO_2$+NAD(P)H+$\alpha$KG
	αKGDH	sucA,B	αKG+NAD+CoA$\Rightarrow$$CO_2$+NADH+SucCoA
	SCS	sucC,D	SucCoA+GDP(ADP)+Pi$\Leftrightarrow$GTP(ATP)+CoA+SUC
	SDH	sdhCDAB	SUC+FAD$\Rightarrow$$FADH_2$+FUM
	Frd	frd	FUM+$FADH_2$$\Rightarrow$SUC+FAD
	Fum	fumA,B,C	FUM$\Leftrightarrow$MAL
	MDH	mdh	MAL+NAD$\Leftrightarrow$NADH+OAA
	Mez	medB	MAL+NADP$\Rightarrow$$CO_2$+NADPH+PYR
	Sfc	sfc	MAL+NAD$\Rightarrow$$CO_2$+NADH+PYR
	Icl	aceA	ICIT$\Rightarrow$GOX+SUC
	MS	aceB	AcCoA+GOX$\Rightarrow$CoA+MAL

Appendix C: A table of genes and their associated functions

(a) Global regulatory genes

Gene names	Description (function as encoded protein)
arcA	Anoxic redox control
cra	Catabolite repressor activator
crp	cAMP receptor protein
fnr	Fumarate and nitrate reductase
rpoS	RNA polymerase sigma factor
rpoD	RNA polymerase, sigma 70 subunit
soxS	Dual transcriptional activator
soxR	Superoxide response protein
mlc	Making large colonies protein

(b) PhoB regulatory genes

Gene names	Description (function as encoded protein)
phoB	Dual transcription regulator
phoR	Sensor kinase of the PhoRB two component signal transduction pathway
phoA	Alkaline phosphatase precursor
phoE	Outer-membrane pore protein
phoH	PhoH protein (phosphate starvation inducible protein PsiH)
phnC	ATP binding component of the alkylphosphonate ABC transporter
pstS	Subunit of phosphate ABC transporter
ugpB	Subunit of glycerol-3-P ABC transporter
phoU	Phosphate transport system regulatory protein
phoM(creC)	Sensor histidine kinase of the CreCB two-component signal transduction system

(c) Metabolic pathway genes

Gene names	Description (function as encoded protein)
ptsH	Phosphohistidinoprotein-hexose phosphotransferase
ptsG	Glucose phosphotransferase enzyme IIBC[Glc]
pfkA	6-phosphofructokinase
pykF	Pyruvate kinase
lpdA	Lipoamide dehydrogenase
gltA	Citrate synthase
icdA	Isocitrate dehydrogenase
sucA	α–ketoglutarate dehydrogenase
sdhC	Succinate dehydrogenase

mdh	Malate dehydrogenase
zwf	Glucose 6-phosphate-1-dehydrogenase
gnd	6-phosphogluconate dehydrogenase
tktA	Transketolase I
tktB	Transketolase II
talA	Transaldolase A
talB	Transaldolase B
edd	6-phosphogluconate dehydratase
eda	Entner-Doudoroff aldolase
ldhA	D-lactate dehydrogenase
yfiD	Stress-induced alternate pyruvate formate-lyase subunit
asr	Acid shock RNA
gadA	Glutamate decarboxylase A
ackA	Acetate kinase
pta	Phosphate acetyltransferase

(d) Nitrogen regulatory genes

Gene names	Description (function as encoded protein)
rpoN	RNA polymerase, sigma 54 (sigma N) factor
gdhA	Glutamate dehydrogenase
gltB	Glutamate synthase, large subunit
gltD	Glutamate synthase, small subunit
glnA	Glutamine synthetase
glnB	Protein PII-control the level and activity of glutamine synthetase
glnD	Uridylyltransferase/uridylyl-removing enzyme
glnE	Glutamine synthetase adenylyltransferase [multifunctional]
glnG	NtrC transcriptional dual regulator
glnK	Nitrogen regulatory protein
glnL	NtrB sensory histidine kinase
nac	Nac DNA-binding transcriptional dual regulator

(e) Respiratory chain genes

Gene names	Description (function as encoded protein)
cyoA	Cytochrome bo terminal oxidase subunit II
cydB	Cytochrome bd-I terminal oxidase subunit II
atpA	ATP synthase α-subunit
ndh	Ubiquinone oxidoreductase II
nuoA	NADH: ubiquinone oxidoreductase
sodA	Superoxide dismutase (Mn)

Appendix D: Gene names

Gene	Product
Sugar uptake	
PEP-carbohydrate phosphotransferase system	
ptsG	PTS system, glucose-specific IIBC component
ptsH	PTS system protein HPr
ptsI	PEP-protein phosphotransferase system enzyme I
crr	PTS system, glucose-specific IIA component
fruR	transcriptional repressor of fru operon and global regulator
fruB	PTS system, fructose-specific IIA/fpr component
fruK	fructose-1-phosphate kinase
fruA	PTS system, fructose-specific transport protein
manA	mannose-6-phosphate isomerase
manX	PTS enzyme IIAB, mannose-specific
manY	PTS enzyme IIC, mannose-specific
manZ	PTS enzyme IID, mannose-specific
malI	repressor of malX and Y genes
malX	PTS system, maltose and glucose-specific II ABC
malY	enzyme that may degrade or block biosynthesis of endogenous mal inducer, probably aminotrasferase
malZ	maltodextrin glucosidase
nagE	PTS system, N-acetylglucosamine-specific enzyme II ABC
nagB	glucosamine-6-phosphate deaminase
nagA	N-acetylglucosamine-6-phosphate deacetylase
nagC	transcriptional repressor of nag operon
nagD	N-acetylglucosamine metabolism
bglX	β-D-glucoside glucohydrolase, periplasmic
bglA	6-phospho-β-glucosidase A
bglB	phospho-β-glucosidase B
bglF	PTS system β-glucosides, enzyme II
bglG	positive regulation of bgl operon
bglJ	2-component transcriptional regulator

Sugar-H⁺ symport and phosphorylation	
glk	glucokinase
galM	galactose-1-epimerase (mutarotase)
galK	galactokinase
galT	galactose-1-phosphate uridylyltransferase
galE	UDP-galactose-4-epimerase
galF	homolog of Salmonella UTP–glucose-1-P uridyltransferase, probably a UDP-gal transferase
galU	glucose-1-phosphate uridylyltransferase
galS	mgl repressor, galactose operon inducer
galR	repressor of galETK operon
galP	galactose-proton symport of transport system
xylB	xylulokinase
xylA	D-xylose isomerase
xylF	xylose binding protein transport system
xylG	putative ATP binding protein of xylose transport system
xylH	putative xylose transport, membrane component
xylR	putative regulator of xyl operon
xylE	xylose-proton symport
fucP	fucose permease
fucO	L-1,2-propanediol oxidoreductase
fucA	L-fuculose-1-phosphate aldolase
fucI	L-fucose isomerase
fucK	L-fuculokinase
fucU	protein of fucose operon
fucR	positive regulator of the fuc operon
araD	L-ribulose-5-phosphate 4-epimerase
araD	L-ribulose-5-phosphate 4-epimerase
araA	L-arabinose isomerase
araB	L-ribulokinase
araC	transcriptional regulator for ara operon
araH	high-affinity L-arabinose transport system membrane protein

(Continued)

Gene	Product
Sugar-H⁺ symport and phosphorylation	
araG	ATP-binding component of high-affinity L-arabinose transport system
araF	L-arabinose-binding periplasmic protein
araE	low-affinity L-arabinose transport system proton symport protein
rhaD	rhamnulose-phosphate aldolase
rhaA	L-rhamnose isomerase
rhaB	rhamnulokinase
rhaS	positive regulator for *rhaBAD* operon
rhaR	positive regulator for *rhaRS* operon
rhaT	rhamnose transport
Direct glucose oxidation and gluconate uptake	
gcd	glucose dehydrogenase
pqqA	PQQ precursor
pqqB	coenzyme PQQ synthesis protein B
pqqC	coenzyme PQQ synthesis protein C
pqqD	coenzyme PQQ synthesis protein D
pqqE	coenzyme PQQ synthesis protein E
pqqF	coenzyme PQQ synthesis protein F
gnl	gluconolactonase
	annotated as gluconolactonase precursor
gntT	high-affinity gluconate permease
gntR	repressor for both *gntT* and *gntK-gntU*
gntK	thermoresistant gluconate kinase
gntU	low-affinity gluconate permease
Glycolysis and gluconeogenesis	
Embden-Meyerhof-Parnas pathway	
pgm	Glucose6P-Glucose1P
fruK	fructose-1-phosphate kinase
pgi	phosphoglucose isomerase
pfkA	6-Phosphofructokinase-I

pfkB	6-Phosphofructokinase-II
fbp	fructose-1,6-bisphosphotase
fba	fructose bisphosphate aldolase
tpi	triose phosphate isomerase
gapA	Glyceraldehyde-3P dehydrogenase-A complex
pgk	phosphoglycerate kinase
gpmA	phosphoglycerate mutase
gpmB	phosphoglycerate mutase
eno	enolase
pykA	pyruvate kinase
pykF	pyruvate kinase
ppsA	phosphoenolpyruvate synthetase
pckA	PEP carboxykinase
sfcA	malic enzyme
maeB	malic enzyme
Penthose phosphate pathway	
zwf	glucose-6-phosphate dehydrogenase
pzwf	glucose-6-phosphate dehydrogenase
ybhE	6-phosphogluconolactonase
gnd	6-phosphogluconate dehydrogenase
rpiA	Ribulose-5P isomerase
rpiB	Ribulose-5P isomerase
rpe	ribulose phosphate 3-epimerase
tktA	transketolase
tktB	transketolase
talA	transaldolase
talB	transaldolase
Entner-Doudoroff pathway	
edd	phosphogluconate dehydratase
eda	multifunctional 2-keto-3-deoxygluconate 6-phosphate aldolase and oxaloacetate decarboxylase
TCA cycle, glyoxylate shunt and ammonia assimilation	
ppc	PEP carboxylase

(Continued)

Gene	Product
pycA	pyruvate carboxylase
gltA	citrate sinthase
acnA	aconitate hydrase 1
acnB	aconitate hydrase 2
icdA	isocitrate degydrogenase
sucA	2-ketoglutarate dehydrogenase
sucB	2-ketoglutarate dehydrogenase
sucC	succinate thiokinase
sucD	succinate thiokinase
lpdA	lipoamide dehydrogenase (NADH); component of 2-oxodehydrogenase and pyruvate complexes; L-protein of glycine cleavage complex
sdhC	succinate dehydrogenase, cytochrome b556
sdhD	succinate dehydrogenase, hydrophobic subunit
sdhA	succinate dehydrogenase, flavoprotein subunit
sdhB	succinate dehydrogenase, iron sulfur protein
fumA	fumarase A = fumarate hydratase Class I
fumC	fumarase C = fumarate hydratase Class II, isozyme, oxygen-stable
mdh	malate dehydrogenase
frdA	fumarate reductase, anaerobic, flavoprotein
frdB	fumarate reductase, anaerobic, iron-sulfur protein subunit
frdC	fumarate reductase, anaerobic, membrane ancho polypeptide
frdD	fumarate reductase, anaerobic, membrane anchor polypeptide
aceA	isocitrate lyase
aceB	malate synthase
aceK	isocitrate dehydrogenase kinase/phosphatase
iclR	repressor of the aceBAK operon
glnA	glutamine synthetase
gltB	glutamate synthase, large subunit (putative)
gltD	glutamate synthase, small subunit
gdhA	NAD+/NADP+-glutamate dehydrogenase
aspC	aspartate aminotransferase
tyrB	aromatic amino acids aminotransferase, tyrosine repressible

	Acetate and fermentative metabolism
poxB	pyruvate oxidase
ackA	acetate kinase
pta	phosphotransacetylase
acs	acetyl-CoA synthetase
ldhA	fermentative D-lactate dehydrogenase, NAD-dependent
adhE	CoA-linked acetaldehyde dehydrogenase and iron-dependent alcohol dehydrogenase
adhP	alcohol dehydrogenase
mhpF	acetaldehyde dehydrogenase
yjgB	Zn-dependent alcohol dehydrogenase
dhaT	1,3-propanediol dehydrogenase
	Iron-dependent alcohol dehydrogenase
budR	transcriptional regulator
budA	alpha-acetolactate decarboxylase
budB	catabolic acetolactate synthase
budC	acetoin(diacetyl) reductase

Appendix E: Precursor requirements (μmol/g DW) for biomass synthesis of *E. coli*

Table. Precursor requirements (μmol/g DW) for biomass synthesis of *E. coli*

Precursor	Amount	Stoichiometry														
		G6P	F6P	R5P	E4P	GAP	3PG	PEP	PYR	ACoA	OAA	AKG	CO2	NADPH	ATP	NADH
Ala	488								1					1		
Arg	281											1	1	4	7	-1
Asn	229										1			1	3	
Asp	229										1			1		
Cys	87						1							5	3	-1
Gln	250											1		1	1	
Glu	250											1		1		
Gly	582						1							1	-1	-2
His	90			1										1	4	-1
Ile	276								1				-1	5	2	
Leu	428								2	1			-2	2		
Lys	326								1		1		-1	4	2	-1
Met	146										1			8	6	-1
Phe	176				1			2					-1	2	1	
Pro	210											1		3	1	
Ser	205						1							1		-1
Thr	241										1			3	2	
Trp	54			1	1			1					-1	2	4	-2
Tyr	131				1			2					-1	1	1	

Published by Woodhead Publishing Limited, 2013

	Demand	G6P	F6P	R5P	E4P	GAP	3PG	PEP	PYR	ACoA	OAA	AKG	CO2	NADPH	ATP	NADH
Val	402								2				-1	2		-2017
Protein	630								2750	428	1447	991	-1940	11338	27144	-1366
RNA	100			144	361		874	668			262		368	1163	6540	-200
DNA	129			100							50		50	274.6	1001.6	64
Lipids	8.4					161	97		1842				-97	2821	2100	-42
LPS	27	34		76			17			302			-17	521	462	27
Peptidoglycan	154		25					27	81	54	27			189	270	
Glycogen	154	154	54												154	
Polyamines	41											41	-41	123	82	
Precursor		G6P	F6P	R5P	E4P	GAP	3PG	PEP	PYR	ACoA	OAA	AKG	CO2	NADPH	ATP	NADH
Total		188	79	950	361	161	1406	695	2831	2626	1786	1059	-1677	16429	37754	-3534

Published by Woodhead Publishing Limited, 2013

Index

ABC transporter: 131, 452
aceBAK: 87, 161, 164, 180, 318, 397, 449, 458
AceK (aceK): 76, 87, 107, 167, 170, 172, 318, 458
acetate formation: 52, 56, 84–7, 112, 115, 139, 255, 297, 369, 394, 423
acetate kinase (Ack): 4, 20, 85, 135, 168, 406, 409, 430–1, 453, 459
acetate overflow: 139
acetyl-CoA synthetase: 20, 74, 458
acetyl phosphate (acP): 20, 73–4, 131
acid resistance: 154–7, 175
aconitase (Acn): 17, 82, 168, 178, 363
ACS (acs): 20, 24, 74, 81, 82, 85, 163–4, 180, 186, 286–8, 382, 385, 388–9, 450, 459
acyl carrier protein (ACP): 48
adenylate cyclase (Cya). 101–2
ADH: 61, 72–3, 146, 244–5, 251, 415, 419–20, 424, 451
adhE mutant: 73, 244, 253–4
alkaline phosphatase (phoA): 132, 135, 452
allolactose: 101
allosteric: 7, 10, 17, 53, 56, 65, 71, 130, 140, 142, 162, 360, 368, 389, 397, 408, 411, 424, 426
γ-amino butyric acid (GABA): 155
ammonia-limited cultures: 322, 329
ammonia assimilation: 38, 122, 223, 457

AmtB: 121
anabolism: 1, 3, 5, 20, 32, 52, 143, 216
anaerobic respiration: 32, 143
anaplerotic pathway: 20–3, 58, 65, 113, 116, 170, 279, 281, 362, 364, 382, 385, 394
anaplerotic metabolism: 303–4, 318
antiporter: 97, 155
antioxidant: 148–9, 411
ArcA (arcA): 31, 72, 75, 80, 83–4, 95, 110–11, 114–15, 117, 125, 134, 137–42, 144–7, 160–1, 164–5, 187, 390, 392, 397, 432, 449, 452
arcA mutant: 140–2
ArcB (arcB): 138–42, 145, 147
arcB mutant: 140–2
ArcA/B (arcA/B): 72, 95, 137–8, 140, 143, 169, 449
arginine decarboxylase: 155
ATase: 128, 129, 155
atom mapping matrices (AMMs): 265, 274
ATPase: 155, 173
ATP binding casset (ABC): 99
[ATP]/[ADP] ratio (ATP/ADP ratio): 315, 361, 410
ATP dissipation: 313
autophosphorylation: 139, 140
autotrophic: 33, 200, 215, 220, 222, 224, 227–8, 230, 231, 232–6, 239–40, 242–3, 346

bifunctional kinase/phosphatase: 76, 397
1, 3-bisphospho grycarate (1, 3BPG): 9
branch chain amino acid: 38

C_1 metabolism: 336, 347
Calvin cycle: 35, 36, 220–2, 225, 235–6, 243, 263, 335, 337, 343–4, 346–50
cAMP-Crp: 59, 75, 95, 102–4, 111, 114, 125, 164
CAP (catabolite gene-activator protein): 102, 145
carbon catabolite repression (CCR): 100–3, 118–20
carbon limitation (C-limitation): 123, 128, 175, 185
catabolism: 1, 3, 5, 20, 50, 52, 104, 119, 122, 139, 143, 186, 216, 241, 263, 327, 330, 337, 347, 410
CcpA (catabolite control protein A): 118–19
CE-MS (capillary electrophoresis mass spectrometry): 265
central metabolic pathway: 1, 57, 66, 95, 121, 146, 180–1, 216, 298, 361, 371, 431
CE-TOF (time of flight) /MS: 265
Chlorella: 215, 220, 224–6, 231, 233
chorismic acid (CM): 42
Citrate synthase (CS): 17, 74, 107, 153, 168, 363, 406, 428, 431, 444, 452
^{13}C metabolic flux analysis: 263, 265, 330, 359, 360
C/N ratio: 123, 125–9
confidence interval: 295
confidence limit: 295
control coefficient: 252

Corynabacterum glutamicum: 100, 119, 299
Cra(*cra*): 65, 70, 72, 79, 80–1, 87, 95, 104–11, 114–15, 117, 122, 125–6, 134, 144, 148, 161, 164, 173–4, 177, 432, 437, 449, 452
cra mutant: 104–8, 110
Crp (*crp*): 95, 102, 111, 114–17, 120, 122–3, 125–8, 135, 144–6, 150, 155–6, 159, 161, 162, 164, 379, 449, 452
crp knockout: 114–16
crp mutant: 116–17
Cya (*cyaA*): 102–3, 111, 113, 115–16, 120, 122, 126, 135, 263, 449
Cyanobacteria: 29, 33–4, 334–7, 346–51
cyclic light-autotrophic/dark-heterotrophic: 220, 230, 243
cydAB: 137, 437, 449
cyoABCD: 137, 449–50
cytochrome bd (Cyd): 26–7, 53, 137, 403, 453
cytochrome bo (Cyo): 26, 29, 137, 453
cytoplasmic membrane: 27, 97–8

DAHPS (deoxyalabino hepturose phosphate synthase): 42–3
deviation index: 252–3, 256
diauxie: 100, 103
doublet: 273, 300–2, 323, 326–7, 329, 339, 348, 379
doublet of doublet: 273

early stationary phase: 20, 55, 79, 81–4, 174, 180–7
EI (enzyme I): 48, 98, 102, 126, 454
EII(enzyme II): 87, 99, 102, 126, 162, 452, 454

Index

EIIBC (EIICB): 48, 59, 99, 102–3, 126, 162, 450
Eda (eda) (phosphor-2-keto-3-deoxy-gluconate): 15, 71–2, 104, 111, 114, 131, 134, 169, 405, 451, 453, 457
Edd (edd)(phosphor-gluconate dehydratase): 14, 71–2, 104, 169, 405, 451, 453, 457
electron acceptor: 25, 27, 137–8, 142–3, 426
electron transport: 19, 27, 32, 104, 153–4, 220–1, 223, 232, 235, 239, 349, 350–1, 411, 426
electrophoresis (2D electrophoresis): 56, 149, 265
Embden-Meyerhof-Parnas (EMP) pathway: 1, 6, 7–8, 10, 16, 25, 32, 44, 50–1, 53, 110, 330, 374, 376
energy charge: 4, 53, 123
enolase (Eno): 9, 181, 287–8, 294, 344–5, 450, 457
erythrose 4-phosphate (E4P): 13–14, 225–6, 277, 287–8, 335, 342, 371, 451
exchange coefficient: 286
exchange flux: 286
exponential growth: 55, 77, 79, 81, 85, 166, 173–4, 180–2, 185–6, 232, 234, 238, 246, 336, 362, 369, 377, 382, 407, 416–17, 428, 432
exponential phase: 79, 82, 84, 168, 170–1, 179, 181–4, 228–9, 237, 240, 246
extracellular flux: 264, 292, 310

fadR gene knockout: 165
fadR mutant: 165-3
fatty acid: 3, 5, 24, 47–8, 87, 106, 108, 165–7, 169, 173–4, 176, 181, 186–7, 224, 310, 318, 327

Fenton reaction: 148
first order washout kinetics: 303
flux ratio (analysis): 303, 439
Fnr (fnr): 31, 72, 75, 95, 110, 114, 125, 128, 137–8, 142–8, 390, 392, 397, 437, 449, 452
fnr mutant: 144–8
focApfl operon: 138
fractional enrichment: 265, 267
fragment mass distribution vector (FMDV): 285
fructose 1, 6-bisphosphate (F16BP,FDP): 7–8, 25, 50, 114, 118–19, 246–7, 250, 314, 368, 386, 392, 408, 420, 437, 450
fructose bisphosphatase (Fbp): 25
fructose diphosphate aldorase (Fba): 8, 405, 431
fructose PTS: 49, 117–18
fructose repressor: 65, 104
FruR (*fruR*): 65, 70, 104, 432, 437, 454
fumarase (Fum): 19, 82, 168, 178, 187, 406, 431, 458
Fur (fur) (ferric uptake regulator): 115, 125, 449
futile cycle: 52–3, 70, 236, 313, 316, 347, 376, 382
f values: 302–4, 323, 325, 327–8

GABA: 155
galactose: 99, 101, 106, 455
GalP (galP): 99, 455
GAPDH: 9, 35, 52, 60, 73, 82, 146, 152, 168, 172, 244–6, 248, 362–3, 392, 410, 415, 418–19, 424–5, 428, 431, 438, 450
GC-MS (gas chromatography mass spectrometry): 263, 265, 272, 274, 285, 291, 294, 334, 336–7, 342, 377

glnALG: 127
glucokinase (Glk, *glk*): 65, 99, 168, 181, 288, 360, 405, 427, 431, 438, 450, 455
gluconate: 55, 57–8, 60–1, 65–6, 70–2, 74, 76, 85, 368, 374, 408, 413, 416–17, 419–20, 422, 425–6, 455–6
gluconeogenesis: 24, 35, 70, 82, 114, 375, 456
gluconolactonase: 11, 456
glucose dehydrogenase: 374, 456
glucose- limited culture(s): 322, 325, 327–8, 332
glucose 6-phosphate (G6P): 6–7, 10, 71, 363
glucose 6-phosphate dehydrogenase (G6PDH): 71, 363, 405, 431, 457
glucose phosphate isomerase (Pgi): 7, 168, 363
glucose repression: 100
glutamate decarboxylase: 154–5, 453
glutamate dehydrogenase (GDH): 38, 120, 223, 453, 458
glutamate synthase (GOGAT): 38, 120, 453
glutamine synthetase (GS): 38, 120, 127, 223, 403, 453
gulutamate synthetase: 223
glyceraldehydes 3-phosphate (GAP): 8, 222
glyceraldehydes 3-phosphate dehydrogenase (GAPDH): 9, 13, 166, 363, 403, 405
glycerol: 51, 55, 57–8, 60–3, 65–6, 70–4, 76, 85, 87, 328–9, 413–14, 416–17, 419–20, 422, 426
glycolytic flux: 65, 123, 232, 246, 248, 250–1, 256–7, 376–7, 397, 415, 423–4
glyoxylate (GOX): 23, 451

glyoxylate cycle: 24, 298, 336
glyoxilate pathway: 23–5, 81, 85, 87, 106, 110, 113, 116–17, 140, 142, 160, 164, 174, 177, 279, 359, 361, 364, 382
glyoxylate shunt: 24, 62, 74–5, 87, 104, 139, 167, 169, 170, 172, 182–3, 186–7, 304–6, 308, 310–12, 316–19, 323–5, 327–8, 334, 361, 364, 367–8, 382, 384, 386, 389, 392–5, 397–8, 409, 457
gnd gene knockout: 368
gnd mutant: 359, 370, 374–6

heat shock: 95, 151–2, 158–9, 161–3, 165, 167, 173, 433, 435
heterotrophic: 100, 215, 220, 223–4, 227, 230–4, 236–40, 242–3, 263, 336–8, 340, 342, 343–50
Hexokinase (Hxk): 4, 6, 363
hexose monophosphate (HMP) (pathway): 6, 10
Hpr kinase/phosphorylase (HprK): 118–19

icdA gene knockout: 398
icdA mutant: 398, 405–7
icd mutant: 399–412
icd gene deletion: 398, 411
inducer exclusion: 101–3
intracellular flux: 218, 263–4, 285, 311–12, 330, 334
isocitrate dehydrogenase (ICDH): 18, 168, 317, 333, 349, 363, 406, 431, 452, 458
isocitrate lyase (Icl): 23, 74, 76, 87, 107, 168, 363, 406, 431, 458
isotopomer balance: 265, 270, 275, 278, 280–1, 286, 291–2
isotopomer distribution vector (IDV): 267–8, 278, 290, 337

isotopomer mapping matrix (IMM): 268
isotopomer representation: 265, 278

KDPG (2-keto-3-deoxy 6-phosphogluconate): 14–15, 71, 451
α-ketoglutarate dehydrogenase (KDGH): 74, 431, 452, 458

labeling experiment (^{13}C-labeling experiment): 263–4, 270, 284, 299, 303, 346
labeling pattern: 275, 278–9, 285, 300, 326, 337, 344, 347
lac repressor: 101
lactose permease (LacY): 101–2
Lactococcus lactis: 119, 253
LamB: 98–9
LDH: 61, 73, 82, 146, 183, 187, 244–5, 250–6, 385, 388–9, 392, 398, 415, 418–19, 424, 426–7, 431, 433, 438, 451
ldhA gene deletion: 428, 433, 438
ldhA gene knockout: 426, 432
ldhA mutant: 426, 428–34, 436–9
lpdA gene knockout: 382
lpdA mutant: 382–3, 385–9
lysine decarboxylase: 155

malate dehydrogenase (MDH): 19, 107, 168, 363, 406, 431, 453, 458
malate synthase (MS): 87, 168, 406, 431, 458
mannose: 50, 99, 108, 162, 454
ManXYZ (*manXYZ*): 50, 99, 162, 450
mass distribution: 274, 285, 288, 290–4, 377
mass distribution vector (MDV): 273

mass spectrometry: 264–5, 285, 403, 404
metabolite activity vector: 265
Mgl (*mgl*): 99, 106, 108, 455
microalgae: 219–21, 232, 235, 240, 242–3
mixotrophic: 215, 220, 222, 224, 229, 231–2, 234, 236–41, 243, 263, 336–8, 341, 342–5, 347–51
Mlc (mlc): 59, 87, 95, 102–3, 111, 115, 117, 122, 125–6, 161–3, 449–50, 452
Monod: 100
multiplet (multiplet pattern): 273, 322, 337–9, 346–8, 367
multiplet intensities: 264, 302–3, 325
multivarent control: 38
mutase: 9, 42, 457

Nac (*nac*): 125, 127–8, 453
NADH oxidase (NOX, *nox*): 139, 140
NADH oxidation: 139, 143
Nar (nar): 142–4
natural abundance: 301, 303
net flux: 286, 294–6, 386, 393
nitrate: 25, 32, 138, 140, 142–4, 223–4, 226, 228–30, 349–50, 426, 452
nitrate reductase: 222, 452
nitrate respiration: 143
nitrite: 32, 144, 223
nitrite reductase: 32, 144, 223
N-limited: 129, 321–2, 324, 328–9, 333
NMR (2D NMR): 263–5, 267, 270, 272–4, 284–5, 299–300, 303, 307, 334, 336–7, 339, 342–3, 367, 377, 379
NMR spectra: 264–5, 300, 307, 336, 339, 342, 367, 379

Non-oxidative PP pathway: 14, 174, 178, 185, 324, 328, 330, 347, 368, 369, 376
non-PTS: 102–3
Ntr gene: 127
nutrient stress: 113, 117, 125

OmpC (*ompC*): 97–8, 158
OmpF (*ompF*): 97–8, 158, 449
outer membrane: 48, 97–8, 452
overflow metabolism: 20, 85, 368, 376, 394
β-oxidation: 47–8, 186
oxidative phophorylation: 3, 5, 227, 235, 237, 239, 243, 329, 350–1
oxidative PP pathway: 14, 71, 152, 276, 298, 312, 326, 328, 333, 343, 347, 367, 369, 374–7, 379, 381–2, 384, 388–9, 398, 404
oxidative stress: 29, 75, 79, 142, 148–9, 153, 165, 174–5, 178
oxygen-limited: 138, 256, 423
oxygen stress: 125

P_{II}: 122–3, 125, 127–30
pck deletion mutant: 309–10, 312, 317–19
pck gene knockout: 361
pck knockout: 307, 311, 318–19
pck mutant: 307–14, 316–19
pentose phosphate (PP) (pathway): 1, 6, 10–11, 71, 79, 82, 106, 182–3, 222, 225, 263, 275, 295, 298, 335, 347, 362
periplasm: 26, 32, 48, 97–9, 109, 131–2, 155, 158
Pfl activase: 138
Pfl deactivase: 138
pfl gene knockout: 137, 412, 420, 423–4

pflA mutant: 244, 246, 251–2, 256, 412–13, 415–20, 422, 425–6
pflB mutant: 412, 420
pfl mutant: 420, 423–4
pgi mutant: 320–8, 330–4
Pgl (*pgl*)(D-glucono- -lactone 6-phosphate): 10, 286, 451
phoB mutant: 132, 134
PhoQ/PhoP: 157
Pho regulon: 130–3, 135, 158
phoR mutant: 133, 135
phosphate starvation: 131–2, 158, 452
phosphoacetyltransferase (Pta): 20
phophoenol pyryuvate (PEP): 3, 7, 9, 102
PEP carboxylase (Ppc): 21, 87, 361, 405, 431, 457
PEP carboxykinase (Pck): 405, 431, 457
phosphofructkinase (Pfk): 4, 7, 25, 50, 53, 60, 65, 70, 82, 110, 115, 117–18, 152, 168, 172, 181, 183, 185, 288, 316, 335–6, 344, 347, 361–3, 368, 377, 379, 381–2, 386, 397, 404–5, 428, 431, 450, 452, 456–7
2-phosphoglycerate (2PG): 9, 10, 36, 183
3-phosphoglycerate (3PG): 3, 33, 35, 44, 278–9, 303, 314, 318–19, 342, 347
6-phosphogluconate (6PG): 11, 169
6-phosphogluconate dehydrogenase (6PGDH): 71, 169, 363, 405, 431, 453, 457
phosphoglycerate kinase (Pgk): 4, 9, 35, 60, 152, 166, 168, 172, 287, 392, 405, 410, 428, 431, 450, 457
phosphoglycerate mutase (Pgm): 9, 450, 456

phosphorelay: 102, 130–1, 139, 156
phosphor transferase (system): 6–7, 10, 48–51, 59, 87, 97, 102, 108, 166, 181, 361, 452, 454
photophosphorylation: 236–7, 239
photorespiration: 36, 222, 347–8
photosynthesis: 33–5, 220, 221–3, 232, 237, 239, 241, 334, 346–7, 349
photosynthetic bacteria: 33, 219
polyP (poly phosphate): 132
porin protein: 97, 158
positional representation: 265, 279
Pox (pyruvate oxidase): 20, 85, 163, 182–3, 187, 384, 386, 388–9, 425, 426, 459
poxB knockout mutant: 386
poxB mutant: 386–7
ppc gene knockout: 319, 361
ppc mutant: 244, 248, 250, 253, 361–5, 367–8
Pps (PEP synthase): 25, 70, 104, 283, 305, 309, 336, 344, 386, 389, 450, 457
precursor-amino acid relationship: 289
prephenic acid (PA): 42–3
proteinogenic amino acid: 301
proton motive force (PMF): 26, 27, 155
PRPP (phosphor ribosilpyrophosphate): 45, 368, 408
Pseudomonas: 100, 119
pstSCAB: 131–2, 449
pta mutant: 85, 244, 246, 251, 253
PTS (phosphotransferase system): 48–50, 59–60, 65, 71, 87, 97–9, 102–3, 108, 113–14, 116–19, 122, 126, 162, 166, 169, 172, 174, 181, 185, 256, 316, 361, 368, 424, 450, 454

ptsG: 50, 87, 99, 102–3, 111, 114–15, 117, 126, 161–2, 368, 432, 449–50, 452, 454
pykF mutant: 244, 246, 250–1, 254, 256, 259, 377–9, 381–2
pyrroloquinoline quinine (PQQ): 374
pyruvate carboxylase (Pyc): 22
pyruvate dehydrogenase (complex) (PDHc): 16, 31, 85, 138, 140, 142, 159, 267, 405, 431, 439
pyruvate kinase (Pyk): 4, 10, 107, 168, 363, 405, 431, 439, 452, 457
pyruvate oxidase (PoxB): 20, 85, 459

quinon (benso-,ubi-, mena-,: ubi-, naphto-,ubisemi-,plast-): 20, 27–9, 137, 140, 145, 147, 165, 374, 453
quinol (ubi-, pyrolo-): 20, 29, 137, 139, 145, 166, 374

reactive oxygen species (ROS): 149, 411
redox ratio (NADH/NAD$^+$): 139
reducing equivalent: 3, 5, 25, 48, 137, 146, 319–20, 349, 390
respiratory chain: 1, 3, 5, 26–7, 30, 52, 73, 113, 116, 125, 128, 134, 137–8, 144, 160, 223, 235, 351, 390, 424, 453
response regulator: 131, 138, 156
reverse CCR: 100
ribose phosphate 3- epimerase (Rpe, rpe): 12, 455, 287–8, 451, 457
ribulose 5-phosphate (R5P): 11–12,
ribose 5-phosphate isomerase (Rpi): 12, 287, 404, 408, 451, 457
RpoD (rpoD): 79–81, 134–5, 150, 156, 173, 177, 433, 450, 452
RpoH (rpoH): 95, 150, 159, 161–2, 433
RpoN (rpoN): 95, 125, 127, 453

RpoS (*rpoS*): 79–81, 83–5, 95, 109–11, 113–15, 117, 125, 135, 144, 150, 155–7, 164, 173–8, 183, 186, 388, 392, 450, 452
RubisCo (ribulose 1, 5-bisphosphocarboxylase oxigenase): 35–6

scalar coupling fine structure: 300, 307, 309–10, 326–7
SCS (succinyl-CoA synthetase): 18, 390, 395, 451
SDH (succinate dehydrogenase): 19, 27, 75, 82, 142, 159, 182, 287, 288, 319, 404, 451, 458
sedoheptulose 7-phosphate (S7P): 12, 178, 225–6, 277, 287–8, 335, 451
sensitivity index: 253, 317
sensor kinase: 138, 156, 452
shikimic acid (Shik): 42
singlet: 273, 300–2, 323, 326, 339, 379
SoxRS (*soxRS*): 75, 79–81, 142, 150, 153, 165, 173, 177–8
SoxR/S (soxR/S): 95, 111, 114–15, 117, 125, 128, 134, 149, 450
SoxR (*soxR*): 148–50, 152–4, 437, 452
soxR mutant: 149–54
SoxS (*soxS*): 142, 148, 149–54, 437, 452
soxS mutant: 149, 150, 152–4
stationary phase: 20, 55, 77–9, 81–5, 109, 117, 155–6, 173–5, 177–87, 228–9, 388
stoichiometric matrix: 216, 263, 289, 297
substrate level phosphorylation: 30, 32, 99, 138, 244, 390

sucA gene knockout: 382–4, 390,
sucA mutant: 390–8
sucC gene knockout: 383–4, 390
sucC mutant: 390–5, 397
succinate-glycine pathway: 223
superoxide: 29, 142, 452
superoxide dismutase: 148–9, 165, 453
symporter: 97, 99, 108
Synechocystis: 334–7, 339, 342–4, 347–8, 351

transaldorase (Tal): 169, 177–8, 326, 453, 457
transaminase: 36, 38, 41–2
transhydrogenase (THD): 6, 127–8, 153, 223, 312, 320, 330, 332–3, 369, 406, 430–1
transketolase (Tkt): 12–14, 107, 177–8, 326, 453, 457
triose phosphate isomerase (Tpi): 8, 60, 168, 172, 181, 303, 377, 379, 392, 450, 457

uniporter: 97
UR: 128, 129
UTase: 121, 128–9

virulence: 100, 120

xylose: 51, 106–7, 426, 455
xylulose 5-phosphate (X5P): 12, 14, 51, 225–6, 287, 335, 346, 371, 451

zwf gene deletion: 320, 375
zwf gene knockout: 263, 319, 330, 368
zwf mutant: 320–2, 324, 328–33, 370, 375–6,